ILSI Human Nutrition Reviews

Series Editor: Ian Macdonald

Thirst:
Physiological and Psychological Aspects

Edited by David J. Ramsay and David Booth

With 91 Figures

Springer-Verlag
London Berlin Heidelberg New York
Paris Tokyo Hong Kong

David Booth, MA,BSc,BA,PhD,DSc,CPsychol,FBPsS
Professor of Psychology, Nutritional Psychology Research Group,
School of Psychology, University of Birmingham, Edgbaston,
Birmingham B15 2TT, UK

David J. Ramsay, MA,DM,DPhil
Professor of Physiology and Senior Vice Chancellor, University of
California - San Francisco,
San Francisco, CA 94103-0400, USA

Series Editor
Ian Macdonald MD,DSc,PhD,FIBiol
Emeritus Professor of Applied Physiology
University of London, UK

QP
139
.T45
1991

ISBN 3-540-19641-2 Springer-Verlag Berlin Heidelberg New York
ISBN 0-387-19641-2 Springer-Verlag New York Berlin Heidelberg
ISSN 0-936-4072

Series Editor's Foreword

Thirst, experienced by everyone, is a sensation that is both essential to life and has many social overtones. The ingestion of fluid may or may not be preceded by the sensation of thirst. Indeed, fluid intake may be stimulated by the desire to acquire food energy, as in the infant.

To expand understanding of the complicated physiological and psychological aspects of thirst, the International Life Sciences Institute (ILSI) convened an international panel of distinguished scientists at a May 1990 workshop in Washington, DC. Well before the workshop, the panelists prepared chapters on a comprehensive range of subjects relating to thirst. Each participant was given the opportunity to read and comment on the other chapters, with the objective of producing a comprehensive, accurate, and up-to-date volume on the subject of thirst.

This volume is one of a series concerned with topics of growing interest to those who want to increase their understanding of human nutrition. Written for workers in the nutritional and allied sciences rather than for the specialist, these volumes aim to fill the gap between the textbook on the one hand and the many publications addressed to the expert on the other. The target readership spans medicine, nutrition and the biological sciences generally and includes those in the food, chemical, and allied industries who need to take account of advances in those fields relevant to their products.

ILSI is a non-profit making, worldwide scientific foundation established in 1978 to advance the understanding of scientific issues relating to nutrition, food safety, toxicology, and the environment. By bringing together scientists from academia, government, and industry, ILSI promotes a balanced approach to solving problems with broad implications for the well-being of the world community.

London   Ian Macdonald
June 1990

Preface

Water is the nutrient that we need most urgently and continuously. The craving for water becomes overwhelming when we are seriously dehydrated; so it seems that there are clear signals of body fluid status to the brain, and there must be neural mechanisms organizing a search for water that can dominate behaviour when necessary.

Often, though, water is in abundant supply. Finding and taking a drink are apparently uncomplicated acts. Yet there are beverages that are so highly appreciated by the palate, or perhaps as symbols, or for contents other than water, that they are sometimes consumed at great cost of effort or money or even to the detriment of health.

Physiological research into the effects of water deficit on fluid intake has made considerable advances in the last quarter century. In contrast, although considerable information on the consumption and the qualities of beverages has now been collected, little psychological research has yet been done on this or other influences on the ingestion of fluids. However, that appears to be changing. There are several scientific monographs and conference volumes on the acceptability of beverages as well as on the physiology of thirst. This book is the first, however, to bring together authoritative reviews of contemporary research on both physiological and psychological aspects of thirst and drinking behaviour.

Each author was asked to deal with a predefined set of topics so that 31 concise chapters covered the main scientific issues in a connected sequence without undue overlap. With limited space, authors have had to assume that readers have a basic acquaintance with the terminology and concepts of the physiological and behavioural sciences. Nevertheless, we have endeavoured to ensure that specialized terms are explained as they arise. Also, we have asked authors, if they use the word thirst to make a scientific point, to specify which of its several possible senses they intend.

The book was put together by an unusual but highly effective procedure, long followed by Dahlem Konferenzen, for example, and used for other books in the series of ILSI Human Nutrition Reviews. Draft chapters were circulated to other authors for comments that were also circulated. All the authors were then brought together for 2¹/₂ days to discuss and edit the final version of each chapter and the commentary that follows it.

As a result of the precirculations and the round-table meeting, the contents of this book have been subjected to a more rigorous and public form of peer review than is applied to many papers in scientific journals. The chapters have been revised in the light of precirculated comments and discussion at the workshop, as well as of editorial suggestions. Points of dispute, further issues and additional information are presented in the comments and replies at the end of some chapters. The book should therefore give good insight into the current state of understanding of thirst and widen the scope for further advance in this vital field of research.

We are greatly indebted to our colleagues for accepting the requirements of this approach and its demands on their time and patience. We thank them and our colleagues on the Planning Committee for their confidence. Our job was made almost enjoyable by the great and calm efficiency of Lili C. Merritt and her staff at the ILSI Washington Office, cheerfully administering the meeting, the precirculations and the production of the final manuscript. We hope that many readers will benefit from this enterprise and from the careful work of all our authors and editorial assistants.

San Francisco, California   David Ramsay
Birmingham, UK   David Booth
May 1990

Contents

Chapter 8. Mineral Appetite: An Overview

SECTION III.   Neural Pathways of Water Deficit Signals

Chapter 9. Central Projections of Osmotic and Hypovolaemic Signals in Homeostatic Thirst

SECTION IV.   Neural Organization of Drinking Behaviour

Chapter 13. Sensory Detection of Water

Chapter 14. Neostriatal Mechanisms Affecting Drinking

Chapter 15. Drinking in Mammals: Functional Morphology, Orosensory Modulation and Motor Control

Chapter 29. Effect of Changes in Reproductive Status on Fluid Intake and Thirst

Chapter 30. Disorders of Thirst in Man

SECTION VIII.   A Concluding View

Chapter 31. Thirst and Salt Intake: A Personal Review and Some Suggestions

Contributors

Dr P.H. Baylis
Endocrine Unit, Royal Victoria Infirmary, Newcastle-upon-Tyne,
NE1 4LP, UK

Dr D.A. Booth
School of Psychology, University of Birmingham, Edgbaston,
Birmingham, B15 2TT, UK

Dr J.M. de Castro
Department of Psychology, Georgia State University, Atlanta,
Georgia 30303, USA

Dr D. Denton
Howard Florey Institute of Experimental Physiology and Medicine,
Melbourne University, Parkville 3052, Victoria, Australia

Dr G.L. Edwards*
The Department of Physiology and Pharmacology, College of
Veterinary Medicine, University of Georgia, Athens, Georgia,
USA

Dr D. Engell
Behavioral Sciences Division, US Army Natick Research,
Development and Engineering Center, Natick,
Massachusetts 01760-5020, USA

Dr A.N. Epstein
Leidy Laboratory 6018, University of Pennsylvania, Philadelphia,
Pennsylvania 19104, USA

Dr J.T. Fitzsimons
The Physiology Laboratory, University of Cambridge, Downing
Street, Cambridge, CB2 3EG, UK

Dr M.J. Fregly
Department of Physiology, Colleges of Medicine and Liberal Arts
and Sciences, University of Florida, Gainesville, Florida 32610, USA

Dr B.G. Galef, Jr
Department of Physchology, McMaster University, Hamilton, Ontario,
L8S 4K1, Canada

Dr M. El Ghissassi*
Laboratoire de Neurobiologie des Régulations, Collège de France,
75231 Paris Cedex 05, France

Dr R.M. Gilbert
Addiction Research Foundation, 33 Russell Street, Toronto,
M5S 2S1, Canada

Dr L. Gray
Department of Geriatrics, University of Melbourne, Austin Hospital,
Melbourne, Victoria 3084, Australia

Dr J.E. Greenleaf
Laboratory for Human Environmental Physiology, Life Science
Division (239-11), NASA Ames Research Center, Moffett Field,
California 94035, USA

Dr W.G. Hall
Department of Psychology, Duke University, Durham,
North Carolina 27706, USA

Dr E. Hirsch*
Behavioral Sciences Division, US Army Natick Research,
Development and Engineering Center, Natick,
Massachusetts 01760-5020, USA

Dr P.C. Holland
Department of Psychology, Duke University, Durham,
North Carolina 27706, USA

Dr A.K. Johnson
Spence Laboratory, Department of Psychology and Pharmacology and
the Cardiovascular Center, University of Iowa, Iowa City, Iowa 52242,
USA

Dr C.I. Johnston
Departments of Medicine and Geriatrics, University of Melbourne,
Austin Hospital, Melbourne, Victoria 3084, Australia

Dr F.S. Kraly
Department of Psychology, Colgate University, Hamilton, NY 13346,
USA

Dr M.J. McKinley
Howard Florey Institute of Experimental Physiology and Medicine,
Melbourne University, Parkville 3052, Victoria, Australia

Dr S. Nicolaïdis
Laboratoire de Neurobiologie des Régulations, Collège de France,
75231 Paris Cedex 05, France

Dr R. Norgren
Department of Behavioral Science, College of Medicine, The
Pennsylvania State University, Hershey, Pennsylvania 17033, USA

Dr B.J. Oldfield
Howard Florey Institute of Experimental Physiology and Medicine,
Melbourne University, Parkville 3052, Victoria, Australia

Dr P.A. Phillips
Departments of Medicine and Geriatrics, University of Melbourne,
Austin Hospital, Melbourne, Victoria 3084, Australia

Dr D.J. Ramsay
Department of Physiology, University of California – San
Francisco, San Francisco, California 94103-0400, USA

Dr G.L. Robertson
Box 131, University of Chicago Hospital, 5841 S. Maryland Avenue,
Chicago, Illinois 60637, USA

Dr B.J. Rolls
Department of Psychiatry and Behavioral Sciences, The Johns Hopkins
University School of Medicine, 600 North Wolfe Street, Baltimore,
Maryland 21205, USA

Dr N.E. Rowland*
Department of Physiology, Colleges of Medicine and Liberal Arts
and Sciences, University of Florida, Gainesville, Florida 32610,
USA

Mr M. Saltmarsh
Mars GB Ltd, Four Square Division, Armstrong Road, Basingstoke,
Hants RG24 0NU, UK

Dr T. Schallert
Department of Psychology and Institute for Neuroscience, University
of Texas at Austin, Austin, Texas 78712, USA

Dr E.M. Stricker
Department of Behavioral Neuroscience, University of Pittsburgh,
479 Crawford Hall, Pittsburgh, Pennsylvania 15260, USA

Dr E. Szczepanska-Sadowska
Department of Clinical and Applied Physiology, Medical Academy of
Warsaw, 00-730 Warsaw, Jazgarzewska 17 str, Poland

Dr S.N. Thornton*
Laboratoire de Neurobiologie des Régulations, Collège de France,
75231 Paris Cedex 05, France

Dr T.N. Thrasher
Department of Physiology, University of California, School of
Medicine, 513 Parnassus Box 0444, San Francisco, California 94143,
USA

Dr J.B. Travers
College of Dentistry, Ohio State University, 305 West 12th
Avenue, Columbus, Ohio 43210, USA

Dr H. Tuorila
Department of Food Chemistry and Technology, University of
Helsinki, Viikki, SF-00710 Helsinki, Finland

Dr J.G. Verbalis
Division of Endocrinology and Metabolism, 930 Scaife Hall,
University of Pittsburgh, School of Medicine, Pittsburgh,
Pennsylvania 15261, USA

Dr H.P. Zeigler
Department of Psychology, Hunter College, 695 Park Avenue, New
York, New York 10021, USA

*Did not attend workshop.

Section 1

Perspectives on Thirst

Chapter 1

Evolution of Physiological and Behavioural Mechanisms in Vertebrate Body Fluid Homeostasis

J.T. Fitzsimons

Body Fluid Homeostasis

Fluid and electrolyte balance is maintained because, as stated by Claude Bernard, constancy of the internal environment is a necessary condition for a free and independent existence (Bernard 1878). L. J. Henderson was instrumental in introducing Bernard's ideas to the English-speaking world by arranging for *Introduction à l'étude de la médecine expérimentale* to be translated into English by H. C. Green. Many others, including W. M. Bayliss, E. H. Starling, J. S. Haldane, J. Barcroft and W. B. Cannon, built on Bernard's ideas in their own work. Cannon introduced the term homeostasis, defined as ". . . the physiological rather than the physical arrangements for attaining constancy". He went on to say why use of a special word was justified: "The co-ordinated physiological reactions which maintain most of the steady states of the body are so complex, and are so peculiar to living organisms, that it has been suggested . . . that a specific designation for these states be employed – *homeostasis*." (Cannon 1929, p. 400).

The body fluids in the vertebrates are subject to close homeostatic control and there are good reasons for this. Elkinton and Danowski (1955) wrote in an influential book, *The Body Fluids*, "That man is a creature of his planetary environment is nowhere more evident than with respect to the fluids of the body, for some of the very elements of one are the basic constituents of the other." (p. 35). The ionic composition of the extracellular fluid bears many resemblances (and important differences) to sea water, and the assumption is that life originated in the oceans because water with its unique properties, dissolved substances and physicochemical stability was a fit environment for the synthesis of organic substances which could form protoplasm. Although the composition of the extracellular fluid in present-day vertebrates may reflect the ionic composition of the ancient seas, it would be misleading to think of it as being identical with the sea in which the particular species first appeared (Elkinton and Danowski 1955). This does not accord with what is known about the probable chemical evolution of the ocean nor with the evolution of physiological mechanisms which enable living organisms to make rapid adjustments to changes in their environment.

The body fluids of many present-day marine invertebrates and some primitive vertebrates tend to be isosmotic with the sea in which they live. Such osmoconformers depend on the stability of the ocean for their continued existence, but even here there are variations in the regulation of individual ions. The ocean is a natural homeostat and for aeons it must have provided the stable conditions necessary for continuation of the limited repertoire of reactions of primitive organisms. But as the composition of the primeval ocean changed and living organisms invaded fresh water and dry land, there came the need to develop physiological mechanisms to preserve the initial favourable physicochemical conditions. The ability to evolve such mechanisms determined the whole pattern of evolution. Constancy of the internal environment is the rule among practically all vertebrates and many invertebrates. The ways in which this is achieved are simple in conception but of great complexity in execution.

Homeostatic Adaptations

In addition to their respiratory and nutritive requirements, the cells of the body need an internal environment of the correct osmolality, ionic composition, pH and temperature in order to function properly. In any advanced organism this means possession of a circulation to enable renewal of the immediate cellular environment, and mechanisms for the acquisition and preservation of essential constituents, and for the disposal of waste products. The circulation imposes its own exigencies on those of fluid and electrolyte homeostasis, of which the one of most direct interest to the present discussion is the requirement that the volume of the extracellular fluid be defended as well as its composition. Fluid and electrolyte homeostasis and thermoregulation are intimately related; the maintenance of heat balance in a hot climate is a heavy drain on the body's reserves of water and salt because of increased evaporative losses and, in some animals, the need to sweat.

The underlying strategy in fluid and electrolyte homeostasis is always the same – to obtain and conserve the water and solutes necessary to maintain the constancy of the internal environment. The tactics vary according to the external environment, and the nature and complexity of the life that the animal leads. The tactical diversity is illustrated by the very wide range of adaptations for fluid and electrolyte homeostasis found in vertebrates, listed as follows:

1. Ingestion of water and salt under the control of mechanisms of thirst and sodium appetite.
2. Restricting water and solute loss from the surface of the body through limited permeability of the body covering.
3. Controlled percutaneous fluid absorption by amphibia when in water. (Some arthropods are able to absorb water from a humid atmosphere.)
4. Production of urine of variable concentration through operation of a countercurrent mechanism.
5. Water sparing by excreting highly soluble urea in high concentration as the nitrogenous waste product.
6. Water sparing by excreting insoluble uric acid as the nitrogenous waste product.
7. Excretion of salts in high concentration by the kidney.
8. Excretion of salts in high concentration by gills or salt glands.
9. Resisting the dehydrating effect of sea water by accumulation of urea and trimethylamine oxide to reduce the osmotic gradient between the body fluids and external environment.
10. Protection of the water content of the cells through increases in the intracellular concentration of free amino acids in response to rises in extracellular osmolality.

11. Preferential maintenance of plasma volume in dehydration.
12. Reduction in evaporative water loss by having a low metabolic rate.
13. Water sparing by allowing the body temperature to rise during the day and dissipating heat at night by conduction and radiation without the use of water.
14. Water sparing by insulating the body with fur or clothing to reduce heat gain by radiation and conduction from the sun and surroundings.
15. Evaporative water loss required to maintain constant body temperature is less in larger animals because of the smaller surface to volume ratio.
16. Reduction in pulmonary water loss by lowering the temperature and therefore the water content of expired air through countercurrent heat exchange in the nasal passages.
17. Behavioural reduction in water loss by seeking shade, remaining in a burrow during the day, or huddling together.
18. Water sparing by hibernation, aestivation and torpor.
19. Water storage, as in the urinary bladder in the desert frog, or the rumen in the Bedouin goat.

Most valuable general accounts of homeostatic adaptations are to be found in Smith (1953), Elkinton and Danowski (1955), Schmidt-Nielsen (1964, 1983) and Bentley (1971). Skadhauge (1981) deals comprehensively with osmoregulation in birds. An excellent summary of the major steps in vertebrate evolution is given by Romer in a presidential address to the AAAS (1967). Comparative aspects of drinking behaviour are considered in detail in Chapter 3 of Fitzsimons (1979). A brief survey of the operation of some of these adaptations in the vertebrates will now be given, followed by a more detailed consideration of the phylogeny of thirst and drinking behaviour, and of the receptor mechanisms of deficit-induced or primary drinking behaviour.

Phylogeny of Homeostasis in Vertebrates

Fish

Animals that live in water require less water than those that live on dry land because there are no thermoregulatory or respiratory water losses, and for the same reasons their water needs are more constant (Elkinton and Danowski 1955). Osmotic stability is the rule among vertebrates and many invertebrates as well, particularly the freshwater forms. All freshwater animals, from protozoa upwards, are hypertonic to their surroundings. The major homeostatic problem is coping with the osmotic inflow of water and diffusional loss of solutes. The means for getting rid of water vary in complexity from the contractile vacuole to the fully developed glomerulotubular kidney. Osmotic inflow of water and loss of solutes are limited by having an impermeable integument.

Animals that live in sea water have the opposite problem. With the exception of the cartilaginous fish or elasmobranchs and one of the two groups of cyclostomes, the hagfish, which have the same or a slightly higher osmolar concentration than sea water, the body fluids of most marine vertebrates are hypotonic to the surrounding sea so that they are continuously losing water by osmosis and gaining solutes by diffusion through the gills and oral membranes. To deal with these challenges to homeostasis, marine animals have evolved a number of strategies. The ancestors of the hagfish reduced their osmotic water loss by allowing the ionic composition of extracellular fluid to rise until it approached that of sea water. The early elasmobranch fishes also reduced osmotic water loss when they were driven by the widespread aridity of the Devonian period to invade the sea, but in a quite different way. They diminished the osmotic gradient between themselves and the surrounding sea by accumulating urea and trimethylamine oxide in the body fluids (Smith 1953). Active reabsorption of urea by the kidney together with the development of gills impermeable to urea resulted in the accumulation of enough urea to make the body fluids slightly more concentrated than sea water. The physiological uraemia reverses the osmotic flow of water across the gills and in this way the elasmobranch fish gains solute-free water from the sea. The elasmobranch does not drink, though it obtains some water from its prey through feeding. The surplus salt that enters the body is excreted through a rectal salt secretory gland, kidneys and perhaps through the gills as well.

The bony fish or teleosts evolved from freshwater fishes much later than the elasmobranchs. When they invaded sea water, they solved the problem of gaining solute-free water differently (Smith 1953). The urea-retention device to extract water from the sea was no longer available. Instead the teleost fish obtains water by drinking the sea water in which it lives, and then disposes of the unwanted sodium chloride by active secretion via the gills, and Mg^{2+} and SO_4^{2-} by excretion via the kidneys. Owing to the difficulty in obtaining solute-free water, urine production is much less than in freshwater fish, and some marine teleosts have aglomerular kidneys.

Particularly interesting fishes from the point of view of fluid and electrolyte homeostasis are the migratory teleosts such as the trout and salmon and the eel, which divide their time between fresh water and sea water. In fresh water, they behave in the way described above, not drinking, and dealing with the osmotic inflow of water by producing copious dilute urine. When they enter sea water, the concentration of the body fluids rises owing to the osmotic withdrawal of water, causing cellular dehydration, partly opposed by an increase in intracellular amino acid concentration. The fish start to drink and the rate of glomerular filtration falls so that they become almost anuric. Some lampreys, belonging to one of the cyclostome groups, behave similarly. They drink while in sea water but stop drinking as soon as they enter fresh water. The hagfish are exclusively marine and, although isosmotic with their environment, they drink sea water.

Amphibia

With colonization of dry land fluid and electrolyte homeostasis becomes more complex. Seasonal drying out of rivers in the late Devonian period may have been responsible for the first appearance of the amphibia almost 300 million years ago (Romer 1967). Fishes that could convert to air breathing using their air sacs, and especially those that could leave the stagnant and stifling water of ponds that were drying out and explore the river bed for other ponds with water still present, would have prospered. These were the ancestors of the present-day amphibia.

The amphibian, spending part of its time in fresh water where breeding and early development take place, and part of its time on dry land, has a much more complicated regulatory problem than a purely aquatic animal. The amphibian kidney is remarkably responsive to this requirement. In fresh water the osmotic inflow of water is dealt with by excretion of large volumes of dilute urine, but on dry land urine flow ceases almost completely through shutdown of glomerular filtration and increased reabsorption of water by the renal tubules and urinary bladder. An example of useful water storage is found in the desert frog which can store up to 30% its body weight of water in the urinary bladder and lymph spaces during the rainy season and survive on this while aestivating in its burrow during the dry season (Bentley 1971).

The need to rehydrate rapidly and automatically may account for development of a skin which actively takes up sodium and which is passively permeable to water, although a waterproof skin would have been an advantage on dry land. Flexibility of control in the face of these widely differing conditions is introduced by varying the rate of secretion of neurohypophysial hormones which act on the kidney and urinary bladder to conserve water and on the skin to promote water uptake. Like other freshwater animals, amphibia do not drink, nor do they need to because percutaneous uptake of water is amply sufficient. However, more surprising in an animal so dependent on water, is the apparent absence in many amphibia of water-seeking behaviour when dehydrated (Adolph 1943). Amphibia still have not adapted to the urgent necessity of truly terrestrial vertebrates of actively ensuring their own hydration in an inhospitable environment.

Reptiles

Reptiles are the first truly terrestrial vertebrates. They appeared in Upper Carboniferous and Permian periods about 220 million years ago and have been unquestionably successful on dry land since they are found in all the deserts of the world. The secret of their success, shared with the birds and mammals, is the ability to enclose their offspring within water impermeable membranes so that early development no longer has to take place in ponds and streams where there is a grave risk of being eaten by a predator. Reptiles lay closed (cleidoic) eggs, each of which contains its own water supply, providing a suitable environment for the developing embryo. A small number of eggs can be deposited in secure places on dry land away from danger (Smith 1953). The aquatic reptiles such as turtles, crocodiles and alligators, are air-breathers. They have evolved from terrestrial ancestors and still lay their eggs on dry land. Because they are air-breathing the problem of fluid and electrolyte exchanges over a large surface area of gills does not exist.

Reptiles are admirably suited to terrestrial life (Bentley 1971). Evaporative losses through the skin are less than in amphibia, and these and pulmonary losses are further reduced at low ambient temperatures by the reptile's poikilothermy. Although the kidneys do not possess loops of Henle, water losses are limited by uricotelism. As water is reabsorbed from the urine in the renal tubules and cloaca, the poorly soluble uric acid and its sodium and potassium salts are precipitated out of solution so that the final product is a semi-solid paste of a mixture of urine and faeces. Uricotelism is a feature of reptiles and birds and this may be associated with embryonic development in a closed egg which has a very limited water supply. Because of its insolubility, the uric acid produced by the developing embryo does not accumulate in an increasingly concentrated solution which would be the case if ammonia or urea were the end-product of nitrogen metabolism. Many reptiles possess nasal salt glands that eliminate excess sodium and potassium, and they are also markedly tolerant to hypernatraemia, completing the picture of water-conserving adaptations.

A major adaptation to life on dry land is the capacity to drink. Reptiles are carnivorous and therefore obtain all or some of the water they need from their prey. However, many reptiles also drink when allowed water after a period of water deprivation. Drinking in response to dehydration by an animal on dry land requires a sequence of motivated behaviour of first seeking water and then ingesting it in appropriate amounts, a much more complicated behaviour pattern than simply opening the mouth and swallowing water as in the case of a fish.

Birds

Birds developed from reptiles in the late Jurassic period about 150 million years ago. They are homeothermic, non-fossorial and usually active during the day, which means that their water-regulatory problems are more severe than those of many reptiles and mammals. Birds have a high body temperature and tolerate quite large increases in body temperature allowing water to be saved which would otherwise have been needed to reduce the temperature. Loops of Henle appear for the first time in the birds, allowing the elaboration of hypertonic urine through a countercurrent mechanism, but the loops are not as well developed as in mammals. The maximum urinary concentration possible is about double that of plasma. Urinary hypertonicity and uricotelism contribute to water conservation, and the possession of nasal salt glands, particularly in marine birds, provides for further economy of water (Skadhauge 1981).

Most birds need to drink but their mobility is such that they can spend a great deal of the time a considerable distance from the source of water. Birds are efficient drinkers but not all have to drink water. Some get all the water they need from succulent plants. Many marine birds are able to obtain their water requirements from the preformed and metabolic water of their diet of teleostean fish, excreting the salt in high concentration via the nasal salt gland. The nasal salt gland also allows some terrestrial birds as well as marine species to take advantage of salty water for their supply of water.

Mammals

More is known about the mechanisms of fluid and electrolyte homeostasis in mammals than in any other vertebrate group, but it should be borne in mind that knowledge is based on the study of very few species. Most mammals are terrestrial and generally they must seek out water and drink appropriate amounts of it in order to remain in water balance. Mammals do not possess any additional adaptations to terrestrial life that are not already present in one or other vertebrate group, but they have highly effective thirst mechanisms, and many of the adaptations listed earlier are particularly well developed (Schmidt-Nielsen 1964). The mammalian kidney is a more versatile organ of fluid and electrolyte homeostasis than the kidneys of the other vertebrate groups and it can produce a much wider range of urinary concentration, depending on whether the need is to conserve or dispose of water. Mammals are ureotelic, and in animals that consume large amounts of protein, urea makes an important contribution to the medullary osmotic gradient, helping to concentrate the non-urea constituents in the filtrate. In herbivores urea makes a much smaller contribution to the concentrating process than it does in rats or dogs which may eat large amounts of protein (Robinson 1988).

Renal water conservation is especially efficient in those mammals that have successfully colonized the most arid regions of the earth or which live in the sea, with the countercurrent mechanism reaching its highest development. Urinary concentrations of up to 25 times the plasma concentration are possible in some desert rodents, which as a consequence have no need to drink, since they can obtain all the water they need from the preformed water in the diet and from water of oxidation. The kangaroo rat is the best known example of the rat that "never drinks", yet it is not exceptionally "dry", its water content being similar to that of other mammals (Schmidt-Nielsen 1964). Schmidt-Nielsen points out that the kangaroo rat can be made to drink by obliging it to feed on soyabeans. This results in an increase in the rate of urea formation requiring increased urine production for its disposal. Such is the urinary concentrating ability of the kangaroo rat that even in these circumstances the additional water requirements can be met by providing the animal with sea water to drink. Another remarkable rodent is the sand rat which lives on succulent halophytes from which it obtains virtually all its water, easily getting rid of the large amounts of salt in highly concentrated urine.

Other inhabitants of the desert produce highly concentrated urine and subsist on relatively little drinking water. During the winter the camel may get by on the water contained in the plants it eats. Even during the heat of summer, the camel can forego drinking for at least two weeks and will withstand a water loss exceeding 25% of the body weight with little apparent ill effect (Schmidt-Nielsen 1964). Camels, and other large mammals found in the desert such as the burros and desert kangaroos (Dawson 1978), maintain their plasma volume during dehydration better than other species, including man, which inhabit temperate regions. This they do by retaining plasma proteins within the circulation. Maintenance of the circulating volume presumably is an important aspect of their tolerance to dehydration.

Marine mammals such as whales and seals also produce highly concentrated urine and have no need to drink. They feed on fish, various invertebrates and plankton, and obtain water from the relatively (to sea water) hypotonic body fluids of their prey. The low Mg^{2+} and SO_4^{2-} levels of their intestinal contents indicate that they do not generally drink sea water (Elkinton and Danowski 1955). As in marine reptiles, the problem of fluid and electrolyte exchanges over a large surface area of gills does not arise because, like all mammals, marine mammals are air-breathing. There are also no thermoregulatory water losses in marine mammals.

Drinking as a Homeostatic Adaptation

Of all adaptations for fluid and electrolyte homeostatis, mechanisms that ensure an adequate intake of water and essential electrolytes are the most important since they are the starting point for all other adaptations. Claude Bernard believed that the central nervous system was responsible for matching intakes and losses in order to achieve the constancy of the milieu intérieur and that the sensation of thirst played a major part in accomplishing this:

> C'est le système nerveux, avons-nous dit, qui forme le rouage de compensation entre les acquits et les pertes. La sensation de la soif, qui est sous la dépendance de ce système, se fait sentir toutes les fois que la proportion de liquide diminue dans le corps à la suite de quelque condition telle que l'hémorr(h)agie, la sudation abondante; l'animal se trouve ainsi poussé à réparer par l'ingestion de boissons les pertes qu'il a faites. Mais cette ingestion même est réglée, en ce sens qu'elle ne saurait augmenter au delà d'un certain degré la quantité d'eau qui existe dans le sang; les excrétions urinaires et autres éliminent le surplus, comme une sorte de trop plein. (Bernard 1878, p.115).

Two Homeostatic Constants, Osmolality and Extracellular Volume, and the Two Thirsts Resulting When These Deviate from Normal

When water is freely available, the normal pattern of drinking generally ensures that water intake exceeds the minimal amount needed to replace the obligatory losses through skin, lungs, kidneys and intestines (Fitzsimons 1979). Drinking in these circumstances is not under the control of deficit-triggered mechanisms. The excess intake is excreted in non-maximally concentrated urine. The amounts of water drunk may vary considerably between individuals, and in some species, preformed water in the diet and metabolic water may obviate the necessity to drink at all. In the rat, there appears to be a basic urge to drink that cannot be suppressed by infusing water at rates far in excess of requirements, by routes that by-pass the oropharynx. Little is known about the mechanism of normal or secondary drinking other than it is probably not caused by cellular dehydration or hypovolaemia. Normal drinking shows the characteristics of a nychthemeral rhythm. The pattern of water intake and amounts drunk depend on habit, diet, ecological conditions and climate. The provision of water in this way enables the various homeostatic adaptations to function efficiently.

But should water intake fall below a certain minimal rate, the increasing fluid deficit that results causes thirst, vasopressin release and the setting in action of other mechanisms that lead to renal fluid and electrolyte conservation. The fluid deficit responsible for these responses may be mainly cellular, or extracellular, or both. The rest of this chapter will be concerned with deficit-induced or primary drinking.

It is a principle of body fluid and electrolyte homeostasis that the composition of the immediate environment of the cells of the body be stabilized. In many of the higher vertebrates the osmolality and ionic composition of extracellular fluid are controlled to within very narrow limits. The osmolality is set by the ratio of water to solute so that controlling either the amount of water or the amount of solute would suffice to set the extracellular osmolality at its desired value. It is clearly simpler to control water than a heterogeneous mixture of solutes, hence control of body water occupies a privileged position in fluid and electrolyte homeostasis. A rise in osmolality of about 1% or 2% results in a conscious desire to drink and near maximal antidiuresis. A fall in osmolality of this order results in water diuresis.

Of less immediate concern in body fluid and electrolyte homeostasis is the volume of extracellular fluid. Percentage decreases in blood volume or pressure of about 10% or more arouse thirst and they also cause a delayed increase in sodium appetite. Vasopressin secretion is little affected by small changes in blood volume or pressure, but decreases of about 10% or more may produce hormone levels many times the level required to produce maximal antidiuresis. The first homeostatic priority is that there should be a layer of fluid of the correct concentration and composition in contact with the cells, but because of metabolism this means that the fluid has to be constantly renewed, which in turn means that there must be enough extracellular fluid for the proper functioning of a circulation. However, in view of the highly variable demands that are made on the circulation, the volume of extracellular fluid is necessarily less tightly controlled than its concentration.

This is well illustrated in the classical study by McCance (1936) in which he caused severe sodium chloride deficiency in himself and two medical students by restricting salt intake and inducing sweating. McCance found that removal of sodium was at first followed by loss of enough water to maintain normal body fluid osmolality. But as the extracellular dehydration and hypovolaemia worsened, further removal of sodium was not followed by an equivalent loss of water. When this degree of sodium deficiency was reached, the extracellular volume started to be increasingly well maintained, but at the expense of falling sodium concentration and osmolality. The subjects experienced a peculiar sensation in the mouth which was described as thirst by two of them and which caused them to drink freely. However "thirst" was not relieved by water but it was by hypertonic sodium chloride.

This experiment illustrates the normal hierarchy for the maintenance of two homeostatic constants, body fluid osmolality and extracellular volume, and how the priority alters as sodium deficiency develops. Changes in intake or excretion of water caused by alterations in cellular or extracellular hydration described above reflect this priority; thirst or water diuresis are aroused by smaller percentage changes in extracellular osmolality than extracellular volume. In McCance's experiment the initial preservation of osmolality as the extracellular solute content fell was accomplished by simultaneous loss of extracellular water in the correct isotonic proportion. Once the extracellular volume deficit became significant, the priority changed, and volume not concentration was preserved. The developing hypovolaemia caused fluid retention, and increasing "thirst" and water intake despite the hypotonicity thereby resulting. The "thirst" included an element of what appeared to be a less well-defined increase in sodium appetite.

Drinking Behaviour in Submammalian Vertebrates

Cyclostomes and teleostean fish in sea water drink continuously owing to the dehydrating effect of the hypertonic environment in which they live. Acquisition of water is a simpler activity for an aquatic organism than for a terrestrial animal because water is immediately available and in order to drink the animal simply has to open its mouth and swallow. The eel continues to drink in sea water after removal of the prosencephalon and the mesencephalon (Takei et al. 1979). Drinking is continuous and it seems likely that it is entirely reflex. It is a moot point whether or not fish experience thirst, a conscious sensation leading to motivated drinking behaviour. When the eel is transferred to fresh water it stops drinking but it can be made to drink again, either by infusing hypertonic sodium chloride intravenously, or by bleeding (Hirano 1974), or by injecting angiotensin systemically. Papaverine caused flounder in fresh water to drink, an effect prevented by simultaneous administration of captopril, suggesting that endogenous renin released by hypotension was responsible (Balment and Carrick 1985). The decerebrate eel drinks as vigorously to angiotensin as the intact eel but vagotomy abolished the response (Takei et al. 1979). It appears, therefore, that some bony fish respond to the same dipsogenic stimuli as terrestrial vertebrates, though the neural organization for these responses shows less encephalization. The immediate onset or cessation of drinking when euryhaline fish move between sea water and fresh water seems to depend on chloride sensitive receptors on the surface of the body (palatal organ, lateral line organ and the olfactory system) and not on the dehydrating effect of sea water. These anticipatory responses which depend on external stimuli recall other anticipatory mechanisms of fluid and electrolyte homeostasis, such as pre-absorptive diuresis or antidiuresis when water or hypertonic sodium chloride is placed on the tongue of a rat (Nicolaïdis 1978), and pre-absorptive satiety in drinking behaviour (Rolls et al. 1980).

The early life history of amphibia does not distinguish them from other freshwater fish. It is only when amphibia undergo metamorphosis and become air-breathing that they can be regarded as terrestrial animals, but they still spend a great deal of time in or near water and must return to water in order to breed. As has been shown, amphibia do not drink but rehydrate themselves by controlled absorption through the skin. This is clearly efficient; there is no particular virtue in being able to drink per se; what is important is to be able to rehydrate when necessary. But it is perhaps unexpected that a dehydrated frog in the laboratory near a pool of water does not show any water-seeking behaviour and may die from water lack a few centimetres from the pool (Adolph 1943). However, should a dehydrated frog stumble on water through random movement, it stays there so that it may be aware of the need for and the benefit to be derived from water (Adolph 1943). Perhaps the additional task of having to locate water before being able to absorb ("drink") it, is still beyond the capabilities of the nervous system of an animal that was so recently a fish, and indeed is still partly one. On the other hand some evolutionary progress has been made by the amphibia to cope with life on dry land, since some more terrestrial Australian frogs which inhabit arid regions burrow in order to reach the damp subsoil, and salamanders seek humidity by moving directly towards water some distance away.

Reptiles are the first truly terrestrial vertebrates. They do not need to return to water in order to breed, so that maintaining body hydration by immersion in a pond, absorbing water through the skin, is no longer a realistic option for an animal that may spend all its time away from water. The vital terrestrial adaptation shown by reptiles is the ability to engage in the motivated behaviour of seeking water and drinking it in appropriate amounts (Adolph 1943; Schmidt-Nielsen 1964). However, the ability of a reptile to seek out water seems limited by comparison with what we know about mammals, and desert lizards can remain in water balance on dietary and metabolic water as the sole source of fluid (Templeton 1972). Bradshaw (1978) studied a group of desert lizards living on a rock where there was a pool of water. He found that the lizards never went to the pool but some might die of dehydration even though the water was less than three yards away. However, when it rained, the lizards ". . . went insane and they ran around and jumped in the air, and they drank the water as it fell." He found that when desert lizards were kept in the laboratory it was necessary to have the water dripping, and then the lizards would regularly come to it and drink. Bradshaw believes that this is because in the normal course of events, the only water that desert lizards ever come across is rain. On the other hand, other lizards are capable of recognizing a stationary source of water and drink regularly from it.

Reptiles drink in response to water deprivation, cellular dehydration, extracellular dehydration and to angiotensin (Fitzsimons and Kaufman 1977; Kobayashi et al. 1979). For example, the iguana drank and retained enough water after systemic injections of hyperosmotic solutions of sodium chloride or sucrose to dilute the injected load to isotonicity. The onset of drinking after an osmotic stimulus was a leisurely affair of between 30 minutes and 4 hours. Once started, drinking was a deliberate action. The animal immersed its snout in water, swallowed for several seconds and then, raising its head in chicken-like fashion for several more seconds, allowed the water to pass down its capacious gullet into the stomach. Excretion of an osmotic load is extremely slow in the iguana compared to a mammal, and it is noteworthy that the iguana appears to compensate for this by relying more on behavioural osmoregulation through drinking. The iguana also drank in response to extracellular dehydration, and to systemic angiotensin. The thirst challenges to which reptiles drink are therefore similar to those to which teleostean fish respond, but clearly for the terrestrial reptile this is much more complicated than it is for the fish, and it presumably demands higher nervous organization, of which we know practically nothing.

Avian drinking is a more robust phenomenon than drinking in reptiles. Many birds drink by sipping water and tilting their heads back to swallow but pigeons and other Columbiforme species immerse their beaks and draw up water with a vigorous pumping action of the gullet. Birds drink in response to water deprivation, cellular dehydration, extracellular dehydration, angiotensin given systemically or intracranially, and to tachykinins and bombesins. The pigeon has been particularly well investigated, and a great deal is known about osmometric thirst and responses to angiotensin in this species. In experiments with different combinations of intracranial and intravascular infusions of hyperosmotic solutions, Thornton (1986a, b) showed that the results were consistent with an osmometric hypothesis of drinking. Many birds drink in response to angiotensin (Kobayashi et al. 1979), and the pigeon is particularly good in this respect, being at least as sensitive as the rat, and drinking about three times as much water although the pigeon and rat have similar body weights (Evered and Fitzsimons 1981a, b).

Birds are unusual in that they also drink vigorously in response to the tachykinins and bombesins, series of peptides which were first isolated from amphibian skin but have since been found in brain tissue and elsewhere (de Caro 1986). These effects on drinking are not mediated through the angiotensin receptor, and it is interesting that these peptides are without any stimulatory effect in rats, and indeed inhibit angiotensin-induced drinking.

Drinking Behaviour in Mammals

There is a great deal of detailed information about drinking behaviour in mammals. In man, need for water is manifest as the sensation of thirst, and need for sodium as the less well-defined sensation of salt hunger or appetite. Both sensations are referred to the oropharyngeal region. It is assumed from their behaviour that mammals, and possibly birds and reptiles, experience similar sensations when dehydrated or sodium deficient.

For most animals, survival in dry hot conditions depends on how strongly they are motivated to seek and ingest adequate amounts of water. Driven by thirst, a human victim of severe water lack will make any effort and run any risk in order to obtain water. In the heat of the desert a man may produce 12 litres of sweat during the day, a rate of sweating of more than 1 litre per hour, and rates as high as 4 litres per hour for 2 or 3 hours may occur. These losses should be compared with the normal plasma volume of about 3.5 litres and the fact that dehydration amounting to 15%–25% body weight is lethal. The rate of sweating is unaffected by moderate dehydration or excessive drinking, but with acclimatization to heat, sweating becomes more profuse and starts at a lower body temperature. Salt loss in sweat and urine is less than it might otherwise be owing to increased secretion of aldosterone, but sweating losses may be considerable. The blood volume also increases, resulting in improved skin blood flow and increased cardiovascular tolerance to work in the heat. However, these adaptations to the increased heat load require increased intakes of water and salt if they are to remain effective.

Man has evolved drinking habits which, in temperate climates, result in his taking more water than needed to maintain normal body water content. The urine is therefore not usually maximally concentrated. On the other hand, in the desert, when sweating is profuse, the amounts of water drunk may be inadequate to replace the losses, a phenomenon which Adolph (1947) called voluntary dehydration, and which may amount to 2%–4% of the body weight. Adolph found that in dehydrated man, neither the temperature of the water, nor added sugar or salt made much difference to immediate drinking, which remained appreciably less than what was needed to restore body water. The usual pattern according to Adolph is that the subject drinks about half of what he needs and then takes the remainder when he sits down to a meal. Schmidt-Nielsen (1964) comments on his own inability to drink 1 litre water over 1 minute compared with the ease with which a donkey can drink 7 or 8 litres per minute. Presumably pre-absorptive satiety mechanisms limit the amounts of water that can be immediately drunk.

Storage of water is an uncommon homeostatic adaptation and in most circumstances little purpose is served carrying a load of unused water. Energy would be wasted and mobility perhaps impaired. Only in a few very limited cases has a mechanism to store water for future use evolved as a biological adaptation to desiccation. A good example is the desert frog's ability to survive while aestivating in its burrow on the large amounts of dilute urine that were stored in its bladder during the rainy season. A more contrived example of water storage is the deliberate over-drinking by a non-thirsting human subject of a litre of water immediately before exertion in the desert. The additional water load taken when levels of antidiuretic hormone are high will be eliminated so rapidly as sweat that little of it will be lost in the urine (Adolph 1947). The amount of water that can be stored by pre-drinking is a small fraction of the future demand, but it is a way of avoiding voluntary dehydration and ensuring that the body is in water surfeit at the start of a day in the desert.

Although anticipatory storage of water in the body for future water needs is uncommon and not particularly useful, it is an extremely valuable option to be able to store water when rehydrating. Temporary storage of water after drinking allows very large amounts of water to be drunk extremely rapidly without running the risk of a rapid fall in body fluid osmolality, haemolysis and water intoxication as water is absorbed. It is claimed that some desert ruminants can store water temporarily in the gastrointestinal tract, enabling them to circumvent the hazards of too rapid rehydration when they drink. A dehydrated camel may drink over 30% of its body weight in a single session, quantities of more than 100 litres being common (Schmidt-Nielsen 1964). Like the camel, zebu cattle tolerate infrequent watering and when eventually allowed to rehydrate also drink huge quantities of water in a very short time without suffering from water intoxication (Nicholson 1985). Bedouin goats are even more remarkable in this respect. They are able to graze in the sun-scorched desert some days' walking distance from any source of water, and then when they eventually reach the watering place, dehydrated animals have been found to drink over 40% their dehydrated weight during their short visit to the water hole (Shkolnik et al. 1980). This is up to half the body water, a truly awesome potential osmotic insult which is eased by the slow release of water from the rumen into the body fluids.

Receptors for Thirst Induced by Cellular Dehydration

Cellular dehydration is an effective stimulus to increased drinking in all vertebrate groups that are able to drink. In mammals, the thirst of cellular dehydration can be aroused by (a) water deprivation, (b) injection of hyperosmotic concentrations of non-penetrating solutes, or (c) inducing potassium deficiency. The classical view is that in the mammal this type of thirst originates in osmoreceptors in the basal forebrain, which share in the general cellular dehydration and are stimulated by diminution in their own water content. The system is extremely sensitive and the amounts of water drunk are such as to maintain normal plasma osmolality. In view of their comparable responses to changes in effective extracellular osmolality, birds and reptiles must have similar osmometric thirst systems. In some mammals, notably goat and sheep, receptors specifically sensitive to the concentration of sodium in the CSF seem to be important (Andersson 1978). The osmoreceptor/sodium receptor controversy is discussed elsewhere (Fitzsimons 1989) but two points should be noted. First, a dual osmoreceptor/sodium receptor mechanism may operate, particularly in herbivorous animals. Second, since most of the extracellular solute is made up of sodium salts, it makes little difference whether changes in osmolality or in sodium concentration underlie cellular dehydration thirst.

Quite apart from the question whether or not there are sodium-sensitive receptors in addition to or instead of osmoreceptors in the basal forebrain, the possibility of receptors elsewhere in the central nervous system or in peripheral sites such as the gastrointestinal tract and liver has not been ruled out. For example, lesions of the area postrema which involve the adjacent nucleus solitarius cause rats to over-drink in response to a number of thirst challenges, including overnight water deprivation and hyperosmotic sodium chloride, as well as to hypovolaemia and angiotensin II, though there are accompanying changes in renal function and feeding which make interpretation hazardous (Miselis et al. 1987). But such results recall the fact that decerebrate fish drink reflexly in sea water, that receptors on the surface of the body are responsible, and that the response is abolished by section of the vagi and glossopharyngeal nerves. As already mentioned, the fish also drinks in response to intravenous hypertonic sodium chloride, suggesting that there are internal osmoreceptors or sodium-sensitive receptors. It is attractive to think that these may be located in the hindbrain but evidence is lacking.

Receptors for Thirst Induced by Extracellular Dehydration

Extracellular dehydration has been shown to be an effective stimulus to increased drinking in all vertebrate groups that are able to drink. In mammals, extracellular dehydration causes an initial thirst and a delayed increase in sodium appetite, an appropriate sequence because sodium is needed as well as water to restore the extracellular fluid volume to normal. Sodium appetite will not be further considered here, but in view of the homeostatic importance of maintaining an adequate extracellular fluid volume, it is a vital component of behaviour in terrestrial vertebrates, particularly in the herbivores whose diet generally lacks adequate amounts of sodium. In mammals, thirst of extracellular origin may be aroused by (a) removing blood, (b) inducing sodium deficiency, (c) causing sequestration of extracellular fluid, or (d) interfering with venous return to the heart by obstructing the inferior vena cava. It also occurs in several clinical conditions such as severe diarrhoea or vomiting, heart failure, and circulatory shock.

The common factor is hypovolaemia, actual or simulated. Since the vascular and interstitial fluid compartments are directly coupled through the Starling capillary filtration/reabsorption system, control of the vascular compartment ensures control of the whole extracellular compartment. Changes in plasma volume approaching

10% are required to arouse hypovolaemic thirst or initiate disposal of excess extracellular fluid by the kidneys. Because the circulation requires such varied rates and distribution of blood flow, the cardiovascular stretch receptors must have an ample margin within which to control the circulation before bringing fluid and electrolyte homeostatic mechanisms into play. Second, interstitial fluid can be quickly moved in and out of the circulation when the need arises through adjustments in the Starling filtration/reabsorption mechanism.

The receptors that initiate hypovolaemic thirst are generally thought to be cardiovascular stretch receptors which respond to underfilling of the circulation with a reduction in inhibitory nerve impulse traffic to the thirst centres. Among the receptors involved are those that register left atrial pressure, which is the filling pressure of the left ventricle, and receptors in the ventricle itself. Nervous impulses travel in the vagi in myelinated fibres and in the mass of unmyelinated C fibres from the atria and ventricles, and in the sympathetic nerves. The relative importance of these different receptors in drinking behaviour is not known.

Suggestive evidence that receptors from the heart may play a role in thirst was provided originally by Sobocinska, who found that blockade of one of the vago-sympathetic nerves in the dog lowered the threshold of drinking in response to osmotic stimulation (Sobocinska 1969). It was suggested that the increased reactivity of the thirst mechanism was explained by removal of the tonic inhibitory impulses that normally travel in these nerves from the volume receptors. Vagosympathectomy also prevented haemorrhage from lowering the osmotic threshold, and isosmotic extracellular expansion from elevating it, because these procedures were no longer able to cause significant changes in tonic inhibitory activity (Kozlowski and Szczepanska-Sadowska 1975). Crushing the left atrial appendage in the sheep prevented the increase in drinking caused by hypovolaemia, presumably because there were too few remaining receptors to register a significant enough reduction in impulses to arouse thirst, but drinking in response to hypertonic sodium chloride was significantly increased though the thirst threshold was not given (Zimmerman et al. 1981).

Reducing the venous return to the heart by obstructing the inferior vena cava results in a fall in central venous pressure and cardiac output and a presumed reduction in inhibitory inputs to thirst neurons from volume receptors. In both rat and dog this causes increased water intake and a marked decrease in urine flow leading to water retention (Fitzsimons 1969; Fitzsimons and Moore-Gillon 1980; Thrasher et al. 1982). Changes in intrathoracic vascular volume also affect drinking behaviour in the primate. When volume receptors are stimulated by increases in central blood volume during weightlessness or when pressure is applied to the lower part of the body, adipsia and increased urinary fluid loss occur in man (Gauer and Henry 1976) and monkey (Kass et al. 1980). This is followed by increased drinking when the central venous volume returns to its usual value as normal conditions are restored. These observations support the view that intrathoracic vascular volume plays a role in controlling drinking behaviour by affecting volume receptor discharge.

Stimulation of a proportion of the receptors in a pulmonary vein-atrial junction by inflating a balloon placed there, caused a reduction in spontaneous water intake without affecting food intake, and an increase in urine flow so that the dog went into negative fluid balance (Moore-Gillon and Fitzsimons 1982). Drinking induced by different thirst challenges was also attenuated by distension of a pulmonary vein-atrial junction. Similar results have been obtained in the rat by stretching receptors at the junction of the superior vena cava and right atrium (Kaufman 1984). In both the rat and dog experiments, inhibition of drinking by maximally stimulating an unknown proportion of venoatrial vascular stretch receptors suggests a role for these particular receptors in hypovolaemic thirst, but the stimulation is an imprecise imitation of the effects of hypervolaemia. Even when the volume receptors are being directly stimulated in this way, it is likely that other cardiovascular receptors, notably arterial baroreceptors, are secondarily affected by the haemodynamic changes that inevitably occur, and these other receptors could be having an effect on drinking. Recent evidence in the dog suggests that this may be the case. It has been found that drinking in response to caval obstruction was greatly reduced by sinoaortic denervation, and completely abolished by combined cardiac and sinoaortic denervation (Quillen et al. 1988).

It is believed, but unproven, that an acute reduction in arterial baroreceptor activity also arouses drinking behaviour. However, after aortic baroreceptor deafferentation (leaving the carotid sinus innervation intact) in the rat, there was an immediate fall in spontaneous water intake, with some recovery over the 5 day experimental period. The fall was thought to be brought about reflexly by the acute rise in blood pressure produced by the partial baroreceptor deafferentation, via intact carotid sinus or cardiopulmonary baroreceptors (Werber and Fink 1981). Supporting this view is the finding that acute changes in arterial blood pressure affect drinking behaviour in the way expected. In the rat, intravenous angiotensin II caused rats to drink more water when the pressor response to angiotensin was prevented than when the arterial pressure was allowed to rise (Robinson and Evered 1986). In the dog, restricting renal blood flow by inflating a balloon placed in the abdominal aorta above the renal arteries caused renin-dependent drinking and a fall in urine flow (Fitzsimons and Moore-Gillon 1981). Arterial blood pressure rose above the obstruction and fell below it. When the balloon was deflated there was a second bout of drinking which was also renin-dependent. Since the arterial pressure returned to normal after deflation of the balloon, the "off response" drinking could have been explained by a reduction in inhibitory baroreceptor stimulation which allowed the persisting renin stimulus to cause more drinking. In the case of the drinking response during inflation, the greater the ratio, rise in pressure upstream of the obstruction to the fall in pressure in the renal arteries downstream of the obstruction, the smaller the water intake. However, in neither the rat nor the dog experiment just described, can it be concluded that the pressor inhibition of water intake was necessarily mediated by arterial baroreceptors. It should also be noted that sustained rises in arterial blood pressure do not themselves have any permanent effect on drinking behaviour, presumably because the baroreceptors are reset to the higher level of pressure.

Chemoreceptors may also affect drinking behaviour. Exposure to low oxygen pressure causes a reduction in spontaneous and induced water intake (Jones et al. 1981a,b). Hypoxia also suppresses increased sodium appetite in spontaneously hypertensive rats (Behm et al. 1984).

It is usually assumed that drinking and antidiuresis in hypovolaemia are explained by removal of the tonic inhibition normally exerted by cardiovascular stretch receptors. However, the increase in vasopressin that occurs in severe haemorrhage may be caused by an increase in the discharge rate of ventricular receptors, not a decrease (Wang et al. 1988). Ventricular denervation, but not sinoaortic denervation, markedly attenuated the increase in vasopressin after haemorrhage. Goetz and his colleagues suggest that vasopressin secretion is not continually held in check by tonic discharge of cardiovascular stretch receptors. It is unlikely that such a hard-working system would ever have evolved since the system would only be switched off by severe haemorrhage or hypovolaemia, which are rare events. They believe that a reflex from ventricular receptors stimulates vasopressin secretion in haemorrhage.

Similar arguments may apply to drinking behaviour. When tonic inhibition has been entirely removed in severe hypovolaemia, additional activation of thirst neurons is required and can only come from a different category of response. In the case of blood pressure control, chemoreceptors in the periphery and brainstem provide extra sympathetic activation when the blood pressure falls below the operating range of the baroreceptors. But the little evidence there is suggests that these receptors inhibit thirst. Another possibility is that inputs from ventricular receptors to the central nervous system, similar to those which may stimulate release of vasopressin, could provide the additional stimulation to thirst neurons. A third possibility is that in severe hypovolaemia, when cardiovascular stretch receptors have ceased firing, continued renal renin secretion owing to the low renal perfusion pressure may be critical for continued stimulation of drinking and vasopressin release.

All the remarks made so far about receptors for hypovolaemic thirst apply to mammals. But birds, reptiles and teleostean fish also drink in response to extracellular fluid deficits and are presumed to possess cardiovascular receptors similar to those found in mammals, though very little is known about such possible receptors (Heymans and Neil 1958). It is thought that birds possess cardiovascular mechanoreceptors and chemoreceptors and that the circulatory reflexes are similar to those in mammals (Sturkie 1976). Circulatory responses to baroreceptor and chemoreceptor stimulation in fish have been described (Johansen 1971), and something is also known about the sensory inputs, including those from baroreceptors, involved in the circulatory and respiratory adjustments that accompany submersion in diving birds (Stephenson and Butler 1987). However, nothing is known about the possible involvement of these receptors in hypovolaemic thirst in the submammalian vertebrates.

Hormones, Neurotransmitters and Other Humoral Factors in Drinking Behaviour

A large number of hormonal and humoral inputs have been implicated in thirst. These are fully discussed by Szczepanska-Sadowska (Chapter 7) and will only be touched on here. Chapter 7 gives an admirable summing up of the present position.

Vasopressin is released in cellular dehydration and in hypovolaemia. The available evidence suggests that it may participate in the control of water intake by affecting the osmosensitivity of the thirst system. In the dog, subpressor amounts of vasopressin lowered the threshold of drinking in response to intravenous infusion of hyperosmotic sodium chloride (Szczepanska-Sadowska et al. 1974) and injection of vasopressin into the third cerebral ventricle caused drinking (Szczepanska-Sadowska et al. 1982). This action of vasopressin could account for the continuing water intake despite hyponatraemia in some cases of the syndrome of inappropriate secretion of vasopressin. It is entirely fitting that the hormone so central to renal water conservation, and perhaps also playing a role in blood pressure control, should contribute to thirst. There is no evidence that neurohypophysial hormones affect drinking behaviour in the submammalian vertebrates.

The renal renin–angiotensin system is activated in extracellular dehydration. Circulating angiotensin contributes to hypovolaemic thirst but its participation is not essential. The more severe the circulatory changes and the quicker they develop, the more important is the contribution from blood-borne angiotensin to hypovolaemic thirst. The possible role of cerebral renin in thirst is unsettled. In the longer-term, angiotensin is also a stimulus to increased sodium appetite, and here it seems more likely that angiotensin generated by cerebral renin is important. Among most bony fish, there is considerable evidence that the renal renin–angiotensin system is involved in the regulation of renal function, drinking, cardiovascular function and adrenocortical secretory patterns (Henderson 1987). It may be significant that in sea-water adapted euryhaline species there appears to be increased adrenocortical activity, relative polydipsia and glomerular antidiuresis, all of which may be caused by angiotensin. Although not all are agreed that angiotensin assists in the adaptation to hyperosmotic environments, drinking in response to exogenous angiotensin II seems to be characteristic of euryhaline fish but not stenohaline fish (Kobayashi et al. 1983). Little is known about the reptilian renin–angiotensin system but angiotensin has been shown to be dipsogenic in a number of species. The avian renin–angiotensin system, and drinking and other responses of birds to angiotensin, suggest many similarities with mammals. The main difference seems to be in the control of blood pressure. Little is known about extrarenal renins in the submammalian vertebrates.

Mineralocorticoids are also increased in extracellular dehydration, and, as Rice and Richter (1943) first showed, they stimulate sodium appetite in the rat. Mineralocorticoids also stimulate sodium appetite in pigeons (Epstein and Massi 1987). Small amounts of mineralocorticoid and angiotensin act synergistically to cause a greater increase in sodium appetite than produced by either alone in the rat (Epstein 1986) and in the pigeon (Massi and Epstein 1990). Mineralocorticoids also cause increased water intake, so that there are some resemblances in this to the dual behavioural effect of angiotensin. The effect, however, is indirect and is probably caused by cellular dehydration resulting from potassium deficiency and sodium retention.

The demands of the fetus during pregnancy and of the newborn baby during lactation result in maternal increases in intakes of both water and sodium chloride. Many of the hormones of pregnancy and lactation have been found to have significant effects on fluid and electrolyte intake. The synergistic action of a number of hormones whose levels increase at this time may be responsible. These hormones include oestrogen, ACTH, cortisol, corticosterone, prolactin and oxytocin (Denton 1982). The ovarian hormone, relaxin, has recently been found to be dipsogenic in male and female rats (Thornton and Fitzsimons 1989).

Both angiotensin and vasopressin may affect drinking behaviour by acting in the brain as neurotransmitters. Other humoral agents could also act in this way. Some of these are stimulatory and others inhibitory, and there may be important species differences. Tachykinins, bombesins and opioid peptides are antidipsogenic in rats, whereas the first two are dipsogenic in birds, and angiotensins are dipsogenic in both species (de Caro 1986). This makes it difficult to draw any firm conclusions about the role, if any, of endogenous brain peptides in water intake and body fluid and electrolyte homeostasis. There are a number of other hormones and humoral factors which have been implicated in thirst, including atrial natriuretic peptide, insulin, thyroid hormones, histamine, neuropeptide Y and neurotensin. The present evidence on each of these is too fragmentary to make any generalizations about their possible role, if any, in the control of water intake. At present the most that can be said is that many hormones whose secretion increases during clearly defined fluid and electrolyte deficiency states have been found to have a stimulatory effect on drinking behaviour.

Conclusions

The mechanisms by which living organisms achieve fluid and electrolyte homeostasis are extraordinarily diverse, but all are critically dependent for their working on adequate intakes of water and electrolytes. The ways in which water and electrolyte intakes are assured must depend on the ecological conditions and on the nature of the life that the animal leads. In this chapter, the phylogeny of vertebrate fluid and electrolyte homeostasis is outlined, with particular emphasis on the arrangements in the different vertebrate groups for the provision of water. The phylogenetic progression from the relatively simple behaviour of reflex drinking in fish to the much more complex patterns of behaviour in terrestrial animals which involve having to seek water before being able to consume it, is reflected in the ontogeny of drinking behaviour. Epstein (Chapter 31) makes an important distinction between suckling, the most recent ingestive behaviour to have evolved among vertebrates, and the more primitive independent ingestive behaviours of feeding and drinking. He describes work which shows that the adult segregation of feeding and drinking is absent in rat pups and that the different drinking and feeding mechanisms appear abruptly and in sequence when certain critical ages are reached.

Analysis of the developmental and comparative aspects of drinking behaviour has consolidated our understanding of the physiology of thirst mechanisms. Deficit-induced water intake and some aspects of the mechanisms involved have been considered in some detail, but normal drinking behaviour and the mechanisms involved, and sodium appetite, have received no more than the briefest mention. The manifold patterns of normal drinking behaviour are obviously of enormous interest and importance but they do not alter in any fundamental way our understanding of the evolution of the physiological mechanisms of thirst. The same is true of sodium appetite. Although the acquisition of water through predation or because of the nature of the diet may mean that the act of drinking is a rare event in a given species, it is often the case that such animals will drink when the need arises. They therefore possess the repertoire of mechanisms that forms part of their evolutionary heritage.

References

Adolph EF (1943) Physiological regulations. Jacques Cattell Press, Lancaster, PA

Adolph EF and associates (1947) Physiology of man in the desert. Interscience Publishers, New York

Andersson B (1978) Regulation of water intake. Physiol Rev 58:582–603

Balment RJ, Carrick S (1985) Endogenous renin–angiotensin system and drinking. Am J Physiol 248:R157–R160

Behm R, Honig A, Griethe M, Schneider P (1984) Sustained suppression of voluntary sodium intake of spontaneously hypertensive rats (SHR) in hypobaric hypoxia. Biomed Biochim Acta 43:975–985

Bentley PJ (1971) Endocrines and osmoregulation: a comparative account of the regulation of water and salt in vertebrates. Springer-Verlag, Berlin, Heidelberg, New York

Bernard C (1878) Leçons sur les phénomènes de la vie communs aux animaux et aux végétaux. Baillière et fils, Paris

Bradshaw SD (1978) Volume regulation in desert reptiles and its control by pituitary and adrenal hormones. In: Barker Jorgensen C, Skadhauge E (eds) Osmotic and volume regulation: proceedings of the Alfred Benzon Symposium XI. Munksgaard, Copenhagen, p 55

Cannon WB (1929) Organization for physiological homeostasis. Physiol Rev 9:399–431

Dawson TJ (1978) Osmotic and volume regulation during dehydration in desert kangaroos. In: Barker Jorgensen C, Skadhauge E (eds) Osmotic and volume regulation: proceedings of the Alfred Benzon Symposium XI. Munksgaard, Copenhagen, pp 22–32 de Caro G (1986) Effects of peptides of the "gut–brain–skin triangle" on drinking behaviour of rats and birds. In: de Caro G, Epstein AN, Massi M (eds) The physiology of thirst and sodium appetite. Plenum Press, New York, pp 213–226

Denton D (1982) The hunger for salt: an anthropological, physiological and medical analysis. Springer-Verlag, Berlin, Heidelberg, New York

Elkinton JR, Danowski TS (1955) The body fluids basic physiology and practical therapeutics. Williams Wilkins, Baltimore

Epstein AN (1986) Hormonal synergy as the cause of salt appetite In: de Caro G, Epstein AN, Massi M (eds) The physiology of thirst and sodium appetite. Plenum Press, New York, pp 395–404

Epstein AN, Massi M (1987) Salt appetite in the pigeon in response to pharmacological treatments. J Physiol (Lond) 393:555–568

Evered MD, Fitzsimons JT (1981a) Drinking and changes in blood pressure in response to angiotensin II in the pigeon *Columba livia*. J Physiol (Lond) 310:337–352

Evered MD, Fitzsimons JT (1981b) Drinking and changes in blood pressure in response to precursors, fragments and analogues of angiotensin II in the pigeon *Columba livia*. J Physiol (Lond) 310:353–366

Fitzsimons JT (1969) The role of a renal thirst factor in drinking induced by extracellular stimuli. J Physiol (Lond) 201:349–368

Fitzsimons JT (1979) The physiology of thirst and sodium appetite. Monographs of the Physiological Society, no 35. Cambridge University Press, Cambridge

Fitzsimons JT (1989) Bengt Andersson's pioneering demonstration of the hypothalamic "drinking area" and the subsequent osmoreceptor/sodium receptor controversy. Acta Physiol Scand 136: Suppl 583, 15–25

Fitzsimons JT, Kaufman S (1977) Cellular and extracellular dehydration, and angiotensin as stimuli to drinking in the common iguana *Iguana iguana*. J Physiol (Lond) 265:443–463

Fitzsimons JT, Moore-Gillon MJ (1980) Drinking and antidiuresis in response to reductions in venous return in the dog: neural and endocrine mechanisms. J Physiol (Lond) 308:403–416

Fitzsimons JT, Moore-Gillon MJ (1981) Renin-dependence of drinking induced by partial aortic obstruction in the dog. J Physiol (Lond) 320:423–433

Gauer OH, Henry JP (1976) Neurohormonal control of plasma volume. In: Guyton AC, Cowley AN (eds) International review of physiology, cardiovascular physiology II, vol 9. University Park Press, Baltimore, pp 145–190

Henderson IW (1987) Phylogeny of renal, cardiovascular and endocrine activity of the renin–angiotensin system. In: Taylor EW (ed) The neurobiology of the cardiorespiratory system. University Press, Manchester, pp 394–414

Heymans C, Neil E (1958) Reflexogenic areas of the cardiovascular system. Churchill, London

Hirano T (1974) Some factors regulating water intake by the eel, *Anguilla japonica*. J Exp Biol 61: 737–747

Johansen K (1971) Comparative physiology: gas exchange and circulation in fishes. Ann Rev Physiol 33:569–612

Jones RM, LaRochelle FT Jr, Tenney SM (1981a) Role of arginine vasopressin on fluid and electrolyte balance in rats exposed to high altitude. Am J Physiol 240:R182–R186

Jones RM, Terhaard C, Zullo J, Tenney SM (1981b) Mechanism of reduced water intake in rats at high altitude. Am J Physiol 240:R187–R191

Kass DA, Sulzman FM, Fuller CA, Moore-Ede MC (1980) Renal responses to central vascular expansion are suppressed at night in conscious primates. Am J Physiol 239:F343–F351

Kaufman S (1984) Role of right atrial receptors in the control of drinking in the rat. J Physiol (Lond) 349:389–396

Kobayashi H, Uemura H, Wada M, Takei Y (1979) Ecological adaptation of angiotensin-induced thirst mechanisms in tetrapods. Gen Comp Endocrinol 38:93–104

Kobayashi H, Uemura H, Takei Y, Itatsu N, Ozawa M, Ichninohe K (1983) Drinking induced by angiotensin II in fishes. Gen Comp Endocrinol 49:295–306

Kozlowski S, Szczepanska-Sadowska E (1975) Mechanisms of hypovolaemic thirst and interactions between hypovolaemia, hyperosmolality and the antidiuretic system. In: Peters G, Fitzsimons JT, Peters-Haefeli L (eds) Control mechanisms of drinking. Springer-Verlag, Berlin, Heidelberg, New York, pp 25–35

Massi M, Epstein AN (1990) Angiotensin/aldosterone synergy governs the salt appetite of the pigeon. Appetite 14:181–192

McCance RA (1936) Experimental sodium chloride deficiency in man. Proc R Soc Lond B 119:245–268

Miselis RR, Shapiro RE, Hyde TM (1987) The area postrema. In: Gross PM (ed) Circumventricular organs and body fluids, vol II. CRC Press, Boca Raton, pp 185–207

Moore-Gillon MJ, Fitzsimons JT (1982) Pulmonary vein-atrial junction stretch receptors and the inhibition of drinking. Am J Physiol 242:R452–R457

Nicolaïdis S (1978) Rôle des réflexes anticipateurs oro- végétatifs dans la regulation hydrominérale et energétique. J Physiol (Paris) 74:1–19

Nicholson MJ (1985) The water requirements of livestock in Africa. Outlook Agric 14:156–164

Quillen EW, Reid IA, Keil LC (I988) Cardiac and arterial baroreceptor influences on plasma vasopressin and drinking. In: Cowley AW Jr, Liard J-F, Ausiello DA (eds) Vasopressin: cellular and integrative functions. Raven Press, New York, pp 405–411

Rice KK, Richter CP (1943) Increased sodium chloride and water intake of normal rats treated with deoxycorticosterone acetate. Endocrinology 33:106–115

Robinson JR (1988) Reflections on renal function, 2nd edn. Blackwell Scientific Publications, Oxford

Robinson MM, Evered M (1986) Angiotensin II and arterial pressure in the control of thirst. In: de Caro G, Epstein AN, Massi M (eds) The physiology of thirst and sodium appetite. Plenum Press, New York, pp 193–198

Rolls BJ, Wood RJ, Rolls ET (1980) Thirst: the initiation, maintenance and termination of drinking. Prog Psychobiol Physiol Psychol 9:263–321

Romer AS (1967) Major steps in vertebrate evolution. Science 158:1629–1637

Schmidt-Nielsen K (1964) Desert animals physiological problems of heat and water. Clarendon Press, Oxford

Schmidt-Nielsen K (1983) Animal physiology: adaptation and environment, 3rd edn. Cambridge University Press, Cambridge

Shkolnik A, Maltz E, Choshniak I (1980) The role of the ruminant's digestive tract as a water reservoir. In: Ruckebusch Y, Thivend R (eds) Digestive physiology and metabolism in ruminants. MTP Press, Lancaster, pp 731–742

Skadhauge E (1981) Osmoregulation in birds. Springer-Verlag, Berlin, Heidelberg, New York

Smith HW (1953) From fish to philosopher. Little, Brown, Boston

Sobocinska J (1969) Effect of vagosympathectomy on osmotic reactivity of the thirst mechanism in dogs. Bull Acad Pol Sci 17: 265–270

Stephenson R, Butler PJ (1987) Nervous control of diving response in birds and mammals. In: Taylor EW (ed) The neurobiology of the cardiorespiratory system. University Press, Manchester, pp 369–393

Sturkie PD (1976) Avian physiology, 3rd edn. Springer-Verlag, Berlin, Heidelberg, New York

Szczepanska-Sadowska E, Kozlowski S, Sobocinska J (1974) Blood antidiuretic hormone level and osmotic reactivity of thirst mechanism in dogs. Am J Physiol 227:766–770

Szczepanska-Sadowska E, Sobocinska J, Sadowski B (1982) Central dipsogenic action of vasopressin. Am J Physiol 242:R372–R379

Takei Y, Hirano T, Kobayashi H (1979) Angiotensin and water intake in the Japanese eel *Anguilla japonica*. Gen Comp Endocrinol 38:466–475

Templeton JR (1972) Salt and water balance in desert lizards. In: Maloiy CMO (ed) Comparative physiology of desert animals Symposia of the Zoological Society of London no 31. Academic Press, London, pp 61–77

Thornton SN (1986a) The influence of central infusions on drinking due to peripheral osmotic stimuli in the pigeon *Columba livia*. Physiol Behav 36:229–233

Thornton SN (1986b) Osmoreceptor localization in the brain of the pigeon *Columba livia*. Brain Res 377:96–104

Thornton SN, Fitzsimons JT (1989) ICV porcine relaxin stimulates water intake but not sodium intake in male and female rats. Appetite 12:242

Thrasher TN, Keil LC, Ramsay DJ (1982) Hemodynamic, hormonal and drinking responses to reduced venous return in the dog. Am J Physiol 243:R354–R362

Wang B, Flora-Ginter G, Leadley RJ, Goetz KL (1988) Ventricular receptors stimulate vasopressin release during hemorrhage. Am J Physiol 254:R204–R2ll

Werber AH, Fink GD (1981) Cardiovascular and body fluid changes after aortic baroreceptor deafferentation. Am J Physiol 240:H685–H690

Zimmerman MB, Blaine EH, Stricker KM (1981) Water intake in hypovolemic sheep: effects of crushing the left atrial appendage. Science 211:489–491

Chapter 2

Water: Distribution Between Compartments and its Relationship to Thirst

D.J. Ramsay

Introduction

Internal homeostatic processes in animals take place in an aqueous environment. Substances are transported between tissues and organs in physical solution, and biochemical and cellular processes use water as the universal solvent. In terrestrial animals, many important exchanges between the internal and external environment take place at an air/water interface, for example, the uptake of oxygen at alveolar membranes in the lung. For some systems, the amount or volume of water is the critical variable. For example, maintenance of cardiovascular function depends upon mechanisms which ensure constancy of blood volume. Other systems depend more critically on the concentrations of solutes. This is generally true of cell transport processes and biochemical reactions (Somero 1986). It is not surprising, therefore, that a critical homeostatic function, found in all animals, is control of water balance to ensure maintenance of the volume and composition of body fluids.

Water is generally lost by a number of routes at variable rates, but continuously. Thus in humans, there is continuous evaporative loss of water through the skin and respiratory tract. The amount lost may increase dramatically during such circumstances as raised body temperature and exercise, but is never reduced to zero. Again, renal elaboration of urine is a continuous process and, although its concentration and volume may vary over a wide range, flow rates do not normally fall to zero. Intake of fluid, however, is an intermittent process. Thus a normal animal is not always drinking, in spite of continuous loss of fluid in a number of ways. It could be argued that the normal physiology of fluid balance is for animals to acquire a water deficit of increasing severity over time which is corrected intermittently by the intake of water. This view is supported by the fact that when water is consumed, an amount appropriate to the deficit is usually taken in (Thrasher et al. 1981). This is discussed later and in other contributions to this volume.

It should be stressed, however, that not all water intake depends on this simple deficit/satiety model. The water content of food and that produced by metabolism can supply significant amounts of intake, particularly in some species. Patterns of intake are influenced by other behaviours and by social factors (Fitzsimons 1972). The amount of water consumed on each occasion depends heavily on learning. The importance of these factors are discussed in many of the contributions to this volume. The situation in humans is even more complex. The allure of alcohol, caffeine, sweetness or flavour in fluid will often obscure physiological control of water intake and result in water consumption unrelated to deficit. However, in the long term, it can be argued that physiological control mechanisms must underpin drinking behaviour. Fluid intake which is inadequate to support fluid balance will result in a change in behaviour and enhanced fluid intake. Excesses of fluid intake due to the desire for some additive rather than the need for water may lead to fluid overload which will diminish thirst, but which only becomes a problem when it exceeds the excretory capacity of the kidneys. In such an overload situation, the survival necessity of thirst mechanisms is not present and they presumably are not triggered.

Fluid Compartments

Once ingested water has been absorbed through the gastrointestinal tract, it is distributed throughout the total body water. Approximately two-thirds of the water is in cells under steady-state conditions and one-third in the extracellular fluid. In order to discuss the general signals which sense the volume and composition of the body fluids and input to the control of water intake, it is important to understand the factors which determine water movements between fluid compartments.

Extracellular and Intracellular Fluid Compartments

There is much evidence that most cells behave as osmometers with acute changes in extracellular osmolality (Dick 1965). Cell volume varies in inverse proportion to extracellular osmolality. The situation is complicated, however, by a number of factors which include the non-solvent volume of the cell and availability of water for osmotic exchange across cell membranes. Even so, over the physiological range of variation to which most cells are exposed in vivo, there is an inverse linear relationship between extracellular osmolality and cell size.

It is apparent, however, that such osmometric behaviour is not maintained chronically. Thus when avian red cells are suspended in hypertonic media, the cells shrink as would be predicted from osmometric principles, but soon swell back towards their original volume (Kregenow 1971b). The mechanism appears to be due to a coupled influx of Na^+ and K^+ in a 1:1 ratio together with Cl^-. Conversely, Kregenow (1971a) demonstrated that such red cells swell in hypotonic media, but return to their original volumes in about 90 minutes by a mechanism which involves extrusion of K^+ and Cl^-. The extrusion of K^+ from red cells had been noted earlier by Davson (1937) as a pre-haemolytic phenomenon when red cells were suspended in hypotonic media. Thus red cells are capable of changing their osmotic content, and adjusting cell size when external tonicity is altered.

A number of studies have extended these studies to many cell types, and it seems clear that most cells contain mechanisms to alter their solute content in order to control cell volume (Rink 1984; Hoffman 1986; Ellory and Hall 1988). Thus under isosmotic conditions there is no net transfer of ions across cell membranes. If, however, cells are subject to a hypotonic environment they swell and then extrude solute. These mechanisms include potassium chloride co-transport, increase in conductive K^+ and Cl^- channels, K^+/H^+ and Cl^-/HCO_3^- exchanges, and taurine and amino acid transport (Hoffman 1986; Ellory and Hall 1988). Cell volume will return to normal. In a hypertonic environment, cells will first shrink, and then gain solute due to activation of $Na^+K^+2Cl^-$ co-transport or Na^+/H^+ and Cl^-/HCO_3^- exchange. Thus cells contain mechanisms which allow volume maintenance under conditions of chronic change in the osmolality of the extracellular fluid.

Shifts of solute between intra- and extracellular fluid have also been shown in whole animals. For example, when dogs were depleted of extracellular solute by peritoneal dialysis with 5% glucose for 4 hours, the extracellular fluid volume fell due to osmometric movement of water into cells. However, within 24 hours, in the absence of fluid exchange with the environment, extracellular fluid volume had been restored approximately half way to normal (Coxon and Ramsay 1968a). As this occurred without change in plasma sodium and potassium concentration, mobilization of these solutes is indicated. More detailed measurements have shown that movement of sodium, potassium and chloride into extracellular fluid, presumably from cells and possibly from bone, is significant following 80–100 minutes of sodium depletion (Rampton and Ramsay 1974).

The first line of defence in the protection of the constancy of cell volume and body fluid osmolality is provided by thirst and renal mechanisms. However, relatively rapid movement of solute between cells and extracellular fluid will occur if these mechanisms fail to keep plasma osmolality within fairly narrow limits. Solute transfer mechanisms allow successful cellular adaptation to a situation of chronic high or low plasma osmolality, particularly when these changes occur slowly.

Interstitial and Plasma Compartments

The partition of extracellular fluid between intravascular and extravascular compartments depends upon the operation of "Starling forces" across capillaries and postcapillary venules (Michel 1984). These membranes have a high hydraulic conductivity and present a large surface area to the blood flowing by them. At the arterial end of the capillary, the balance of hydrostatic pressures and colloid osmotic pressures favours ultrafiltration of fluid into the interstitial space, whereas at the venular end, the reverse is usually true. The effect of considerable protein leak into the interstitial fluid is minimized by the presence of macromolecules in the interstitial fluid which effectively excludes protein from solution and by lymphatic drainage (Bert and Pearce 1984). Thus the direction of net fluid movement across the capillary wall normally depends on the balance of precapillary and postcapillary resistances.

In situations of hypovolaemia, there is a reduction in mean capillary pressure in many tissues. Less ultrafiltrate is lost from the circulation in capillary beds, and net absorption can occur. This mechanism allows rapid mobilization of fluid from the interstitial compartment, and allows rapid restoration of plasma and blood volumes, and thus cardiovascular function.

It is not surprising, therefore, that receptors sensitive to volume and pressure are present in the circulation and provide inputs to thirst. There is evidence that such receptors are also sensitive to interstitial fluid volume. For example, Sonnenberg and Pearce (1962) showed that blood volume expansion will only cause diuresis and natriuresis if the interstitial compartment is normal or expanded. Our work on the control of water diuresis in salt depleted dogs emphasized that replacement of interstitial fluid volume was critical in the restoration of the ability of these volume deficient animals to develop water diuresis (Coxon and Ramsay 1968b). It would appear reasonable to suggest that the behaviour of any receptors sensitive to vascular wall stretch would be modified by conditions in the interstitial fluid. A change in volume in the interstitial fluid in the wall of a blood vessel might be expected to alter intramural tension. Both passive effects, due to a change in the elastic elements, and active effects, due to response of smooth muscle elements, would contribute to the change in wall tension. Increases in interstitial fluid volume have only small effects on interstitial fluid pressure. However, interstitial fluid pressure is exquisitely sensitive to small reductions in volume below the normal set point (Guyton 1965). Thus receptors in a vascular wall would be sensitive to changes in wall tension, and this is determined by intra- and extravascular pressures.

The Hypothalamus and Thirst

The relationship between the hypothalamus and forebrain, drinking and antidiuresis was demonstrated by Andersson (1953, 1978). Electrical stimulation or injections of hypertonic saline in the anterior hypothalamus caused copious drinking. Additionally, large lesions in this area rendered the animal adipsic and unresponsive to stimuli which normally cause thirst (Andersson and McCann 1955). Later, Buggy and Johnson (1977) drew attention to the importance of structures in the anterior wall of the third ventricle. Lesions of this region in rats caused adipsia and, although animals developed severe dehydration, vasopressin secretion remained unstimulated.

Intensive study has highlighted the importance of forebrain circumventricular organs in water balance, in particular the organum vasculosum laminae terminal (OVLT) and subfornical organ (SFO) (Thrasher 1989). These organs contain fenestrated capillaries and thus lack the normal blood–brain barrier. They are ideally suited to sense the chemical composition of the blood. In dogs, the OVLT is important in sensing raised plasma osmolality and the SFO plasma volume via the renin–angiotensin system, but in other species these regions seem less well differentiated (Thrasher et al. 1982a). The rich connectivity between these regions and other brain structures involved in the control of water balance, such as the supraoptic nucleus, underlines the importance of these structures (Miselis 1982; Ramsay and Thrasher 1991).

Regulation of Plasma Osmolality

From the preceding discussion, the importance of the control of plasma osmolality is apparent. Indeed, in mammals, plasma osmolality is one of the most tightly regulated variables (Ramsay et al. 1988). It rarely changes in normal animals by more than 2%–3%, largely due to thirst and control of renal water loss via regulation of vasopressin secretion. The correlation between plasma "concentration" and drinking was accurately described by Mayer (1900), but confirmation of precise causal relationships awaited the work of Gilman (1937). He showed that intravenous infusions of hypertonic sodium chloride solutions into dogs caused drinking, whereas similarly hypertonic infusions of urea did not. These findings have been extended by a number of workers in the last 50 years (Fitzsimons 1979; Ramsay and Thrasher 1991).

The broken line in Fig. 2.1 shows the relationship between the change in plasma osmolality and water intake in dogs following infusion of hypertonic sodium chloride solutions (Wood et al. 1977). In these experiments, the saline was infused bilaterally via the common carotid arteries exteriorized in skin loops. This method allows the hypertonic stimulus to be delivered to the fore- and midbrain without significantly affecting the volume and concentration in the systemic circulation. As a comparison, the effects of raising plasma osmolality on plasma vasopressin concentration are also shown (Wade et al. 1982a, b).

The characteristics of the control system which maintains plasma osmolality constant are clear from this treatment. At the normal "set point" of plasma osmolality,

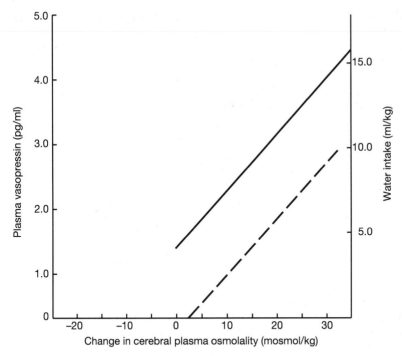

Fig. 2.1 The relationship between plasma vasopressin (solid line) and water intake (broken line) in dogs and change in cerebral plasma osmolality. (From Ramsay (1989), with permission.)

a basal rate of secretion of vasopressin will ensure the elaboration of a concentrated urine and there will be no drinking. As water continues to be lost through the skin, lungs and urine, plasma osmolality increases, stimulating further secretion of vasopressin, concentrating the urine more and thus slowing down the rate of development of dehydration. As the osmotic stimulus increases, so presumably does the intensity of thirst until drinking occurs. The appropriate volume of fluid is consumed to return plasma osmolality to normal.

Much of the development of use of plasma vasopressin/osmolality relationships is due to the elegant work of Robertson in humans (Robertson 1985; Robertson et al. 1977). Instead of measuring volumes of water consumed, individuals were asked to rate how thirsty they were (Robertson 1984). Rolls and her co-workers have used a series of more detailed questions to assess the degree of thirst. Although there is some disagreement about the precise relationship between the thirst and vasopressin thresholds or set point, the control of plasma osmolality by the dual mechanisms of thirst and vasopressin secretion has been shown in a wide variety of animals and is well accepted (Robertson 1984; Ramsay 1989).

As Gilman first noted, intravenous infusion of hypertonic saline stimulates drinking whereas urea does not. The experiments of Verney (1947) also showed that only solutes such as sodium chloride which do not readily penetrate cell membranes caused pituitary-induced antidiuresis. In later experiments in dogs (Thrasher et al. 1980) and sheep (McKinley et al. 1978) this was found to be true for thirst and vasopressin secretion. Thus only solutes which can cause osmotic withdrawal of water from cells, and presumably their shrinkage, can stimulate drinking and vasopressin secretion. The term cellular dehydration drinking seems appropriate in these circumstances of raised plasma osmolality.

However, this simple view of osmometric drinking as proposed by Wolf (1950) should be approached with caution in the light of the rapid adjustments in cell size described earlier. The finding that acute increases in plasma osmolality cause predictable linear decreases in cell size – with cell swelling during hypotonicity – led to the development of the concept of osmoreceptors operating as cells sensitive to their own volume. Thus the osmoreceptive elements in circumventricular tissue, or elsewhere, would generate sensory information in proportion to their volume and input this via neural connections to thirst and vasopressin release systems. The finding that cells adapt their volumes quite rapidly to changed osmolality implies that osmoreceptor cells must lack mechanisms which adjust their volumes if the simple model of an osmoreceptor cell is to be retained. Alternatively, the transduction

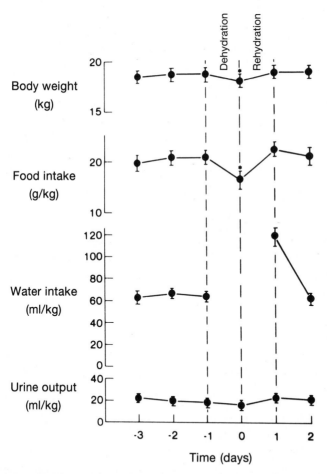

Fig. 2.2 Daily means ± SE of body weight, food and water intake and urine output during a 3-day control period, 24 h of water, but not food, deprivation and a 2-day recovery period. (From Thrasher et al.(1984), with permission.)

mechanism may include input from the ion membrane currents involved with osmoreceptor cell volume maintenance. Biophysical investigation of identified osmoreceptor cells should provide further elucidation.

Dehydration Natriuresis

Recent evidence has shown that control of sodium balance is also involved in maintenance of plasma osmolality (McKinley et al. 1983a, b; Thrasher et al. 1984;

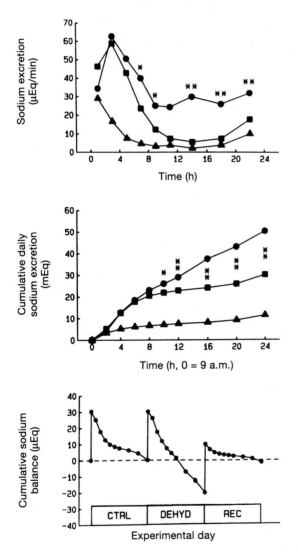

Fig. 2.3. Rate of sodium excretion (top), daily sodium excretion (middle) and cumulative sodium balance (bottom) during 3 experimental days. Filled squares, control days (CTRL); filled circles, dehydration day (DEHYD); and filled triangles, recovery day (REC). *$P<0.05$ between control and dehydration day. (From Metzler et al. (1986), with permission.)

Metzler et al. 1986). The renal response to reduced extracellular fluid volume is sodium retention and antidiuresis, due to stimulation of the renin–angiotensin–aldosterone system and vasopressin release respectively. During dehydration, however, extracellular fluid volume falls but sodium concentration rises. In a wide variety of species, water deprivation is accompanied by natriuresis, not an antinatriuresis.

Data from a population of dogs during dehydration are shown in Fig. 2.2. It is clear that during 24 hours of water, but not food, deprivation, urine volume does not decrease. This is due to increased sodium excretion (Fig. 2.3) which obligates continued excretion of water. During the 24 hours of water deprivation a negative sodium balance of 1.9 ± 0.2 mEq/kg was accumulated. When access to water was restored, dogs drank rapidly and showed a profound antinatriuresis until sodium balance was restored (Fig. 2.3). It is of great interest that a salt appetite has been demonstrated in dehydrated rats when allowed access to fluids (Weisinger et al. 1985).

Thus, part of the response to raised plasma osmolality is renal excretion of sodium. As plasma osmolality rises, cells gain solute and ameliorate changes in their volumes, and the kidney loses sodium which ameliorates the rise in plasma sodium. In the hierarchy of homeostatic control mechanisms, the maintenance of plasma osmolality seems pre-eminent.

The quantitative importance of the dehydration natriuresis should be emphasized. Dogs with OVLT lesions (Thrasher et al. 1983) and sheep with anterior third ventricle lesions which include the OVLT (McKinley et al. 1983) no longer show a dehydration natriuresis. Plasma sodium concentration rises 2–3 times as much during water deprivation compared with normal animals. It is difficult to escape the conclusion that control of sodium balance by the brain is intimately involved in the maintenance of plasma osmolality.

Reduction of Extracellular Fluid Volume

The general relationship between reduction in body fluid volume and thirst has been appreciated in the clinical literature since the last century (Fitzsimons 1979). Experimentally, it is clear that reduction of extracellular fluid volume without change in its osmolality leads to stimulation of drinking – extracellular dehydration thirst. For example, reduction in extracellular fluid volume by subcutaneous or intraperitoneal injection of polyethylene glycol causes extravascular sequestration of protein-free fluid at the site of injection, reduction in blood volume and thirst. After 8–10 hours of injection, salt appetite can be demonstrated (Stricker 1981; Fitzsimons 1961). Thirst can also be demonstrated following haemorrhage (Fitzsimons 1961; Ramsay and Thrasher 1986; Thrasher and Keil 1987) and mechanical restriction of venous return (Thrasher et al. 1982b).

There is good evidence that reduction in thoracic blood volume is a key element in the production of thirst and vasopressin secretion during hypovolaemia. Low pressure cardiopulmonary receptors in the right and left atria and high pressure arterial baroreceptors are involved. Reduction in stimulation of these receptors in hypovolaemia leads to thirst and vasopressin secretion via identified pathways which relay through the nucleus of the tractus solitarius. Additionally, hypovolaemia and arterial hypotension directly and reflexly cause stimulation of renal renin secretion and raised plasma angiotensin II. Angiotensin II interacts with the brain via forebrain circumventricular organs – in many species the subfornical organ – and causes drinking. Thus neural and humoral systems can act in concert to produce thirst.

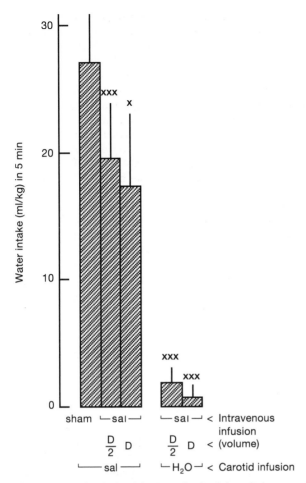

Fig. 2.4. Water intake in eight water-deprived dogs during a combination of carotid and intravenous infusion. All carotid infusions are split between the two carotids. D = fluid deficit as calculated from change in body weight during water deprivation. Intravenous infusion D = volume of saline equal to fluid deficit infused; D/2 = volume of saline equal to half the fluid deficit infused. Carotid infusions of saline (sal) and water (H_2O) were at a rate of 0.6 ml/kg/min. Significances: $*P < 0.05$, $** P < 0.01$, $*** P < 0.001$ when compared to control. (From Ramsay et al. (1977), with permission.)

Interaction of Cellular and Extracellular Dehydration

As has been stressed earlier, a normal cycle of water balance in terrestrial animals is continuous loss of fluid with intermittent drinking to satiety. In periods between drinking, plasma osmolality will rise and blood (or extracellular fluid) volume will fall. Both mechanisms can contribute to the development of thirst.

Ramsay et al. (1977) investigated the interaction between the two inputs in dogs deprived of water for 24 hours (Fig. 2.4). When the raised plasma osmolality stimulus to hypothalamic osmoreceptors was removed by infusions of water bilaterally via the common carotid arteries, subsequent drinking was reduced by 70%. On a separate occasion, restoration of extracellular fluid volume by intravenous infusion of artificial extracellular fluid reduced drinking by 30%. A combination of the two treatments eliminated drinking by these dehydrated dogs. Similar quantitative contributions of raised plasma osmolality and reduced extracellular fluid volume are also found for vasopressin secretion during dehydration (Wade et al. 1983). Although the precise contributions of the two inputs vary between species during dehydration, where it has been studied, both are involved.

Summary

The similarity between the physiological inputs to the control of thirst and those to vasopressin secretion is striking. These represent the controlled input and output to water balance, and emphasize the importance of the regulatory aspects of drinking. The translation of these basic inputs into learned behaviour, selection of fluids to be ingested and hedonic effects of fluid additives in the complex human situation is a challenge. The success of even the human organism in keeping plasma osmolality and its volume within such narrow limits, however, provides eloquent testimony to the dominant influence of physiological regulation of thirst over all other inputs in the control of water balance.

References

Andersson B (1953) The effect of injections of hypertonic NaCl solutions into different parts of the hypothalamus of goats. Acta Physiol Scand 28:188–201

Andersson B (1978) Regulation of water intake. Physiol Rev 58:582–603

Andersson B, McCann SM (1955) The effect of hypothalamic lesions on the water intake of the dog. Acta Physiol Scand 35:312–320

Bert JL, Pearce RH (1984) The interstitium and microvasculary exchange. In: Renkin EM and Michel CC (eds) Handbook of physiology, section 2, vol IV, Part I. American Physiological Society, Baltimore, pp 521–547

Buggy J, Johnson AK (1977) Preoptic hypothalamic periventricular lesions: thirst deficits and hypernatremia. Am J Physiol 233:R44–R52

Coxon RV, Ramsay DJ (1968a) The effect of sodium depletion on the renal response to water loading in dogs. J Physiol (Lond) 197:617–630

Coxon RV, Ramsay DJ (1968b) The effect of infusions of plasma and of saline on the renal response to water loading in sodium depleted dogs. J Physiol (Lond) 197:631–638

Davson H (1937) The loss of potassium from the erythrocyte in hypotonic saline. J Cell Comp Physiol 10:247–264

Dick DAT (1965) Cell water. Butterworth, London

Ellory JC, Hall AC (1988) Human red cell volume regulation in hypotonic media. Comp Biochem Physiol 90A:533–537.

Fitzsimons JT (1961) Drinking by rats depleted of body fluid without increase in osmotic pressure. J Physiol (Lond) 159:297–309

Fitzsimons JT (1972) Thirst. Physiol Rev 52:468–561

Fitzsimons JT (1979) The physiology of thirst and sodium appetite. Cambridge University Press, Cambridge

Gilman A (1937) The relation between blood osmotic pressure, fluid distribution and voluntary water intake. Am J Physiol 120:323–328

Grantham JJ (1977) Pathophysiology of hypo-osmolar conditions: a cellular perspective. In: Andreoli TE, Grantham JJ, Rector FC, Jr (eds) Disturbances in body fluid osmolality. American Physiological Society, Bethesda, pp 217–226

Guyton AC (1965) Interstitial fluid pressure: II. Pressure volume curves of interstitial space. Circ Res 16:452–460

Hoffman EK (1986) Anion transport mechanisms in the plasma membrane of vertebrate cells. Biochim Biophys Acta 864:1–31

Kregenow FM (197la) The response of duck erythrocytes to non-hemolytic hypotonic media. Evidence for a volume controlling mechanism. J Gen Physiol 58:372–395

Kregenow FM (1971b) The response of duck erythrocytes to hypertonic media. Further evidence for a volume controlling mechanism. J Gen Physiol 58:396–412

Mayer A (1900) Variations de la tension osmotique de sang chez les animaux privés de liquides. Comp Ren Biol 52:153–155

McKinley MJ, Denton DA, Weisinger BS (1978) Sensors for antidiuresis and thirst – osmoreceptors or CSF-sodium detectors? Brain Res 141:89–103

McKinley MJ, Denton DA, Nelson JF et al (1983a) Dehydration induces sodium depletion in rats, rabbits and sheep. Am J Physiol 245:R287–R292

McKinley MJ, Denton DA, Park RC et al (1983b) Cerebral involvement in dehydration induced natriuresis. Brain Res 263:340–343

Metzler CH, Thrasher TN, Keil LC, Ramsay DJ (1986) Endocrine mechanisms regulating sodium excretion during water deprivation in dogs. Am J Physiol 251:R560–R568.

Michel CC (1984) Fluid movements through capillary walls. In: Renkin EM and Michel CC (eds) Handbook of physiology, section 2, vol IV, Part I. American Physiological Society, Baltimore, pp 375–409

Miselis RR (1981) The efferent projection of the subfornical organ of the rat: a circumventricular organ within a neural network subserving water balance. Brain Res 1–23

Rampton DS, Ramsay DJ (1974) The effects of production of sodium depletion by peritoneal dialysis with 5% glucose of the volume and composition of extracellular fluid of dogs. J Physiol (Lond) 237:535–554

Ramsay DJ, (1989) The importance of thirst in maintenance of fluid balance. In: Bayliss PH (ed) Water and salt homeostasis in health and disease. Bailliere Tindall, London, pp 371–392. (Clinical Endocrinology and Metabolism, vol 3, no 2)

Ramsay DJ, Thrasher TN (1986) Hyperosmotic and hypovolemic thirst. In: deCaro G, Epstein AN, Massi M (eds) The physiology of thirst and sodium appetite. Plenum Press, New York, pp 83–96

Ramsay DJ, Thrasher TN (1991) Thirst and water balance. In: Stricker EM (ed) Neurobiology of food and water intake. Plenum Press, New York, pp 353–387 (Handbook of behavioral neurobiology)

Ramsay DJ, Rolls BJ, Wood RJ (1977) Thirst following water deprivation in dogs. Am J Physiol 232: R93–R100

Ramsay DJ, Thrasher TN, Bie P (1988) Endocrine components of body fluid homeostasis. Comp Biochem Physiol 90A:777–780

Rink TJ (1984) Aspects of the regulation of cell volume. J de Physiol (Paris) 79:388–394

Robertson GL (1984) Abnormalities of thirst regulation. Kidney Int 25:460–469.

Robertson GL (1985) Osmoregulation of thirst and vasopressin secretion: functional properties and their relationship to water balance. In: Schrier R (ed) Vasopressin. Raven Press, New York, pp 202–213

Robertson GL, Athar S, Shelton RL (1977) Osmotic control of vasopressin function. In: Andreoli TE, Grantham JJ, Rector FC, Jr (ed) Disturbances in body fluid osmolality. American Physiological Society, Bethesda, pp 125–148

Somero GN (1986) Protons, osmolytes and fitness of internal milieu for protein function. Am J Physiol 251:R197–R213 (Regulatory, integrative and comparative physiology vol 20)

Sonnenberg H, Pearce JW (1962) Renal response to measured blood volume expansion in differently hydrated dogs. Am J Physiol 206:1–7

Stricker EM (1981) Thirst and sodium appetite after colloid treatment in rats. J Comp Physiol Psychol 95:1–25

Thrasher TN (1989) Role of forebrain circumventricular organs in body fluid balance. Acta Physiol Scand 136 Suppl 583:l4l–150.

Thrasher TN, Keil LC (1987) Regulation of drinking and vasopressin secretion: role of the organum vasculosum laminae terminalis. Am J Physiol 253:R108–R120 (Regulatory, integrative and comparative physiology vol 22)

Thrasher TN, Brown CJ, Keil LC et al (1980) Thirst and vasopressin release in the dog: an osmoreceptor or sodium receptor mechanism? Am J Physiol 238:R333–R339 (Regulatory, integrative and comparative physiology vol 7)

Thrasher TN, Nistal-Herrera JF, Keil LC et al (1981) Satiety and inhibition of vasopressin secretion after drinking in dogs. Am J Physiol 240:E394–E401 (Endocrinology and Metabolism vol 3)

Thrasher TN, Keil LC, Ramsay DJ (1982a) Lesions of the organum vasculosum of the lamina terminalis (OVLT) attenuate osmotically-induced drinking and vasopressin secretion in the dog. Endocrinology 110:1837–1839

Thrasher TN, Keil LC, Ramsay DJ (1982b) Hemodynamic, hormonal and drinking responses to reduced venous return in the dog. Am J Physiol 243:R354–R362 (Regulatory, integrative and comparative physiology vol 12)

Thrasher TN, Simpson JB, Ramsay DJ (1982c) Lesions of the subfornical block angiotensin induced drinking in the dog. Neuroendocrinology 35:68–72.

Thrasher TN, Keil LC, Ramsay DJ (1983) Altered responses to dehydration in dogs with lesions of the organum vasculosum laminae terminalis (OVLT). Proceedings of the 29th International Congress Physiological Sciences, p 49.

Thrasher TN, Wade CE, Keil LC, Ramsay DJ (1984) Sodium balance and aldosterone during dehydration and rehydration in the dog. Am J Physiol 247:R76–R83 (Regulatory, integrative and comparative physiology vol 16)

Verney EB (1947) The antidiuretic hormone and the factors which determine its release. Proc R Soc Lond B 135:25–106.

Wade CE, Bie P, Keil LC et al (1982a) Osmotic control of vasopressin in the dog. Am J Physiol 243:E287–E292 (Endocrinology and Metabolism vol 6)

Wade CE, Bie P, Keil LC et al. (1982b) Effect of hypertonic intracarotid infusions on plasma vasopressin concentration. Am J Physiol 243:E522–E526 (Endocrinology and Metabolism vol 6)

Wade CE, Keil LC, Ramsay DJ (1983) Role of volume and osmolality in the control of plasma vasopressin in dehydrated dogs. Neuroendocrinology 37:349–353.

Weisinger RS, Denton DA, McKinley MJ, Nelson JF (1985) Dehydration induced sodium appetite in rats. Physiol Behav 34:45–50.

Wolf AV (1950) Osmometric analysis of thirst in man and dog. Am J Physiol 161:75–86.

Wood RJ, Rolls BJ, Ramsay DJ (1977) Drinking following intracarotid infusions of hypertonic solutions in dogs. Am J Physiol 232:R93–R100 (Regulatory, integrative and comparative physiology vol 1)

Commentary

Robertson: There may be important species differences in the way that sodium homeostasis affects osmoregulation. In humans, fluid deprivation results in a gradual fall in the rate of urine sodium excretion, presumably as a result of a slight reduction in blood volume. Usually, this is associated with a slightly positive or neutral sodium balance. In some cases, however, fluid deprivation results in a negative sodium balance because, like eating in general, the intake of sodium decreases even more than output. It is unclear whether this avoidance of intake is due to an innate inhibitory effect of fluid deprivation on appetite for food or is simply the result of the subject's knowledge that eating is likely only to increase the intensity of thirst. However, it does indicate that natriuresis is not a significant element of osmoregulation in humans. The absence of this capability may be the result of evolution to an upright posture since the exacerbation of the hypovolaemia that would result from a natriuretic response to fluid deprivation would pose a greater threat to maintenance of blood pressure in a bipedal than in a quadrupedal animal. Because of this and their insensitivity to volume stimuli, the thirst and vasopressin responses to fluid deprivation in humans appear to be mediated almost totally by the osmoregulatory mechanism.

Ramsay: Robertson raises an interesting question. If dogs are producing a urine of moderate concentration before water deprivation occurs, they also will show reduction in urinary volume and maximal concentration. However, when that point is reached, natriuresis begins to occur. We have shown that the development of negative sodium balance does not depend on reduction in food intake during water deprivation. In one set of studies, food intake did not decrease, and the natriuresis which occurred produced as profound a negative sodium balance as in other studies where some reduction in food intake did occur. The suggestion that assumption of the upright posture might have caused an evolutionary blunting of the water deprivation natriuretic response is an interesting one. However, I still feel that careful investigations of human – or primate – responses to water deprivation would be worthwhile. It is difficult to imagine that a mechanism which seems to have such a widespread distribution in the animal kingdom would be completely missing in humans.

Chapter 3

The Ontogeny of Drinking

W.G. Hall

Issues in the Development of Drinking

By attaching to their mothers' nipples and sucking, mammalian neonates obtain all their necessary fluids and nutrients. This behaviour, a defining characteristic of Mammalia known as "suckling", shows a remarkably similar form across species (Peiper 1963), from rodent pups to young human infants (Fig. 3.1). Though specific features may differ, suckling serves the ingestive requirements of the preweaning period.

Suckling is most frequently studied in terms of its role in nutrition and food-getting, but from our present interest in fluid procurement and the developmental origins of thirst, we can also ask: "Is suckling drinking?" The answer to this question is not so straightforward as one might first imagine. Suckling is clearly drinking in the sense of being fluid consumption. The issue can be raised in a more instructive way, though, by asking whether infants detect fluid needs and respond to them with changes in their suckling behaviour. For rodent infants, in which the issue has been explicitly tested, the answer appears to be 'no'. These findings, reviewed in the following sections, lead to further questions about the ontogeny of drinking. Is there any early behaviour that is activated by dehydration, or is there any evidence that the CNS of developing animals is sensitive to this condition? When does hydrationally relevant drinking normally develop and what processes underlie this development? When does discrimination of different fluids emerge and how does this relate to the onset of spontaneous fluid ingestion?

Such questions are of obvious importance for our appreciation of the biology and behaviour of developing infants but, beyond this, their resolution can provide considerable information on how appetitive ingestive systems are controlled and organized, and may provide new ways to study ingestive behaviour. Exploring such issues will reveal a few of the unique experimental opportunities for understanding the organization of appetitive behaviour that are provided by developmental analysis.

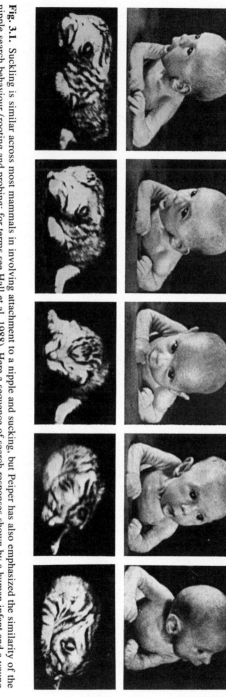

Fig. 3.1. Suckling is similar across most mammals in involving attachment to a nipple and sucking; but Peiper has also emphasized the similarity of the nipple search behaviour (rooting and probing; for terms see Hall et al. 1988). Here a sequence of search responses shown by a human infant and a young kitten are compared. (From Peiper (1963), with permission.)

The Ontogeny of Ingestive Responses to Dehydration

Infant Mammals are Highly Vulnerable to Fluid Loss, Yet...

Vulnerabilities

At birth, most altricial mammals have poorly developed pelage and high ratios of skin surface to body volume. These characteristics make them highly vulnerable to both thermal stress and to evaporative fluid loss. Elaborate systems of maternal care and infant appetitive behaviour have evolved to help protect infant mammals from these vulnerabilities. Nonetheless, in infant rodents for example, even a brief separation from the mother can result in a substantial loss of body fluid (e.g. Heller 1949; Friedman 1975, 1979). The extracellular fluid compartment is particularly vulnerable to fluid loss during early development. Loss of extracellular fluid becomes a special problem for any human infants suffering from gastrointestinal disorders or illnesses which result in vomiting or diarrhoea (Winters 1982).

Protection

At birth, most mammals already have some limited abilities to concentrate urine and retain fluid, and to deal with hypotonic loads (e.g. Heller 1947; Falk 1955; Adolph 1957; Dlouha et al. 1963). In particular, neonatal capabilities are usually adequate to deal with the normal variability of the fluid and electrolytes received in mothers' milk. Renal function does not fully mature, however, until weaning age in most mammals. Notably, concentrations of artificial milk formulas used for human infants before the middle of this century appear to have been beyond the capabilities of many infants. In fact, modern understanding of the history of formula rearing indicates that the onset of successful rearing practices depended critically on the establishment of diets with appropriate osmotic properties in addition to appropriate balances of other components and production under improved sanitation conditions (Cone 1984). Because formula-reared infants are very vulnerable to dehydration from diarrhoea owing to other features of artificial diets, the osmotic and electrolytic properties of such formulas are now recognized as all the more important. In some situations, even slight deviation from normal concentration can cause problems. For example, a number of clinical reports describe infants who became critically dehydrated from hyperosmotic formula which was made up from powdered stock that had accidentally been measured using rounded spoonfuls rather than level ones (e.g. Taitz and Byers 1972).

In general, though, it is clear that maternal behaviour and maternal production of milk have evolved to ensure adequate hydration. Hydration is achieved as a result of multiple adaptations. Maternal nesting behaviour reduces evaporative loss of fluid. Maternal nursing responses anticipate pups' fluid and caloric needs and are synchronized by a number of mechanisms (see reviews by Galef 1981a; Alberts and Gubernick 1983) such as thermally mediated control of maternal behaviour (Leon 1979). In primates, the thoroughness of maternal attention further relieves neonates from needing to exhibit many of the appetitive suckling responses required of non-primate mammals. Finally, lactation provides a fluid supply appropriately balanced and largely buffered against short-term changes in fluid availability.

Table 3.1. Approximate timing of events in rat pup development[a]

Age (days)	Weight (g)	Morph./Motor/Physiol.	Thirst related		
			Normal ingestion	Independent ingestion	Effective dipsogens
				Mouth/swallow oral infusions	
birth 0	6	Poikilothermic	SUCKLING	Flavour-related learning	NaCl
		Crawl		Will lick from floor Some taste discrimination	
6	14	Some fur Walk			Angiotensin II Renin PG (all non-selective)
		Teeth erupt			Isoproterenol Water intake preferentially stimulated by some dipsogens
12	26	Ears open Eyes open	Nibble solids Food found in stomach Spontaneous drinking	Full taste preference/ aversion	
18	40	Full fur Approach homeothermic Onset intestinal sucrase activity	(NaCl inhibits suckling) NORMAL WEANING		NaCl induced anorexia
24	60	Full renal response to water and salt loads			
60–90		Puberty			

[a] See text for references; also see Epstein (1976), Hall (1990), Leshem et al. (1988).

The centrality of hydrational factors in the adaptive nature of maternal behaviour has recently been shown in the impressive fluid recovery and conservation demonstrated by rodent mothers. In the process of licking and cleaning their young, these females are able to recover 65%–70% of the fluid lost through lactation and are able to recycle it for future milk production (Baverstock and Green 1975; Friedman and Bruno 1976). Their behaviour is sensitive to dehydration and, in fact, they treat their young as sources of fluid (Friedman et al. 1981). This system provides an example of the way in which interlocking needs and adaptations contribute to organizing complex mammalian behaviour and behavioural interactions.

... Yet Suckling Behaviour is not Stimulated by Dehydration

Absence of a Suckling Response

In the face of hydrational vulnerabilities, there is remarkably little responsiveness to dehydration in the suckling behaviour of infants themselves. In infant rodents, the animals focused on here because their drinking development has been most thoroughly studied (Table 3.1 gives a general developmental timetable of this species), dehydration induces little additional enthusiasm for suckling or inclination to increase fluid consumption at the nipple.

Studies of the approach and nipple-attachment component of suckling behaviour, the appetitive portion of the behavioural sequence, provide a first kind of evidence that suckling is not affected by dehydration. When young rat pups are both dehydrated and deprived of food and are later placed with the mother to suckle, they attach to their mother's nipples no more quickly or frequently than pups that have not been dehydrated (see review by Blass et al. 1979). In addition, when pups are experimentally dehydrated by injection of sodium chloride solution (producing a cellular stimulus) or polyethylene glycol (PG; producing an extracellular stimulus), their nipple attachment behaviour is not enhanced at any age (Drewett and Cordall 1976; Drewett 1978a; Bruno et al. 1982, 1983) nor is suckling stimulated by central renin injection (Leshem et al. 1988). In fact, after 15 days of age, both sodium chloride and PG actually inhibit nipple attachment (Bruno et al. 1982).

Dehydration also has little influence on the intake component of suckling. Intake at the nipple is not increased at any age in response to dehydration produced by sodium chloride injection (Friedman 1975; Bruno et al. 1982, 1983) nor by the dipsogen angiotensin (Leshem et al. 1988). Hypovolaemia produced by PG injection in pups did not increase intake in investigations by Drewett (1978a) and Bruno et al. (1982). Drewett argues that an earlier report (Friedman 1975) of PG-stimulated intake in suckling pups was an artifact resulting from the effect of PG on fluid retention that distorted the weight-gain based intake measures. Bruno et al. further found that, by 20 days of age, either PG or sodium chloride injection actually inhibited suckling intake.

These data provide considerable converging evidence that infant rodents show little or no response to dehydration in their suckling behaviour, in either nipple attachment or intake components. When sensitivity to dehydration finally develops as pups become older, suckling is inhibited, not enhanced, by dehydration.

Suckling and Feeding

Could this absence of response to dehydration be understood by thinking of suckling as feeding rather than drinking? Probably not. Early in development, there is little resemblance between suckling and feeding, either in the sensory-motor features of the behaviours or in their control and modulation (see review by Hall and Williams 1983; also Drewett 1978b; Epstein 1984, 1986). Thus, early suckling and later feeding are not homologous. It is only as weaning approaches that pups come to view the mother as a food source (e.g. Kenny et al. 1979; Lichtman and Cramer 1989). This late-emerging perception of suckling as feeding may explain why dehydration comes to inhibit suckling intake in older pups.

Suckling was initially studied by psychobiologists as a precursor to adult ingestive systems, but the lack of evidence for continuity between this infant mode of ingestion and the adult modes of ingestion is now recognized. Epstein (1986) puts it, "Suckling is not the kindergarten for adult feeding and drinking." Suckling remains a fascinating biological system worthy of considerable study in its own right, but it is not the origin for drinking.

To summarize this perspective on suckling and drinking: infant rodents consume fluid at the nipple, and this fluid serves hydrational needs. They are not drinking in the sense of preferring a hypotonic fluid or having a specific response to fluid deprivation or fluid need. They do drink in the sense of using the unique mammalian adaptation of suckling to consume fluids (for discussion of "early adaptations" see Galef 1981a; Hall and Oppenheim 1987; Alberts and Cramer 1988). It is the robustness of this suckling response and not its sensitivity to dehydration that ensures adequate hydration. Since mothers do not supply more fluid or calories than infants can handle, there has been little need for mammalian infants to evolve specific responses to dehydration or protection against surfeit.

If Drinking Doesn't Come From Suckling, Where Does it Come From?

On the basis of the information reviewed above, it should be clear that the study of the development of drinking and its controls must be different from the analysis of suckling. Although a number of authors have argued similarly regarding the development of feeding, the case is more compelling for drinking – and more interesting. Unlike the case for feeding, we are in an excellent situation to explore drinking development because we have a good understanding of stimuli to drink and well explored hypotheses about how they are transduced (at least relative to our understanding of feeding, e.g. Stricker, 1984; see Section II). Also, even though

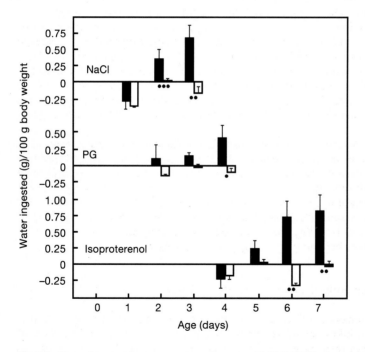

Fig. 3.2. The early onset of oromotor responsiveness to NaCl (1.25% body weight, 1.25 M NaCl, s.c.), PG (1.25% body weight, 40% polyethylene glycol, s.c.) and isoproterenol (500 μg/kg, 1.25 ml/100 g volume, s.c.) as measured by water intake (weight gain as percentage of body weight) in rat pups held at a flowing water spout. Open bar, vehicle injected; solid bar, dipsogen. (From Wirth and Epstein (1976).)

pups do not drink by suckling, systems for thirst are present from early in life, in ontogenetic anticipation of their later roles in appetitive behaviour.

Dehydration Modulates Early Oromotor Responses

In recent years, researchers interested in the developmental origins of drinking have exploited oromotor systems for fluid lapping and swallowing that are present early in development. When infant mammals receive oral stimulation with fluids, they make lapping and swallowing movements as well as show facial responsiveness dependent on the taste of the solution. This oromotor responsiveness has permitted analysis of gustatory development, as well as the analysis of the development of ingestive systems and their controls (Hall 1990). In fact, all mammals that have been studied show such oromotor responsiveness well before birth and are known to swallow amniotic fluid actively (e.g. Bradley and Mistretta 1973; Smotherman and Robinson 1988). Such ingestion may even play a role in homeostasis (e.g. Lev and Orlic 1972; McCance 1959).

Research using a variety of techniques to study the ingestive responses of newborns to orally presented fluids has shown young mammals, particularly young rodents, to be exquisitely sensitive to dehydration from early in their development. Rat pups, from the time of birth, will consume solutions infused into their mouths and will lap solutions spread on the floor beneath them. This form of "independent ingestion", unlike suckling, resembles adult ingestion in its motor topography and physiological control. Dehydration clearly stimulates intake in these tests of ingestion which are independent of suckling behaviour. For example, in the first demonstration of this type, Wirth and Epstein (1976) showed that cellular dehydration produced increased intake of solutions which flowed into neonates' mouths from an external tube (Fig. 3.2). Based on further work of Epstein and collaborators and Bruno and collaborators, it has been possible to construct a timetable for the onset of responsiveness to various dipsogenic stimuli (Table 3.1). In general, cellular dehydration produces increased intake from 1–3 days; extracellular dehydration stimulates intake as early as 4–6 days.

Dehydration, in fact, may be the only stimulus for increased oromotor activity in young rodents. These animals appear to have just two controls of independent ingestion: an inhibition of ingestion produced by gastric fill and a modulation of ingestion by dehydration (Hall 1990). Neither nutritive deprivation nor nutritive loads influence oromotor responses, beyond their contribution to the gastric fill signal, until 9 days of age. Thus, despite dehydration's lack of effect on suckling, it is ontogenetically the first physiological signal regarding body state to have an influence on oromotor behaviour.

Maturation of Responses to Dehydration Stimuli

Although responsiveness to dehydration is present from early in development, it is not suggested that drinking or thirst systems are fully mature. During the course of development, (1) full responsiveness to specific dipsogens gradually emerges, (2) discrimination of hydrationally appropriate ingesta develops, and (3) appetitive behavioural competence appears. Development in each of these areas is important to the ultimate expression of hydrationally related adaptive behaviour in adult animals.

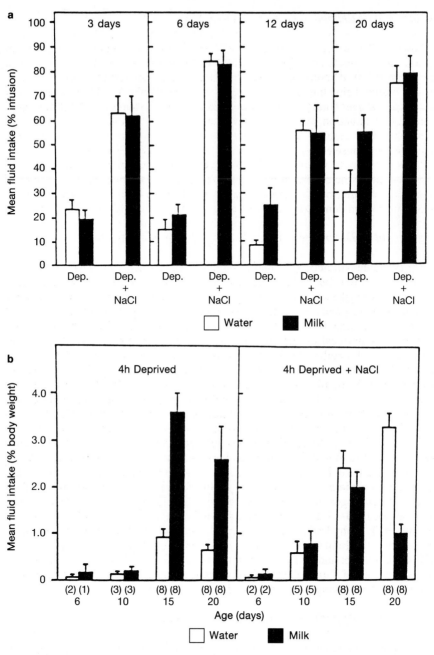

Fig. 3.3. Emergence of selectivity in water and milk saline intake after dehydration, for oral infusion tests (**a**) and ingestion from cups (**b**). Rat pups were either 4 h deprived (baseline condition) or were deprived and injected with NaCl (1% body weight, 1 M NaCl, s.c.). Note that in the oral infusion tests (**a**) dehydration begins to have a relatively greater effect on water intake by 12 days of age, though milk intake is still stimulated as well. In tests with fluids in cups (**b**), selective stimulation of water intake is apparent at 15 days, and dehydration significantly inhibits intake of milk at 20 days of age. (From Bruno (1981), with permission.)

Development of Responsiveness to Dehydration

As shown in Table 3.1, sensitivity to all the dipsogenic stimuli that have been studied is not complete until 6–8 days (Epstein 1984; also see discussion in Leshem and Epstein 1988). It should be noted that the response induced by sodium chloride injection, besides being the earliest to mature, has an adult-like threshold for activation from the earliest ages studied (e.g. Almli 1973; Wirth and Epstein 1976). Interestingly, in the natural situation pups are much more likely to suffer from extracellular dehydration than cellular dehydration (see p.37 and Friedman 1979); yet, the extracellular response emerges a bit later. It seems likely that this gradual emergence of full responsivity to various dipsogens depends on maturation of neural systems involved in either the detection of dehydration or in its behavioural expression. In this regard, a maturation of discrete neurotransmitter systems involved in ingestive behaviour has been demonstrated (Ellis et al. 1984).

Development of Discrimination of Thirst Appropriate Ingesta

When they are dehydrated, animals need to consume osmotically or electrolytically appropriate solutions, but such selective ingestion is slow to develop. Before 6 days of age, pups can make some rudimentary discriminations between solutions in their ingestion and do respond differentially to water compared to milk and water compared to sucrose (Hall and Bryan 1981; Johanson and Shapiro 1986). Also, despite considerable anatomical immaturity, the olfactory system functions well enough to serve as the basis for both unconditioned and conditioned discriminations (Coopersmith and Leon 1988). Nonetheless, before 6 days of age, dehydration indiscriminately increases the intake of all solutions in oral infusion tests, even fluids that will further exacerbate dehydration. For example, in oral infusion tests using pups under 6 days of age, dehydration produced by sodium chloride injection equivalently increased intake of water, milk, and concentrated sucrose solutions that were themselves dehydrating (Fig. 3.3). After this age rat pups begin to show preferential intakes of water, compared to milk or hypertonic saline solutions. Yet, even so, during most of the preweaning period, dehydration still has a stimulatory effect on milk intake. Angiotensin and renin administration also indiscriminately increase intake of all solutions at early ages, not showing a completely selective effect on water versus milk intake until 16–17 days (Leshem et al. 1988; Misantone et al. 1980). For young pups, this lack of hydrationally appropriate discrimination is no doubt partially attributable to the immaturity of the gustatory system (e.g. Mistretta and Bradley 1986) but may reflect immaturity of the drinking system as well. A further contribution to the emergence of fluid selection is discussed after the description of appetitive features of early drinking in the next section.

Development of Appetitive Components of Drinking

The oromotor responses that have been discussed so far constitute the final component of the drinking response sequence. Yet, the developmental emergence of a dehydrated animal's ability to direct and maintain contact with a fluid source and to extend its behaviour to searching the environment for fluids is also of interest. Considerably less is known about these capabilities. Initially it was thought that the independent ingestive behaviour of young rats was poorly directed in the sense that pups could not search out or maintain contact with a topographically restricted source of diet in their test containers. For example, when 6-day-old pups were required to ingest when diet was restricted to a limited area on the floor of the test container, they were very inefficient in maintaining contact with the solution and consumed relatively little of it (Hall and Bryan 1981). It was thus believed that young pups could not orient their ingestion to restricted sources of ingesta in their environment. However, it has recently been shown that under different conditions even 3-day-old pups will ingest efficiently from localized diet sources and consume amounts similar to pups

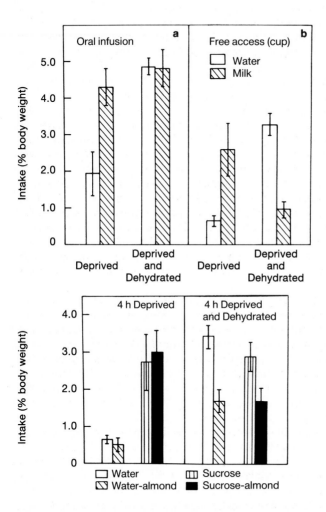

Fig. 3.4. Upper panel, comparison of the effect of dehydrating 20-day-old rat pups then measuring intake in two types of ingestive tests: (**a**) oral infusion, (**b**) free access from a cup. Water intake is increased in the oral infusion tests, but anorexia (inhibition of milk intake) is seen only in intake from a cup (deprivation and dehydration procedures same as in Fig. 3.3). Such dehydration anorexia may depend on olfaction because (as seen in lower panel) adding odour (almond) to a sucrose solution results in a suppressed intake after dehydration; a similar effect is seen in the case of simply adding odour to water. (From Bruno and Hall (1982).)

ingesting from the entire floor surface (Hall et al. 1990). More importantly, dehydration produced by sodium chloride injection caused increased intake that was also efficiently maintained when solutions were restricted to a small area.

In the above experiments, pups were placed with their mouths near the test solutions, and therefore they only had to maintain their contact with fluids, they did not have to seek fluids. To test a type of water-seeking behaviour, Bruno (1981) placed dehydrated pups, initially, at water and then moved them away so that they had to return to drink. In this case, pups showed efficient intake only after 10 days of age. Thus, ability to direct drinking towards an existing and known fluid source seems to mature after the ability to maintain behaviour at a fluid source.

A further primary feature of adult appetitive behaviour is the animal-initiated search or exploration for appropriate ingesta, a behavioural component generally believed to depend on an increased locomotor activity as well as an increased responsiveness to potential fluid-signalling cues. This component of the appetitive sequence has not been studied for drinking development. It is known, that by 23 days, dehydrated rat pups are able to initiate drinking in a novel test environment (Almli 1973), but it is not known when or how pups first begin to search their environment in response to dehydration. Earlier conclusions that ingestive responses to dehydration were late-maturing (e.g. Adolph 1968; Friedman and Campbell 1974) probably resulted from tests in which the poorly developed search and recognition capabilities of pups precluded their contacting fluid and ingesting (Wirth and Epstein 1976). While some potentially relevant information is available on the effects of maternal deprivation (thus social, nutritional, and hydrational deprivation) and on locomotor activity (e.g. Moorcroft et al. 1973; Campbell and Raskin 1978; Hofer 1975, 1978), such effects are still poorly understood and the relation of search activity to hydrational state has not been specifically addressed. In weanling age pups, however, analysis of the approach components of the drinking sequence has proven important in revealing how the full distinction between drinking and feeding eventually emerges.

When adult animals are dehydrated they do not eat, even when they are hungry. Indeed, one of the most common features of the mammalian response to dehydration is an inhibition of food ingestion; this inhibition is known as "dehydration anorexia" (Lepkovsky et al. 1957). A portion of the phenomenon of dehydration anorexia is attributable to differential effects of dehydration on oromotor responsiveness to different tastes, which emerges towards weaning age as described above. However, there is a more significant mechanism in the production of dehydration anorexia that appears to operate on appetitive or approach behaviour and that was uncovered in comparisons between two types of ingestive tests.

The responses of dehydrated pups to milk and water in tests of lapping from the floor and oral-infusion tests were found to be similar before 15 days of age. In both types of test, intake of water and milk was increased by cellular dehydration. However, at 20 days of age the tests gave different results (Bruno 1981). Dehydration increased the ingestion of orally infused milk but blocked the intake of milk when animals were ingesting from the floor of their test containers (Fig. 3.4). In other words, when fluid food was placed in the mouth, ingestion was enhanced by dehydration (though as noted above, not as much as water intake), but when milk had to be acquired from the environment, intake was blocked by dehydration. The oromotor consummatory response component in the ingestive sequence thus appeared to be stimulated by dehydration, irrespective of diet or age, although responses of the preceding approach and contact component in the sequence were inhibited.

This control exerted at the food-contact component of ingestion appears to be based on a hydrational modulation of a pup's response to the odour of the solutions before they reach the mouth. When 20-day-old pups were dehydrated and offered a diet with little or no odour, such as sucrose, their intake was as enhanced as it was with plain water, even though the intake of such a solution can produce further dehydration (Bruno and Hall 1982). However, when odour was added to the sucrose, ingestion was inhibited by dehydration. Indeed, anorexia was shown to water when it was simply given an added odour. No anorexia was expressed to any solution infused directly into the mouth. Thus, reaction to an olfactory signal appears to cause reduction of intake after dehydration, largely by limiting pups' approach to or

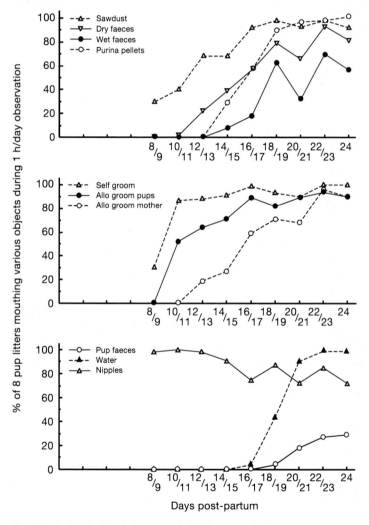

Fig. 3.5. A description of the onset of nibbling and drinking at weaning age in rats as determined by daily observation. Note that mouthing of various objects occurs earlier than the onset of drinking activity (i.e. mouthing water) though water was only available from a drinking spout and thus may have been more difficult for pups to sample than the other items. (From Galef (1979), with permission.)

contact with the solution and not primarily by affecting oromotor responses to solutions.

The Normal Emergence of Drinking in Rodents

Despite pups' clear sensitivity to dehydration and their ability to express it in ingestion in experimental situations, fluid consumption independent of the mother and suckling does not normally emerge until the weaning process begins, around two weeks of age in the laboratory situation. Rat pups begin weaning by nibbling at items in their environment: bedding, faeces, chow, and each other (Fig. 3.5). By about 16 days they actually ingest chow and some caecotrophe (Babicky et al. 1973; Galef 1979), a semi-solid but nutritively rich component of the maternal excreta (Leon 1974). Suckling gradually wanes during the course of weaning, but if pups are prevented from gaining access to the food source and a lactating mother remains available, they will continue to suckle (e.g. Lichtman and Cramer 1989).

Drinking (from water spouts) does not appear to occur until a day or two after solid food ingestion is well established (e.g. Bolles and Woods 1964; Adolph 1968; Babicky et al. 1972; Redman and Sweney 1976). This precedence of feeding is a phenomenon that many investigators have noted. It has been taken as an indication that hunger precedes thirst in development (Teitelbaum 1971), and is puzzling in the context of the findings described above. Pups respond to dehydration from early in development, but do not spontaneously begin to drink water until after weaning is well underway. While this phenomenon may partially reflect the manner in which drinking fluids have been made available, it also seems to indicate that spontaneous licking and lapping of fluids does not have the same need-dependent priority as gnawing and chewing. A rat's first drinking might, in fact, not occur until enough solid food has been ingested to produce a frank dehydration. Such a dehydration-stimulated initiation of drinking might normally be delayed because pups are rarely in fluid deficit since they are still obtaining large volumes of fluid from the mother. Another potential factor in the onset of drinking may be that pups begin to drink in order to wet their mouths to aid in eating chow. Kissileff (1971) has reported that such a pattern of "prandial drinking" appears early in ontogeny though somewhat after the time drinking is first seen. It should be noted that both these hypotheses could be taken to imply that much of the episodic consumption of water in adult rats is learned during these initial experiences and is thus anticipatory in nature. While the role of learning in adult drinking may be difficult to evaluate (Holland, Chapter 17), hypotheses for the normal onset of water ingestion might be more readily investigated.

Weaning occurs in a broader context of developing independence on the part of young (Galef 1981a; Alberts and Gubernick 1983). For rodents, this is the normal time of emergence from the nest. In experimental settings that are naturalistic, the course of weaning is made more obvious than in the standard experimental cage and the factors that influence weaning are more easily observed. Studies of weaning carried out in such settings have revealed that three factors contribute to what pups eat (Galef 1981b):

1. The physical presence of adult rats at a feeding site attracts pups to that feeding site and markedly increases the probability of young rats initiating weaning on the foodstuff located there.

2. Adult rats deposit residual olfactory cues in areas that they visit and pups prefer to explore and eat in an area soiled by conspecifics rather than in a clean area.

3. The milk of a lactating female rat contains gustatory cues from her diet and, at weaning, an isolated pup exhibits a preference for a diet of the same flavour as the diet that its mother has been eating during lactation (Galef, 1981b).

Similar influences do not seem to occur for drinking. Although pups do their first nibbling and feeding in the vicinity of adults and thus wean onto foods that are judged safe and familiar by more experienced animals, conspecific presence appears to have little influence on water intake imperatives (Galef 1978) or even at later ages (Timberlake 1983). Pups make the transition to drinking independently of parental guidance. Such self-guidance in the transition to drinking may occur because by weaning age pups are well prepared to discriminate rehydrating fluids from foods, though in contrast they may not be knowledgeable about the best foods and thus rely on others' experience. In any case, this difference in social influence on the two modes of ingestion represents a major and unexplored difference between drinking and feeding.

At the time of weaning, the diurnal pattern of drinking is quickly established. By 25–30 days, pups are consuming an adult proportion of both food and water during the night hours (Levin and Stern 1975; Redman and Sweney 1976). Furthermore, although considerable prandial water intake may occur around weaning age, this pattern of feeding punctuated by drinking is lost as rats approach puberty (Kissileff 1971) though much drinking still occurs near the time of feeding (Kraly 1984, and Chapter 18).

Weaning in Other Species

Studies of weaning typically focus on transitions from suckling to feeding as opposed to the transition from suckling to drinking. Thus, although it is known that for most species drinking begins at weaning with the onset of feeding, little is known about the factors responsible for initial fluid intake independent of the mother. Because the timing of weaning is usually well correlated with other maturational events, its occurrence is closely related to the relative development of a species at birth. The fact that there is great variability in developmental status at birth across different species would lead one to expect the postnatal development of drinking to be highly variable, despite the overall mammalian similarity in the form of suckling behaviour. Altricial marsupials exemplify the extreme of immaturity with an extensive developmental period required prior to weaning. On the other hand, the precocial guinea pig that can be weaned from the mother immediately after birth provides an example of the well-developed ingestive capabilities of a species born mature. Relative maturity probably determines the time at which dehydration influences suckling and the time at which dehydration influences oromotor responsiveness to fluids, but this question has not been examined in animals other than rodents.

Human infants offer an example of a highly evolved suckling system. Altricial in most ways (though not as much so as the rat), human infants are highly dependent on maternal care, with the mother considerably more active in nursing than for non-primates. An infant human's ingestive behaviour is actually quite flexible, however. Unlike rodent infants, humans will readily suckle from artificial nipples, and will feed from cups at early ages. Human infants also show relatively well-developed taste discriminative ability (e.g. Desor et al. 1973; Crook 1978). In addition, anecdotal information indicates that dehydration increases intake during suckling early in infancy (e.g. Hutchinson and Cockburn 1986). As noted above, this can have dangerous results in young infants when dehydrating solutions are not discriminated. Beyond about 4 months, though, the common impression of parents and paediatricians is that dehydrated human infants generally chose to suckle bottles containing water or juices rather than milk (though I have found no direct test of this question). This finding would suggest that the distinction between feeding and drinking is made relatively early in our species and long before language has had an influence. Considerably more information on the ontogeny of drinking in human infancy is needed, however. This information may not be too difficult to obtain, given the frequency of dehydration in clinical situations.

Opportunities of Ontogenetic Analysis: Behavioural and Neurobiological

In outlining what is known about the development of thirst, I have tried to emphasize the value of developmental analysis. While a number of interpretive difficulties exist in the analysis of early ingestion (e.g. what is suckling?), developmental analysis still provides an important means for learning about how the system is put together and for understanding the factors that modulate and control components of ingestive behaviour.

Developmental analysis creates at least four types of opportunity for the study of ingestive behaviour.

First, at early stages of development, ingestive systems may be immature, simplified, and thus more easily understood. As such they are available for study in isolation from factors that normally confound a dissection of their mechanism. Six-day-old rat pups, as an example, provide an excellent preparation in which to study neural mechanisms for the production of ingestion stimulated by cell dehydration.

Second, complexities that are added during development can be appreciated one by one, and newly appearing substrates can be analysed and contrasted to their absence at earlier stages. In this vein, the gradual appearance of components in the appetitive sequence of drinking represents a fascinating opportunity for analysis.

Third, developmental analysis permits control, manipulation, and evaluation of the significance of the history of an animal's ingestion-related experience. Only by such analysis can we hope to identify early events of major relevance in shaping later behaviour.

Finally, to the extent that aspects of the organization and control of ingestion change in relation to neurobiological development, we may be able to associate behavioural changes with events in neural maturation and thus establish brain–behaviour relationships in ways not possible in the absence of the natural experiment of development.

The opportunities for ontogenetic study have not yet been fully exploited. The new perspectives on behavioural organization created by observing the dynamics of developmental change promise to make it one of our most powerful tools for understanding the biology of appetitive behaviour.

Acknowledgements. I am grateful to N. Burka, A.N. Epstein, S. Frankmann and S. E. Swithers for helpful comments on earlier versions of this chapter. Some of the work described here and preparation of the chapter were supported by NICHD Grant HD17457 and by NIMH RSD Award K02-MH00571.

References

Adolph EF (1957) Ontogeny of physiological regulations in the rat. Q Rev Biol 32:89–137
Adolph EF (1968) The origins of physiological regulations. Academic Press, New York
Alberts JR, Cramer CP (1988) Ecology and experience: sources of means and meaning of developmental change. In: Blass EM (ed) Handbook of behavioral neurobiology, (vol 9). Developmental psychobiology and behavioral ecology. Plenum Press, New York, pp 1–39
Alberts JR, Gubernick DJ (1983) Reciprocity and resource exchange: a symbiotic model of parent–offspring relations. In: Moltz J, Rosenblum L (eds) Symbiosis in parent–young interactions. Plenum Press, New York
Almli CR (1973) The ontogeny of the onset of drinking and plasma osmotic pressure regulation. Dev Psychobiol 6:147–158
Babicky A, Pavlik L, Ostadalova I, Kolar J (1972) Determination of the onset of spontaneous water intake in infant rats. Physiol Bohemoslov 21:467–471
Babicky A, Parizek J, Ostadalova I, Kolar J (1973) Initial solid food intake and growth of young rats in nests of different sizes. Physiol Bohemoslov 22:557–566
Baverstock P, Green B (1975) Water recycling in lactation. Science 187:657–658
Blass EM, Hall WG, Teicher MH (1979) The ontogeny of suckling and ingestive behaviors. In: Sprague JM, Epstein AN (eds) Progress in psychobiology and physiological psychology, (vol 8). Academic Press, New York, 243–299
Bolles RC, Woods PJ (1964) The ontogeny of behavior in the albino rat. Anim Behav 12:427–441
Bradley RM, Mistretta CM (1973) Swallowing in fetal sheep. Science 179:1016–1017
Bruno JP (1981) Development of drinking behavior in preweanling rats. J Comp Physiol Psychol 95:1016–1027
Bruno JP, Hall WG (1982) Olfactory contributions to dehydration-induced anorexia in weanling rats. Dev Psychobiol 15:493–505
Bruno JP, Craigmyle L, Blass EM (1982) Dehydration inhibits suckling behavior in weanling rats. J Comp Physiol Psychol 96:405–415
Bruno JP, Blass EM, Amin F (1983) Determinants of suckling versus drinking in weanling albino rats: influence of hydrational state and maternal contact. Dev Psychobiol 16:177–184
Campbell BA, Raskin LA (1978) Ontogeny of behavioral arousal: the role of environmental stimuli. J Comp Physiol Psychol 92:176–184
Cone TE Jr (1984) Infant feeding: a historical perspective. In: Howard RB, Winter HS (eds) Nutrition and feeding of infants and toddlers. Little, Brown and Company, Boston
Coopersmith R, Leon M (1988) The neurobiology of early olfactory learning. In: Blass EM (ed) Handbook of behavioral neurobiology. Plenum Press, New York, pp 283–308
Crook CK (1978) Taste perception in the newborn infant. Infant Behav Dev 1:52–59
Desor JA, Maller O, Turner RE (1973) Taste in the acceptance of sugars by human infants. J Comp Physiol Psychol 83:496–501
Dlouha H, Krecek J, Kreckova J (1963) Water diuresis and the effect of vasopressin in infant rats. Physiol Bohemoslov 12:443–451
Drewett RF (1978a) Gastric and plasma volume in the control of milk intake in sucking rats. Q J Exp Psychol 30:755–764
Drewett RF (1978b) The development of motivational systems. Prog Brain Res 48:407–417
Drewett RF, Cordall KM (1976) Control of feeding in suckling rats: effects of glucose and osmotic stimuli. Physiol Behav 16:711–718
Ellis S, Axt K, Epstein AN (1984) The arousal of ingestive behaviors by the injection of chemical substances into the brain of the suckling rat. J Neurosci 4:945–955
Epstein AN (1976) Feeding and drinking in suckling rats. In: Novin D, Wyrwicka W, Bray GA (eds) Hunger: basic mechanisms and clincal implications. Raven Press, New York, pp 193–202
Epstein AN (1984) The ontogeny of neurochemical systems for control of feeding and drinking. Proc Soc Exp Biol Med 175:127–134

Epstein AN (1986) The ontogeny of ingestive behaviors: control of milk intake by suckling rats and the emergence of feeding and drinking at weaning. In: Ritter R, Ritter S, Barnes CD (eds) Neural and humoral controls of food intake. Academic Press, New York, pp 1–25

Falk G (1955) Maturation of renal function in infant rats. Am J Physiol 181:157–170

Friedman MI (1975) Some determinants of milk ingestion in suckling rats. J Comp Physiol Psychol 89:636–647

Friedman MI (1979) Effects of milk consumption and deprivation on body fluids of suckling rats. Physiol Behav 23:1029–1034

Friedman MI, Bruno JP (1976) Exchange of water during lactation. Science 197:409–410

Friedman MI, Campbell BA (1974) Ontogeny of thirst in the rat: effects of hypertonic saline, polyethylene glycol, and vena cava ligation. J Comp Physiol Psychol 87:37–46

Friedman MI, Bruno JP, Alberts JR (1981) Physiological and behavioral consequences in rats of water recycling during lactation. J Comp Physiol Psychol 95:26–35

Galef BG Jr (1978) Differences in affiliative behavior of weanling rats selecting eating and drinking sites. J Comp Physiol Psychol 92:431–437

Galef BG Jr (1979) Investigation of the functions of coprophagy in juvenile rats. J Comp Physiol Psychol 93:295–305

Galef BG Jr (1981a) The ecology of weaning: parasitism and the achievement of independence by altricial mammals. In: Gubernick DJ, Klopfer P (eds) Parental care in mammals. Plenum Press, New York, pp 211–242

Galef BG Jr (1981b) The development of flavor preferences in man and animals: the role of social and nonsocial factors. In: Aslin RN, Alberts JR, Peterson MR (eds) Development of perception: psychobiological perspectives. Academic Press, New York, pp 411–431

Hall WG (1990) The ontogeny of ingestive behavior: changing control of components in the feeding sequence. In: Stricker EM (ed) Handbook of behavioral neurobiology, (vol 10). Food and water intake. Plenum Press, New York, in press

Hall WG, Bryan TE (1981) The ontogeny of feeding in rats. IV. Taste development as measured by intake and behavioral responses to oral infusions of sucrose and quinine. J Comp Physiol Psychol 95: 240–251

Hall WG, Oppenheim RW (1987) Developmental psychobiology: prenatal, perinatal, and early postnatal aspects of behavioral development. Ann Rev Psychol 38:91–128

Hall WG, Williams CL (1983) Suckling isn't feeding, or is it? A search for developmental continuities. In: Rosenblatt JS, Hinde RA, Beer RA, Busnell M (eds) Advances in the study of behavior, (vol 13). Academic Press, New York, pp 219–254

Hall WG, Hudson R, Brake SC (1988) Terminology for use in investigations of nursing and suckling. Dev Psychobiol 21: 89–91

Hall WG, Denzinger A, Phifer CB (1990) Behavioral activation and the guidance of early feeding. Submitted

Heller H (1947) The response of newborn rats to administration of water by the stomach. J Physiol Lond 106:245–255

Heller H (1949) Effect of dehydration on adult and newborn rats. J Physiol Lond 108:303–314

Hofer MA (1975) Studies on how early maternal separation produces behavioral change in young rats. Psychosom Med 37:245–264

Hofer MA (1978) Hidden regulatory processes in early social relationships. In: Bateson PPG, Klopfer PH (eds) Perspectives in ethology. Plenum Press, New York, pp 135–163

Hutchison JH, Cockburn F (1986) Practical paediatric problems. Lloyd-Luke, London

Johanson IB, Shapiro EC (1986) Intake and behavioral responsiveness to taste stimuli in infant rats from 1 to 15 days of age. Dev Psychobiol 19:593–606

Kenny JT, Stoloff ML, Bruno JP, Blass EM (1979) Ontogeny of preference for nutritive over non-nutritive suckling in Albino rats. J Comp Physiol Psychol 93:752–759

Kissileff HR (1971) Acquisition of prandial drinking in weanling rats and in rats recovering from lateral hypothalamic lesions. J Comp Physiol Psychol 77:97–109

Kraly FS (1984) Physiology of drinking elicited by eating. Psychol Rev 91:478–490

Leon M (1974) Maternal pheromone. Physiol Behav 13:441–453

Leon M (1979) Mother–young reunions. In: Sprague JM, Epstein AN (eds) Progress in psychobiology and physiological psychology, (vol 8). Academic Press, New York, pp 301–334

Lepkovsky S, Lyman R, Fleming D, Nagumo M, Dimick MM (1957) Gastrointestinal regulation of water and its effect on food intake and rate of digestion. Am J Physiol 188:327–331

Leshem M, Epstein AN (1988) Thirst induced anorexias and the ontogeny of thirst in the rat. Dev Psychobiol 21:651–662

Leshem M, Boggan B, Epstein AN (1988) The ontogeny of drinking evoked by activation of brain angiotensin in the rat pup. Dev Psychobiol 21:73–76

Lev R, Orlic D (1972) Protein absorption by the intestines of the fetal rat in utero. Science 177:522–524

Levin R, Stern JM (1975) Maternal influences on ontogeny of suckling and feeding rhythms in the rat. J Comp Physiol Psychol 89:711–721

Lichtman AH, Cramer CP (1989) Relative importance of experience, social facilitation, and availability of milk in weaning of rats. Dev Psychobiol 22:347–356

McCance RA (1959) The maintenance of stability in the newly born. Arch Dis Child 34:361–370, 459–470

Misantone LJ, Ellis S, Epstein AN (1980) Development of angiotensin-induced drinking in the rat. Brain Res 186:195–202

Mistretta CM, Bradley RM (1986) Development of the sense of taste. In: Blass EM (ed) Handbook of behavioral neurobiology, (vol 8). Developmental psychobiology and developmental neurobiology. Plenum Press, New York

Moorcroft WH, Lytle LD, Campbell BA (1973) Ontogeny of starvation-induced behavioral arousal in the rat. J Comp Physiol Psychol 75:50–67

Peiper A (1963) Cerebral function in infancy. Consultants Bureau, New York

Redman RS, Sweney LR (1976) Changes in diet and patterns of feeding activity of developing rats. J Nutr 106: 615–626

Smotherman WP, Robinson SR (1988) The uterus as environment: the ecology of fetal behavior. In: Blass EM (ed) Handbook of behavioral neurobiology, (vol 9). Developmental psychobiology and behavioral ecology. Plenum Press, New York, pp 149–196

Stricker EM (1984) Biological basis of hunger and satiety: therapeutic implications. Nutr Abstr Rev 42:333–340

Taitz L, Byers HD (1972) High calorie/osmolar feeding and hypertonic dehydration. Arch Dis Child 47:257–260

Teitelbaum P (1971) The encephalization of hunger. In: Stellar E, Sprague JM (eds) Progress in physiological psychology, (vol 4). Academic Press, New York, pp 319–350

Timberlake W (1983) The functional organization of appetitive behavior: behavior systems and learning. In: Zeiler MD, Harzem P (eds) Advances in analysis of behavior, vol 3. Biological factors in learning. Wiley, Chichester, pp 177–221

Winters RW (1982) Principles of pediatric fluid therapy. Little, Brown and Company, Boston

Wirth JB, Epstein AN (1976) The ontogeny of thirst in the infant rat. Am J Physiol 230:188–198

Commentary

Nicolaïdis: By the time mammals are born they have already experienced a lot of drinking. Prenatal induction of drinking preferences may arise from drinking of amniotic fluid. In fact, the offspring from rats which have experienced extracellular dehydration during pregnancy show an enhancement of their preference for salty water when they are given a choice. Extracellular dehydration during pregnancy increases salt appetite of offspring (Nicolaïdis et al. 1990).

Reference

Nicolaïdis S, Galaverna O, Metzler CH (1990) Extracellular dehydration during pregnancy increases salt appetite of offspring. Am J Physiol 258:R281–283

Schallert: Although pups themselves may not respond to dehydration by suckling, it remains possible that a non-suckling behaviour such as emitting ultrasound or making orofacial or other movements in response to dehydration may stimulate the dam to nurse.

Chapter 4

Influences on Human Fluid Consumption

D.A. Booth

Scientific Study of Thirsty Behaviour

The primary meaning of words like thirst and hunger is dispositional. They refer to causal structures in the behaviour of the moment. This is a logical prerequisite of their various usages in ordinary speech (Ryle 1949; Wittgenstein 1953; Peters 1954). It is also the key to measurement of influences on drinking.

Thus, being thirsty is having an appetite (avidity, desire) for water or, more loosely, the overt tendency to seek fluids and to ingest them (Fig. 4.1). This only roughly delimits an aspect of the momentary organization of behaviour, as psychological performance to do with water consumption. So the term thirst makes a good title for a book but it is not likely to give scientists a tight theoretical construct.

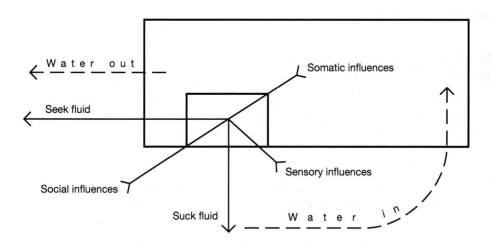

Fig. 4.1. A sketch of the processing of information in thirst that is observable in human beings and other animals, and the affected exchange of water between the external and internal environments. This causal model lumps discrete influences on fluid intake into three categories and the sequence of search and consummatory behaviour into one function. The small compartment of the body is neural circuitry that engineers this mental processing evident within drinking behaviour.

Sensations, Signals and Behavioural Neuroscience

Dynamics within behaviour resulting in fluid intake should not be confused with subjective experiences in conscious awareness or signals from body to brain that at times are also involved, nor with the physicochemical processes in neural networks that underlie any behaviour and consciousness.

Such distinctions are a conceptual necessity. The private phenomenology, the physiological signals and the neural activity are relevant to thirst only insofar as they are involved with acts of seeking and ingesting fluid. Mouth dryness would not be a sensation of thirst unless we recognized it as the experience that can arise when we begin to want some fluid (perhaps after a period without drinking), and that normally disappears again after intake of water, along with the desire to drink. Similarly, a neural or hormonal signal of thirst can be identified as such only by its facilitation specifically of water-drinking behaviour, not by its origin in water deficit nor by just the words someone utters when the signal arises.

The distinctions are also empirically sound. The desire to have a drink can occur in the absence of dehydration, of a bodily sensation such as of dryness of the mouth and throat, or of a conscious sense that one may be in real need of water. Physiologists rightly insist that somatic variables be measured, not just speculated on. Psychologists should insist similarly on measurement of the causal processes within the organization of behaviour, and indeed on analysis of the ontogeny of this structure (Hall, Chapter 3). Moreover, all should recognize that the scientific study of a private sensation or pleasure requires it to have a distinct causal role, and that this is far trickier to disconfound than traditionally assumed (Booth 1990c; Booth et al. 1987). Ratings per se do not scale or directly measure subjective experiences: unless the scores on one wording are uncorrelated with the scores using another phrase, there is no evidence that different private experiences underly the two ratings: the rater may be "referring" a craving for water to dryness of the mouth (or vice versa).

Objective phenomena such as plasma osmolality, salivary secretion rate, or gastric distension are not being discriminated by ratings of thirst, mouth dryness or lack of epigastric fullness if all these scores are highly correlated with each other (or with fluid intake). That is, to the extent that a set of ratings all predict water intake, they cannot be distinguishing sense of need, sensation of thirst and volume of water in the drink that is consumed.

Drinking Behaviour, Thirst Ratings and Fluid Volumes

Fluid consumption is behaviour. It is sipping, sucking, lapping or gulping, controlled by information about water (in the case of thirst in the narrow sense). The momentary disposition to ingest fluid is most fully expressed in selective ingestion. It can also be expressed in selective approach and in symbolic behaviour. Rats may communicate to each other about water sources (Galef, Chapter 20). A rat can signal a desire for water by selecting the lever that generally delivers water, not strong saline. A dog can beg us to refill its water bowl. We can show our thirst to someone else by a gesture indicating the drink we prefer or by a spoken complaint of oral discomfort that a drink would relieve.

A tradition in biomedical research divides research on eating between food intake and food choice. This would be as fallacious in the context of thirst as it is for hunger. Choice is often the choice of amount as well as a choice between alternatives.

Preferences change during ingestion, as a result of habituation or satiation conditioning. Even when the decision to stop is made while plenty of a drink remains, what has happened is the choice of another ingestate or to switch to non-ingestive behaviour (Schallert, Chapter 14).

Thirst and satiation can be rated by any verbal behaviour implying a wish to drink that can be translated into numbers (scored). What exact sort of tendency to drink a particular wording expresses for a person in given circumstances is not known until validated on that person's actual behaviour in situations that are similar and where the influences on consumption have been measured. Cumulated intake of plain water is commonly used to validate thirst ratings but it can be a poor criterion of most human dispositions to drink, for two reasons. In most affluent societies, few people drink water often. Also, the volume consumed at a drink confounds a whole series of changes in performance; the quality of thirst at the drink's start is likely to differ greatly from that towards its end.

To track this changing structure of motivation, we need ratings that perform discriminatively (Booth 1981) or renewed choices between ingestates (Booth 1977, 1980). A verbal score by itself is no more a measure of a causal process than is an intake volume. Use of the words "thirst sensation" and "pleasantness", for example, must be empirically shown and cannot be assumed to distinguish dryness of the mouth from palatability of the beverage, and either of these from time since the last drink.

The sorts of instructions and response formats commonly used to get scores out of people include so-called visual analogue scales, Likert formats, semantic differentials, magnitude estimation, and the 9-phrase hedonic "scale". All these are only ways of rating; the mere procedure does not necessarily succeed in scaling, i.e. in measuring anything subjective or objective. To make an effective series of quantitative judgements, the rater must be referring to one single qualitative attribute (like wanting a drink) and be using two anchor levels of it on some analogue dimension such as a line, maybe sectored into a row of unlabelled boxes or dashes. "Not at all" is a fairly exact anchor level, if it can be used, and it is good for one end of a rating dimension. Generally, the problem is choosing a decent high anchor. "As extreme as you can imagine" would avoid end effects. A midpoint (and no anchor at the high end) based on some familiar standard is better, which is why exploitation of learned preferences and normal drinking situations gives such precise measurements – from intake tests no less than from ratings.

Controls of Human Fluid Intake

The disappearance of fluid is thus a very indirect measure of thirst. Yet it is much more easily measured than drinking behaviour. Furthermore, measurements of water intake are indispensable to the study of water balance. They may be sufficient to make considerable advances in physiology. Intake volumes are of rather limited help, however, in relating the mechanisms of body fluid control to the mechanisms of human drinking behaviour. To get causally relevant information, we must also measure the influences on intake, wherever they come from. The briefer the period of intake measurement the better, too.

Influences on Daily Water Intake

Water is one of the most difficult nutrients to estimate in the intake of free-living human beings by traditional dietary methods. Ershow and Cantor (1989) produced the first representative estimates of a nation's water intake from all sources. They observed effects of age (cp. Phillips et al., Chapter 26) and sex (cp. Baylis, Chapter 29) and made a case for attributing them entirely to requirements related to body size. Small regional differences and even smaller seasonal variations were seen but none of the interactions between regions and seasons that would be expected from thermal effects on invisible losses. The age-group median water intakes paralleled the recommended US baseline allowances, such as 30–40 ml/kg/day for adults. However, multiple regression of the above variables accounted for "only about 15% of the variance in total water intake of most age groups" (Ershow and Cantor 1989); it is not entirely clear whether this was analysis across individuals within groups or across group averages, but in either case, as the authors say, it is very limited evidence that drinking is influenced by physiological requirements for water. They note that we do not know how much better the prediction would have been if it had included strenuous activity (Greenleaf, Chapter 27), smoking (Gilbert, Chapter 23) and relevant medical conditions (Robertson, Chapter 30).

The clearest indication that human drinking is not controlled by water balance, at least on the side of excess, is the skew to high daily intakes in all subgroups (Ershow and Cantor 1989). Of course, balance is still achieved by excretion, while there would be limits to the conservation of body water. The long tail of water intakes above twice the median is very reminiscent of the population distributions of alcohol intake over which epidemiologists have long argued (Gilbert, Chapter 23). Being a case of secondary insipidus myself, I can believe that there is at least a substantial minority who sip through much of the day, simply because there is a drink always to hand, if not to keep the mouth wet or the brain caffeine level up.

Thus, grouped daily water intake data are not useful scientifically. Their practical uses seem few if any. Yet, average daily intake is regarded as the primary datum in dietary assessment. This seems to be an unfortunate carryover from the success of nutritional science in understanding long-term deficiencies in the intake of essential nutrients.

Daily water intake may be a conceptual index of thirst. Nevertheless, it is a most unsatisfactory measure of the organization of thirst, because a great deal of potentially varied drinking behaviour goes into a day's water intake.

Temporal Pattern of Drinks

The pattern of fluid intake bouts around the clock should have a better chance of relating to physiological and environmental changes than the 24-hour intake measure. However, ingestive bout sizes and time intervals from one bout to the next cannot by themselves distinguish influences on behaviour. They are only useful as indices when one or more putative influences on bout size and/or interbout interval are varied in a controlled or unconfounded manner (Booth 1972a, 1990b; de Castro, Chapter 21).

That is, bout size, considered without regard to conditions, cannot be the measure of satiation (or of appetite or palatability), nor can bout-to-bout interval be the measure of satiety. In order to find the determinants of the behaviour that initiates a drink, maintains it or terminates it, we have to measure the processes influencing the behaviour that occurs during each of those parts of the drink. Equally, we shall not understand the influences on continued postdipsic satiety if it is assumed that not-drinking is always the same behaviour. The end of the bout must be compared with various times later, when water processing is at different stages and also fluid availability may well have changed; the social context may preclude thirst or encourage need-free drinking.

The Single Drink

Most physiologists and biological psychologists have used the discrete drinking occasion as the primary measure of drinking. Often, however, the sequence of behaviour resulting in the whole drink is not elucidated.

A primitive measure of changes in the influences on intake during the drink is comparison of bout size with initial rate of ingestion (Booth 1972b, 1974; Booth and Davis 1973; Van Vort and Smith 1987). Clearly it is better to monitor modulators of ingestion during and after the bout (e.g. sensory preferences: Booth 1980; Gibson and Booth 1989). Sensory adaptation or "saturation" (such as the cloying effect of a sweetener), stimulus-specific habituation of ingestion ("sensory satiety"), counting mouthfuls, and learned or unlearned dependence of sensory facilitation on internal state, can all be identified by this technique.

The temporal pattern of drinking movements or intake rate within the bout can be recorded. However, identifying two uncorrelated parameters of the cumulative intake curve shows only that there are two distinct influences; it does not tell us the cause of either (Booth 1990a).

Postingestional influences cannot be deduced from an ingestive response change coinciding with a change in the gut, blood or tissues. The sensory attributes of a beverage can create an expected sating effect based on past experience. If rated expectations account for the size of that drink and for intake for some time following, the evidence for direct physiological control is lacking (Booth 1990a). That is, only controlling out prior expectations can disconfound cognitive from physiological influences (Booth 1987a); thirst ratings as such cannot.

A good match of expectancy and outcome is evidence that satiation and post-bout satiety were learned. Such acquired controls of intake may still depend on physiological effects occurring in an anticipated manner, not just on sensory cues from the fluid and perhaps environmental cues. Experiments should use moderately unexpected physiological consequences of drinking; resulting deviations from expected satiety would be evidence that learned physiological control was operative.

Fluid portion volumes are constrained by the sizes of drinking vessels and beverage packs. A drink's traditional size though may be adapted to the average physiology, as conventional meal sizes appear to be (Booth and Mather 1978).

Quantitating Causes of Drinking

To provide evidence of the causal processes within a momentary disposition to drink, observed outputs must be systematically related to inputs, measured concurrently or inferred via a validation procedure. These are of course the conditions for any causal science, applicable to psychological phenomena as much as they have long been applied to physics, physiology and macroeconomics.

A decent concept of psychological causation clarifies how to investigate behavioural and cognitive processes. As for any other sort of mechanism, a powerful cause is one that reliably produces a big effect. The quantitative evidence for causal strength is the association of small differences in a stimulus with differences in a response that are large relative to uncontrolled variation; output is discriminating and finely tuned or sensitive to the input (Booth et al. 1987, 1989).

This tightness of the functional relationship has to be specific. Qualitative disconfounding is achieved by double dissociation. It can be done statistically by orthogonal canonical analysis, for example. Quantitative identification of a causal process requires an output that is more sensitive to an input than are other candidate outputs or inputs. Hence, in order to investigate a thirst mechanism, we must measure at least either two strong influences on the process or two of its effects on what the organism is doing (or the human being is saying). When a network of cognitive processes is involved, sufficient numbers of unconfounded inputs and/or outputs have to be measured to build up and to test a realistic model of the organization of the behaviour.

A corollary is that a thirst mechanism is identified by showing that a particular water-relevant input specifically affects an ingestion-related output. This simply works in reverse the principle that expressing a desire for water is an act that successfully discriminates water-relevant stimuli in the diet.

These requirements for causal analysis also apply to interventions that affect the controls of fluid intake, whether physical such as a lesion, a drug or electrical stimulation, or informational such as verbal instructions. To tell us about the neural or cognitive engineering of thirst, the drinking test must isolate one or more of the behavioural processes affected (Booth 1990b).

Integration Among Discrete Influences on Drinking

The measured sensitivities of discrete causal processes can then be used to characterize their exact interactions in the individual's overall response to a measured situation (Booth 1987b; Booth and Blair 1989). The strengths of the major contributors to variations in behaviour may well combine linearly over the range of that person's tolerance for deviations from the most motivating version of the situation (Fig. 4.2).

Such individualized analysis of the structure of drinking motivation reveals any idiosyncrasies (Fig. 4.3). Personal specifications of determinants of drinking in common situations can also be summed over whatever population sample the sociological, medical or commercial interest requires. Individualized measurement is very sensitive to group differences. For example, small differences could be detected between sexes and age groups in the distributions of individually preferred levels of sugar in a fruit drink and to show that these differences produced an artifactual relationship to body weight (Booth and Conner 1988).

The beverage industries have enormous amounts of practical expertise and statistical data concerning sales, consumer preferences and expert descriptions of the sensed constituents of drinks. There are, however, almost no data on the strengths and interactions of the sensory and contextual influences on drinking. Yet these are the real causal processes within the individual consumer's mind (Booth 1990d).

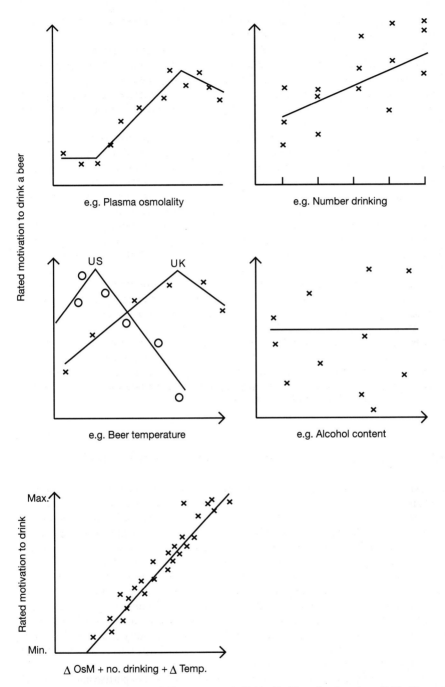

Fig. 4.2. Notional data on major influences on an individual's thirst when beer is available. The preferred temperature of the beer is likely to differ between an American and a Briton. Drinking behaviour is rather precisely predicted by a personal integration formula that averages the distances of the inputs' levels from maximum or minimum motivating power, in units of the sensitivity of thirst to each input ("just tolerable differences": cf. Conner et al. 1988).

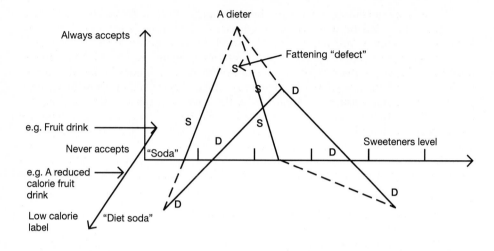

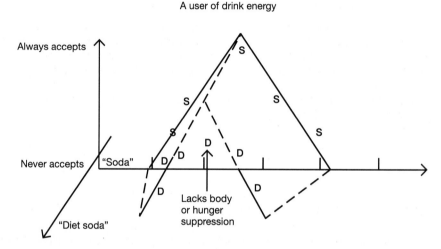

Fig. 4.3. Tolerance triangles for sweetness at two levels of perceived calories in a soft drink, for two notional users (a dieter and a user of soda calories to assuage hunger). S and D are raw data points for regular (sugar) and diet sodas respectively.

An Outline Causal System of Thirst

The apparently simple and biologically basic human behaviour of taking a drink must in fact be a very complex psychobiosocial system. A highly co-ordinated network of mechanisms has to be active in order for a person to have a thirst. This chapter is an attempt to illustrate the potential for the above rather advanced approach to the identification and measurement of causal interactions in fluid consumption, particularly by human beings, and to show that the approach applies equally to sensory, somatic and social factors in thirst. Examples of causal analysis of thirst in clinical situations are given by Robertson (Chapter 30), although measuring only physiological inputs, not disconfounding oral or temporal inputs (cp. Robinson 1990).

A simple model based on commonsense can provide a framework for such quantitative mechanistic research into influences on drinking behaviour (Fig. 4.1). The observed outputs are ingestive or approach movements, or relevant symbolic communications such as a thirst rating. The observed inputs are stimuli from certain potential ingestates – the sensory factors in beverages. Sometimes inputs also come from signals of the hydromineral economy of the body, and/or from occasions perceivable as appropriate for drinking, be they watering places, the activities of conspecifics or times in one's personal routine. Evidence for existence of a thirst is a momentary sensitivity of those particular outputs to such inputs (compare Booth 1987d, 1990a for human ingestive motivation in general).

Influences on the search for and selection of fluids and the initiation, maintenance and termination of drinking can thus be split into three categories (Fig. 4.1):

1. The sensed characteristics of the fluid itself
2. The internal environment (including the physical effects of the external environment on water balance)
3. The interpreted external environment, particularly of other people and socially significant activities

However, the many distinct sensory factors in drinking do not act independently; they modulate each other. Also, physiological and cultural influences are mediated through the sensory controls. Furthermore, the discrete influences operate within and across these three categories in an integrated rather than an additive fashion, and distinctively to the individual.

Nevertheless, we can try to focus on each category, before turning to overall integration and its origins.

Sensory Sensitivity of Drinking

The exact nature of the motivation to drink in any particular individual in specified circumstances can be decided only by analysing the behavioural structure. Observing water intake or a thirst rating does not tell us, for example, how high a tonicity would be accepted or whether an oil might be consumed.

What is more or less palatable at any given moment is the dietary object of the appetite. A real thirst could be a desire for solid foods that yield lots of water.

What is wanted though may be solely the sensory effects of the beverage, such as cooling, wetting or chemospecific stimulation of the oropharynx; perhaps sometimes swallowing is incidental. If it were technically feasible to rig a sensory match, a solid or a non-aqueous fluid might momentarily be equally acceptable. More practicably, some quite different orosensory impact might easily become regarded as refreshing. Fizziness could be a case in point, expanding the market for sodas once they become readily available (Tuorila, Chapter 22). Heavy carbonation has its limitations, however. The current boom in mineral waters, both natural and flavoured, may reflect subtler palates, plus concern about health or safety, and fashion or faddism. Water in this form has recently been selling well enough in the UK to support magazine advertising using "smart" humour in two-page photographic spreads.

The Tolerance Triangle

Improved measurement of the palatabilities of foods and drinks (Booth et al. 1983, 1987) has shown that unlearned and learned preferences have different shapes of hedonic function (Booth 1990d). When acquired stimulus control is exerted in familiar surroundings, there is the same decrement in learned response at perceptually equal distances on either side of the trained level of the stimulus; what in learning theory is called the intradimensional generalization gradient will be symmetric. That is, if the test is sufficiently realistic, a sensory preference established by habit yields an isosceles "tolerance triangle" (Fig. 4.2 temperature, Fig. 4.3 sweetness). An innate sensory preference or aversion, on the other hand, produces a monotonic function from liked to disliked levels or a mixed function, such as the traditional asymmetric and flat inverted-U given e.g. by sodas tested with a wide range of sweetness (Booth et al. 1987).

A rat's short-term choices among familiar saline solutions could provide an example. If the hedonic function shows a descending limb that is flatter or steeper than the ascending limb when preference is plotted against salt concentration in units of equal discriminability, then that is evidence for an innate component to that salt preference (or salt appetite if the preference depends on sodium deficiency).

We first analysed the cognitive processes in a person's behaviour for cases of single gustatory influences on choices among fluids, such as the saltiness of soup (Booth et al. 1983), the sweetness of fruit-flavoured drinks or vegetable soup (Conner et al. 1986, 1988), and the bitterness of coffee (Booth et al. 1989).

In this last example, the taste of caffeine is a powerful quality factor in thirst for coffee. The rated preference for a sample of instant coffee was sensitive to differences in concentration of added caffeine in all 60 consumers tested, whether they usually drank coffee caffeinated or decaffeinated, with or without sugar or in a mild or a regular roast (Booth et al. 1989). Yet half the assessors (from each drink type, age band and sex) could not discriminate those same differences in caffeine level in the same coffee by the bitterness of the samples, rated in the same way as the preferences.

Everybody experienced all the coffees as bitter. Also, the taste of caffeine is unambiguously bitter at the higher levels used. Yet in the context of roast coffee bean flavour the contribution of caffeine to the preferred flavour overall was subliminal. An important research implication is that the conventional technique of rating descriptions using imposed vocabulary cannot be relied on to identify all the major influences on drinking, eating or other behaviour.

Somatic Sensitivity of Drinking

Several chapters in this book deal with the sensitivity of human drinking to imposed dehydration. We have little idea yet of when or even whether water deficit contributes to drinking in normal circumstances.

The wish for a drink when in need of water is likely to be affected by palatability and by appropriateness to the occasion. A desire for effects of water on the mouth or circulation may interact with sensory factors to initiate drinking, and perhaps to maintain it for longer than with a less palatable fluid. Even a thirst for Adam's ale from the mains tap or the river may be more than the desire for the effects of water on tissues (Engell and Hirsch, Chapter 24).

An absolute threshold for osmotic thirst, or the slope of a dose–response line (Robertson, Chapter 30) or its regression coefficient (Weiffenbach 1989), are not sufficient to tackle such issues. We need a parameter of suprathreshold sensitivity that enables calculation of the exact interactions of the physiological influences with sensory and social influences. The required measure of causal strength is the relationship of slope to response variance that represents the difference threshold for the relevant range of the stimulus (Booth 1987b; Booth et al. 1989; Conner et al. 1988). We could then calculate interactions among somatic and sensory or social influences on drinking, because they are all reduced to one and the same metric of sensitivities. Integration of dehydration signals could also be quantitated, just as sensory integration has been.

If we could measure somatic sensitivity functions in habitual surroundings, we might find evidence that the physiological signalling is learned. At plasma osmolalities above the levels to which drinking had been conditioned, the function would start flattening relative to the innate function (elicited by artificial tests in the laboratory), and might even descend (as does learned gastrointestinal-volume satiety of hunger; Gibson and Booth 1989) (Fig. 4.2).

Some human thirst, in the looser sense of the term, may be sensitive to postingestional action of constituents other than water (Gilbert, Chapter 23). Indeed, drinking might not be facilitated at all by the effects of the water content of the drink; there may even be an inverse sensitivity. A drinker's thirst for his or her favourite tipple is no less real if the object for ingestion is ethanol molecules and perhaps the subsequent compulsion to get rid of fluid is regarded as a penalty for indulging. Similarly, if caffeine addiction existed, the thirst for a drink of cola, coffee or tea would be for the drug and not for the nutrient in which it is dissolved.

Social Sensitivity of Drinking

Personal attitudes and the nature of the occasion for the drink are also likely to interact with sensory influences on fluid consumption. In addition or even alternatively to sensory and somatic gratification, a drink may provide the emotional satisfaction of participating in social rituals such as taking refreshment during a break from work, accepting the friendly offer of a beverage, or making a symbolic act of celebration. The water ingested may be quite incidental to the conventional use of a beverage. A candy might do instead of a coffee at a work break, the offer could be a cigarette instead of a beer, and a gesture of the hand might be a sufficient symbol. That is, although the person is thirsty in at least the broad sense, the wanted drink can be switched for something else appropriate.

Nevertheless, sensory characteristics and physiological after-effects may contribute to a drink's conventional appropriateness to a situation. The fact that a drink was socially triggered carries no implication in itself whether or not a water deficit was also having an effect; measures of sensitivity of the drink consumption to intracellular and extracellular depletion are needed to decide that issue. It is essential to measure somatic and sensory influences in a defined situation in which the beverage is usually consumed. Hot, soft and alcoholic beverages have a great diversity of uses, as also tap or bottled water itself appears to (e.g. Elmadfa and Huhn 1986).

Some existing social influences on drink choice or drink size quite possibly used to relate to relief of need for water, whether in the individual's past experience (see below and Holland, Chapter 17) or in the history of the species (Fitzsimons, Chapter 1).

Nutritional conditioning of the social reinforcers has not been excluded for eating (Booth 1990a) and so the same issue needs addressing for drinking.

Individual Analysis of Sensory and Social Interactions

The usual methods of analysing survey responses and attitude scores (including so-called causal modelling; Breckler 1990) group the data from people who may integrate influences quite differently. These idiosyncratic structures can be characterized by individualized cognitive analysis of two or more interacting influences, whether sensory, somatic or social (Booth 1987d; Fig. 4.2) and whether each operates consciously or not (Booth et al. 1987).

This has been illustrated by the integration of sensory and social influences within individual British consumers' preferences among variants of a drink of instant coffee (Booth 1987b; Booth and Blair 1989; L. Lähteenmäki, A.J. Blair and D.A. Booth, unpublished data). The social factors are the emotional meanings, health attributes and marketing images of the coffee; each consumer names these prior to the quantitative tests. Regression analysis of the levels of drink constituents and labels with ratings of preferences and of sensory and attitudinal ascriptions yields a cognitive network for each person (Fig. 4.4).

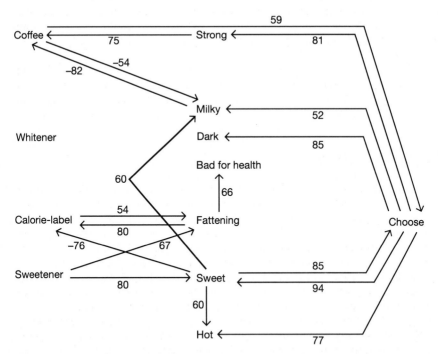

Fig. 4.4. Cognitive path analysis of an individual person's thirst for a hot drink at a break from work (Booth and Blair 1989). Each node represents a measured input or output variable. Each line represents statistically significant ($P < 0.01$) linear covariation between the two nodes it connects (the numbers are partial regression coefficients x 100).

This many-noded system of coffee drinking choices involved the effects of varied levels of coffee solids, whitener and sweetener but also the caloric image of the sweetener, manipulated by veridical labelling as "low-calorie sweetener", "sugar" or half and half. The label made a substantial difference to the preference for a given level of sweetness in some but not all assessors. Those affected usually gave evidence of conscious mediation by an attribution of fattening qualities to the drink. Also there were a few consumers who preferred the caloric sweetener and attributed refreshing or alerting benefits to the drink.

These tests were for the drinks taken by themselves. Perhaps even these consumers would not have been so concerned about the caloric attributes of the drink if it were to be consumed as part of a meal. Would such drinking be thirst? This leads to the general issue of thirst/hunger overlap.

Thirst and Hunger

If someone takes a drink between meals purely for calories, the distinction between thirst and hunger is lost. A beverage that is consumed only with meals would also unite the motives. This ambiguity arises naturally for water-rich foods such as ripe fruit and raw flesh.

The overlap is much broader, however. Many animals have causal links from eating to drinking (Kraly, Chapter 18). Human ingestive motivation, though, is commonly directed towards meals and snacks which usually consist of both food and drink. Indeed, the drink may be more taken for granted than the food; it may be omitted if the food is wet enough. We have a general appetite for both food and drink at meals.

This is not to deny that sometimes we wish to express an appetite for a fluid in particular and hence to distinguish that desire from appetite for food. That is what gives us uses for two terms, thirst and hunger, as well as the generic term appetite, applying to motivation based on any need.

Fluid Foods not Affecting Eating

This part-unity of drinking and eating motivations is reflected in traditional drinking habits and hence in much of the beverages market. Most types of drink are calorific, with alcohol, or sugar and/or cream(er). Moreover, small but highly calorific food items are often served with the drink. Yet most consumers of beverages away from meals probably want no more than liquid refreshment (Booth 1987a).

Such drinking entails energy consumption that bypasses the mechanisms to compensate by eating less at the next meal (Booth 1988). It would therefore help to prevent obesity if it were easier to separate mere water intake from energy intake (Blair et al. 1989). Low-calorie crunchy items that fitted the sensory characteristics of popular drinks better than apple or celery would also help. The concern to control shape (often beyond any health needs) has opened a market for low-calorie drinks. It may not be helpful, however, to identify these foodless drinks by a "Diet" or "Slim" brand name, since some consumers susceptible to obesity treat this as the promise of a slimming medicine or as an excuse to eat more energy than the product saves (Booth 1987a).

Commonalities of Hunger and Thirst

The fact that we want beverages with our meals also has some fundamental theoretical implications. It means that much of the psychology of the appetite for foods can be used as a source of working hypotheses for the psychology of the appetite for fluids.

The cognitive processes in drinking (at least at meals and snacks) may be similar to those in eating. This has indeed been confirmed whenever food attitude research has been extended to drinks (cf. above; Tuorila, Chapter 22).

The ontogeny of ingestion may also be similar for foods and drinks. It has been proposed that eating for energy in mature omnivores is entirely under learned control (Booth et al. 1972, 1974), unlike salt appetite in the rat for example (Krieckhaus and Wolf 1968). Therefore it should not be taken for granted that dehydration evokes thirst innately.

Origins of Influences on Human Fluid Consumption

Much biologically basic and physiologically excitable human behaviour is indeed learned. A tolerance triangle in an influence on drinking would be diagnostic of its learned origin (see above). Changes in the controls of human fluid consumption from birth or before should also be studied directly (Hall, Chapter 3). Are drinking movements congenitally influenced by water-relevant stimuli such as coolness or dryness in the mouth? Can human infants who have not experienced ingestive rehydration recognize water specifically as what they want when they need it?

There have long been data for another omnivorous species (Booth et al. 1972, 1974) indicating that virtually all ingestive appetite is acquired. Subsequently, even stronger evidence has been obtained that this is true for babies (Harris and Booth 1987; Harris et al. 1990), young children (Birch et al. 1980; Birch 1987) and adults (Booth et al. 1976, 1982; Conner et al. 1988).

Learned Preferences for Fluids

After babies begin to detect the low salt level of breast milk, exclusive exposure to it induces a preference for its blandness over light salting (Harris et al. 1990). Repeated sampling of unfamiliar fruit drink flavours similarly increases students' liking for them (Pliner 1982).

Even the smooth-brained rat learns to like fluid foods (Booth et al. 1972). Flavour preferences can be reinforced by sensory characteristics (Holman 1975), caloric yield (Booth et al. 1972; Booth and Davis 1973; Elizaldi and Sclafani 1988; Sherman et al. 1983 ; Tordoff and Friedman 1986) or protein repletion (Booth 1974; Baker et al. 1987). Even more to the point, flavour preferences can be also conditioned by water repletion (Revusky 1968; Booth 1979). Therefore, it requires direct behavioural analysis to exclude associative learning as the origin of human liking for the texture of water, its cooling or anti-irritant effect or its taste or smell (Norgren, Chapter 13).

Learned Water-Specific Dietary Selection

Integration of sensory and somatic information is logically necessary to succeed in selecting among diets specifically to meet a nutrient need (Booth 1987c). Sodium appetite (Denton, Chapter 8) is the facilitation of ingestion by the conjunction of sodium need signals and otherwise aversively strong salty tastes; it does not depend on learning in the rat. On the other hand, protein appetite can be acquired; repletion with protein conditions ingestive behaviour to the combination of mild protein need and a fluid flavour (Baker et al. 1987; Gibson and Booth 1986) or a solid texture (Booth and Baker 1990). Thus, to the extent that there is a preference for water texture, temperature or taste during water deficit, it is an open question whether this is innate or acquired or whether both sorts of ontogeny contribute (Hall, Chapter 3).

Such learned use of conjunctions of cues is well known in forms of configural conditioning and cognitive "chunking" (Wickelgren 1979). Such a basic mechanism could enable not only specificity of dehydration-induced intake for water (Booth 1979, 1985; Norgren, Chapter 13) but also drinking of an amount appropriate to need (Verbalis, Chapter 19).

Learned Control of Fluid Intake Volume

The simplest mechanism for anticipatory "oral metering" would be to stop drinking in response to stimuli from the gut that had proved appropriate to that fluid. Making a correct count of ingestive movements (Mook 1988) or of sensed sips or swallows seems a much more complicated task (Booth 1976).

Associative conditioning of the conjunction of fullness cues with flavours has indeed been shown to control the selection between liquid foods at both the start and the finish of meals in rats (Booth 1972b, 1980; Booth and Davis 1973; Gibson and Booth 1989) and in people (Birch and Deysher 1985; Booth et al. 1976, 1982; Booth and Toase 1983). Such a learned beverage-specific satiation could end a drink at an appropriate gut fill ahead of tissue restoration.

The idea that thirst and its satiation when water and food are freely available are innate could be an illusion generated by two theoretical mechanisms that might regulate non-specific intakes during severe water deficit. One is the increase in general drive with increasing dehydration; this energizes all behaviour and so is likely to increase drink sizes. The other mechanism is the punishing effect of extreme gastrointestinal distension, which will put increasingly strong inhibition on drink size before tissue repletion. Thus, no oral metering, learned or unlearned, may be needed to explain apparently anticipatory control of drink size after long water deprivations or large osmotic loads. To demonstrate water-specific satiation, drink-size control has to be disconfounded from general arousal and from abdominal discomfort.

Is Control of Human Fluid Intake Entirely Acquired?

Thus, there is no reason to assume that the influences on human drinking in normal circumstances have an instinctively fixed underpinning – in the choices between food and drink, among beverages or of the amounts drunk. All the immediate influences on the selection, initiation or termination of a drink could have gained their power as a result of past personal experience. People would react out of familiarity with the drink and the context of drinking (Booth and Conner 1988; Tuorila, Chapter 22). This could result partly from sheer habituation. A variety of associations to the consequences of drinking in such situations is also likely (Holland, Chapter 17), with the result that there is a desire to drink or not to drink certain beverages in those circumstances subsequently.

Summary

Appetite for water is the objective psychological phenomenon of being disposed to consume fluids or solids expected to have effects that in fact result from water being yielded to the tissues. The capacity of human behaviour to discriminate those ingestates most capable of rectifying body fluid balance has yet to be demonstrated, despite (or perhaps because of) our confidence it's so easy. Any such appetite specifically for water is, however, integrated in normal thirst with a much broader range of dietary, environmental and sometimes pharmacological objects of the desire for fluids, and is often closely related to appetite for foods as well. This in itself is good reason to believe that all the control of human drinking behaviour in familiar surroundings has been learned, like other biologically important human appetites. Hence the innateness of dehydration-induced water consumption in people should not be presumed in the absence of conclusive ontogenetic analysis. The issue is due critical psychological and physiological investigation.

Albeit qualitative or even anecdotal, the sensory and social psychology of human fluid consumption turns out to be quite extensive. Characterization of the causal bases of beverage palatability and cultural practices and attitudes would enable the physiological psychology of normal human thirst to advance further and faster as well. To understand *Gestalten* of the sensory, somatic and social influences on human thirst, the organization of drinking behaviour needs to be analysed in diverse individuals in each distinct circumstance where it is liable to occur, from the earliest development of beverage consumption to the changes in thirst in normal and abnormal circumstances throughout life.

References

Baker BJ, Booth DA, Duggan JP, Gibson EL (1987) Protein appetite demonstrated: learned specificity of protein-cue preference to protein need in adult rats. Nutr Res 7:481–487

Birch LL (1987) The acquisition of food acceptance patterns in children. In: Boakes RA, Popplewell DA, Burton MJ (eds) Eating habits. Wiley, Chichester, pp 107–130

Birch LL, Deysher M (1985) Conditioned and unconditioned caloric compensation: evidence for self-regulation of food intake in young children. Learn Motiv 16:341–355

Birch LL, Zimmerman S, Hind H (1980) The influence of social affective context on preschool children's food preferences. Child Dev 51:856–861

Blair AJ, Booth DA, Lewis VJ, Wainwright CJ (1989) The relative success of official and informal weight reduction techniques: retrospective correlational evidence. Psychol Health 3:195–206

Booth DA (1972a) Satiety and behavioral caloric compensation following intragastric glucose loads in the rat. J Comp Physiol Psychol 78:412–432

Booth DA (1972b) Conditioned satiety in the rat. J Comp Physiol Psychol 81:457–471

Booth DA (1974) Acquired sensory preferences for protein in diabetic and normal rats. Physiol Psychol 2:344–348

Booth DA (1976) Approaches to feeding control. In: Silverstone T (ed) Appetite and food intake. Abakon/ Dahlem Konferenzen, Berlin, pp 417–478

Booth DA (1977) Appetite and satiety as metabolic expectancies. In: Katsuki Y, Sato M, Takagi SF, Oomura Y (eds) Food intake and chemical senses. University of Tokyo Press, Tokyo, pp 317–330

Booth DA (1979) Is thirst largely an acquired specific appetite? Behav Brain Sci 2:103–104

Booth DA (1980) Conditioned reactions in motivation. In: Toates FM, Halliday TR (eds) Analysis of motivational processes. Academic Press, London, pp 77–102

Booth DA (1981) How should questions about satiation be asked? Appetite 2:237–244

Booth DA (1985) Food-conditioned eating preferences and aversions with interoceptive elements: learned appetites and satieties. Ann NY Acad Sci 443:22–37

Booth DA (1987a) Evaluation of the usefulness of low-calorie sweeteners in weight control. In: Grenby TH (ed) Developments in sweeteners – 3. Elsevier Applied Science, London, pp 287–316

Booth DA (1987b) Individualised objective measurement of sensory and image factors in product acceptance. Chem Ind (Lond) 1987 (13):441–446

Booth DA (1987c) Central dietary "feedback onto nutrient selection": not even a scientific hypothesis. Appetite 8:195–201

Booth DA (1987d) Objective measurement of determinants of food acceptance: sensory, physiological and psychosocial. In: Solms J, Booth DA, Pangborn RM, Raunhardt O (eds) Food acceptance and nutrition. Academic Press, London, pp 1–27

Booth DA (1988) Mechanisms from models – actual effects from real life: the zero-calorie drink-break option. Appetite 11:94–102

Booth DA (1990a) Summary: Concluding Session. Ann NY Acad Sci 575:466–471

Booth DA (1990b) The behavioral and neural sciences of ingestion. In: Stricker EM (ed) Handbook of behavioral neurobiology, vol 10. Food and water intake. Plenum Press, New York

Booth DA (1990c) Learned ingestive motivation and the pleasures of the palate. In: Bolles RC (ed) Hedonics of taste. Erlbaum, Hillsdale, NJ

Booth DA (1990d) Learned role of tastes in eating motivation. In: Capaldi ED, Powley TL (eds) Taste, experience and feeding. American Psychological Association, Washington DC.

Booth DA, Baker BJ (1990) dl-Fenfluramine challenge to nutrient-specific textural preference conditioned by concurrent presentation of two diets. Behav Neurosci 104:226–229

Booth DA, Blair AJ (1989) Objective factors in the appeal of a brand during use by the individual consumer. In: Thomson DMH (ed) Food acceptability. Elsevier Applied Science, London, pp 329–346

Booth DA, Conner MT (1988) Preferred sweetness of a lime drink and preference for sweet over non-sweet foods, related to sex and reported age and body weight. Appetite 10:25–35

Booth DA, Davis JD (1973) Gastrointestinal factors in the acquisition of oral sensory control of satiation. Physiol Behav 11:23–29

Booth DA, Mather P (1978) Prototype model of human feeding, growth and obesity. In: Booth DA (ed) Hunger models: computable theory of feeding control. Academic Press, London, pp 279–322

Booth DA, Toase AM (1983) Conditioning of hunger/satiety signals as well as flavour cues in dieters. Appetite 4:235–236

Booth DA, Lovett D, McSherry GM (1972) Postingestive modulation of the sweetness preference gradient in the rat. J Comp Physiol Psychol 78:485–512

Booth DA, Stoloff R, Nicholls J (1974) Dietary flavor acceptance in infant rats established by association with effects of nutrient composition. Physiol Psychol 2:313–319

Booth DA, Lee M, McAleavey C (1976) Acquired sensory control of satiation in man. Br J Psychol 67:137–147

Booth DA, Mather P, Fuller J (1982) Starch content of ordinary foods associatively conditions human appetite and satiation, indexed by intake and eating pleasantness of starch-paired flavours. Appetite 3:163–184

Booth DA, Thompson AL, Shahedian B (1983) A robust, brief measure of an individual's most preferred level of salt in an ordinary foodstuff. Appetite 4:301–312

Booth DA, Conner MT, Marie S (1987) Sweetness and food selection: measurement of sweeteners' effects on acceptance. In: Dobbing J (ed) Sweetness. Springer-Verlag, London, pp 143–160

Booth DA, Conner MT, Gibson EL (1989) Measurement of food perception, food preference, and nutrient selection. Ann NY Acad Sci 561:226–242

Breckler SJ (1990) Applications of covariance structure modeling in psychology: cause for concern? Psychol Bull 107:260–273

Conner MT, Haddon AV, Booth DA (1986) Very rapid, precise measurement of effects of constituent variation on product acceptability: consumer sweetness preferences in a lime drink. Lebens-Wiss-Technol 19:486–490

Conner MT, Haddon AV, Pickering ES, Booth DA (1988) Sweet tooth demonstrated: individual differences in preference for both sweet foods and foods highly sweetened. J Appl Psychol 73:275–280

Elizaldi G, Sclafani A (1988) Starch-based conditioned flavor preferences in rats: influence of taste, calories and CS-US delay. Appetite 11:179–200

Elmadfa I, Huhn W (1986) Consumption and preference of drinking water. In: Diehl JM, Leitzmann C (eds) Measurement and determinants of food habits and food preferences. University Department of Human Nutrition, Wageningen, pp 129–135

Ershow AG, Cantor KP (1989) Total water and tapwater intake in the United States: population-based estimates of quantities and sources. FASEB, Bethesda, MD

Gibson EL, Booth DA (1986) Acquired protein appetite in rats: dependence on a protein-specific need state. Experientia 42:1003–1004

Gibson EL, Booth DA (1989) Dependence of carbohydrate-conditioned flavor preference on internal state in rats. Learn Motiv 20:36–47

Harris G, Booth DA (1987) Infants' preference for salt in food: its dependence upon recent dietary experience. J Reprod Infant Psychol 5:97–104

Harris G, Thomas A, Booth DA (1990) Development of salt taste preference in infancy. Dev Psychol 26:534–538

Holman EW (1975) Immediate and delayed reinforcers for flavor preferences in rats. Learn Motiv 6: 91–100

Krieckhaus EE, Wolf G (1968) Acquisition of sodium by rats: interaction of innate mechanisms and latent learning. J Comp Physiol Psychol 65:197–201

Mook DG (1988) 0n the organization of satiety. Appetite 11:27–39

Peters RS (1954) The concept of motivation. RKP, London

Pliner P (1982) The effects of mere exposure on liking for edible substances. Appetite 3:283–290

Revusky SH (1968) Effects of thirst level during consumption of flavored water on subsequent preferences. J Comp Physiol Psychol 66:777–779

Robinson PH (1990) Gastric function in eating disorders. Ann NY Acad Sci 575:456–465

Ryle G (1949) Concept of mind. John Murray, London

Sherman JE, Hickis CF, Rice AG, Rusiniak KW, Garcia J (1983) Preferences and aversions for stimuli paired with ethanol. Anim Learn Behav 11: 101–106

Tordoff MG, Friedman MI (1986) Hepatic portal glucose infusions decrease food intake and increase food preference. Am J Physiol 251:R192–196

Van Vort W, Smith GP (1987) Sham feeding experience produces a conditioned increase of meal size. Appetite 9:21–29

Weiffenbach JM (1989) Assessment of chemosensory functioning in aging: subjective and objective procedures. Ann NY Acad Sci 561:56–65

Wickelgren WA (1979) Chunking and consolidation: a theoretical synthesis. Psychol Rev 86:44–60

Wittgenstein L (1953) Philosophical investigations. Blackwells, Oxford

Commentary

Szczepanska-Sadowska: The notion of thirst as a disposition to have a drink encompasses a broad spectrum of circumstances relevant to human life. Some reservations arise about its application to animal studies, however. How is it possible to define precisely when the animal is actually disposed to have a drink?

Greenleaf: How can the disposition to have a drink, i.e. thirst, occur without a sensation of thirst? Is this not a contradiction in terms? The first definition of thirst in Webster's Unabridged Dictionary (1981) is: "a sensation of dryness in the mouth and throat associated with a desire for liquids; also the bodily condition that induces this sensation; 1b: a desire for potable liquids or to drink; 2: to have a vehement desire; crave … ". Perhaps the term thirst is so ambiguous that it should not be discussed.

Booth: In my view, it is a waste of time trying to agree a proper scientific use of the word thirst (or of any other everyday psychological word). Thank you for the support from a dictionary for my claim that the major ordinary use of the word thirst is to refer to "a desire for potable liquids or to drink", not a desire specifically for water. Note also that Webster defines the sensation only by association and the physiological state only by what it induces. A mental process that has the nature of a disposition does not have to be verbalizable or even conscious. Thirst is very adequately measurable in people or non-verbal animals, by seeing how responsive any ingestion-related behaviour is to cues to sources of water. This is something like testing whether a reflex mechanism has been selectively facilitated. It is a perfectly objective matter, whether verbal, behavioural or intake data are used. Both expressing a feeling of thirst and drinking a fluid have to be tested further in order to have definite evidence of appetite for water.

Epstein: I don't understand why you use the phrase "thirsty behaviour." It's like calling olfaction smelly behaviour. I agree that the sensation of thirstiness is an epiphenomenon for those of us who are trying to understand its mechanisms. If we keep insisting that it is a sensation, then the only science that can be done with it is human psychophysics.

The use of the word "organism" in this context is naive and pretentious. Most organisms are plants, and even among animals thirst and drinking behaviour are not universal.

Booth: On all these fronts, you've missed the logical implications of what you agree with in my argument. Olfaction precisely is successful smelling behaviour. Thirst, I point out, is an organism's momentary power to execute (water-) drinking behaviour. The thirst in the thirsty behaviour is the ability of the movements to be sensitive to cues of water, if we're using the word thirst in its narrow sense, or to fluids or beverages if in the sense that is broader except that it excludes watery solids.

Yes, this is psychophysics, but not the old-fashioned psychophysics of sensations somehow introspected on by human beings. It's the objective psychophysics of discriminative performance (Torgerson 1958), investigatable in non-verbal animals as well as in humankind. This can identify causally real cognitive mechanisms, not mere epiphenomena.

Thus if a plant performs the right way, it's thirsty. So, when I use the word, I do mean "organism". I agree we should say animal if we mean it. Yet it would be empirically incorrect and ethically egregious to say animal when we are referring primarily to people (as Chapter 4 does), because a person has important characteristics in addition to the qualities of any other animal we know. Should you say animal when you mean vertebrate, though? Do invertebrates get thirsty (Fitzsimons, Chapter 1)?

Reference

Torgerson WS (1958) Theory and methods of scaling. Wiley, New York

Denton: The suggestion that thirst is learned is difficult to reconcile with a Darwinian view. A highly organized behavioural pattern like water drinking has emerged phylogenetically because it carried great survival advantage for land animals in the face of constant threat of desiccation. This propensity to drink water with dehydration, but not the motivation to drink salt solution if depleted of sodium, is abolished by lesion, not of higher areas such as cortex, but of the anterior wall of the third ventricle. There appears to be a hardwired organization involving sensors of ionic and hormonal changes, as well as receptors for volume of blood vessels. Learning occurs on this genetically determined base, and it would appear that a large amount of drinking behaviour in humans is socially conditioned and without motivation by changes in the milieu interieur. Water intake might be regarded as having a hierarchical organization such as described by Tinbergen for a number of innate behavioural patterns.

Hall (Chapter 3) describes the maturation of drinking responses to dehydration, and shows that this behaviour is clearly distinct from suckling. Thirst is innate.

Booth: What I am indicating is that it has yet to be shown by appropriate measurement of the physiological and behavioural history of the individual that dehydration-induced drinking is innate in any mammal. This is an example of a very live issue particularly in human biology: what in our behaviour involves reflex mechanisms that get wired up independently of learning? (Much unlearned behaviour is not truly "hardwired": for example, visual discrimination of vertical from horizontal in kittens and the strength of sodium appetite in the rat are greatly affected by neural activity induced postnatally.)

You recognize the need for behavioural evidence to establish the innateness of sodium appetite. Water appetite is no different. Disruption by lesions is beside the point of the behavioural criteria. In fact, it is Hall and his colleagues themselves who have shown classical and instrumental conditioning in rats considerably earlier than water-selection responses to dehydration. So there are all the mechanistic possibilities outlined by Holland (Chapter 17) even in that species. *Rattus* and *Homo* are such extraordinary biological successes because of their great adaptability of behaviour to environmental opportunities. Maybe losing compulsive reflexes to desiccation kept our ancestors sane enough to survive when only the desert specialists otherwise had a chance.

No amount of functional speculation can settle a mechanistic issue. Hierarchical schemes of motivation also stand or fall on the facts of behaviour. Nobody can yet cite signal detection experiments that show a disguised osmotic load to give adult human subjects a water-specific appetite, as we all presume it does.

Robertson: I must take issue with the suggestion that the controls of selection, initiation or termination of a drink could have gained their power as a result of past experience. A naive human subject receiving an intravenous infusion of hypertonic saline almost invariably develops a thirst and a desire to drink that cannot be attributed to any past experience. Patients with diabetes insipidus develop thirst and polydipsia immediately upon the development of polyuria, even though there is nothing in their past to prepare them for this experience. Similarly, when started on antidiuretic therapy, their thirst and polydipsia cease almost immediately even though these symptoms may have been present all of their life. These dipsogenic stimuli, at least, appear to be innate and not a product of experience.

Booth: A learned osmotic thirst would disappear when the osmotic stimulus disappears; instant recovery from diuresis would be no evidence either way. Clinically "immediate" development of polydipsic thirst need not signify innateness either;

learning of control by the physiological signal could in principle track the polyuria very closely. After all, caloric conditioning of ingestion occurs in a single meal. How do you know that extra drinking is not reinforced at the very first drink after polyuria begins?

Then how do you know that infused subjects, and indeed the diabetes patients, never had any past experience of dehydration and indeed of some effects of drinking on it to generalize from? Moreover, if osmotic thirst is innate, what is going on in those who do not get thirsty after saline infusion?

Those sorts of issue have to be addressed before we could have scientific evidence whether or not these controls of drinking are unlearned. Experimental tests of dipsogenic stimuli would also be more illuminating if the pattern of sensory preferences among fluids were ascertained at both the start and the finish of the first drink after the first-ever hypertonic saline infusion, by means of validated ratings or real drink choices. This would extend the range of outputs (such as differently worded ratings) over which we could do intercorrelations. When we measure all major inputs too, we can identify causal relationships, such as effects of sensory or social influences on the sensitivity of ingestive output to plasma osmolality. Water and salt intakes do not tell you what the ingestion is directed at sensorily and cognitively, any more than carbohydrate, protein and fat intakes tell you why foods are preferred (Booth 1987a,b; Booth et al. 1989). A thirst rating affected by dehydration does not tell you what it is that the somatic variable has influenced. All the psychological questions remain to be answered. Is it ice cream that's really wanted, not water? Or a coffee or a beer? And would it be a large beer or a small one?

References

Booth DA (1987a) Central dietary "feedback onto nutrient selection": not even a scientific hypothesis. Appetite 8:195–201

Booth DA (1987b) Objective measurement of determinants of food acceptance: sensory, physiological and psychosocial. In: Solms J, Booth DA, Pangborn RM, Raunhardt O (eds) Food acceptance and nutrition. Academic Press, London, pp 1–27

Booth DA, Conner MT, Gibson EL (1989) Measurement of food perception, food preference, and nutrient selection. Ann NY Acad Sci 561:226–242

Section II
Physiological Influences on Drinking

Chapter 5

Osmoreceptors for Thirst

M.J. McKinley

Osmoregulatory Responses

The osmolality of human plasma is normally maintained within a quite narrow range around a set point which may vary between different individuals from 280–295 mosm/kg (this set point is lowered in pregnant women). If plasma osmolality does begin to increase from the normal set point so that the plasma tonicity (effective osmotic pressure) alters, a number of physiological mechanisms become operative. As a result, thirst develops, vasopressin is released from the posterior pituitary and renal water losses diminish because of the formation of more concentrated urine (Verney 1947, Wolf 1950, Robertson and Athar 1976). While the kidneys are partially able to correct the hypertonicity, water intake is usually necessary to restore the composition and volume of body fluids to their original state. Therefore thirst, the desire to ingest water, may be viewed as an important component of a co-ordinated system dedicated to the maintenance of constant body fluid tonicity.

For such a homeostatic system to function effectively and efficiently, it would seem essential for animals to possess some mechanism for detecting small changes in plasma tonicity, which then engages the body's osmoregulatory mechanisms. It is commonly thought that osmoreceptors subserve this function in mammals (Verney 1947; Wolf 1950; Robertson and Athar 1976), but it should be stated at the outset, that nobody has actually been able to give a convincing description of the appearance of such sensors. However, we do know something of their properties and modus operandi, and there are clues to the bodily location of the osmoreceptors. These topics will be discussed in the subsequent paragraphs.

Cellular Dehydration and Thirst

On observing a greater depression of the freezing point of plasma when animals became dehydrated, Mayer (1900) concluded one of the early reports of an investigation of thirst, thus: "Les expériences précédentes tendent à démontrer que le soif est liée a l'état hypertonique du milieu intérieur". While Mayer thought that thirst was generated when the osmotic pressure of body fluids increased, his hypothesis only partly explains the phenomenon of osmotically stimulated thirst.

Addition of solute to the blood will certainly stimulate thirst or water drinking in many cases. However, it is the effective osmotic pressure (tonicity) of the body fluids, rather than the absolute osmotic pressure of the solution relative to an ideal semi-permeable membrane which determines water intake. Molecules such as sodium chloride, sucrose or sorbitol may be quite different chemically, but, mainly because of their size and polar nature, they do not permeate through cell membranes very quickly. On the other hand, smaller molecules like urea or glycerol do permeate the plasma membrane of cells. When concentrated solutions of sodium chloride, sucrose or sorbitol (hyperosmolar to plasma) are infused into the bloodstream, an osmotic gradient is established across cell membranes, and water then moves out of cells by osmosis, causing cellular dehydration. When such solutions are infused into their bloodstream, experimental animals (e.g. dogs, goats, sheep), drink considerable quantities of water (Fig. 5.1) , and we presume they experience thirst. On the other hand, increasing the osmolality of the plasma of these animals by addition of hyperosmolar solutions of urea, isomannide or glycerol results in much less water drinking (Gilman 1937; Holmes and Gregerson 1950; Fitzsimons 1961; Olsson 1972; McKinley et al. 1978). Because these small molecules (urea, isomannide, glycerol) permeate into cells relatively quickly, an osmotic gradient is not maintained across cell membranes with the latter infusions, so that there is little movement of water out of cells by osmosis. In other words, the osmotic stimuli that have been found experimentally to be the most effective in stimulating drinking are also those which cause cellular dehydration (Gilman 1937). From these observations, the notion has developed that thirst results from cellular dehydration of a particular group of cells in the body (Wolf 1950).

Since Gilman first proposed the cellular dehydration theory of thirst as a result of his investigations in dogs, several other species have also been studied in this regard. It is now clear that increasing the osmolality of plasma with solutes such as sodium chloride, sucrose, mannitol, sorbitol or fructose, induces water drinking in such diverse species as dogs, goats, sheep, cattle, pigeons and iguanas. Systemic infusions of hyperosmolar urea, glycerol or isomannide are much less effective dipsogens in these animals (Holmes and Gregerson 1950; Wolf 1950; Olsson 1972; Fitzsimons

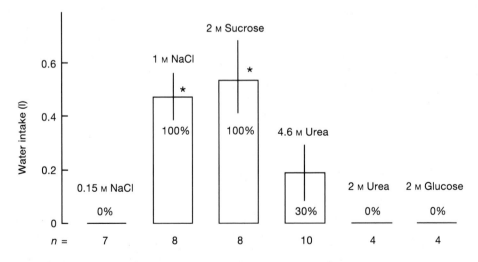

Fig. 5.1. Water drinking (mean,SEM) in sheep induced by infusion of various hypertonic solutions at 1.3 ml/min into the carotid artery of sheep for 25 min. *Significantly different from infusion of 0.15M NaCl, $P<0.01$; % = percentage of trials in which drinking response occurred. (Data from McKinley et al. (1978).)

and Kaufman 1977; McKinley et al. 1978; Thornton 1984; McKinley et al. 1987). With laboratory rats, this is not quite the case, because hypertonic fructose and sorbitol as well as urea are weaker dipsogens than hypertonic sodium chloride or sucrose (Fitzsimons 1961). It is not clear why rats are different in this regard, although it has been suggested that it may be related to the fact that rats are slower drinkers than the other species investigated (Fitzsimons 1961).

An important consideration, of course, is how the thirst mechanism in man responds to hypertonicity (also see Chapter 25). We are all aware of thirst which follows the eating of salty foods, and a number of studies have shown that ingestion of or intravenous infusions of hypertonic sodium chloride induces thirst in men and women (Arden 1934; Wolf 1950; Zerbe and Robertson 1983; Phillips et al. 1985). The average increase in plasma sodium concentration and osmolality necessary to stimulate thirst in young men is 4 mmol/l and 7 mosm/kg (baseline of 284 mosm/kg) respectively (Phillips et al. 1985). In pregnant women also, an average increase of 7 mosm/kg in plasma osmolality from a lower baseline (average 280 mosm/kg) was needed to stimulate thirst. Post-partum, the same women had a baseline plasma osmolality of 289 mosm/kg and the thirst threshold was reached at 298 mosm/kg (Davison et al. 1984). Unfortunately there are few studies on man of the thirst-inducing effectiveness of hypertonic solutions other than sodium chloride. However, Zerbe and Robertson (1983) have observed that intravenous infusion of hypertonic mannitol was as effective as hypertonic sodium chloride in stimulating thirst, whereas hypertonic urea was less effective. Based on these data, it seems likely that osmoreceptor stimulation in man is associated with cellular dehydration, and the osmoreceptor probably behaves similarly in man, dog, goat and sheep.

It should be kept in mind that this concept of the osmoreceptor responding to cellular dehydration is based on the results of experiments in which systemic infusions of hyperosmolar solutions were given over periods of no more than an hour. In the circumstance of dehydration occurring over several hours or days, it is possible that a different mechanism for detecting hypertonicity becomes operative in the longer term. This is because there is evidence that brain cells regain cell water as the duration of the hypertonicity increases (Arieff et al. 1977), and therefore thirst and vasopressin secretion are probably not directly related to the absolute degree of cellular dehydration. Evidence showing that the secretion of vasopressin in chronically hydrated rats which were acutely infused with hypertonic saline was not directly related to the volume of the cells in the brain also supports this idea (Verbalis et al. 1986). Of course, the acceptability of this conjecture depends on the proviso that osmoreceptor cells behave like other brain cells in the chronic regulation of their volume, and this is not known.

Location of the Osmoreceptors

Osmoreceptors in the Brain

Theories regarding the location of osmoreceptors have been greatly influenced by the work of E. B. Verney. He showed that infusions of hypertonic sodium chloride or sucrose were much more effective in stimulating vasopressin (ADH) secretion if they were infused into the carotid artery rather than into a peripheral vein (Verney 1947). Because the internal carotid artery supplies the forebrain, Verney proposed that the selective increase in tonicity of carotid blood was detected by osmoreceptors in the diencephalon, with resultant vasopressin secretion. Intracarotid infusion of hyperosmolar urea was less effective than hypertonic sodium chloride or sucrose, and Verney proposed that cellular dehydration was the modus operandi of these osmoreceptors. The obvious similarity to the induction of water drinking by systemic infusion of hypertonic solutions and the observation by Bengt Andersson that injections of hypertonic sodium chloride into the hypothalamus of goats caused drinking (Andersson 1952), resulted in the belief that hypothalamic osmoreceptors controlled both thirst and vasopressin secretion. Interestingly, the Verney-type experiment of comparing intracarotid with intravenous infusions of hypertonic solutions, but this time measuring water intake, was not carried out until the 1970s. The results from studies in goats and dogs were quite clear. Selectively increasing the tonicity of the carotid blood stimulated water drinking (Olsson 1972; Wood et al. 1977). These results indicate an osmoreceptor in the CNS .

Sites in the CNS

There are conflicting data in the literature regarding the existence of osmoreceptors in the liver regulating vasopressin secretion (Haberich 1968; Glasby and Ramsay 1974), and no conclusive studies on hepatic osmoreceptors subserving thirst. By far the most extensively studied osmoreceptors are those that are thought to be situated in the brain.

The Medial Hypothalamus

Considerable effort has been expended in searching for the exact location of the cerebral osmoreceptors which subserve thirst in animals. The initial and most direct method of approach has been to make microinjections of hypertonic solutions into particular brain regions and to observe the resultant drinking behaviour. Andersson's influential early study (Andersson 1952) has already been mentioned, but its details (Andersson 1953) are important. The solution used was usually sodium chloride at 330 mmol/l, and volumes of 50–150 µl were injected into the goat's brain. The injections effective in causing drinking were spread throughout the hypothalamus, but tended to be close to the midline. The most effective sites were in the rostral medial hypothalamus. No effect was obtained from the lateral hypothalamus or supraoptic nucleus (Andersson 1953). Andersson advised caution in the interpretation of these results, in that only hypertonic sodium chloride solution was injected (no other osmotic agent for comparison), and the concentrations were supraphysiological.

The Lateral Preoptic Region

The next site to come to notice in this regard was the lateral preoptic region, where it was found that microinjection of hypertonic solutions of sodium chloride or sucrose induced water drinking in rats and rabbits (Peck and Novin 1971; Blass and Epstein 1971). Ablation of the same region caused inhibition of osmoregulatory drinking, and injection of water into the lateral preoptic region prevented the water intake normally induced by systemic hypertonicity (Peck and Novin 1971; Blass and Epstein 1971). However, further studies in rats by these authors showed that microinjection of hypertonic solutions into several other sites (eg bed nucleus of the stria terminalis and anterior hypothalamic region) could also elicit drinking (Peck and Blass 1975). It is not easy to estimate the extent of spread of these injected solutions whose hypertonicity needed to be supraphysiological in order to elicit drinking. However, they do support the notion that preoptic/hypothalamic sites are involved. In another study, drinking was elicited in rats with injections into the lateral preoptic region of hypertonic sucrose with the osmolality increased by only 30 mosm/kg (Blass 1974).

Other evidence which is consistent with the existence of osmoreceptors in the lateral preoptic region comes from electrophysiological studies. Neurons there have been found that increase their firing rate in response to systemic infusions of hypertonic solutions (Tondat and Almli 1976; Malmo and Malmo 1979). Unfortunately this information alone does not allow the conclusion that these neurons are in fact osmoreceptive. While they may well be part of a neural pathway involved in osmoregulatory mechanisms, it is quite possible that they are activated indirectly from osmoreceptors which are in fact located elsewhere. Indeed, neurons have been found in other preoptic and hypothalamic sites which increase their activity in response to hypertonicity (Hayward and Vincent 1970; Sayer et al. 1984; Malmo 1976; Gutman et al. 1988),

The Circumventricular Organs

These neuronal structures have the special feature that they lack a normal blood–brain barrier, due to the presence of fenestrations in the capillary endothelium (Weindl 1973). When evidence was obtained that some central osmoreceptors may be situated in a region lacking the blood–brain barrier (McKinley et al. 1978; Thrasher et al. 1980a), interest focused on the circumventricular organs as possible sites of osmoreceptors.

The experimental evidence for this view comes from investigations in which the sodium concentration of cerebrospinal fluid (CSF) was monitored during systemic infusions of various hyperosmolar solutions. It was found that infusions into the bloodstream of hyperosmolar urea could be as effective as hyperosmolar sodium chloride or sucrose in increasing the sodium concentration (and therefore the tonicity) of CSF (Fig. 5.2). In this respect, therefore, urea is like sodium chloride and sucrose, in that it is effectively excluded by the blood–brain and blood–liquor barriers (Javid and Settlage 1956; Crone 1965; Oldendorf 1971). When any of these three solutes is used to increase the osmolality of plasma, water moves out of the brain by osmosis across these barriers into the bloodstream, causing both the sodium concentration and tonicity of cerebral fluids to increase. The question then arises as to how a sensor in the brain driving thirst could distinguish an infusion of urea from one of hypertonic sodium chloride or sucrose (see above), while these agents all cause similar changes behind the blood–brain barrier. Thus the suggestion was made that osmoreceptors may reside in two of the forebrain circumventricular organs, the organum vasculosum of the lamina terminalis (OVLT) and/or the subfornical organ (McKinley et al. 1978), two midline structures found in the anterior wall of the third ventricle.

Consistent with this suggestion is evidence that ablation of the OVLT reduces water drinking by sheep and dogs in response to systemic infusion of hypertonic saline (McKinley et al. 1982; Thrasher et al. 1982a). Despite the fact that ablation of the subfornical organ in the rat or dog does not inhibit water drinking to systemic hypertonicity (Simpson et al. 1978; Thrasher et al. 1982b), it does reduce drinking and vasopressin release in response to subcutaneously injected hypertonic saline

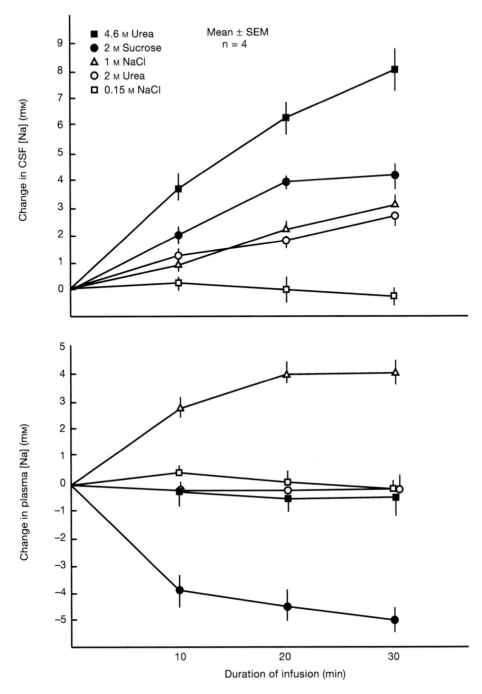

Fig. 5.2. Increases in the sodium concentration of CSF of conscious sheep induced by infusions of concentrated solutions of sodium chloride, sucrose or urea into the carotid artery of sheep. The infusion rate was 1.6 ml/min and the CSF was obtained from the lateral ventricle. (From McKinley et al. (1978), with permission.)

(Mangiapane et al. 1984). It also reduces drinking to small increases in plasma sodium concentration (Hosutt et al. 1981), and microinjection of 0.2 mol/l sodium chloride into the subfornical organ has been reported to stimulate water drinking in the rat (Comargo et al. 1984). Electrophysiological experiments have shown that some cells in the rat OVLT (in vitro brain slices) increase their activity in response to increased tonicity of their bathing medium (Sayer et al. 1984; Vivas et al. 1990). Some subfornical organ neurons also increase their activity in response to hypertonicity (Sibbald et al. 1988; Gutman et al. 1988). It is still possible, of course, that the osmoresponsive neurons in these two circumventricular organs were being indirectly stimulated by synaptic input from some other region.

Although the results from infusions of urea suggested osmoreceptors were located in the circumventricular organs, the problem remains, however, that while hyperosmolar urea is a relatively weak dipsogen (Fig. 5.1), many investigators have found that it does induce some water drinking when infused into the bloodstream (Gilman 1937; Holmes and Gregerson 1950; Fitzsimons 1961; Olsson 1972; McKinley et al. 1978; Zerbe and Robertson 1983). It seems unlikely that this water intake was caused by dehydration of osmoreceptors in the circumventricular organs, because such cellular dehydration would be expected to show a rapid onset and be quite transient, since the urea should permeate into such cells and the cellular dehydration should not be maintained. In fact the drinking response to hyperosmolar urea has quite a long latency and persisted for some time (Gilman 1937; Holmes and Gregerson 1950; Wolf 1950; Fitzsimons 1961). Therefore, it seems probable that the thirst and water drinking induced by systemic infusions of concentrated urea results from dehydration of brain sites with a blood–brain barrier. Here, the cellular dehydration may be maintained for hours because urea is excluded by the blood–brain barrier (Javid and Settlage 1956). Thus it is likely that in addition to osmoreceptors in circumventricular organs, there are also sensors which mediate osmotically-induced water drinking at sites elsewhere in the brain.

The Anteroventral Third Ventricle Region (AV3V)

In regard to the anterior wall of the third ventricle, studies in goats, rats and sheep have shown that water drinking can either be temporarily or permanently abolished if this region is ablated (Andersson et al. 1975; Buggy and Johnson 1977; Lind and Johnson 1983; McKinley et al. 1986). In the rats and sheep with AV3V lesions, that recover some water-drinking behaviour, this drinking now fails to respond to systemic hypertonicity (Buggy and Johnson 1977; McKinley et al. 1986). Rats and sheep with AV3V lesions fail adequately to increase vasopressin secretion in response to systemic hypertonicity, and are also unable to suppress vasopressin release when plasma tonicity falls to quite low levels (Johnson et al. 1978; McKinley et al. 1984). This may indicate that the detection of both increases and decreases in plasma tonicity is disrupted by ablation of the AV3V region. In normal rats and sheep, small injections of hypertonic sodium chloride or sucrose into the AV3V elicit drinking (McKinley et al. 1974; Buggy et al. 1979), although the hypertonicity of the solutions injected in these experiments was supraphysiological. The AV3V region is composed of several distinct cerebral structures: the preoptic periventricular region, ventral part of the median preoptic nucleus, anterior hypothalamic periventricular area and the organum vasculosum of the lamina terminalis (OVLT). The OVLT has already been discussed in regard to osmoreceptor location. Of the other components of the AV3V, the median preoptic nucleus (synonym: nucleus medianus) is noteworthy because discrete ablation of this site disrupts osmotically induced water drinking in the rat (Mangiapane et al. 1983), and it is likely that this is the crucial tissue that has to be ablated for the AV3V-lesion to disrupt osmoregulatory drinking (Lind et al. 1981) (see also Chapter 30). The median preoptic nucleus receives afferent neural input from both the OVLT and subfornical organ (Miselis 1981; Chapter 10) and it is possible that information from osmoreceptors there (and in other regions) is relayed via the median preoptic nucleus. Single unit recordings in this nucleus in anaesthetized sheep have shown that median preoptic neurons increase their activity in response to intracarotid infusions of hypertonic solutions which increase cerebral blood osmolality by 1%–2% (McAllen et al. 1990) consistent with the idea that these neurons are part of an osmoregulatory pathway. However, ablation of the median preoptic nucleus, while disrupting drinking in response to intraperitoneal hypertonic sodium chloride injection during daylight hours, does not do so if the experiment is made at night (Gardiner and Stricker 1985), indicating that some osmoregulatory neural pathways must exist outside of the median preoptic nucleus.

Overview

The aforementioned studies in experimental animals have produced evidence that osmoreceptors subserving thirst could be situated in one or more of the following forebrain sites: the medial hypothalamus, the lateral preoptic nucleus, bed nucleus of the stria terminalis, anterior hypothalamus, OVLT and subfornical organ. There is also evidence suggesting that the osmoreceptors are probably not confined to just one of these regions. For example, while ablation of the lateral preoptic nucleus severely disrupts the water drinking normally induced by intraperitoneal injection of hypertonic saline, it does not reduce drinking in response to intravenous injection of hypertonic saline (Coburn and Stricker 1978). Ablation of the OVLT reduces, but does not abolish, water drinking in response to systemic hypertonicity (McKinley et al. 1982; Thrasher et al. 1982a) indicating that osmoreceptors must also exist outside the OVLT. Similarly, ablation of the subfornical organ, bed nucleus of the stria terminalis or anterior hypothalamic region does not abolish osmoregulatory drinking (Blass and Epstein 1971; Simpson et al. 1978; McKinley et al. 1982). From this evidence, it is likely that osmoreceptors are distributed in a number of preoptic/hypothalamic sites, a view that has also been expressed by others (Peck and Novin 1971; Blass 1974). Whether they are all of equal sensitivity or physiological importance is unknown. Whether the osmoreceptors subserving thirst also regulate vasopressin secretion from the neurohypophysis is also unclear at this stage. It is unlikely that the vasopressin-containing neurons of the supraoptic nucleus are osmoreceptors for thirst because injection of hypertonic solutions into this site does not stimulate water drinking in the rat (Peck and Blass 1975).

Osmoreceptors and/or Sodium Sensors

Evidence for Sodium Receptors

Assertions have been made that neurons responding to changes in the sodium concentration of CSF, rather than cellular dehydration, are the sensors which signal alterations in the tonicity of blood (Andersson and Olsson 1973; Andersson 1978).

First, they showed that water drinking could be induced in goats by infusing small amounts of hypertonic sodium chloride or sodium bicarbonate into the third cerebral ventricle. Equivalent infusions of hypertonic saccharides into the third ventricle were ineffective (Andersson et al. 1967; Andersson and Olsson 1973). Second, infusions into the lateral ventricle of isotonic or hypertonic solutions of non-electrolytes such as fructose, glucose or mannitol, which reduced the sodium concentration but not the tonicity of CSF, inhibit water drinking responses to intracarotid infusions of hypertonic sodium chloride in goats (Olsson 1973). They inhibit water drinking by dehydrated goats and also sheep (Olsson 1975; Leksell et al. 1981). Third, intracerebroventricular infusions of drugs which modify transmembrane sodium transport were found to disrupt water drinking in goats (Andersson 1978; Rundgren et al. 1979).

To account for these observations Andersson and colleagues propose that when the tonicity of the circulating plasma increases, an increase in the sodium concentration of CSF also occurs, and the sodium sensors in the vicinity of the anterior wall of the third ventricle are stimulated.

Some variations on the sodium sensor hypothesis have been advanced by others. It has been suggested that both osmoreceptors and sodium sensors are involved in osmotically stimulated water drinking (McKinley et al. 1974; 1978) and the sodium sensors respond to the sodium concentration of cerebral extracellular fluid, rather than the cerebrospinal fluid per se (McKinley et al. 1980). Another variation on the sodium sensor theme is that an adequate ambient sodium concentration is permissive for osmoreceptors to function (Thornton 1984).

Criticisms of the Sodium Receptor Hypothesis

While the sodium sensor hypothesis has not gained overwhelming support, many of the criticisms can be refuted. A frequent objection to this hypothesis has been that the dipsogenic effects of intracerebroventricular hypertonic saline are due to non-specific neural stimulation of the hypothalamus. This seems unlikely because no other behavioural changes are observed when dipsogenic infusions of hypertonic saline are made into the lateral ventricle. This contrasts with the hyperventilation, rage, gnawing and eating which have been reported to accompany drinking in the goat in response to intraventricular infusions of potassium chloride, a known general excitant of neural tissue (Olsson 1969). In the sheep, we have also observed that intraventricular infusions of non-specific neuronal excitants such as glutamate or potassium chloride cause many other behavioural changes such as bleating, hyperventilation, scratching, eating and muscle twitches before drinking is induced. By contrast, the quiet, unhurried, deliberate sipping and drinking of water, which occurs during intraventricular infusion of hypertonic sodium chloride, is not accompanied by any of the other behaviours mentioned (M.J. McKinley and M. Gannon, unpublished observations). A further reason for suspecting that the effects of sodium in the CSF are not non-specific, is the observation that the concentration of sodium in the lateral ventricle needed to induce drinking in the sheep is an increase of 10–20 mmol/l (McKinley et al. 1980) which falls within the range of physiological change of this parameter. Another criticism which was made (McKinley et al. 1974) of the ineffective intracerebroventricular infusions of hypertonic saccharide solutions that Andersson and colleagues used, was that the solutions were not prepared in an artificial CSF solution. This would result in the concentrations of important ions such as sodium,

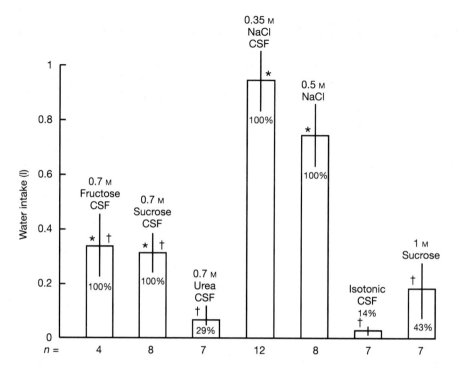

Fig. 5.3. Water drinking by sheep in response to infusions of various hypertonic solutions into the lateral ventricle. The infusion rate was 0.05 ml/min for 20 minutes Note the greater water intake induced by infusions of 0.35M NaCl prepared in artificial CSF compared with that caused by infusion of equiosmolal 0.7M fructose or sucrose prepared in an artificial CSF. Mean and SEM are given and the percentage of sheep which responded with drinking are given at the base of each block.
* Significantly greater than with isotonic CSF (*P*<0.01).
† Significantly less than with 0.35M NaCl CSF (*P*<0.01). (From McKinley et al. (1978).

potassium, calcium, magnesium and chloride not being maintained at adequate levels in the CSF, possibly influencing the drinking behaviour. Indeed, when hypertonic solutions of sucrose or fructose prepared in an artificial CSF were infused into the third or lateral ventricle of sheep, drinking responses were obtained (Fig. 5.3), (McKinley et al. 1974; 1978). However they were always less than the response to infusion of equiosmolar hypertonic sodium chloride (Fig. 5.3), leading us to postulate a dual sodium sensor–osmoreceptor regulation of thirst (McKinley et al. 1978). However a further problem arises with intraventricular infusion of hyperosmotic solutions of saccharides dissolved in artificial CSF. Even though the sodium concentration of the infusate is adequate, water will move by osmosis into the ventricle causing the sodium concentration of the CSF to fall below normal. In order to prevent this occurring, hypertonic solutions of mannitol combined with sodium chloride at supranormal concentration were infused into the lateral ventricle and maintained the sodium concentration of CSF normal. Water drinking in response to this infusion was still considerably less than the response to an intraventricular infusion of equiosmolar hypertonic sodium chloride alone (McKinley et al. 1980). Subsequent experiments have also shown evidence that intraventricular infusions of hypertonic sodium chloride are more dipsogenic than hypertonic saccharides in the rat, pig and pigeon (Buggy et al. 1979; Thornton 1984; Baldwin and Thornton 1986; Osborne et al. 1987a). In addition, water drinking by water-deprived rats is inhibited if the sodium concentration of CSF is reduced by means of intraventricular infusions of isotonic mannitol (Osborne et al. 1987b). The dog appears to be an exception in that intracerebroventricular hypertonic sucrose is equipotent with sodium chloride at stimulating drinking (Thrasher et al. 1980b). The suggestion that sodium sensors may only be of importance in herbivores (Fitzsimons 1979), therefore seems unlikely, and the evidence supports the view that sodium sensors for drinking may be widespread among mammalian species. With regard to primates, there has been no study of sodium sensors for drinking, but evidence of cerebral sodium sensors mediating vasopressin secretion has been produced (Swaminthan 1980).

Evidence which shows that a periventricular sodium receptor could not be the sole sensor for osmotically induced drinking, was the observation (previously discussed above) that infusions of concentrated solutions of urea into the bloodstream could elevate the sodium concentration of CSF as much as infusions of hypertonic sodium chloride or sucrose in sheep (Fig. 5.2), dogs and rats (McKinley et al. 1978; Thrasher et al. 1980a; Epstein 1980), but hyperosmolar urea infusions were less dipsogenic (Fig. 5.1). Therefore a sodium sensor alone could not account for the greater drinking seen in response to the infusions of sodium chloride or sucrose. It is feasible however, that the lesser drinking which occurs in response to systemic infusions of hyperosmolar urea is mediated by sodium receptors (or osmoreceptors as discussed above) behind the blood–brain barrier (McKinley et al. 1978).

Summary

In summary, it is unlikely that sodium receptors responding to changes in CSF sodium concentration are the sole mediators of thirst associated with the tonicity of the extracellular fluid. It is possible however, and even likely, that cerebral sodium receptors (inside the blood–brain barrier) together with osmoreceptors (both inside and outside the blood–brain barrier) participate in detecting changes in the tonicity of blood and engage neural circuitry subserving thirst.

Physiological Significance of Osmoreceptors

The experiments which have established the existence of cerebral osmoreceptors have usually utilized infusions of different solutes directly into the peritoneum, bloodstream or brain. None of these procedures are normal physiological phenomena, and it is important to consider the physiological conditions when these cerebral sensors might operate. Phillips et al. (1984) studied changes in plasma variables associated with the normal water drinking of young men during the working day. The subjects appeared to drink before any fluid deficit developed and before the tonicity of plasma increased, suggesting the osmoreceptors were not activated. Of course with ingestion of salt, thirst develops and the increase in plasma sodium concentration (Arden 1934; McCance and Young 1944) can exceed that needed for the thirst threshold to be reached in man (Phillips et al. 1985), suggesting osmoreceptors initiate this thirst. With water deprivation for 1–3 days, exercise or thermal dehydration, the plasma osmolality and sodium concentration increase to levels greater than the thirst threshold also (Black et al. 1944; Costill and Fink 1974;

Rolls et al. 1980). In these conditions it is likely that osmoreceptors are activated and signals from them combine with influences associated with the concommitant hypovolaemia to stimulate thirst. In the dehydrated dog, it has been shown that reducing the tonicity of cerebral blood considerably reduces (by an average of 72%) subsequent drinking. Such drinking can be abolished if the hypovolaemia is simultaneously corrected (Ramsay et al. 1977). Thus, it is likely that the cerebral osmoreceptors play an important role in the induction of thirst in dehydrated individuals.

Acknowledgement. M. J. McKinley is supported by the National Health and Medical Research Council of Australia.

References

Andersson B (1952) Polydipsia caused by intrahypothalamic injections of hypertonic NaCl solutions. Experientia 8:157–158

Andersson B (1953) The effect of injections of hypertonic NaCl solutions into different parts of the hypothalamus of goats. Acta Physiol Scand 81:188–201

Andersson B (1978) Regulation of water intake. Physiol Rev 58:552–603

Andersson B, Olsson K (1973) On central control of body fluid homeostasis. Condit Reflex 8:147–160

Andersson B, Olsson K, Warner RG (1967) Dissimilarity between the central control of thirst and of the release of antidiuretic hormone. Acta Physiol Scand 71:57–64

Andersson B, Leksell LG, Lishajko F (1975) Perturbations in fluid balance produced by medial placed forebrain lesions. Brain Res 99:261–275

Arden F (1934) Experimental observations upon thirst and on potasssium overdosage. Aust J Exp Biol Med Sci 12:121–122

Arieff AI , Guisado R, Lazarowitz VC (1977) Pathophysiology of hyperosmolar states. In: Andreoli TE, Grantham JJ, Rector FC (eds) Disturbance in body fluid osmolality. American Physiological Society, Bethesda MD, pp 227–250

Baldwin BA, Thornton SN (1986) Operant drinking in pigs following intracerebroventricular injections of hypertonic solutions and angiotensin II. Physiol Behav 36:325–328

Black DAK, McCance RA, Young WF (1944) A study of dehydration by means of balance experiments. J Physiol (Lond) 102:406–414

Blass EM (1974) Evidence for basal forebrain osmoreceptors in the rat. Brain Res 82:69–76

Blass EM, Epstein AN (1971) A lateral preoptic osmosensitive zone for thirst in the rat. J Comp Physiol Psychol 76:378–394

Buggy J, Johnson AK (1977) Preoptic-hypothalamic periventricular lesions: thirst deficits and hypernatremia. Am J Physiol 233:R44–R52

Buggy J, Hoffman WE, Phillips MI, Fisher AE, Johnson AK (1979) Osmosensitivity of rat third ventricle and interactions with angiotensin. Am J Physiol 236:R76–R82

Coburn PC, Stricker EM (1978) Osmoregulatory thirst in rats after lateral preoptic lesions. J Comp Physiol Psychol 92:350–361

Comargo LAA, Menani JV, Saad WA, Saad WA (1984) Interaction between areas of the central nervous system in the control of water intake and arterial pressure in rats. J Physiol (Lond) 350:1–8

Costill DL, Fink WJ (1974) Plasma volume changes following exercises and thermal dehydration. J Appl Physiol 37:521–525

Crone C (1965) The permeability of brain capillaries to non-electrolytes. Acta Physiol Scand 67:407–417

Davison JM, Gilmore EA, Durr J, Robertson GL, Lindheimer MD (1984) Altered osmotic thresholds for vasopressin secretion and thirst in human pregnancy. Am J Physiol 246:F105–F109

Epstein AN (1980) Consensus, controversies, and curiosities. Fed Proc 37:2711–2716

Fitzsimons JT (1961) Drinking by nephrectomized rats injected with various substances. J Physiol (Lond) 155:563–579

Fitzsimons JT (1979) The physiology of thirst and sodium appetite. Cambridge University Press, Cambridge

Fitzsimons JT, Kaufman S (1977) Cellular and extracellular dehydration and angiotensin as stimuli to drinking in the common iguana, *Iguana iguana*. J Physiol (Lond) 265:443–463

Gardiner TW, Stricker EM (1985) Impaired drinking responses of rats with lesions of nucleus medianus: circadian dependence. Am J Physiol 248:R224–R230

Gilman A (1937) The relation between blood osmotic pressure, fluid distribution and voluntary water intake. Am J Physiol 120:323–328

Glasby MG, Ramsay DJ (1974) Hepatic osmoreceptors? J Physiol 243:765–777

Gutman MB, Ciriello J, Mogenson GJ (1988) Effects of plasma angiotensin II and hypernatremia on subfornical organ neurons. Am J Physiol 254:R746–R754

Haberich FJ (1968) Osmoreception in the portal circulation. Fed Proc 27:1137–1141

Hayward JN, Vincent JD (1970) Osmosensitive single neurones in the hypothalamus of unanaesthetized monkeys. J Physiol (Lond) 210:947–972

Holmes JM, Gregerson MI (1950) Observations on drinking induced by hypertonic solutions. Am J Physiol 162:326–337

Hosutt JA, Rowland N, Stricker EM (1981) Impaired drinking responses of rats with lesions of the subfornical organ. J Comp Physiol Psychol 95:104–113

Javid M, Settlage P (1956) Effect of urea on cerebrospinal fluid pressure in human subjects. JAMA 160:943–949

Johnson AK, Hoffman WE, Buggy J (1978) Attenuated pressor responses to intracranially injected stimuli and altered antidiuretic activity following preoptic-hypothalamic periventricular ablation. Brain Res 157:161–166

Leksell LG, Congiu M, Denton DA et al. (1981) Influence of mannitol-induced reduction in CSF sodium on nervous and endocrine mechanisms involved in control of fluid balance. Acta Physiol Scand 112:33–40

Lind RW, Johnson AK (1983) A further characterization of the effects of AV3V lesions on ingestive behavior. Am J Physiol 245:R83–R90

Lind RW, Hartle DK, Brody MJ, Johnson AK (1981) Separation of the functional deficits induced by lesions of the AV3V. Fed Proc 40:390 (Abstr)

Malmo RB (1976) Osmosensitive units in the rat's dorsal midbrain. Brain Res 105:105–120

Malmo RB, Malmo HP (1979) Responses of lateral preoptic neurons in the rat to hypertonic sucrose and NaCl. Electroenceph Clin Neurophysiol 46:401–408

Mangiapane ML, Thrasher TN, Keil LC, Simpson JB, Ganong WF (1983) Deficits in drinking and vasopressin secretion after lesions of the nucleus medianus. Neuroendocrinology 37:73–77

Mangiapane ML, Thrasher TN, Keil LC, Simpson JB, Ganong WF (1984) Role for the subfornical organ in vasopressin release. Brain Res Bull 13:43–47

Mayer A (1900) Variations de la tension osmotique du sang chez les animaux prives de liquides. C R Soc Biol (Paris) 52:153–155

McAllen RM, Pennington GL, McKinley MJ (1990) Osmoresponsive units in the sheep median preoptic nucleus. Am J Physiol 259:R593–R600

McCance RA, Young WF (1944) The secretion of urine during dehydration and rehydration. J Physiol (Lond) 102:415–428

McKinley MJ, Blaine EH, Denton DA (1974) Brain osmoreceptors cerebrospinal fluid electrolyte composition and thirst. Brain Res 70:532–537

McKinley MJ, Denton DA, Weisinger RS (1978) Sensors for antidiuresis and thirst – osmoreceptors or CSF sodium detectors. Brain Res 141:89–103

McKinley MJ, Denton DA, Leksell L, Tarjan E, Weisinger RS (1980) Evidence for cerebral sodium sensors involved in water drinking in sheep. Physiol Behav 25:501–504

McKinley MJ, Denton DA, Leksell L et al. (1982) Osmoregulatory thirst in sheep is disrupted by ablation of the anterior wall of the optic recess. Brain Res 236:210–215

McKinley MJ, Congiu M, Denton DA et al. (1984) The anterior wall of the third cerebral ventricle and homeostatic responses to dehydration. J Physiol (Paris) 79:421–427

McKinley MJ, Denton DA, Leventer M et al. (1986) Adipsia in sheep caused cerebral lesions. In: de Caro G, Massi M, Epstein AN (eds) The physiology of thirst and sodium appetite. Plenum Press, New York, pp 321–326

McKinley MJ, Denton DA, Gellatly D, Miselis RR, Simpson JB, Weisinger RS (1987) Water drinking caused by intracerebroventricular infusions of hypertonic solutions in cattle. Physiol Behav 39:459–464

Miselis RR (1981) The efferent projections of the subfornical organ of the rat: a circumventricular organ within a neural network subserving fluid balance. Brain Res 230:1–23

Oldendorf WH (1971) Brain uptake of radiolabelled amino acids, amines, and hexoses after arterial injection. Am J Physiol 221:1629–1639

Olsson K (1969) Effects of slow infusions of KCI into the third brain ventricle. Acta Physiol Scand 77:465–474

Olsson K (1972) Dipsogenic effects of intracarotid infusions of various hyperosmolar solutions. Acta Physiol Scand 85:517–522

Olsson K (1973) Further evidence for the importance of CSF Na concentration in central control of fluid balance. Acta Physiol Scand 88:183–188

Olsson K (1975) Attenuation of dehydrative thirst by lowering of the CSF Na. Acta Physiol Scand 94:536–538

Osborne PG, Denton DA, Weisinger RS (1987a) Inhibition of dehydration induced drinking in rats by reduction of CSF Na concentration. Brain Res 412:36–42

Osborne PG, Denton DA, Weisinger RS (1987b) Effect of variation of the composition of CSF in the rat upon drinking of water and hypertonic NaCl solutions. Behav Neurosci 101:371–377

Peck JW, Blass EM (1975) Localization of thirst and antidiuretic osmoreceptors by intracranial injections in rats. Am J Physiol 228:1501–1509

Peck JW, Novin D (1971) Evidence that osmoreceptors mediating drinking in rabbits are in the lateral preoptic area. J Comp Physiol Psychol 74:134–147

Phillips PA, Rolls BJ, Ledingham JGG, Morton JJ (1984) Body fluid changes, thirst and drinking in man during free access to water. Physiol Behav 33:357–363

Phillips PA, Rolls BJ, Ledingham JGG, Forsling ML, Morton JJ (1985) Osmotic thirst and vasopressin release in humans: a double-blind crossover study. Am J Physiol 248:R645–R650

Ramsay DJ, RollS BJ, Wood RJ (1977) Thirst following water deprivation in dogs. Am J Physiol 232:R93–R100

Robertson GL, Athar S (1976) The interaction of blood osmolality and blood volume in regulating plasma vasopressin in man. J Clin Endocrinol Metab 42:613–620

Rolls BJ, Wood RJ, Rolls ET, Lind H, Lind W, Ledingham JGG (1980) Thirst following water deprivation in humans. Am J Physiol 239:R476–R482

Rundgren M, McKinley MJ, Leksell LG, Andersson B (1979) Inhibition of thirst and apparent ADH release by intracerebroventricular ethacrynic acid. Acta Physiol Scand 105:123–125

Sayer RJ, Hubbard JI, Sirret NE (1984) Rat organum vasculosum laminae terminalis in vitro: responses to neurotransmitters. Am J Physiol 247:R347–R379

Sibbald JR, Hubbard JI, Sirret NE (1988) Responses from osmosensitive neurons of the rat subfornical organ in vitro. Brain Res 461:205–214

Simpson JB, Epstein AN, Camardo JS (1978) Localization of receptors for the dipsogenic action of angiotensin II in the subfornical organ of the rat. J Comp Physiol Psychol 92:581–608

Swaminthan S (1980) Osmoreceptors or sodium receptors: an investigation into ADH release in the rhesus monkey. J Physiol (Lond) 307:71–83

Thornton SN (1984) A central Na receptor and its influence on osmotic and angiotensin II induced drinking in the pigeon Columba livia. J Physiol (Paris) 79:505–510

Thrasher TN, Brown CJ, Keil LC, Ramsay DJ (1980a) Thirst and vasopressin relase in the dog: an osmoreceptor or sodium receptor mechanism. Am J Physiol 238:R333–R339

Thrasher TN, Jones Keil LC, Brown CJ, Ramsay DJ (1980b) Drinking and vasopressin release during ventricular infusions of hypertonic solutions. Am J Physiol 238:R340–R345

Thrasher TN, Keil LC, Ramsay DJ (1982a) Lesions of the organum vasculosum of the lamina terminalis (OVLT) attenuate osmotically induced drinking and vasopressin release in the dog. Endocrinology 110:1837–1839

Thrasher TN, Simpson JB, Ramsay DJ (1982b) Lesions of the subfornical organ block angiotensin-induced drinking in the dog. Neuroendocrinology 35:68–72

Tondat LM, Almli CR (1976) Evidence for independent osmosensitivity of lateral preoptic and lateral hypothalamic neurons. Brain Res Bull 1:241–249

Verbalis JG, Baldwin EF, Robinson AG (1986) Osmotic regulation of plasma vasopressin and oxytocin after sustained hyponatremia. Am J Physiol 250:R444–R451

Verney EB (1947) The antidiuretic hormone and the factors which determine its release. Proc R Soc B 135:25–106

Vivas L, Chiaraviglio E, Carrer HF (1990) Rat organum vasculosum laminae terminalis in vitro: responses to changes in sodium concentration. Brain Res 519:294–300

Weindl A (1973) Neuroendocrine aspects of circumventricular organs. In: Ganong WF, Martini L (eds) Frontiers in neuroendocrinology. Oxford University Press, London, pp 3–32

Wolf AV (1950) Osmometric analysis of thirst in man and dog. Am J Physiol 161:75–86

Wood RJ, Rolls BJ, Ramsay DJ (1977) Drinking following intracarotid infusions of hypertonic solutions in dogs. Am J Physiol 232:R88–R92

Zerbe RL, Robertson GL (1983) Osmoregulation of thirst and vasopressin secretion in human subjects: effect of various solutes. Am J Physiol 244:E607–E614

Commentary

Epstein: McKinley has done us all a service by doing the best that can be done with a confusing field. I believe that the nature and location of the brain's osmosensors for thirst will not be revealed until we (1) give up the idea that Na^+ (as sodium chloride) is the prototypical osmotic stimulus and (2) do more of our work in the brain parenchyma. Sodium is not just an osmotic stimulus. It is a neural excitant and an osmotic stimulus. Non-electrolyte, non-sodium solutes like sucrose and mannitol are the prototypical osmotic agents. They are excluded from cells and do not excite them. When Blass and I found the thirst osmosensors in the rat lateral preoptic we did so by requiring that candidate sites yield thirst when they were treated with sucrose or Na^+. Sites that yielded thirst when they were treated with Na^+ but that were behaviourally silent when treated with sucrose were rejected. These were more widespread and included the hypothalamic sites that were first studied by Andersson. Second, a great deal can happen between blood and brain and between CSF and brain, and a return to studies of the brain substance itself would be welcome.

Thrasher: Epstein recommends increased attention to administration of non-sodium salt osmotic agents into brain parenchyma in future studies to localize and elucidate the nature of the cerebral osmoreceptors. It should be noted that this proposal by Epstein, as well as the continued use of intraventricular (lateral or third) administration of hypertonic sodium chloride by McKinley and colleagues derives from a preconceived concept of the anatomical structure of the osmoreceptive area.

The fact that McKinley et al. utilize infusions into the ventricular space to stimulate "sodium" or "osmo" receptive elements is a clear indication that they do not know the precise location of the receptors themselves, or even whether they are discretely localized. In this context, the anterior chamber of the third ventricle or the entire lamina terminalis are not discrete structures. On the other hand Epstein conceives of a point-source for the receptors in brain parenchyma such as the lateral preoptic area for osmoreception as described in the rat by Blass and Epstein (1971).

The crucial, unresolved issue is whether the behaviour stimulated by infusion of hypertonic sodium chloride into the third ventricle of sheep is due to activation of specific sensory elements or to excitation of neuronal elements near the ventricular ependyma which are non-specifically excited by the acute rise in sodium concentration. The apparent "specificity" of the responses to increased CSF sodium could be correct or could be elicited because the neurons reached by the high sodium concentrations all have to do with mechanisms regulating water balance. Unfortunately, the check prescribed by Epstein of infusing a hypertonic saccharide or other non-sodium solution to test for non-specific activation of these neurons does not yield a clear distinction. That is, intraventricular administration of hypertonic saccharide solutions cause very little stimulation followed by a subsequent decline in responsiveness of mechanisms which stimulate water intake and secretion of vasopressin. The issue comes full circle when McKinley et al. say the reason the non-sodium solutions inhibit is because the relevant receptors are responding to sodium concentration. Thus, if the receptors are in a diffuse periventricular localization, discrete parenchymal injections of stimuli will probably not elicit responses. In contrast, if the receptors are localized discretely, it is likely that they will be activated by either discrete injections or ventricular infusions. Thus, both techniques seem to arrive at a correct answer.

Reference

Blass EM, Epstein AN (1971) A lateral preoptic osmosensitive zone for thirst in the rat. J Comp Physiol
    Psychol 76:378–394

Robertson: Animals in which the osmoregulation of thirst and vasopressin secretion
have been selectively abolished by electrolytic lesions also afford an excellent model
in which to define the role of osmoreceptors in prandial and other types of drinking.
Have any studies of this type been undertaken?

McKinley: Ablation of the midline lamina terminalis (which includes the subfornical
organ, OVLT and median preoptic nucleus) does not reduce water drinking associated
with feeding in sheep although drinking in response to acute intravenous infusion of
hypertonic sodium chloride is abolished. However, if the lesion is extended to include
periventricular tissue a further 1–2 mm around the coronal midline on either side of
the lamina terminalis, total adipsia has been observed for several months and it is
necessary to administer water into the rumen at regular intervals to prevent severe
dehydration and death (McKinley et al. 1986).

Reference

McKinley MJ, Denton DA, Leventer M et al. (1986) Adipsia in sheep caused by cerebral lesions. In: de
    Caro G, Massi M, Epstein AN (eds) The physiology of thirst and sodium appetite. Plenum Press, New
    York, pp 321–326

Johnson: The change in the daily pattern of water intake under ad libitum conditions
in "recovered" rat with lesions of the periventricular tissue of the anteroventricular
third ventricle (AV3V) is subtle. As reviewed by Kraly (Chapter 18), normal rats
drink a large percentage of their daily intake in association with feeding. Furthermore,
the intact animal exhibits a high correlation between the quantity of food meals and
the volume of water injected. Rats with chronic AV3V lesions still show the majority
of their daily drinking in association with meals. However, the high correlation
between meal size and the meal associated drinking bout volume is abolished in the
lesioned rat (Bealer and Johnson, 1980). Rats with AV3V lesions are chronically
impaired in their drinking responses to hormonal stimuli (i.e., angiotensin II and
hyperosmolality). Therefore, one interpretation of the effect of AV3V lesions is that
animals sustaining such damage are "blind" to hormonal stimuli that arouse thirst
through an action on forebrain circumventricular tissues. Thus, AV3V lesions may
remove the fine tuning of drinking responses by abolishing sensitivity to angiotensin
II and osmotic input.

Reference

Bealer SL, Johnson AK (1980) Preoptic-hypothalamus periventricular lesions alter food-associated drinking
    and circadian rhythms. J Comp Physiol Psychol 94: 547–555

Chapter 6

Volume Receptors and the Stimulation of Water Intake

T.N. Thrasher

Introduction

It is now well established that loss of volume from the intracellular compartment and loss of volume from the extracellular compartment represent primary and independent stimuli to cerebral mechanisms which initiate drinking and secretion of arginine vasopressin (AVP). Chapter 5 is devoted to a discussion of receptors and mechanisms which stimulate thirst in response to loss of intracellular fluid (ICF) volume. The purpose of this chapter is to examine the receptors which stimulate thirst in response to a deficit in extracellular fluid (ECF) volume. The ECF is composed of interstitial and plasma compartments. As there are no known receptors monitoring the interstitial fluid volume per se the mechanisms considered here are limited to those which respond to changes in vascular volume. Thus, the terms hypo- and hypervolaemia will be used to indicate decreases and increases, respectively, in relation to normal circulating blood volume. The first aim of this chapter is to consider the evidence for and the relative importance of signals arising from peripheral cardiovascular receptors and the renal renin–angiotensin system in drinking caused by hypovolaemia. The second aim is to consider interactions between volume stimuli and osmotic stimuli arising from dehydration as this is a natural event and, in the world outside the laboratory, probably a frequent occurrence. A limited selection of references is given.

Integrated Responses to Acute Hypovolaemia

Experimentally induced hypovolaemia as a stimulus to thirst is usually brought about by withdrawing blood (i.e. haemorrhage); subcutaneous or intraperitoneal injection of a hyperoncotic solution of colloid which causes withdrawal of a protein-free filtrate of plasma into the injection site (Fitzsimons 1961; Stricker 1968); complete ligation of the abdominal vena cava (Fitzsimons 1969); partial constriction or obstruction of the thoracic inferior vena cava (Fitzsimons and Moore-Gillon 1980; Thrasher et al. 1982a). Manoeuvres affecting flow in the inferior vena cava do not actually reduce blood volume but decrease venous return and thus, cause a relative hypovolaemia downstream of the constriction site. All of these manoeuvres will result in the same general sequence of events although the rates of onset differ. That is, ligation of the abdominal vena cava causes an abrupt decrease in venous return and thus a rapid fall in both arterial and atrial pressure (Lee et al. 1981); haemorrhage leads to a slower fall in venous return causing an initial fall in atrial pressure and, as blood loss progresses, a fall in arterial pressure; whereas administration of colloid requires 1 – 2 hours to cause significant hypovolaemia and may not depress mean arterial pressure at all (Stricker 1968, 1977). The loss of volume from the vascular system leads to a well-defined sequence of events which in turn, stimulates a series of homeostatic responses. A scheme of the known receptors which stimulate homeostatic responses to a haemorrhage, for example, will be used to provide the background for further discussion (Fig. 6.1).

As blood is lost from the vascular system the rate of venous return decreases and thus, pressure in the atria falls which leads to an unloading of atrial receptors. Similarly the narrowing of arterial pulse pressure and eventual fall in mean arterial pressure decreases input from baroreceptors in the carotid sinus and arch of the aorta. As the input from atrial and arterial receptors to medullary cardiovascular control centres is tonically inhibitory, the decreased input leads to a reflex increase in sympathetic outflow and decreased parasympathetic tone to the heart (Abboud and Thames 1983). The increased sympathetic outflow causes increased cardiac performance, increased smooth muscle tone in both arterial and venous vascular beds (Abboud and Thames 1983) and increased secretion of renin by the kidneys (Davis and Freeman 1976).

The afferent input from arterial and atrial receptors is also tonically inhibitory to hypothalamic mechanisms controlling secretion of corticotropin releasing factor (CRF) and AVP (Share 1988). Thus, during haemorrhage, unloading these receptors leads to reflex increases in AVP and ACTH secretion. The rise in plasma AVP reduces urinary water loss and, via its vasoconstrictor action, plays a role in the maintenance of blood pressure during haemorrhage (Share 1988). The rise in ACTH causes increased adrenal secretion of glucocorticoids, which appear to play an important role in the ability to withstand the stress of acute haemorrhage.

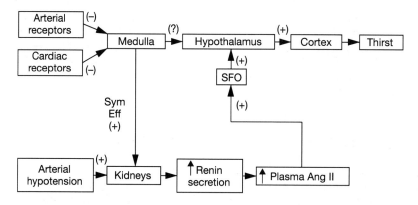

Fig. 6.1. Schematic illustration of the two proposed mechanisms to stimulate thirst in response to hypovolaemia. Unloading arterial and cardiac receptors reduces tonic inhibitory input (−) to medullary centres which send projections to the hypothalamus. It is unclear whether the ascending pathways are themselves inhibitory or stimulatory (?) to hypothalamic thirst centres. Alternatively, hypovolaemia can stimulate renin secretion reflexly via sympathetic efferents (Sym Eff) or directly via intrarenal mechanisms. The increase in renin secretion causes a rise in circulating Ang II which can stimulate thirst via stimulation of receptors in the subfornical organ (SFO).

As noted above, hypovolaemia can cause increased secretion of renin via reflex mechanisms. However, intrarenal receptors which respond to a decrease in renal perfusion pressure, the "renal baroreceptor mechanism", and a decrease in sodium delivery to the distal tubule, "the macula densa mechanism", can stimulate renin release independently of renal nerves (Davis and Freeman 1976). Therefore, there are multiple mechanisms to ensure that the renin–angiotensin system is activated in response to hypovolaemia. The increase in plasma levels of angiotensin II (Ang II) leads to a number of important responses. Over the short term, the potent vasoconstrictor effects of Ang II help to maintain arterial blood pressure during the hypovolaemic state (Davis and Freeman 1976). Over the longer term Ang II stimulates increased synthesis and secretion of aldosterone which leads to increased renal sodium reabsorption. In addition, an increase in plasma Ang II stimulates secretion of AVP and ACTH and stimulates drinking (Simpson 1981).

Restoration of ECF Volume

The loss of ECF volume stimulates longer-acting mechanisms to ensure replacement of the volume deficit. In order to re-expand ECF volume, both solute (primarily sodium chloride) and water must be replaced. Activation of the renin–angiotensin-aldosterone system ensures maximal renal reabsorption of sodium but, until sodium is ingested, there is no extra sodium to reabsorb. Similarly, the rise in plasma AVP will ensure maximal renal reabsorption of water, but until the animal drinks, there is no extra water to reabsorb. Thus, to replace the volume lost, mechanisms must exist to stimulate intake of both sodium and water. One mechanism which ensures ingestion of sodium is the metabolic demand for calories. That is, all animals must eat and, in the process, ingest sodium. However, some species, such as rats and sheep, also display a specific appetite for sodium in response to sodium depletion and hypovolaemia (Fitzsimons 1979).

Two mechanisms have been proposed to account for the stimulation of drinking in response to hypovolaemia (Fitzsimons 1979). The first is that the stimulus to drink arises from unloading peripheral cardiovascular receptors in the heart and systemic arteries and thus drinking is reflexly mediated. The second is that the increase in circulating Ang II, acting at receptors in the subfornical organ (SFO), is the stimulus to drink. Thus the drinking response is hormonally mediated. A third possibility is that both mechanisms participate in the response and neither constitutes an essential event for drinking to occur.

Evidence for Reflex Stimulation of Drinking

Fitzsimons (1961) provided the first solid evidence that loss of extracellular fluid volume is a primary stimulus to drink. He induced hypovolaemia by haemorrhage or by injecting rats with colloid (i.p.) and observed that both groups drank significantly more water over a 6-hour period than control rats. In order to ensure that the drinking was not secondary to increased renal excretion of water, the experiments were repeated in acutely nephrectomized animals and the results were the same (Table 6.1). Nephrectomized rats given colloid (i.p.) or subjected to haemorrhage drank more water than nephrectomized controls. Although the volumes drunk by intact and nephrectomized rats to colloid or haemorrhage were not compared statistically by

Table 6.1. Effect of hypovolaemia and haemorrhage on water intake in intact and nephrectomized rats

Treatment	6 h Water intake (ml/100 g body weight)			
	Intact rats (n)		Nephrectomized rats (n)	
Sham hypovolaemia	0.5 ± 0.2	(10)	1.6 ± 0.3	(20)
Hypovolaemia[a]	3.9 ± 0.6	(24)	4.7 ± 0.8	(10)
Sham haemorrhage	0.6 ± 0.2	(5)	1.1 ± 0.3	(12)
Haemorrhage[b]	3.3 ± 0.6	(12)	2.5 ± 0.4	(12)

Data taken from Fitzsimons (1961).
[a]Hypovolaemia induced by gum acacia administered i.p. Dose range: 0.7 – 1.7 g/100 g body weight in intact rats and 0.6 – 1.3 g/100 g body weight in nephrectomized rats.
[b]Haemorrhage ranged from 1.7 to 3.6 ml/100 g body weight in intact rats and from 2.5 to 3.4 ml/100 g body weight in nephrectomized rats.

Fitzsimons (1969), it is apparent from the means shown in Table 6.1 that they were not markedly different. Therefore, these observations clearly show that hypovolaemia stimulated drinking in the rat. The fact that nephrectomy had little or no effect on the response suggests that the drinking is mediated primarily by reflexes; and that a rise in plasma Ang II is not essential for drinking in response to hypovolaemia.

Acute hypovolaemia is also a stimulus to drink in dogs. Fitzsimons and Moore-Gillon (1980) have reported that reduced venous return (caused by inflation of a balloon within the abdominal vena cava) stimulated drinking and that the volumes consumed correlated with the reduction in both central venous and arterial pressure. They also reported that infusion of an Ang II receptor antagonist (saralasin, i.v.) reduced drinking in this model by about 50%. Since a significant drinking response remained in the presence of Ang II blockade, these results are compatible with an important reflex component in the response.

Thrasher et al. (1982a) have also observed that an acute reduction in venous return (by inflating a cuff around the thoracic inferior vena cava) is a potent stimulus to drink in dogs. Furthermore, there is a very good correlation ($r = 0.86$; $n = 8$; $P < 0.01$) between the fall in arterial blood pressure and the volumes consumed in response to vena caval constriction. However, infusing an Ang II receptor antagonist (saralasin) either i.v. or intracerebroventricularly (i.c.v.) had no effect on drinking to vena caval constriction. This observation differs from that of Fitzsimons and Moore-Gillon (1980) in that no evidence could be found to support an essential role for circulating Ang II in the drinking response. Nevertheless, both studies agree that input from peripheral cardiovascular receptors is an important source of stimulation to drink in response to hypovolaemia.

The failure to find evidence for circulating Ang II as an essential component of the drinking response in the caval constriction model of hypovolaemia is difficult to reconcile with the 50% reduction during saralasin blockade reported by Fitzsimons and Moore-Gillon (1980). One possible explanation is methodological. In their experiments a balloon inside the vena cava was inflated to lower mean arterial pressure to the same level both in the control and saralasin infusion experiments. Since it is well known that Ang II plays an important role in the maintenance of arterial pressure in this model of hypovolaemia, the required decrease in venous return must have been less during saralasin infusion. Therefore, the falls in atrial pressures would be less, thus contributing less reflex input to drink. That is, the reduced drinking may not have been due to blockade of Ang II receptors, but due to a reduced signal from atrial receptors. In the study by Thrasher et al. (1982a) a constant volume was instilled into a cuff surrounding the vena cava. During saralasin infusion the fall in mean arterial pressure was greater compared to the control experiment but the drinking response was not different. These results illustrate the importance of Ang II in blood pressure maintenance but do not reveal a role in the drinking response. However, the lack of effect of saralasin cannot be used to discard a role for Ang II in the normal response to hypovolaemia, when all mechanisms are functional. All that can be said is that in the caval constriction model, there was no evidence that Ang II was required for the response.

Receptors Mediating Reflex Effects on Drinking

As indicated in Fig. 6.1, sinoaortic baroreceptors and/or cardiopulmonary receptors with afferents in the vagus nerve could serve as the source of stimuli to cerebral mechanisms controlling drinking. If the input from these receptors is tonically inhibitory to thirst mechanisms, as is the case for secretion of AVP, one would predict that any manoeuvre which reduced the firing rate of the receptors would be stimulatory to drinking. Observations in the literature generally lend support to this prediction.

Sobocinska (1969a) has reported that surgical section of either the left or right cervical vagosympathetic trunk in the dog significantly reduces the osmotic threshold to drink (that is, the theoretical reduction in cell volume required to stimulate osmoregulatory drinking). Furthermore, the same effect was observed if the nerves were acutely blocked with local anaesthetic or cooling (Sobocinska 1969a). Also inducing hypovolaemia prior to infusing hypertonic sodium chloride solution reduces the osmotic threshold to drink in dogs and sectioning either the left or right vagosympathetic trunk abolishes the effect of hypovolaemia on the osmotic thirst threshold (Sobocinska 1969b). In contrast, increasing vascular volume by infusing isotonic dextran solution elevated the osmotic thirst threshold and this effect was abolished by left vagosympathectomy (Kozlowski and Szczepanska-Sadowska 1975). The observation that left (10 of 14 dogs) or right (10 of 11 dogs) vagosympathectomy alone reduced the osmotic thirst threshold (Sobocinska 1969a) suggests that, in the euvolaemic state, input from receptors with vagal afferents is inhibitory to thirst centres. Furthermore, the inverse relationship between vascular volume and the osmotic thirst threshold suggests that input from peripheral volume receptors with vagal afferents is an important regulatory input to control of drinking.

More recently Fitzsimons and Moore-Gillon (1980) have reported that acutely anaesthetizing the left cervical vagosympathetic nerve enhanced the drinking response to acute abdominal caval obstruction in the dog. They suggested that the afferent input was tonically inhibitory and that removing it allowed a more robust drinking response to occur. However, anaesthetizing the nerve alone caused mean arterial pressure (MAP) to increase from 98 ± 3 mmHg to 122 ± 6 mmHg. Thus caval obstruction in the control experiment lowered MAP from 99 ± 2 to 80 ± 2 mmHg but during left vagosympathetic blockade, MAP fell from 122 ± 6 to 86 ± 1 mmHg. The larger fall in MAP during nerve blockade (36 mmHg) compared to the control experiment (19 mmHg) offers an alternative explanation for the increased drinking, that is, greater unloading of arterial baroreceptors. Unless the signal to the arterial receptors is controlled, it is impossible to differentiate between responses caused by blocking inhibitory afferent input from the vagus and increasing input from arterial baroreceptors.

In theory, if input from cardiopulmonary receptors with vagal afferents tonically inhibit drinking, as is known for AVP secretion (Share 1988), then acutely blocking these impulses could stimulate a drinking response. Acute blockade of either the left or right vagosympathetic trunk alone apparently is not sufficient to stimulate drinking (Kozlowski and Szczepanska-Sadowska 1975; Fitzsimons and Moore-Gillon 1980). However, Fitzsimons and Moore-Gillon (1980) reported that bilateral blockade of both nerve trunks (injection of xylocaine into skin loops containing nerves plus carotid arteries) did cause spontaneous drinking in four of five dogs with a latency of about 10 min and amounting to 43 ± 13 ml. The authors also reported that bilateral blockade caused severe respiratory distress requiring anaesthesia. Given the limited description of the experimental conditions it is difficult to be sure that the bilateral interruption of vagal impulses was specific for drinking (i.e. would the dogs have eaten if food were present?). Nevertheless, the results are compatible with the concept of a tonic inhibition to drinking mediated by vagal afferents.

We have developed the technique of infusing procaine into the pericardial sac as a means of selectively blocking cardiac nerve transmission in conscious dogs. Thus, instead of interrupting all nerve transmission in the cervical vagosympathetic trunks, the pericardial infusion technique blocks only afferent and efferent nerves supplying the heart. During pericardial procaine infusion reflex changes in heart rate in response to increases or decreases in MAP and the reflex hypotension caused by administration of veratridine into the left atrium (Bezold–Jarisch reflex) were completely eliminated indicating blockade of both efferent and afferent nerves to the heart. The blockade has been maintained for up to 4 hours and causes no apparent discomfort in the dogs. In six dogs which have been tested, spontaneous drinking has never been seen following blockade of the cardiac nerves. This observation does not necessarily contradict previous observations (Fitzsimons and Moore-Gillon 1980) but it does suggest the possibility that drinking induced by bilateral blockade of the vagi may not be due to blocking afferent input from cardiac receptors alone.

Direct evidence of regulatory input from specific regions of the heart itself has also been reported. Moore-Gillon and Fitzsimons (1982) have investigated the effect of inflating a balloon in the junctional region between a pulmonary vein and the left atrium in dogs. They report that balloon inflation reduced water intake in response to isoproterenol and the effect could be prevented by acutely anaesthetizing the left cervical vagosympathetic trunk. They also reported a significant reduction in water intake in response to hypertonic sodium chloride and to 24-h water deprivation during acute inflation of the balloon. Remarkably, when the balloon was inflated for 24 h there was a 48% reduction in water intake, a significant increase in urine volume and doubling of urinary sodium excretion over the same period but no change in food intake.

Kaufman (1984) investigated the influence of stretching receptors in the junctional region of the right atrium and superior vena cava in the rat. Inflating a balloon in this region had no effect on arterial or central venous pressures but attenuated spontaneous night-time water intake, and the acute drinking responses to 24-h water deprivation and isoproterenol (s.c.). The response to hyperoncotic colloid was virtually abolished during balloon inflation but drinking to hypertonic sodium chloride was unaffected. These observations suggest that stretching (loading) right atrial receptors in rats produce many of the same effects on drinking that Moore-Gillon and Fitzsimons (1982) observed when left atrial receptors were activated in dogs. Unfortunately, it is not known whether these inhibitory effects on drinking are unique to receptors on the left side of the heart in dogs and right side of the heart in rats or whether receptors on both sides of the heart can inhibit drinking to hypovolaemic stimuli .

At present, it is difficult to put the reports of Moore-Gillon and Fitzsimons (1982) and Kaufman (1984) into a physiological context because there is no way to compare the stimulus intensity generated by the balloon on stretch receptors in a single veno-atrial junction to naturally occurring stretch of the receptors during hypervolaemia. Nevertheless, the results illustrate that powerful inhibitory signals to central mechanisms regulating water intake arise from receptors located in the left and right atrium.

Few studies have specifically addressed the issue of whether chronically removing input from cardiac and/or arterial baroreceptors alters drinking responses to hypovolaemia. In one study in rats Rettig and Johnson (1986) reported that chronic sinoaortic denervation reduced drinking to an osmotic stimulus on the first or second trial but did not alter responses to subsequent trials. On the other hand, there was no difference in the drinking responses between control and sinoaortic denervated rats to Ang II, hyperoncotic colloid administration, isoproterenol or water deprivation regardless of time of testing. Thus, it appears that the initial loss of sinoaortic baroreceptors specifically affects drinking to an osmotic stimulus, but the effect is transitory. The recovery of responsiveness to hypertonic sodium chloride in the sinoaortic denervated animal remains unexplained.

In contrast, Quillen et al. (1987) reported that surgical denervation of either sinoaortic baroreceptors or cardiac receptors significantly decreased drinking in response to vena caval constriction in dogs. Furthermore, they observed that combined sinoaortic and cardiac denervation virtually eliminated the drinking response to caval constriction even though the increase in plasma Ang II was not different from the sham group. Thus, these observations suggest that it is the unloading of receptors in the heart and the arterial system that constitutes the essential stimulus to initiate drinking to hypovolaemia in the dog. However, as these results are reported as part of a conference proceedings and have not as yet been published in a reviewed journal they should be interpreted cautiously.

Evidence for Ang II Stimulation of Drinking

Evidence linking activation of the renal renin–angiotensin system to drinking was first reported by Fitzsimons (1969). He showed that procedures which decrease renal perfusion pressure, e.g. ligation of the abdominal vena cava or constriction of the abdominal aorta above the origin of the renal arteries, stimulated drinking in rats. Furthermore, nephrectomy markedly reduced drinking induced by these manoeuvres but bilateral ureteric ligation did not. Since all of the effective procedures stimulated renin secretion and since administration of renin or Ang II stimulated drinking (Fitzsimons 1979) the concept of an endocrine mechanism of thirst was born.

The role of Ang II in thirst is complex because drinking can occur in response to both peripheral and intraventricular administration of the hormone. The receptors which mediate drinking in response to an increase in circulating Ang II are located in the subfornical organ (SFO) in rat (Simpson 1981) and dog (Thrasher et al. 1982b). The SFO is one of the circumventricular organs (CVO), a unique group of brain structures with fenestrated capillary endothelia (Simpson 1981). Thus, Ang II and other peptides which are excluded from brain ECF by the blood–brain barrier (BBB) and blood–CSF barrier, can reach receptors on neuronal elements in the SFO to stimulate drinking or other neurally mediated responses. Electrolytic ablation of the SFO (Simpson 1981; Thrasher et al. 1982b) or surgical deafferentation of its rostral projections (Eng and Miselis 1981) markedly reduce or abolish drinking to peripherally administered Ang II but do not affect drinking to i.c.v. Ang II. The receptors which stimulate drinking in response to i.c.v. administration of Ang II appear to be in the nucleus medianus (aka median preoptic nucleus). Lind and Johnson (1982) have proposed that circulating Ang II stimulates neuronal elements in the SFO with efferent projections to the nucleus medianus and these efferent fibres release Ang II as a neurotransmitter in the drinking pathway. The concept of Ang II acting as a stimulus to drink at the SFO and as a neurotransmitter in the neural circuitry of drinking resolved many conflicting observations (Lind and Johnson 1982). However, the dual role of Ang II as both dipsogen and neurotransmitter makes it much more difficult to separate drinking responses dependent on activation of the renal renin–angiotensin system from drinking responses arising from neurally generated formation of Ang II. For example, does centrally generated Ang II always reflect activation by circulating Ang II or does it subserve drinking to other inputs as well?

Blockade of the Renin–Angiotensin System and Hypovolaemic Drinking

Pharmacological blockade of the renin–angiotensin system can be accomplished with antagonists such as saralasin, which compete with Ang II for the receptor site and by converting enzyme inhibitors (CEI) which prevent the conversion of angiotensin I (Ang I) into Ang II. Both methods would appear to offer easily quantifiable results and definitive conclusions as to the role of Ang II in drinking. In practice however, this does not seem to be the case. One problem is that receptor blockers cause varying degrees of agonistic effects. More importantly, infusing saralasin or other Ang II receptor antagonists peripherally blocks all responses to an increase in circulating Ang II including constriction of vascular smooth muscle, aldosterone secretion, negative feedback on renin secretion, as well as the drinking response. Thus, in a hypovolaemic animal in which Ang II is helping to maintain MAP, administering the antagonist may result in dramatic falls in MAP rendering the animals behaviourally incompetent. In this case, it may be impossible to distinguish between a reduction in drinking due to blockade of Ang II-mediated drinking from a reduction in drinking caused by behavioural incapacitation.

To avoid this problem the technique of i.c.v. administration of saralasin to block Ang II-induced drinking was developed (Lee et al. 1981; Thrasher et al. 1982a). The data in Table 6.2 show that the drinking response to an infusion of Ang II (i.v.) could be blocked by a simultaneous i.c.v. infusion of saralasin in both rat and dog. The same dose of saralasin (i.c.v.) had no significant effect on water intake in response to caval ligation (rat) or thoracic caval constriction (dog) compared to the control intake during i.c.v. infusion of artificial CSF. Also the i.c.v. infusion of saralasin had no significant effect on the ability to maintain blood pressure following reduced venous return in either species (Lee et al. 1981; Thrasher et al. 1982a). Thus, these results suggest that circulating Ang II does not play an essential role in the drinking response to these two models of hypovolaemia.

In contrast, Fitzsimons and Elfont (1982) used CEI and arrived at different conclusions. They observed that low doses of captopril (0.5 mg/kg) given peripherally to rats immediately prior to caval ligation resulted in a marked enhancement of the

Table 6.2. Effect of saralasin (i.c.v.) on drinking induced by Ang II (i.v.) and hypovolaemia in rat and dog

Treatment	Water intake[a]			
	Rat	(n)	Dog	(n)
Ang II[b] (i.v.) + Veh[c] (i.c.v.)	3.3 ± 0.5 ml	(11)	250 ± 40 ml	(6)
Ang II (i.v.) + saralasin[d] (i.c.v.)	0.1 ± 0.1 ml[f]	(11)	40 ± 15 ml[f]	(6)
Hypovolaemia[e] + Veh (i.c.v.)	20.0 ± 2.6 ml/kg	(14)	9.7 ± 3.4 ml/kg	(4)
Hypovolaemia + saralasin (i.c.v.)	15.2 ± 2.8 ml/kg	(20)	9.2 ± 2.9 ml/kg	(4)

Data taken from Lee et al. (1981) and Thrasher et al. (1982a).
[a]Values represent mean ± 1 SE.
[b]Ang II dose was 100 ng/min for 30 min in rats; and 20 ng/kg/min for 60 min in dogs.
[c]Vehicle (Veh) was saline in rats and artificial CSF in dogs.
[d]Saralasin dose was 5. 2 μg/h in rats and 0. 2 μg/kg/min in dogs.
[e]Hypovolaemia was induced by caval ligation in rats and intake was measured for 6 h; hypovolaemia was induced by thoracic caval constriction in dogs and intake was measured for 1 h.
[f]Significantly different from Ang II + Veh (i.c.v.) infusion at $P < 0.01$.

drinking response. However increasing the dose by 2 orders of magnitude (50 mg/kg) significantly delayed drinking relative to control, but did not reduce the total amount drunk over the 6-h experiment. In a different approach, the low dose of captopril (0.5 mg/kg) was combined with multiple i.c.v. doses of captopril both before and after caval ligation and this procedure delayed but, again, did not reduce the total volume drunk compared to controls during the 6 h of measurement.

The totally different responses to differing doses of CEI suggests that use of these drugs to dissect the mechanism has many potential pitfalls. The enhanced response to "low" doses of captopril is presumed to be due to blockade of peripheral conversion of Ang I into Ang II, which leads to a build up of Ang I in plasma which, "somehow", gets into the brain and is converted into Ang II. On the one hand there is no evidence to show that Ang I can pass through the blood–brain barrier any better than Ang II. On the other hand, it is possible that the high concentration of converting enzyme in the SFO is not blocked by low doses of captopril and thus, the enhanced drinking is due to formation of high levels of Ang II acting at receptors in the SFO. Furthermore, it has been reported that ablation of the SFO prevents the enhanced drinking response to captopril (Thunhorst et al. 1987). Thus, it seems quite likely that the enhancement of drinking is a peripheral phenomenon requiring an intact SFO.

The reduced drinking response to high doses of CEI is a more difficult issue. The high dose of captopril could reduce drinking by completely blocking the converting enzyme in the SFO, an outcome which would suggest participation of renal Ang II in the response. However, Evered et al. (1980) have argued that high doses of captopril (50 mg/kg, oral or s.c.) block intracerebral formation of Ang II because this dose causes a reduction in the drinking response to i.c.v. Ang I. Fitzsimons and Elfont (1982) reached similar conclusions and noted that i.c.v. administration of microgram doses of captopril eliminated the enhancing effect of peripheral captopril (0.5 mg/kg) on drinking responses. Thus, it is possible that the critical action of CEI to reduce drinking is due to blockade of Ang II formation inside the brain acting at receptors in the nucleus medianus and not peripheral formation of Ang II which acts at receptors in the SFO. If the former is true, the results of Fitzsimons and Elfont (1982) provide more evidence in support of a role for brain Ang II, rather than circulating Ang II, in caval ligation-induced drinking. Alternatively, without knowing whether the high doses of CEI affected blood pressure or other behaviours, it is difficult to separate specific effects on water intake from non-specific effects on physiological control systems. In this regard, CEI interferes not only with the conversion of Ang I into Ang II, but also with the metabolism of bradykinin and the endorphins. Whether alterations in bradykinin or endorphin levels alter mechanisms regulating thirst remains largely unknown.

Effect of SFO Ablation

As noted above the SFO appears to be the receptor site where circulating Ang II can reach neural circuits to stimulate drinking in rats and dogs (Simpson 1981; Thrasher et al. 1982b). Thus, lesions of the SFO would appear to offer an ideal means of evaluating the role of circulating Ang II in drinking without the confusion arising from the use of pharmacological tools. Furthermore, it has been reported that rats with lesions of the SFO drink less to administration of hyperoncotic colloid suggesting that the drinking deficit reflects a component mediated by circulating Ang II (Simpson 1981).

However, Hosutt et al. (1981) have shown that the drinking deficits following SFO lesions are not specific for hypovolaemic stimuli. They and others have reported that SFO lesions also reduce drinking in response to moderate but not severe osmotic challenge (Lind et al. 1984; Mangiapane et al. 1984). Hosutt et al. (1981) suggested that destroying the SFO reduced the overall level of responsiveness of the animals and thus the deficit in drinking was not necessarily indicative of a role for circulating Ang II. However, it is equally possible that both osmoreceptors and Ang II receptors are located in the SFO in the rat and thus, lesions of the SFO reduce responsiveness to both types of stimuli. Based on the available literature, it does not seem possible to favour one or the other possibility. Therefore, the observation that ablation of the SFO reduces drinking to hypovolaemia represents evidence supporting a role for circulating Ang II in the response.

Effect of Nephrectomy on Hypovolaemic Drinking

Various investigators have used acute removal of the kidneys to evaluate the role of circulating Ang II in drinking. On the surface, it would seem that interpretation of the results would be straightforward, i.e., "reduced drinking" implying a role for circulating Ang II and "no effect" implying no role for Ang II in the response. However, Stricker (1977) has argued that nephrectomy combined with either caval ligation or isoproterenol treatment (Hosutt et al. 1978) leads to severe hypotension and thus rats do not drink because they are behaviourally incompetent. In support of this argument, administration of agents which elevate blood pressure (such as renin, pitressin or adrenaline) to nephrectomized rats, also increases water intake toward the level observed in control rats. Thus, Stricker and colleagues have suggested that circulating Ang II is important in a "permissive" manner, but is not the direct stimulus to drink in response to hypovolaemia (Stricker 1977; Hosutt et al. 1978).

The acutely nephrectomized rat is a powerful model to evaluate the role of circulating Ang II on drinking. In the first place, there are many reasons to explain why an animal may not drink in response to hypovolaemia following nephrectomy, e.g. lingering effects of anaesthesia, pain associated with incisions, sudden loss of a major organ system and loss of the renin–angiotensin system which plays a role in many homeostatic responses etc. Therefore, a reasonably normal drinking response to hypovolaemia by the nephrectomized rat is very good evidence that circulating Ang II is not essential. On the other hand, failure to drink requires a much more detailed analysis. In this regard, it is unfortunate that no investigation has documented that the MAP of nephrectomized, cavally ligated conscious rats are really markedly reduced over the usual 6-h experimental period to ensure that behavioural incompetence can be explained by hypotension and not lack of Ang II.

Is Either the Reflex or the Ang II Stimulus to Drink Essential?

In my opinion, there is a considerable body of evidence, only a brief sampling of which has been considered here, to show that drinking in response to a decrease in ECF volume does not depend on the participation of circulating Ang II. However it should be noted that virtually all of these studies manipulate the renin–angiotensin system in one way or another (e.g. nephrectomy, pharmacological antagonists, etc.). Only the study by Quillen et al. (1987) set out purposefully to eliminate all reflex mechanisms and examine whether loss of these reflexes affected drinking to hypovolaemia. They observed that circulating Ang II increased to similar levels in all experiments but that dogs with combined sinoaortic and cardiac denervation did not drink in response to caval constriction. Thus, the database indicating that Ang II does not play an essential role in hypovolaemic drinking is very broad but data showing that reflex mechanisms are essential amount to one report.

To put it another way, the importance of reflex mechanisms has been established primarily by showing that drinking still occurs if Ang II participation is blocked. The important question which remains is, would an increase in circulating Ang II stimulate drinking in most models of hypovolaemia if reflex mechanisms were blocked? In the opinion of the author, the only certain conclusion is that reflex mechanisms are sufficient to account for drinking in response to hypovolaemia but whether they are necessary is a question which requires additional study.

Dehydration as a Stimulus to Drinking

Mammals continuously lose water to the environment via urine, in expired air, in faeces and via sweat or other mechanisms regulating evaporative heat loss. The water that is lost comes from both the intracellular and extracellular compartments. Over a 24-h period of water deprivation in dogs an observed rise in plasma osmolality of 13 ± 2 mosmol/kg (a 4.4% increase above control) and plasma protein of 0.8 ± 0.1 g/dl (a 13.1% increase) indicated significant decreases in ICF and ECF volume (Thrasher et al. 1984). The combination of cellular and extracellular dehydration caused a fourfold increase in plasma AVP, a twofold increase in plasma renin activity and, when water was offered, the dogs ingested 47 ± 4 ml/kg within 6 min. The 4.4% rise in plasma osmolality is well above the usual threshold for drinking in dogs and the 13% decrease in ECF Volume (inferred from the rise in plasma protein) is also above the threshold for drinking in response to hypovolaemia (Kozlowski and Szczepanska-Sadowska 1975; Ramsay and Thrasher 1986). Therefore, both mechanisms are activated by water deprivation and could contribute to the drinking observed.

Contribution of Osmotic and Volaemic Factors to Dehydration-Induced Thirst

Ramsay et al. (1977b) investigated the relative contributions of osmotic and volaemic factors to the drinking response of 24-h water-deprived dogs. They found that intracarotid infusions of water, which reduced the osmolality of blood perfusing cerebral osmoreceptors to predeprivation levels, led to a 72% reduction in water intake. It is important to note that the amount of water infused into the cerebral circulation was not sufficient to change systemic plasma osmolality, therefore, it was possible to equate the reduction in water intake to removal of the cerebral osmotic stimulus. Infusing isotonic saline (i.v.) in amounts sufficient to restore the estimated loss of ECF volume to control led to a 27% reduction in water intake. Finally, combining the intracarotid infusion of water with intravenous infusion of saline reduced the intake of 24-h water-deprived dogs to 93% of their control intake. Thus, this experimental design made it possible to estimate the relative contribution of osmotically (72%) and volaemically (27%) induced drinking in response to water deprivation. Using a similar approach Ramsay et al. (1977a) observed that water-deprived rats consumed 64% – 69% of fluid in response to the osmotic stimuli and 20% – 26% in response to the volume depletion. The main conclusions of these two studies in dogs and rats is that the osmotic component is the more important but that the volaemic component was also responsible for a significant portion of the intake in both species.

Does Ang II Contribute to Dehydration-Induced Drinking?

Dehydration of 24 h duration does not reduce MAP and causes only a small decrement in atrial pressure. However, it does cause reliable increases in plasma renin activity in both rats and dogs (Lee et al. 1981; Thrasher et al. 1984). Therefore, it is possible that circulating Ang II could play a role in the volaemic component of drinking in this model. In a study of this possibility Malvin et al. (1977) reported that i.c.v. infusion of saralasin significantly delayed drinking in water-deprived rats. Also, Barney et al. (1980) observed that administration of captopril reduced water intake in dehydrated rats. Thus, both reports supported a role for Ang II in this form of physiological drinking. However, we attempted to replicate the study of Malvin et al. (1977) but could not find a significant effect of i.c.v. saralasin on drinking in water-deprived rats. In another approach Rolls and Wood (1977) observed that nephrectomized rats previously deprived of water for 21 h drank as much as controls. The fact that nephrectomy did not alter the drinking response to dehydration strongly indicates that circulating Ang II is not an essential component of the drinking response. However, as noted previously, there are no data which show positively that reflex mechanisms are an essential component in this combined model of hypovolaemia and hyperosmolality in the rat.

Dehydrated dogs drink to satiety usually within 3–6 minutes and show no interest in water 60 minutes later (Thrasher et al. 1981). Over the 60 min following ingestion of water, plasma osmolality, protein and AVP all return to predeprivation levels but plasma renin activity remains elevated. Thus, if Ang II were stimulating a component of the drinking response to dehydration in the dog, it is clear that elevated plasma levels 60 min after drinking are not sufficient to stimulate further drinking. Thus, whereas hypovolaemia does contribute to drinking in dehydrated dogs, there is no convincing evidence that circulating Ang II is a component of the response.

Atrial Natriuretic Peptides and Drinking

The last decade has ushered in a new peptide hormone, atrial natriuretic peptide (ANP), which has important implications in the control of body fluid balance (Goetz 1988; Ramsay and Thrasher 1989). Although initially localized to the atria of the heart, ANP has also been found in the brain and in particular, in hypothalamic areas implicated in the regulation of water balance (Standaert et al. 1988). Samson (1987) has reported that peripheral or i.c.v. administration of ANP will inhibit drinking and AVP secretion in rats deprived of water for 3 days or made hypovolaemic by haemorrhage. It is also known that i.c.v. administration of ANP antagonizes drinking induced by Ang II (Goetz 1988). Although the effects reported by Samson have been confirmed by others, the physiological significance of the inhibitory action of ANP on thirst is unknown. This is because the only established physiological stimulus to ANP release by atrial myocytes is increased stretch, which normally accompanies hypervolaemia (Goetz 1988). On the other hand hypovolaemia leads to reduced atrial wall stretch and consequently, decreased secretion of ANP. Thus, it is highly unlikely that circulating ANP would be involved in the control of drinking in response to hypovolaemia except perhaps, by its absence.

In contrast, certain pathological states such as congestive heart failure are characterized by elevated levels of circulating Ang II and ANP (Goetz 1988). Thus, it is possible that inhibition of the dipsogenic effects of Ang II, as well as promoting increased sodium excretion, are beneficial in pathological states of volume overload. Whether ANP actually plays such a role remains to be established.

Concluding Remarks

The aim of this chapter was to sketch out the basic evidence for drinking in response to depletion of the extracellular fluids and to consider the evidence pointing toward reflex and Ang II contributions to this drinking response. It should be evident to the reader that the author is convinced of the primacy of reflex mechanisms as the initiator of drinking responses to hypovolaemia. However, given the rapidity with which neural mechanisms reset to new conditions, it is considered quite likely that Ang II may play an important role in the final replacement of the lost volume, both by way of stimulating increased intake of water and renal reabsorption of sodium. In the past most experimental designs have attempted to force mechanisms controlling water intake to rely on one or another stimulus, the object being to say this or that stimulus was essential. However, redundancies in mechanisms controlling water intake have yielded many of the controversies discussed in this chapter. It may be time to consider an alternative approach to the problem. That is, making many more measurements of the factors which are known to influence drinking in animals in various states of water balance. These would include atrial and arterial pressures, circulating Ang II and osmolality across the full response to a perturbation in body fluid volumes. This approach, although much more laborious, at least allows for a clear definition of the conditions under which drinking occurs and may point to the relative importance of the various factors as they operate in an intact animal.

Acknowledgements. Work from the author's laboratory was conducted in collaboration with Dr D.J. Ramsay. The author would also like to thank Phyllis Colbert for her excellent assistance in the preparation of the manuscript. The studies were supported by HL 29714 and HL 41313.

References

Abboud FM, Thames MD (1983) Interactions of cardiovascular reflexes in circulatory control. In: Handbook of physiology, Am Physiol Soc, Sec 2, Vol III, part 2, chapter 19, pp 675–753

Barney CC, Katovich MJ, Fregly MJ (1980) The effect of acute administration of an angiotensin converting enzyme inhibitor, captopril (SQ 14, 225), on experimentally induced thirst in rats. J Pharmacol Exp Ther 212:53–57

Davis JO, Freeman RH (1976) Mechanisms of renin release. Physiol Rev 56:1–54

Eng R, Miselis RR (1981) Polydipsia and abolition of angiotensin-induced drinking after transections of subfornical organ efferent projections in the rat. Brain Res 225:200–206

Evered MD, Robinson MM, Richardson MA (1980) Captopril given intracerebroventricularly, subcutaneously or by gavage inhibits angiotensin-converting enzyme activity in the rat brain. Eur J Pharmacol 68:443–449

Fitzsimons JT (1961) Drinking by rats depleted of body fluid without increase in osmotic pressure. J Physiol (Lond) 159:297–309

Fitzsimons JT (1969) The role of renal thirst factor in drinking induced by extracellular stimuli. J Physiol (Lond) 201:349–368

Fitzsimons JT (1979) The physiology of thirst and sodium appetite. Cambridge University Press, Cambridge

Fitzsimons JT, Elfont RM (1982) Angiotensin does contribute to drinking induced by caval ligation in rat. Am J Physiol 243 (Regulatory, Integrative Comp Physiol 12):R558–R562

Fitzsimons JT, Moore-Gillon MJ (1980) Drinking and antidiuresis in response to reductions in venous return in the dog: neural and endocrine mechanisms. J Physiol (Lond) 308:403–416

Goetz KL (1988) Physiology and pathophysiology of atrial peptides. Am J Physiol 254 (Endocrinol Metab 17):E1–E15

Hosutt JA, Rowland N, Stricker EM (1978) Hypotension and thirst in rats after isoproterenol treatment. Physiol Behav 21:593–598

Hosutt JA, Rowland N, Stricker EM (1981) Impaired drinking responses of rats with lesions of the subfornical organ. J Comp Physiol Psychol 95:104–113

Kaufman S (1984) Role of right atrial receptors in the control of drinking in the rat. J Physiol (Lond) 349:389–396

Kozlowski S, Szczepanska-Sadowska E (1975) Mechanisms of hypovolaemic thirst and interactions between hypovolaemia, hyperosmolality and the antidiuretic system. In: Peters G, Fitzsimons JT, Peters-Haefeli L (eds) Control mechanisms of drinking. Springer-Verlag, Berlin, pp 25–45

Lee M-C, Thrasher TN, Ramsay DJ (1981) Is angiotensin essential in drinking induced by water deprivation and caval ligation? Am J Physiol 240 (Regulatory Integrative Comp Physiol 9):R75–R80

Lind RW, Johnson AK (1982) Central and peripheral mechanisms mediating angiotensin-induced thirst. In: Ganten D, Printz M, Phillips MI, Scholkens BA (eds) The renin angiotensin system in the brain. Springer-Verlag, Berlin, Heidelberg, pp 353–364 (Experimental Brain Research, Suppl 4)

Lind RW, Thunhorst RL, Johnson AK (1984) The subfornical organ and the integration of multiple factors in thirst. Physiol Behav 32:69–74

Malvin RL, Mouse D, Vander AJ (1977) Angiotensin: physiological role in water-deprivation-induced thirst of rats. Science 197:171–173

Mangiapane ML, Thrasher TN, Keil LC, Simpson JB, Ganong WF (1984) Role for the subfornical organ in vasopressin release. Brain Res Bull 13:43–47

Moore-Gillon MJ, Fitzsimons JT (1982) Pulmonary vein-atrial junction stretch receptors and the inhibition of drinking. Am J Physiol 242 (Regulatory Integrative Comp Physiol 11):R452–R457

Quillen Jr EW, Reid IA, Keil LC (1987) Baroreceptor influences on plasma vasopressin and drinking. In: Cowley Jr AW, Liard JF, Ansiello DA (eds) Vasopressin cellular and integrative functions. Raven Press, New York, pp 405–411

Ramsay DJ, Thrasher TN (1986) Hyperosmotic and hypovolemic thirst. In: de Caro G, Epstein AN, Massi M (eds) The physiology of thirst and sodium appetite. Plenum Press, New York, pp 83–96

Ramsay DJ, Thrasher TN (1989) A physiological role for atrial peptides in endocrine control mechanisms. In: Brenner BM, Laragh JH (eds) Progress in atrial peptide research. Vol III. Raven Press, New York, pp 65–75

Ramsay DJ, Rolls BJ, Wood RJ (1977a) Body fluid changes which influence drinking in the water deprived rat. J Physiol (Lond) 266:453–469

Ramsay DJ, Rolls BJ, Wood RJ (1977b) Thirst following water deprivation in dogs. Am J Physiol 232 (Regulatory Integrative Comp Physiol 3) R93–R100

Rettig R, Johnson AK (1986) Aortic baroreceptor deafferentation diminishes saline-induced drinking in rats. Brain Res 370:29–37

Rolls BJ, Wood RJ (1977) Role of angiotensin in thirst. Pharmacol Biochem Behav 6:245–250

Samson WK (1987) Atrial natriuretic factor and the central nervous system. Endocrinol Metab Clin North Am 16:145–161

Share L (1988) Role of vasopressin in cardiovascular regulation. Physiol Rev 68:1248–1284

Simpson JB (1981) The circumventricular organs and the central actions of angiotensin. Neuroendocrinology 32:248–256

Sobocinska J (1969a) Effect of cervical vagosympathectomy on osmotic reactivity of the thirst mechanism in dogs. Bull Acad Pol Sci 17:265–270

Sobocinska J (1969b) Abolition of the effect of hypovolemia on the thirst threshold after cervical vagosympathectomy in dogs. Bull Acad Pol Sci 17:341–346

Standaert DG, Needleman P, Saper CB (1988) Atriopeptin: neuromediator in the central regulation of cardiovascular function. In: Martini L, Ganong WF (eds) Frontiers in neuroendocrinology, vol 10. Raven Press, New York, pp 63–78

Stricker EM (1968) Some physiological and motivational properties of the hypovolemic stimulus for thirst. Physiol Behav 3:379–385

Stricker EM (1977) The renin–angiotensin system and thirst: a reevaluation. II. Drinking elicited in rats by caval ligation or isoproterenol. J Comp Physiol Psychol 91:1220–1231

Thrasher TN, Nistal-Herrera JF, Keil LC, Ramsay DJ (1981) Satiety and inhibition of vasopressin secretion after drinking in dehydrated dogs. Am J Physiol 240 (Endocrinol Metab 3):E394–E401

Thrasher TN, Keil LC, Ramsay DJ (1982a) Hemodynamic, hormonal, and drinking responses to reduced venous return in the dog. Am J Physiol 243 (Regulatory Integrative Comp Physiol 12):R354–R362

Thrasher TN, Simpson JB, Ramsay DJ (1982b) Lesions of the subfornical organ block angiotensin-induced drinking in the dog. Neuroendocrinology 35:68–72

Thrasher TN, Wade CE, Keil LC, Ramsay DJ (1984) Sodium balance and aldosterone during dehydration and rehydration in the dog. Am J Physiol 247 (Regulatory Integrative Comp Physiol 13):R76–R83

Thunhorst RL, Fitts DA, Simpson JB (1987) Separation of captopril effects on salt and water intake by subfornical organ lesions. Am J Physiol 252 (Regulatory Integrative Comp Physiol 21):R409–R418

Commentary

Stricker: Aside from Quillen et al. (1987) the author should inspect the data of Zimmerman et al. (1981) who destroyed the atrial appendage in sheep and eliminated hypovolaemic thirst, even though PRA was elevated normally. It is important to note that this surgical treatment did not disturb osmotic thirst.

Reference

Zimmerman MB, Blaine EH, Stricker EM (1981) Water intake in hypovolaemic sheep: effects of crushing the left atrial appendage. Science 211:489–491

Thrasher: I did not cite Zimmerman et al. (1981) for a number of reasons. First, McKinley and the group at Melbourne have claimed for years that sheep do not drink to infusions of angiotensin in the physiological range. Thus, in this species, arguments over the role of angiotensin in drinking may be moot. Second, the anatomical and electrophysiological investigations of atrial receptors in rat, cat, and dog have indicated that most of the atrial receptors are located around the junctional region of atrial tissue and veins on both the left and right sides and in the interatrial septum. Very few receptors have been located in the atrial appendages. I have never been able to reconcile damage to the atrial appendages, which should not markedly reduce the population of atrial receptors, with the abolition of hypovolaemic drinking, unless sheep have their atrial receptors in the appendages. Nevertheless, the phenomenon, if not the mechanism, is clear.

Fitzsimons: I am attracted by the idea that an increase in discharge from some category of cardiovascular receptor may stimulate drinking, by analogy with the work of Goetz and his colleagues, who found that ventricular denervation markedly attenuated haemorrhage-induced vasopressin secretion (Wang et al. 1988). It would make sense to have stimulatory receptors working below the range at which the inhibitory baroreceptors function. What does the author think?

Reference

Wang B, Flora-Ginter G, Leadley RJ, Goetz KL (1988) Ventricular receptors stimulate vasopressin release during hemorrhage. Am J Physiol 254:R204–R211

Thrasher: I quite agree that some sort of a stimulatory system arising in the heart as opposed to unloading of an ascending inhibitory system is an attractive idea as a means of stimulating hypovolaemic drinking. However, I have tried to replicate the results of Wang et al. (1988) using a slightly different approach and cannot find any evidence of stimulatory input from the heart to effect vasopressin secretion or drinking in response to haemorrhage in dogs.

Reference

Thrasher TN, O'Donnell CP, Keil LC (1989) Role of cardiac receptors in the stimulation of vasopressin secretion in response to hypovolemia. Proc Int U Physiol Sci 17:490

Stricker: I agree that blood pressure measurements in conscious caval ligated animals would be useful information. However, the issue is whether animals who do not drink are behaviourally competent, not whether the behavioural incompetence results from hypotension. Thus, I believe that it is significant that nephrectomized rats did not drink when given isoproterenol despite a prior period of water deprivation (Stricker 1977), regardless of whether or not this failure to respond to water deprivation results from the severe hypotension that was induced or some other consequence of the isoproterenol treatment.

Reference

Stricker EM (1977) The renin–angiotensin system and thirst: a reevaluation. II. Drinking elicited in rats by caval ligation or isoproterenol. J Comp Physiol Psychol 91:1220–1231

Thrasher: I believe that you, as well as others, have suggested that failure of nephrectomized rats to drink in response to hypovolaemic stimuli may be due to behavioural incompetence secondary to hypotension. The point I was trying to make is that blood pressure measurements in conscious, nephrectomized rats following caval ligation would allow one to determine whether the animals were behaviourally incompetent due to "hypotension". If they were not severely hypotensive, then one could eliminate this as a factor in the reduced drinking response. Until someone provides hard data, we will continue to argue about the importance of hypotension as a cause of the failure to drink.

Chapter 7

Hormonal Inputs to Thirst

E. Szczepanska-Sadowska

Introduction

Thirst is one of the most important homeostatic mechanisms within the complex system of the body fluid control. In healthy subjects operation of the thirst system and of the other systems brought into action by changes in body fluid tonicity or volume is, with some exceptions such as voluntary dehydration, precisely adjusted to smooth and effective restoration of normal conditions. This phenomenon strongly implies existence of a highly integrated co-ordination of thirst with the other effectors. During recent years extensive research has been focused on the role of hormonal inputs. In particular, the possibility of direct effects of hormones on water intake has been debated.

The studies on hormonal control of water intake were initiated at the beginning of the 1950s with a report by Barker et al. (1953) on the effect of pitressin on water intake. Spectacular acceleration of research in this field began with the momentous studies by Fitzsimons (1969) describing the potent thirst stimulating action of components of the renin–angiotensin system (see also Fitzsimons 1979).

The primary objective of this chapter is to present current evidence which implicates some hormones in the control of thirst, with special reference to those hormonal systems which are primarily and directly involved in regulation of body fluid osmolality, volume and blood pressure. The putative mechanisms of action and pathophysiological significance of particular hormones are emphasized. The classical neurotransmitters and sex hormones are not dealt with, as they are reviewed in Chapters 10 and 29. Information on the role of hormones in human thirst may be also found in the other chapters.

Hormones: Their Access to the Thirst System and Their Mechanism of Action

Hormones: General Comments

The past two decades have been a period of vigorous change in the concept of the hormone as a physiological messenger. The classical notion of the substance produced by an endocrine gland and transported to the target organs by the blood has been expanded to encompass virtually every biologically active substance produced by cells and altering cell functions by classical endocrine as well as by paracrine, autocrine and synaptic contacts. The central nervous system (CNS) is the principal site of hormonal influences on the thirst system. In this regard, it is now well established that many classical hormones are synthesized and may act at the level of the CNS through specific receptors. The question then arises whether the hormonal input to the thirst system derives from the blood, from the brain or from both sources.

Access of the Blood-Borne Hormones to the Thirst System

The brain microenvironment is protected efficiently by barriers that limit and regulate passage of substances to and from the CNS (Weindl 1983). In general, the blood–brain barrier (BBB) creates a major hindrance for peptide hormones, unless a specific carrier transport system operates. Peptides may interact with the brain through the specialized BBB deprived regions in circumventricular organs (CVOs). The total surface of this free exchange area is only a small fraction of the BBB. This implies that the direct access of peptide hormones to brain receptors is substantially limited. However, it includes the potential possibility of their powerful influences on many

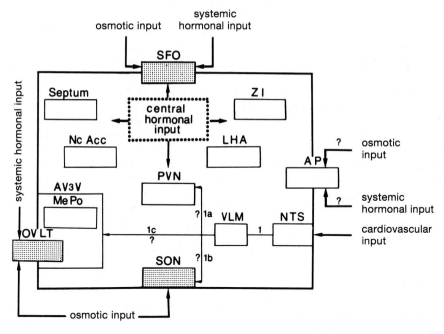

Fig. 7.1. Putative brain areas of relevance to thirst control as indicated by studies with lesions and electrical, osmotic and pharmacological stimulations. The dotted areas indicate putative osmoreceptive regions. AP, area postrema; AV3V, anteroventral third ventricle; LHA, lateral hypothalamic area; MePo, median preoptic nucleus; NcAcc, nucleus accumbens; NTS, nucleus tractus solitarius; OVLT, organum vasculosum laminae terminalis; PVN, paraventricular nucleus; SFO, subfornical organ; SON, supraoptic nucleus; VLM, ventrolateral medulla; ZI, zona incerta; 1, 1a, 1b, and 1c; putative pathways transmitting inputs from the cardiovascular receptors. Not indicated are extensive connections between the structures of the "thirst zone". Systemic hormones may enter the brain through the blood–brain barrier lacking circumventricular organs (AP, OVLT, SFO) or penetrate the blood–brain barrier (not shown). Hormone generated in the central nervous system may influence thirst independently or in concert with systemic hormones and osmotic and cardiovascular inputs.

vital areas, namely the subfornical organ (SFO) and organum vasculosum laminae terminalis (OVLT) which form extensive afferent and efferent connections with many other brain structures within the BBB.

Site and Mechanism of Action

Drinking of water by a thirsty individual results from activation of a complex network responsible for initiation, maintenance and extinction of thirst and of the associated motor activity which enables intake of an appropriate amount of water. Theoretically hormones may alter drinking behaviour through actions on various components of this network. Structures that are presumed to play an important role in regulation of water intake from experimental and clinical evidence are depicted in Fig 7.1. (see also Chapters 9, 11 and 12). Hormones may affect thirst by (a) primary effects on the neurons of the thirst system acting as neurotransmitters at the synaptic clefts, (b) secondary neuromodulatory effects on transmission by other neurotransmitters, (c) secondary effects through modification of the neural microenvironment by affecting systems transporting ions and organic osmolytes at the brain barrier systems and glial cells as well as by metabolic, trophic and vasoactive effects. Several lines of evidence indicate that many hormones that may contribute to the control of water intake fulfil the criteria of neurotransmitters and neuromodulators. Paracrine effects may be exerted by those hormones which are released to the interstitial fluid (ISF) and transported by cerebrospinal fluid (CSF). Presence of the hormone in CSF, long latency and prolonged effects after central administration as well as mismatch in the localization of neurons synthesizing the hormone and of its receptors may suggest (but does not prove) the paracrine action.

Evidence Suggesting Involvement of a Hormone in the Control of Thirst

The following observations may be used to support a contribution of a specific hormone to the control of water intake:

1. Demonstration that administration of exogenous hormone influences intake of water
2. Demonstration that stimulation or blockade of endogenous hormone interferes with spontaneous or stimulated water intake
3. Demonstration of the presence of the hormone and of its receptors in the neural substrate for thirst
4. Demonstration that disturbances in body fluid homeostasis which are known to stimulate thirst influence content of the hormone or number of its receptors within the neural substrate for thirst
5. Electrophysiological evidence that the hormone alters activity of neurons within the neural substrate for thirst.

It should be emphasized however, that none of the above circumstances is conclusive. The potential pitfalls which may lead to inappropriate evaluation of the observed effects may be summarized as follows:

First, administration of exogenous hormone cannot faithfully imitate naturally occurring events that result in release of the endogenous hormone and may be important for its action. Second, it is not possible to reproduce the normal pattern of release of the hormone which would result in optimal concentrations in a specific place. It is important to realize that all known hormones exert manifold, often contradictory effects mediated by distinct subsets of receptors. Under natural conditions the predominant action is presumably determined by the signal accounting for the increased release. It is logical to assume that specific conditions will result in increased amounts of a neurotransmitter or neuromodulator in the specific region. On the other hand, administration of high, pharmacological doses of an exogenous hormone may be required in order to create an appropriate concentration gradient that would enable its diffusion to the specific target site. Thus, the exogenous hormone may exert some unwanted side effects that may interfere with its action on the thirst system. Third, many central effects elicited by neuropeptides are depicted by the U-shaped dose–response curve. Therefore a broad range of concentration should be investigated before the conclusions about the effect of the hormone may be formulated. The concentrations corresponding to those at which the hormone is known to exert peripheral effects or at which it occurs in the CNS are probably the most relevant. Finally, the immunohistochemical data showing presence of the hormone and of its receptors and electrophysiological data disclosing excitatory or inhibitory effects of the hormone on the neurons within areas of the brain known to be involved in thirst should be confronted with the evidence that many structures of the same region are also important for blood pressure, temperature and glucoregulation as well as the control of many neuroendocrine functions, and specifically for secretion of vasopressin that is also sensitive to body fluid osmolality and volume regulation.

Renin–Angiotensin System (RAS)

In 1969 Fitzsimons demonstrated that ligation of the inferior vena cava either above or below the renal veins and constriction of the aorta above but not below the renal arteries caused rats in normal water balance to drink water. As drinking of water could also be initiated by administration of saline extracts of the renal cortex possessing pressor activity, he suggested that the renal dipsogen may be identical with renin. Based on results of these and other investigations he concluded that "the renin–angiotensin system may play a role in genesis of the thirst which follows certain extracellular stimuli" (Fitzsimons 1969). In subsequent studies Fitzsimons and Simons (1969) and Epstein et al (1970) described copious drinking of water in water-sated rats after intravenous (i.v.) or intracranial (i.c.) administration of angiotensin II (Ang II). These studies, which indicated a direct involvement in the control of thirst by an endocrine system with established importance for regulation of electrolyte metabolism and blood pressure, inspired extensive work in many laboratories.

Peripheral and Brain Renin–Angiotensin System

In 1970 Epstein et al. supported convincing evidence that Ang II stimulates thirst by a direct action at the level of the CNS. In subsequent studies it has been demonstrated that systemic Ang II may affect thirst at the level of the SFO (Simpson and Routtenberg 1975). The role of the target site for the blood-borne Ang II has also been ascribed to the OVLT (see Chapter 6). The discovery of the brain renin–angiotensin system allowed explanation of the rapid and powerful dipsogenic effects of Ang II, renin, angiotensinogen and angiotensin I applied directly into the brain tissue (Epstein et

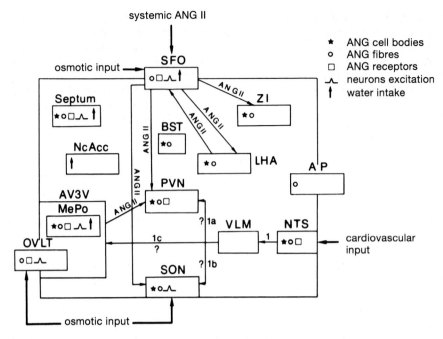

Fig. 7.2. Angiotensinergic system relevant to the thirst system depicted in Fig. 7.1. ANG arrows, angiotensinergic pathways within the areas controlling thirst as indicated by immunohistochemical and electrophysiological evidence cited in the text. For other explanations see Fig. 7.1.

al. 1970; Fitzsimons and Kucharczyk 1978). It is now generally accepted that the central dipsogenic effect of Ang II is executed through the central Ang II receptors that normally mediate the action of the brain-borne Ang II. The presence of all compounds of the renin–angiotensin system in the central nervous system and the possibility of de novo formation of brain Ang II is now well documented (for references see Ganong 1983; Ganten et al. 1985). Immunocytochemical evidence indicates occurrence of Ang II in cell bodies and fibres in the brain regions which overlap with those involved in thirst regulation (Lind et al. 1985; Ihamandas et al. 1989) (Fig. 7.2). Many areas innervated by Ang II-containing fibres exhibit high affinity binding sites for this peptide (Plunkett et al. 1987; Castren and Saavedra 1989). Angiotensin II receptors appear to be regulated depending on the state of body water and sodium content, and the presence of other hormones (Thomas and Sernia 1985; Feldstein et al. 1986; Wilson et al. 1986; Castren and Saavedra 1989).

Site and Mechanism of Action

Initial studies with central administration of Ang II and ablation of various structures indicated SFO and OVLT as important avenues for the systemically generated Ang II and the anteroventral part of the third ventricle (AV3V; including the area preoptica) and septum as the most sensitive sites of action for the brain-borne Ang II (Fitzsimons 1979; Simonnet et al. 1979; Chapter 6). The significance of the paraventricular nucleus (PVN) for the dipsogenic effect of Ang II mediated by the SFO has been emphasized by Gutman et al. (1988b). Recent immunohistochemical and electrophysiological data are in agreement with previous findings. In vitro and in vivo extracellular recordings revealed direct, in most instances excitatory, effects of Ang II on neurons in the SFO, OVLT, AV3V, septum and in the supraoptic nucleus (SON) (Huwyler and Felix 1980; Sayer et al. 1984; Okuya et al. 1987; Tanaka et al. 1987; Gutman et al. 1988a, 1989; Ihamandas et al. 1989). In the study of Ihamandas et al. (1989) the SFO neurons projecting to the SON were excited both by microiontophoretically applied and i.v. Ang II; the effect being blocked by saralasin, an Ang II antagonist.

Neurons of the median preoptic nucleus (MePo) projecting to PVN were found to be excited by electrical stimulation of the SFO and by microiontophoretic application of Ang II (Tanaka et al. 1987). Application of saralasin (an angiotensin antagonist) has also been found to abolish an excitatory input from the lateral hypothalamic area to the SFO (Tanaka et al. 1986a, b). Some of the antidromically identified neurons projecting from the SFO directly to PVN and SON responded both to intracarotid administration of Ang II and to hypertonic saline (Gutman et al. 1988a) which suggests integration of osmotic and hormonal inputs in the SFO. Thus, the electrophysiological and immunohistochemical data together with dipsogenic effects observed after direct application of Ang II to the same brain regions or absence of this effect after their destruction strongly suggest the possibility of a close connection between the peripheral and central RAS with a link between these two components in the SFO (see below and Fig. 7.2). Close coincidence of Ang II-containing cell bodies and fibres with specific binding sites for this peptide (Fig. 7.2) strongly suggests that Ang II may act as a neurotransmitter or neuromodulator. Direct excitatory effects of Ang II on neurons located within the "thirst zone" and short latency of dipsogenic responses after intracranial administration further support this conclusion. Ang II may also modulate transmission in the neural circuits subserving the thirst system. Several studies indicate that the dipsogenic effect of Ang II may be affected by a blockade of cholinergic, dopaminergic and noradrenergic receptors (for references see Fitzsimons 1979). Recent studies supported evidence for the important role of the noradrenergic innervation in thirst induced by centrally and systemically applied Ang II. It has been proposed that ascending noradrenergic projections may provide the MePo with information from baroreceptors which could be integrated in this region with the input from the SFO (see Johnson and Cunningham 1987 and Chapter 9). Although much evidence argues for direct neurotransmittory or neuromodulatory role of Ang II its paracrine role should also be considered. The presence of Ang II receptors in the glial cells (Raizada et al. 1987) which are known to play an important role in maintenance of the brain ionic homeostasis may speak in favour of such a possibility.

Regulatory Significance of Dipsogenic Properties of Ang II

In one of the earliest studies suggesting involvement of renin in drinking induced by extracellular stimuli Fitzsimons (1969) cautiously expressed his idea about the role of angiotensin in the control of thirst by saying:

> . . .even though lower and more physiological concentrations of angiotensin may have no direct stimulating effect on the hypothalamus, the possibility remains that angiotensin may lower the threshold of the nervous centres concerned in drinking sufficiently to ensure that enough water is drunk in response to thirst stimuli arising elsewhere in the body.

Since that time, the importance of the RAS for the control of thirst under physiological and pathological conditions has been the subject of discussion (Epstein 1978; Fitzsimons 1979; Rolls and Rolls 1982; Stricker 1978) which has focused on elucidation of the following objectives:

1. Whether the blood concentrations of Ang II are relevant to blood concentrations of this peptide occurring under physiological and pathological conditions
2. Whether stimulation of the RAS is associated with increased intakes of water
3. Whether blockade of the central or peripheral RASs interferes with water intake under physiological and pathological conditions.

Several studies performed on rats indicate that during systemic administration of Ang II plasma levels of this peptide at the moment of drinking fell within the range of concentrations occurring under physiological and pathological circumstances (Mann et al. 1980; Anke et al. 1988). The dog also appears to be sensitive to dipsogenic action of Ang II (Fitzsimons 1979; Cowley et al. 1981), especially when the peptide is administered together with an osmotic load (Kozlowski et al. 1972). Intravenous administration of Ang II has been found to be much less effective in the sheep and the goat and not effective in the rabbit (Andersson and Westbye 1970; Abraham et al. 1975; Tarjan et al. 1988b). In the goat the dipsogenic action of Ang II has been found to be markedly potentiated by increased CSF sodium concentration (Andersson 1978). Systemic infusion of Ang II has been found to be not effective in eliciting thirst in man (Chapter 25). Recently, it has been shown that the dipsogenic potency of Ang II is markedly enhanced when the hypertensive effect of this peptide is prevented (Robinson and Evered 1987).

Increased secretion of renin has been reported in many disturbances of body fluid balance, pathological states and experimental procedures that are associated with enhancement of thirst (Fitzsimons 1979; Fitzsimons and Moore-Gillon 1980; Thrasher et al. 1982). Two clinical reports describing relief of severe thirst after nephrectomy in patients with hyperreninaemia suggested that excessive release of renin might have been involved in abnormal thirst in these patients (for references see Fitzsimons 1979). Controversial results have been obtained in the experimental studies. Nephrectomy did not abolish or only partly eliminated increased water intake following intraperitoneal administration of hyperoncotic colloid (Fitzsimons 1979; Stricker 1978). Increased intake of water in rats induced by caval ligation (simulating haemodynamic perturbations resulting from inadequate filling of the cardiovascular system) has been effectively abolished by bilateral nephrectomy (see Fitzsimons 1979). In the dog peripheral administration of saralasin partly attenuated thirst induced by caval ligation (Fitzsimons and Moore-Gillon 1980). The same tendency can be noticed in the study of Thrasher et al. (1982) when the water intake is referred to the changes in blood pressure. However, the absolute amounts of water ingested with and without saralasin did not differ in the latter study and intracerebroventricular (i.c.v.) administration of saralasin did not affect thirst induced by vagal ligation, although it completely abolished the effect of exogenous Ang II. Similarly, Lee et al. (1981) were unable to eliminate increased thirst in the rat after caval ligation by prolonged i.c.v. administration of saralasin which abolished the dipsogenic effect of centrally applied Ang II. Therefore, the authors concluded that Ang II is not essential for the enhanced thirst induced by caval ligation (Lee et al. 1981; Thrasher et al. 1982). Thirst induced by water deprivation in rats has been found to be significantly attenuated by prior i.c.v. blockade of Ang II receptors with P-113 (Malvin et al.

1977). These results have not been confirmed by Lee et al. (1981). In addition, Ramsay and Reid (1975) failed to prevent stimulation of drinking in the dog elicited by injection of hyperosmotic sodium choride into the third ventricle by prior administration of saralasin which abolished intake of water induced by i.c.v. applied Ang II. On the other hand, Hoffman et al. (1978) reported that thirst induced by water deprivation in the rat could be effectively inhibited by combined Ang II and cholinergic blockade. Recently, Franci et al. (1989) reported that application of Ang II antiserum into the third ventricle results in significant reduction of water intake in water-deprived rats. The authors suggested that the previous failures to block dehydration-induced drinking by angiotensin antagonists might have been related to incomplete saturation of Ang II receptors. Significant dipsogenic action of central endogenous angiotensin is also suggested by the recent study by Harding et al. (1989) who found that blockade of degradation of the endogenous angiotensin with bestatin (aminopeptidase inhibitor) elicits thirst and elevation of blood pressure in water-sated rats. Thus, the last two studies suggest that central endogenous angiotensin may be involved in the control of thirst under physiological conditions. Some data indicate that the central RAS may be affected by disturbances of body fluid osmolality and volume, the conditions which are known to enhance release of the peripheral renin (Hoffman et al. 1982; Cameron et al. 1985; Lind et al. 1985; Simon-Oppermann et al. 1986).

It is also noteworthy that the central RAS system overlaps the osmosensitive areas and that the brain regions housing the RAS receive the input from the brainstem areas which are involved in blood pressure regulation (McAllen and Blessing 1987; Morris et al. 1987; Shioya and Tanaka 1989). Thus, there exists a potential possibility of modulation of the central RAS by changes in body fluid tonicity and in the loading of the cardiovascular receptors. Action of these stimuli may be potentiated by systemic Ang II arriving at the CVOs (see also Johnson and Cunningham 1987; Chapter 9).

In summary, the available evidence strongly suggests involvement of Ang II in the control of thirst, especially in the sense proposed originally by Fitzsimons (1969, see also Fitzsimons 1979), that angiotensin is, as one of the regulators of thirst, acting synergistically with or potentiating effects of other factors. The question that remains to be elucidated is whether and to what extent Ang II is essential for maintenance of body fluid homeostasis.

Vasopressin

Vasopressin (VP) is a neuropeptide synthesized in the supraoptic, paraventricular and suprachiasmatic (SCN) nuclei, in the bed nucleus of the stria terminalis (BST) and other hypothalamic and extrahypothalamic neurons (for references see Buijs 1987; Ganten et al. 1985). Figs 7.1 and 7.3 show that many structures involved in the control of water intake are innervated by vasopressinergic fibres and possess vasopressin receptors. Ablation of some of these areas and in particular of the AV3V results in hypodipsia and in deficient release of VP (Andersson et al. 1975; Brody and Johnson 1980).

Many factors that cause release of vasopressin are also effective stimulants of thirst and vice versa; inhibition of VP release is usually associated with inhibition of thirst. Specifically, unidirectional changes in activity of the thirst system and of the vasopressinergic system are observed during disturbances of body fluid osmolality

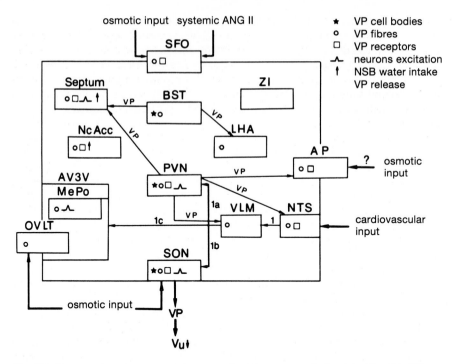

Fig. 7.3. Vasopressinergic system relevant to the thirst system depicted in Fig. 7.1. VP arrows, vasopressinergic pathways within the "thirst zone" as indicated by immunohistochemical and electrophysiological evidence cited in the text. Vu, urine output; NSB, non-stimulus-bound. For other explanations see Fig. 7.1.

or volume (for references see Kozlowski and Szczepanska-Sadowska 1975; Fitzsimons 1979; Robertson et al. 1982; Ramsay and Thrasher 1984; Baylis 1987; Robertson 1987a). Moreover, numerous experimental treatments that stimulate thirst such as cholinergic, dopaminergic, noradrenergic, histaminergic and serotoninergic agents, hypoglycaemia and angiotensin have also been found to release vasopressin (for references see Baylis 1987; Fitzsimons 1979; Sladek and Armstrong 1987).

Effect of Peripheral and Central VP on Water Intake

In 1953 Barker et al. reported that vasopressin is not dipsogenic in normally hydrated dogs but enhances osmotic thirst in dehydrated animals and that enhancement of osmotic thirst occurs with moderate but not with excessive doses of the hormone. The authors submitted also the first trace suggesting that neurohypophysectomy may increase availability of vasopressin for the control of thirst. This assumption, became comprehensible in the light of subsequent studies showing that neurohypophysectomy does not eliminate but may even enhance central release of vasopressin (Mens et al. 1982). The finding that the effect of VP may be reversed with higher doses has been neglected in many subsequent studies in which very high doses were administered and in which no effect or suppression of osmotic thirst have been observed (for references see Fitzsimons 1979). Systemic infusions of vasopressin (pitressin) resulting in more physiological blood concentrations of this peptide, comparable to those that occur during non-hypotensive hypovolaemia were found to cause a significant reduction of the osmotic threshold for thirst. On the other hand, very high blood VP concentration that was associated with a significant elevation of the central blood volume resulted in a clear cut suppression of the osmotic thirst. The latter effect was substantially reduced by vagosympathectomy. Systemic administration of VP did not stimulate water intake at any concentration. It was therefore concluded that lowering of the osmotic thirst threshold is a primary action of vasopressin, which, in concert with its renal antidiuretic effect would allow intake and retention of an additional amount of the fluid. On the other hand, inhibition of thirst which appears at supraphysiological levels is probably secondary to the increase of an inhibitory input to the vasopressinergic system from the cardiovascular receptors (Kozlowski and Szczepanska-Sadowska 1975; Szczepanska-Sadowska et al. 1974).

Central administration of vasopressin did not give unequivocal results (Epstein et al. 1970, see also Kozlowski and Szczepanska-Sadowska 1975, Fitzsimons 1979). Nicolaïdis and Fitzsimons (1975) observed stimulation of spontaneous water intake after bolus injections of 0.1–100 mU (0.24–240 pmol) of VP to the anteroinferior part of the third ventricle. The effect was not consistent, although sometimes very impressive (10.3 ml after administration of 5 mU). Szczepanska-Sadowska et al. (1982) found a significant increase in spontaneous water intake in the dog after bolus injections of very small doses of VP pitressin (30–300 μU=72–720 fmol) into the very rostral part of the third ventricle through the cannulas implanted 1 mm in front of the bregma. Drinking started with a latency of 10–20 min and usually terminated within 60 min. Water intake (150 ml/h) that followed the most effective dose of VP (100–150 μU=240–360 fmol=10^{-14} mol) was comparable with water intake after 10^{-11} mol of Ang II injected intracranially in the dog (Fitzsimons 1979). However, in contrast to Ang II, the range of concentrations resulting in a dose-dependent drinking was much lower. Injection of 3 mU of VP was less effective and plasma osmolality decreased by no more than 1.5%. This suggests that the spontaneous water intake in these experiments might have also been a result of reduction in the osmotic threshold which enabled elimination of small deficits of water that were not sensed prior to administration of vasopressin. In the same study lack of effect or inconsistent responses were observed after injection or infusions of vasopressin into the left ventricle or after injection to the more posterior part of the third ventricle. Eriksson et al. (1987) did not observe spontaneous intake of water after infusion of VP to the anterior part of the third ventricle of the dog at a slow rate of 46–138 fmol/min and at a very high rate (100 pmol/min; Simon-Oppermann, personal communication) through the cannulas located more posteriorly. It appears from these studies that the place and mode of administration may be crucial determinants of appearance of the dipsogenic effect of vasopressin. Pulse injections suddenly increasing concentration of the peptide in the very rostral part of the third ventricle may be more effective than constant rate infusions. The possibility that the constant delivery of the peptide is more efficient in induction of the inactivation or desensitization processes should be taken into account.

Central administration of VP antagonists to the third ventricle effectively suppressed the osmotic thirst in the dog induced by i.v. infusion of hyperosmotic sodium choride (Szczepanska-Sadowska et al. 1987). Furthermore, significant attenuation of water intake after 24-h water deprivation by administration of specific vasopressin antiserum has been recently reported by Franci et al. (1989).

Site and Mechanism of Action

Thus far, the site of a putative dipsogenic effect of vasopressin cannot be defined. It is appropriate to assume that during the natural stimulation the peptide released centrally from the vasopressinergic terminals in the brain will be more relevant for regulation of thirst than that released systemically. With this regard various manoeuvres which are known to stimulate thirst have been found to affect the content of immunoreactive vasopressin in the brain areas related to thirst regulation or in CSF in a manner suggesting central release of this peptide in concert with peripheral release (Epstein et al. 1983; Szczepanska-Sadowska et al.1984a; Cameron et al. 1985; Demotes-Mainard et al. 1986; Landgraf et al.1988). Water intake following 24 h dehydration was correlated with blood VP and CSF vasopressin concentrations prior to drinking (Szczepanska-Sadowska et al. 1984b). The available evidence suggests that the peptide may act in the very rostral part of the third ventricle. Unfortunately, systematic studies with local application of low amounts of VP directly to the brain tissue have not been performed. Previous studies from our laboratory have found enhanced release of vasopressin and non-stimulus-related drinking, correlated with blood vasopressin level and with its antidiuretic action during electrical stimulation of the lateral septum and placements located along the projection of the bed nucleus of the stria terminalis (BST) (Szczepanska-Sadowska et al. 1982). The septum is densely innervated by vasopressinergic fibres originating in the BST. Connections of this structure with PVN are also suggested. It has been also shown that osmotic and hypovolaemic stimuli, as well as electrical stimulation, enhance release of vasopressin in both the blood and the lateral septum (Demotes-Mainard et al. 1986; Neumann et al. 1988). Moreover, numerous specific VP binding sites have been detected in the lateral septum (Raggenbass et al. 1987; Dorsa et al. 1988; Tribollet et al. 1988; Shewey et al. 1989) and a direct excitatory effect of this peptide on lateral septal neurons has been reported (Raggenbass et al. 1988). Therefore, the lateral septum may be a site of action. The other possible sites are SFO and AV3V that are bilaterally connected with the SON and PVN (Morris et al. 1987) and presumably together with SON play an important role in expression of the brain osmosensitivity (for references see Burque 1989; Chaudhry et al. 1989).

Increasing amounts of data indicate that VP may act in the brain as neurotransmitter, neuromodulator or paracrine substance. The peptide has been found to excite neurons in the lateral septum and in the paraventricular nucleus (Inenaga and Yamashita 1986; Raggenbass et al. 1988). In both sites the effect persisted after blockade of the synaptic transmission induced by change in the ionic medium. In the septum, high concentrations of VP elicited inhibition of firing rate after initial excitation (Raggenbass et al. 1988). Whether this finding could be relevant to inhibition of thirst by high doses of vasopressin is not known. Excitatory effects of VP on neurons in the medial preoptic area and in the diagonal band were observed by Nicolaïdis and Jeulin (1984). Interestingly, the same neurons responded also to Ang II and the simultaneous application of these two peptides resulted in clear potentiation of the response (see also Chapter 12). Excitation of the same neurons by Ang II and VP has been also observed in the lateral septum but their combined action has not been tested (Huwyler and Felix 1980). These findings suggest that VP and Ang II may co-operate, the more so as there exists a striking resemblance in distribution of vasopressinergic and angiotensinergic innervations and of their receptors (Figs 7.2 and 7.3). Moreover, colocalization of these peptides has been reported (for references see Ganten et al. 1985; Gardiner and Bennett 1989).

The paracrine effects of vasopressin should also be taken into consideration. The peptide appears in CSF in increased concentration during disturbances of body fluids and may exert some generalized effects on the CNS. In support of this possibility are findings showing an increase of brain and ependymal permeability to water after central administration of vasopressin (Raichle and Grubb 1978; Rosenberg et al. 1986). Furthermore, De Pasquale et al. (1989) reported the impaired appearance of the volume regulatory phase during the acutely induced hyperosmotic state in the vasopressin deficient Brattleboro rat. In agreement with the latter finding are results from our laboratory indicating that i.c.v. applied VP significantly influences sodium handling in the CSF and may affect the volume regulatory phase in acute hyperosmotic state (Szczepanska-Sadowska et al. 1988; Szmydynger-Chodobska et al. unpublished data). In this regard it is worth noting that VP generates organic osmotically active substances (osmolytes) in the kidney (Blumenfeld et al. 1989). Brain ependyma and microvessels are innervated by vasopressin fibres and presence of VP receptors in brain microvessels has been demonstrated (Jojart et al. 1984; Kretzschmar et al. 1986; Pearlmutter et al. 1988). Thus, VP may potentially exert multiple effects on the CNS; some of them being secondary to alterations in the brain microenvironment.

Physiological and Pathophysiological Significance

As reviewed above, VP may be important for expression of the normal osmotic sensitivity under physiological conditions. This assumption is strongly reinforced by numerous studies showing significantly elevated plasma osmolality and sodium concentration in homozygous and heterozygous Brattleboro rats drinking water ad libitum (Möhring et al. 1978; Mann et al. 1980; De Pasquale et al. 1989) in spite of high renin levels in this strain (Mann et al. 1980). Moreover, hypernatraemia has been found to be corrected by treatment with vasopressin (Möhring et al. 1978).

The question arises whether disorders in secretion of vasopressin may contribute to abnormal thirst in some clinical states. Inappropriate thirst is observed in the syndrome of inappropriate secretion of vasopressin (SIADH), and in patients with essential hypernatraemia associated with a partial diabetes insipidus. In both situations parallel disorders of both systems may result in severe disturbances of body fluid osmolality (see De Rubertis et al. 1971; Robertson et al. 1982). Evidence that dDAVP treatment in patients with essential hypernatraemia improved their thirst sensation suggests that the deficient thirst might have been due to inadequate secretion of VP (Dunger et al. 1987). On the other hand, normal vasopressin levels have been observed in some psychotic patients with polydipsia and hyponatraemia (Goldman et al. 1988; Robertson 1987b). However, the possibility of abnormal central release has not been excluded in this study. In patients with SIADH water intake is inappropriate with regard to low plasma osmolality and sodium concentrations and some patients suffer from severe thirst when water intake is restricted to an appropriate level (Whitaker et al. 1979). Thus, it may be inferred that the osmostatic control of thirst is reset in this syndrome; however, systematic investigations of osmotic thirst before and after elimination of SIADH are not available. Nevertheless, it appears that disorders in secretion or action of vasopressin should be taken into consideration as a factor which may contribute to defective osmoregulation of thirst.

In summary, the above survey indicates that although there are still many unexplained findings concerning the role of vasopressin in the regulation of thirst, the available evidence suggests that vasopressin may play a role in determining the osmosensitivity of the thirst system. Its relevance to hypovolaemic thirst remains to be elucidated.

Other Hormones

Atrial Natriuretic Peptide

Several recent publications point to a putative involvement of the atrial natriuretic peptide (ANP) in the control of thirst. The atriocytes of the heart are the primary, peripheral source of ANP. Synthesis of this peptide and presence of ANP binding sites in the brain regions associated with the control of water intake have been also demonstrated (for references see Samson 1988). Electrophysiological studies give evidence for the direct, in most cases inhibitory, effect of ANP on single neuron firing rates (Samson 1987). Central administration of ANP has been found to suppress dehydration-induced and Ang II-induced water intake (Antunes-Rodrigues et al. 1985; Lappe et al. 1986). This action would be in concert with diuretic effects of peripheral ANP. However, it should be emphasized that inhibition of thirst was observed after administration of very high doses of ANP (1 nmol = 3 µg) that appear to be clearly supraphysiological. Evidence from our laboratory indicates that infusion of ANP into the third ventricle of the dog at a rate of 120 ng/min (about 13 pmol/min) and lower did not suppress the osmotic thirst induced by i.v. infusion of hyperosmotic load; on the contrary a tendency for reduction of the osmotic thirst threshold was observed (Kowalik-Borowka et al., unpublished observations).

Attempts were also made to find out whether the blood-borne ANP may be relevant for the control of water intake. Kauffman and Monckton (1988) found that i.v. infusion of ANP did not reduce but rather increased intake of water. The authors suggested that the latter effect might have been due to reduction in blood pressure which was observed after administration of ANP. Intracarotid infusion of ANP in the goat eliciting a 175-fold increase of its plasma concentration did not elicit significant suppression of water intake elicited by 24-h water deprivation (Olsson et al. 1989). However, significant reduction of water intake was observed in the same study during i.c.v. infusion of ANP at three times higher rate. Finally, it should be mentioned that i.c.v. administration of specific ANP antiserum was found to suppress thirst induced by water deprivation in the rat (Franci et al. 1989). The above data indicate that inhibition of thirst is observed after administration of pharmacological doses of ANP. It seems that by itself ANP is probably not an effective inhibitor of thirst, although it cannot be excluded that under certain circumstances it may suppress thirst in conjunction with other inhibitory stimuli (for instance it may act synergistically with an inhibitory input from the cardiovascular receptors under conditions of hypervolaemia). It also appears that under certain circumstances ANP may have a stimulatory effect on the thirst system.

Tachykinins (TCKs)

Three mammalian tachykinins, substance P (SP), neurokinin A and neurokinin B as well as their binding sites are present in the brain regions associated with regulation of water–electrolyte balance and cardiovascular functions (for references see Gardiner and Bennett 1989). It has been found that various members of the family of tachykinins administered intracranially influence water intake in different ways in different species. Eledoisin has been found to elicit copious, short latency drinking in the pigeon and to increase water intake in the rabbit and the sheep (Fitzsimons 1979; Tarjan et al. 1988a). On the other hand, in the rat it effectively inhibited thirst induced by Ang II, water deprivation and hyperosmotic load. Similar effects, though at much higher dose, were observed after administration of substance P. Neurokinin A and its non-mammalian counterpart kassinin suppressed thirst induced by cellular dehydration in the rat, whereas their effects on Ang II-induced thirst were not consistent (de Caro et al. 1988). Thus, it appears that tachykinins effectively inhibit water intake in the rat while they may have stimulatory effects in other species. Whether they are endogenous regulators of thirst and what is the mechanism of their action remains to be elucidated.

Oxytocin

Little is known about the relevance of oxytocin to the control of thirst. Oxytocin-containing neurons and receptors are widespread in the CNS (for references see Gardiner and Bennett 1989). Szczepanska-Sadowska et al. 1975 (cited by Fitzsimons 1979) did not observe any alteration of osmotic thirst during infusion of oxytocin into the lateral ventricle in the dog but the full range of doses has not been investigated. Recently Franci et al. (1989) reported that administration of specific antiserum to oxytocin attenuated water intake in dehydrated rats. Thus, it is likely that this peptide may contribute to the control of thirst together with other compounds. The role of oxytocin in the control of sodium and water intake is discussed in Chapters 11 and 19.

Neuropeptide Y

Recent evidence indicates that neuropeptide Y which is widely distributed in the CNS and colocalized with noradrenaline and many other neuropeptides (for references see Gardiner and Bennett 1989) stimulates water intake in the rat but inhibits drinking induced by water deprivation and associated with food intake in mice (Stanley and Leibowitz 1984; Morley and Flood 1989). Very high doses of neuropeptide Y were applied in both studies. The regulatory significance of these effects awaits elucidation.

Corticotropin Releasing Factor and Pro-opiomelanocortin Derived Peptides. Opioid Peptides

Corticotropin releasing factor (CRF), ACTH, ß-endorphin and a-melanocyte-stimulating (a-MSH) hormone as well as their receptors exist in the CNS in the regions of relevance to thirst. At present there are not enough data to evaluate their role in the control of water intake. Tarjan et al (1987) reported that i.c. applied CRF, ACTH and a-MSH did not influence water intake in wild rabbit, whereas ß-endorphin had an inhibitory effect. On the other hand, peripherally administered ACTH has been repeatedly shown to increase water intake (Blaine et al. 1975; Weisinger et al. 1980). However, these effects appear to be mediated by adrenal steroids.

Opioid peptides and their receptors are widely represented in the brain regions associated with regulation of water–electrolyte balance. The effects of opiate receptor agonists and antagonists on water intake have been extensively investigated; however, the results did not yield a comprehensive picture. Blockade of opiate receptors with naloxone or diprenorphine resulted in hypodipsia (Brown and Holtzman 1981; Czech et al. 1983; Cooper 1984). Accordingly, it has been speculated that endogenous peptides may stimulate thirst. On the other hand, administration of opiate receptor agonists did not provide conclusive results (Cooper 1984; Konecka et al. 1984; Tarjan et al. 1987), presumably because of complex action of opioid peptides through different kinds of receptors. Future studies employing specific agonists and antagonists of opiate receptors should add to understanding of the role of the endogenous opioids in the control of water intake.

Other Peptide Hormones

Insulin and relaxin have been recently reported to stimulate water intake (Kozlowski et al. 1980; Vijande et al. 1989; Thornton and Fitzsimons 1989, see also Baylis 1987). It appears that the thirst-stimulating action of insulin may be related to hypoglycaemia (see Baylis 1987), although the central effects should be also taken into account (Kozlowski et al. 1980). The relevance of insulin and relaxin to physiological control of thirst remains to be elucidated.

Steroid Hormones

In general, steroids can readily penetrate the BBB (Weindl 1983). Moreover, there is evidence that the brain is capable of synthesizing steroids (Le Goascogne et al. 1987). Specific binding sites for gonadal and adrenal steroids exist in the brain regions related to the control of water intake (for references see Miller et al. 1989; Luttge and Emadian 1988). Steroid hormones exert powerful effects on water and electrolyte balance and energy metabolism through their peripheral actions and thus their central action through the regulation of water and salt intake may be of particular interest. Indeed, it has been demonstrated that the number of Ang II and VP receptors as well as biosynthetic capacity of vasopressin neurons in specific brain regions may critically depend on the presence of steroid hormones (Wilson et al. 1986; Kiss et al. 1988; Miller et al. 1989). Weisinger et al. (1980) reported that adrenalectomy abolishes ACTH-induced increase in water intake, whereas treatment of adrenalec-tomized animals with adrenal steroids results in stimulation of water intake which suggests that enhancement of thirst by ACTH was mediated by adrenal steroids. Furthermore, glucocorticoids were found to potentiate the dipsogenic action of Ang II (Gansen and Sumners 1989).

Significant changes in water balance occur during menstruation, pregnancy and lactation. Attempts were made, therefore, to find out whether the gonadal steroid hormones may be involved in increased intakes of water in these states. Denton and Nelson (1978) reported that combined administration of ACTH, prolactin and oxytocin in female rabbits pretreated with oestradiol and progesterone successfully replicated increased intake of water and sodium choride seen in lactating animals. On the other hand, the gonadal steroids do not seem to account for the resetting of osmotic thresholds for thirst and VP release observed during gestation (Durr et al. 1981; see also Chapter 29).

Thyroid Hormones

Deficit or excess of thyroid hormones results in appreciable disturbances of water–electrolyte balance. Thyroid hormones can enter the brain by the carrier-mediated transport system (see Weindl 1983) and some data indicate that they may stimulate water intake (for references see Fitzsimons 1979). However, despite their apparent importance for water–electrolyte metabolism, their role in the control of thirst has not been systemically investigated and it cannot be excluded at present that the increase in water intake induced by thyroid hormones is secondary to the metabolic effects resulting in raised body temperature or increased loss of water due to osmotic diuresis.

Summary

Maintenance of a constant body fluid volume and osmolality depends on co-ordinated actions of thirst and neural and hormonal mechanisms regulating salt intake and excretion of water and electrolytes. Several lines of evidence indicate that changes in hormonal milieu may be important for proper functioning of the thirst system. Spontaneous or stimulated water intakes have been found to be affected by administration of angiotensin II, vasopressin, tachykinins, atrial natriuretic peptide, ACTH, ß-endorphin, glucocorticoids and neuropeptide Y. Although the physiological relevance of these findings is not always clear the currently available evidence strongly suggests that Ang II and VP may play an important role in regulation of thirst. Thirst may be affected by blood-derived and brain-borne hormones. It is likely that the brain-derived hormones may be released by neurons encompassed within the neural network subserving thirst. With regard to Ang II and VP some evidence indicates that their central release may be affected by disturbances in body fluid volume and osmolality, independently from peripheral release. Hormones may act as neurotransmitters, neuromodulators or paracrine regulators. The paracrine action may include passage of ions and other osmotic substances through the brain barriers and between neural and glial elements, trophic functions or regulation of number or affinity of receptors to the other neuroactive substances. The neurotransmitter–neuromodulatory functions of hormones may be important for short-term regulation of water intake, whereas the paracrine effects may play a role in adaptation to prolonged disorders in body fluid osmolality. It is likely that a wide spectrum of hormones may be necessary for redundancy and plasticity of various elements of the thirst system, enabling precise regulation under a variety of conditions and compensation of some isolated morphological and biochemical defects.

Acknowledgements. The author wishes to acknowledge financial support from the Alexander von Humboldt Foundation and from the Research Project CPBR 11.6 during preparation of this chapter as well as to express special thanks to Professor Eckhart Simon and to Dr Simon-Oppermann from the Max Planck Institute für Physiologische und Klinische Forschung in Bad Nauheim for valuable discussion of the manuscript. Preparation of photographs by Mrs Roswitha Bender is gratefully appreciated.

References

Abraham SF, Baker RM, Blaine EH, Denton DA, McKinley MJ (1975) Water drinking induced in sheep by angiotensin–a physiological or pharmacological effect? J Comp Physiol Psychol 88:503–518

Andersson B (1978) Regulation of water intake. Physiol Rev 58:582–603

Andersson B, Westbye O (1970) Synergistic action of sodium and angiotensin on brain mechanisms controlling fluid balance. Life Sci 9:601–608

Andersson B, Leksell G, Lishajko F (1975) Perturbations in fluid balance induced by medially placed forebrain lesions. Brain Res 99:261–275

Anke J, Van Eekelen M, Phillips MI (1988) Plasma angiotensin II levels at moment of drinking during angiotensin II intravenous infusion. Am J Physiol 255:R500–R506

Antunes-Rodrigues J, McCann SM, Rogers LC, Samson WK (1985) Atrial natriuretic factor inhibits dehydration and angiotensin II-induced water intake in the conscious unrestrained rat. Proc Natl Acad Sci USA 82:8720–8723

Barker JP, Adolph EF, Keller AD (1953) Thirst tests in dogs and modifications of thirst with experimental lesions of the neurohypophysis. Am J Physiol 173:233–245

Baylis PH (1987) Osmoregulation and control of vasopressin secretion in healthy humans. Am J Physiol 253:R671–R678

Blaine EH, Covelli MD, Denton DA, Nelson JF, Skulkes AA (1975) The role of ACTH and adrenal glucocorticoids in the salt appetite of wild rabbits (*Oryctolagus cuniculus* (L)). Endocrinology 97: 793–801

Blumenfeld JD, Hebert SC, Heilig CW, Balschi JA, Stromski ME, Gullans SR (1989) Organic osmolytes in inner medulla of Brattleboro rat: effects of ADH and dehydration. Am J Physiol 256:F916–922

Brody MJ, Johnson AK (1980) Role of the anteroventral third ventricle (AV3V) region in fluid and electrolyte balance, atrial pressure regulation, and hypertension. In: Martini L, Ganong WF (eds) Frontiers in neuroendocrinology, vol 6. Raven Press, New York, pp 249–292

Brown DR, Holtzman SG (1981) Narcotic antagonists attenuate drinking induced by water deprivation in a primate. Life Sci 28:1287–1294

Buijs RM (1987) Vasopressin localization and putative functions in the brain. In: Gash DM, Boer GJ (eds) Vasopressin. Principles and properties. Plenum Press, New York, London, pp 91–115.

Burque Ch W (1989) Ionic basis for the intrinsic activation of rat supraoptic neurones by hyperosmotic stimuli. J Physiol (Lond) 417:263–277

Cameron V, Espiner EA, Nicholls MG, Donald RA, MacFarlane MR (1985) Stress hormones in blood and cerebrospinal fluid of conscious sheep: effect of hemorrhage. Endocrinology 116:1460–1465

Castren E, Saavedra JM (1989) Angiotensin II receptors in paraventricular nucleus, subfornical organ, and pituitary gland of hypophysectomized, adrenalectomized and vasopressin deficient rats. Proc Natl Acad Sci USA 86:725–729

Chaudhry MA, Dyball REJ, Honda K, Wright NC (1989) The role of interconnection between supraoptic nucleus and anterior third ventricular region in osmoregulation in the rat. J Physiol (Lond) 410: 123–135

Cooper S (1984) Benzodiazepine and endorphinergic mechanism in relation to salt and water intake. In: de Caro G, Epstein AN, Massi M (eds) The physiology of thirst and sodium appetite. Plenum Press, New York, London, (Series A, Life Sciences, vol 105) pp 239–244

Cowley AW, Switzer SJ, Skelton MM (1981) Vasopressin, fluid and electrolyte response to chronic angiotensin II infusion. Am J Physiol 240:R130–R138

Czech DA, Stein EA, Blake MJ (1983) Naloxone-induced hypodipsia: a CNS mapping study. Life Sci 33:797–803

de Caro G, Perfumi M, Massi M (1988) Tachykinins and body fluid regulation. In: Epstein AN, Morrison AR (eds) Progress in psychobiology and physiological psychology, vol 13. Academic Press, San Diego, pp 31–36

Demotes-Mainard J, Chauveau J, Rodriguez F, Vincent JD, Poulain DA (1986) Septal release of vasopressin in response to osmotic, hypovolemic and electrical stimulation in rats. Brain Res 381:314–321

Denton DA, Nelson JF (1978) The control of salt appetite in wild rabbits during lactation. Endocrinology 103:1880–1887

De Pasquale M, Patlak CS, Cserr HF (1989) Brain ion and volume regulation during acute hypernatremia in Brattleboro rats. Am J Physiol 256:F1059–F1066

De Rubertis FR, Michelis MF, Beck N, Field, Davis BB (1971) "Essential" hypernatremia due to ineffective osmotic and intact volume regulation of vasopressin secretion. J Clin Invest 50:97–111

Dorsa DM, Brot MD, Shewey LM, Meyers KM, Szot P, Miller MA, (1988) Interaction of vasopressin antagonist with vasopressin receptors in the septum of the rat brain. Synapse 2:205–211

Dunger DB, Seckl JR, Lightman SL (1987) Increased renal sensitivity to vasopressin in two patients with essential hypernatremia. J Clin Endocrinol Metab 64:185–189

Durr JA, Stamoutsos B, Lindheimer MD (1981) Osmoregulation during pregnancy in the rat. Evidence for resetting of the threshold for vasopressin secretion during gestation. J Clin Invest 68:337–346

Epstein AN (1978) Consensus, controversies and curiosities. Fed Proc 37:2711–2716

Epstein AN, Fitzsimons JT, Rolls BJ (1970) Drinking induced by injection of angiotensin into the brain of the rat. J Physiol (Lond) 210:457–474

Epstein Y, Castel M, Glick SM, Sivan N, Ravid R (1983) Changes in hypothalamic and extra-hypothalamic vasopressin content of water-deprived rats. Cell Tissue Res 233:99–111

Eriksson S, Simon-Oppermann Ch, Simon E, Gray DA (1987) Interaction of changes in the third ventricular CSF tonicity, central and systemic AVP concentrations and water intake. Acta Physiol Scand 130: 575–583

Feldstein J, Sumners C, Raizada M (1986) Sodium increases angiotensin II receptors in neural cultures from brains of normotensive and hypertensive rats. Brain Res 370:265–272

Fitzsimons JT (1969) The role of a renal thirst factor in drinking induced by extracellular stimuli. J Physiol (Lond) 201:349–368

Fitzsimons JT (1979) The physiology of thirst and sodium appetite. Monographs of the physiological society no 35, Cambridge University Press, Cambridge, pp 128–265

Fitzsimons JT, Kucharczyk J (1978) Drinking and hemodynamic changes induced in the dog by intracranial injection of components of the renin–angiotensin system. J Physiol (Lond) 276:419–434

Fitzsimons JT, Moore-Gillon MJ (1980) Drinking and antidiuresis in response to reductions in venous return in the dog: neural and endocrine mechanisms. J Physiol (Lond) 308:403–416

Fitzsimons JT, Simons BJ (1969) The effect on drinking in the rat of intravenous angiotensin, given alone or in combination with other stimuli of thirst. J Physiol (Lond) 203:45–57

Franci G, Kozlowski GP, McCann SM (1989) Water intake in rats subjected to hypothalamic immunoneutralization of angiotensin II, atrial natriuretic peptide, vasopressin or oxytocin. Proc Natl Acad Sci USA 86:2952–2956

Ganong WF (1983) The brain renin–angiotensin system. In: Krieger D, Brownstein MJ, Martin JB (eds) Brain peptides. Wiley, New York, pp 805–826

Gansen R, Sumners C (1989) Glucocorticoids potentiate the dipsogenic action of angiotensin II . Brain Res 499:121–130

Ganten D, Unger Th, Lang RE (1985) The dual role of angiotensin and vasopressin as plasma hormones and neuropeptides in cardiovascular regulation. J Pharmacol (Paris) 16 (Suppl II):51–68

Gardiner SM, Bennett T (1989) Brain neuropeptides:actions on central cardiovascular control mechanisms. Brain Res Rev 14:79–116

Goldman MB, Luchins DJ, Robertson GL (1988) Mechanisms of altered water metabolism in psychotic patients with polydipsia and hypernatremia. N Engl J Med 318:397–403

Gutman MB, Ciriello J, Mogenson GJ (1988a) Effects of plasma angiotensin II and hypernatremia on subfornical organ neurons. Am J Physiol 254:R746–754

Gutman MB, Jones DL, Ciriello J (1988b) Effect of paraventricular nucleus lesions on drinking and pressor responses to Ang II. Am J Physiol 255:R882–R887

Gutman MB, Douglas L, Jones DL, Ciriello J (1989) Contribution of nucleus medianus to the drinking and pressor responses to angiotensin II acting at subfornical organ. Brain Res 485:49–56

Harding JW, Jensen LL, Quirk WS, Dewey AL, Wright JW (1989) Brain angiotensin: critical role in the ongoing regulation of body fluid homeostasis and cardiovascular function. Peptides 10:261–264

Hoffman WE, Ganten U, Phillips MI, Schmid PG, Schelling P, Ganten D (1978) Inhibition of drinking in water deprived rats by combined central angiotensin II and cholinergic receptor blockade. Am J Physiol 234:F41–F47

Hoffman DL, Krupp L, Schrag D, Nilaver G, Valiquette G, Kilcoyne MM, Zimmerman EA (1982) Angiotensin immunoreactivity in vasopressin cells in rat hypothalamus and its relative deficiency in homozygous Brattleboro rats. Ann NY Acad Sci 394:135–141

Huwyler T, Felix D (1980) Angiotensin II sensitive neurons in septal areas of the rat. Brain Res 195: 187–195

Ihamandas JH, Lind RW, Renauld LP (1989) Angiotensin II may mediate excitatory neurotransmission from the subfornical organ to the hypothalamus supraoptic nucleus: an anatomical and electro-physiological study in the rat. Brain Res 487:52–61

Inenaga K, Yamashita H (1986) Excitation of neurones in the rat paraventricular nucleus in vitro by vasopressin and oxytocin. J Physiol (Lond) 370:165–180

Johnson AK, Cunningham JT (1987) Brain mechanisms and drinking: the role of lamina terminalis-associated systems in extracellular thirst. Kidney Int 32:S35–S42

Jojart J, Joo F, Miklos L, Laszlo FA (1984) Immunoelectro-histochemical evidence for innervation of brain microvessels by vasopressin-immunoreactive neurons in the rat. Neurosci Lett 51:259–264

Kauffman S, Monckton EA (1988) Effect of peripherally administered atriopeptin III on water intake in rats. J Physiol (Lond) 396:379–387

Kiss Z , van Eekelen JAM, Reul JMHM, Westphal HM, De Kloet ER (1988) Glucocorticoid receptor in magnocellular neurosecretory cells. Endocrinology 122:444–449

Konecka AM, Sadowski B, Jaszczak J, Panocka I, Sroczynski J (1984) Suppression of food and water intake after intracerebroventricular infusion of morphine and naloxone in rabbits. Arch Int Physiol Biochim 92:219–226

Kozlowski S, Szczepanska-Sadowska E (1975) Mechanisms of hypovolemic thirst and interactions between hypovolemia, hyperosmolality and the antidiuretic system. In: Peters G, Fitzsimons JT, Peters-Haefeli L (eds) Control mechanisms of drinking. Springer-Verlag, Berlin, Heidelberg, New York, pp 25–35

Kozlowski S, Drzewiecki K, Zurawski W (1972) Relationship between osmotic reactivity of the thirst mechanism and the angiotensin and aldosterone level in the blood of dogs. Acta Physiol Pol 23:417–425

Kozlowski S, Szczepanska-Sadowska E, Sobocinska J, Bak M, Czyzyk A (1980) Stimulation of thirst and ADH release after intracranial injection of insulin in the dog. Verh Dtsch Ges Inn Med 86:1423–1426

Kretzschmar R, Landgraf R, Gjedde A, Ermisch A (1986) Vasopressin binds to microvessels from rat hippocampus. Brain Res 380:325–330

Landgraf R, Neumann J, Schwarzberg H (1988) Central and peripheral release of vasopressin and oxytocin in the conscious rat after osmotic stimulation. Brain Res 457:219–225

Lappe RW, Dinish JL, Bex F et al. (1986) Effects of atrial natriuretic factor on drinking responses to central angiotensin II. Pharmacol Biochem Behav 24:1573–1576

Le Goascogne C, Robel P, Gouezou M, Sonanes N, Baulieu E, Waterman M (1987) Neuro-steroids:cytochrome P-450scc in rat brain. Science 237:1212–1215

Lee MC, Thrasher TN, Ramsay DJ (1981) Is angiotensin essential in drinking induced by water deprivation and caval ligation? Am J Physiol 240:R75–R80

Lind RW, Swanson LN, Ganten D (1985) Organization of angiotensin II immunoreactive cells and fibers in the rat central nervous system. Neuroendocrinology 40:2–24

Luttge WG, Emadian SM (1988) Further chemical differentiation of Type I and Type II adrenocorticosteroid receptors in mouse brain cytosol: evidence for a new class of glucocorticoid receptors. Brain Res 453:41–50

Malvin RL, Mouw D, Vander AJ (1977) Angiotensin: physiological role in water deprivation-induced thirst of rats. Science 197:171–173

Mann JEF, Johnson AK, Ganten D (1980) Plasma angiotensin II: dipsogenic levels and angiotensin-generating capacity of renin. Am J Physiol 238:R372–R377

McAllen RM, Blessing WW (1987) Neurons (presumably A1-cells) projecting from the caudal ventrolateral medulla to the region of the supraoptic nucleus respond to baroreceptor inputs in the rabbit. Neurosci Lett 73:247–252

Mens WBJ, Van Dam AF, Van Wimersma Greidanus TB (1982) Influence of histamine and pentobarbitone on plasma and CSF vasopressin levels of hypophysectomized rats. Brain Res Bull 8:555–557

Miller MA, Urban JH, Dorsa DM (1989) Steroid dependency of vasopressin neurons in the bed nucleus of the stria terminalis by in situ hybridization. Endocrinology 125:2335–2340

Möhring J, Kohrs G, Möhring B, Petri M, Homsy E, Haack D (1978) Effects of prolonged vasopressin treatment in Brattleboro rats with diabetes insipidus. Am J Physiol 234:F106–F111

Morley JE, Flood JF (1989) The effect of neuropeptide Y on drinking in mice. Brain Res 494:129–137

Morris F, Chapman DB, Sokol HW (1987) Anatomy and function of the classic vasopressin-secreting hypothalamus–neurohypophysial system. In: Gash DM, Boer GJ (eds) Vasopressin. Principles and properties. Plenum Press, New York, pp 1–89

Neumann J, Schwarzberg H, Landgraf R (1988) Measurement of septal release of vasopressin and oxytocin by push–pull technique following electrical stimulation of paraventricular nucleus of the rat. Brain Res 462:181–184

Nicolaïdis S, Fitzsimons JT (1975) La dependance de la prise d'eau induite par l'angiotensine II envers la fonction vasomotrice cerebrale locale chez le rat. C R Acad Sci 281D:1417–1420

Nicolaïdis S, Jeulin AC (1984) Converging projections of hydromineral imbalances and hormonal co-action upon neurons surrounding the anterior wall of the third ventricle. J Physiol (Paris) 79:406–415

Okuya S, Inenaga K, Kaneko T, Yamashita H (1987) Angiotensin II sensitive neurons in the supraoptic nucleus, subfornical organ and anteroventral third ventricle of rats in vitro. Brain Res 402:58–67

Olsson K, Dahlborn K, Nygren K, Kalberg BE, Anden NE, Eriksson L (1989) Fluid balance and arterial blood pressure during intracarotid infusions of atrial natriuretic peptide (ANP) in water-deprived goats. Acta Physiol Scand 137:249–257

Pearlmutter AF, Szkrybalo M, Kim Y, Harik SI (1988) Arginine vasopressin receptors in pig cerebral microvessels, cerebral cortex and hippocampus. Neurosci Lett 87:121–126

Plunkett LM, Shigematsu K, Kurihara M, Saavedra JM (1987) Localization of angiotensin II receptors along the anteroventral third ventricle area of the rat brain. Brain Res 405:205–212

Raggenbass M, Tribollet E, Dreifuss JJ (1987) Electrophysiological and autoradiographical evidence of V1 vasopressin receptors in the lateral septum of the rat brain. Proc Natl Acad Sci USA 84: 7778–7782

Raggenbass M, Dubois-Dauphin M, Tribollet E, Dreifuss JJ (1988) Direct excitatory action of vasopressin in the lateral septum of the rat brain. Brain Res 459:60–69

Raichle ME, Grubb RL (1978) Regulation of brain water permeability by centrally released vasopressin. Brain Res 142:191–194

Raizada MK, Phillips J, Crews FT, Sumners C (1987) Distinct angiotensin II receptor in primary cultures of glial cells from rat brain. Proc Natl Acad Sci USA 84:4655–4659

Ramsay DJ, Reid JA (1975) Some central mechanisms of thirst in the dog. J Physiol (Lond) 253:517–525

Ramsay DJ, Thrasher TN (1984) The defence of plasma osmolality. J Physiol (Paris) 79:416–420

Robertson GL (1987a) Physiology of ADH secretion. Kidney Int 32:(Suppl 21)S-20–S-26

Robertson GL (1987b) Dipsogenic diabetes insipidus: a newly recognized syndrome caused by a selective defect in the osmoregulation of thirst. Tran Assoc Am Physicians 100:241–249

Robertson GL, Aycinena P, Zerba RL (1982) Neurogenic disorders of osmoregulation. Am J Med 72: 339–353

Robinson M, Evered MD (1987) Pressor action of intravenous angiotensin II reduces drinking response in rats. Am J Physiol 252:R754–R759

Rolls BJ, Rolls E (1982) Thirst. Problems in the behavioral sciences, Cambridge University Press, Cambridge

Rosenberg GA, Kyner WT, Fenstermacher JD, Patlak CS (1986) Effect of vasopressin on ependymal and capillary permeability to tritiated water in cat. Am J Physiol 251:F485–F489

Samson WK (1987) Atrial natriuretic factor and the central nervous system. Endocrinol Metab Clin North Am 16:145–161

Samson WK (1988) Central nervous system actions of atrial natriuretic factor. Brain Res Bull 20: 831–837.

Sayer J, Hubbard J, Sirett N (1984) Rat organum vasculosum laminae terminalis in vitro: responses to transmitters. Am J Physiol 247:R374–R379

Shewey LM, Boer J, Szot P, Dorsa M (1989) Regulation of vasopressin receptors and phosphoinositide hydrolysis in the septum of heterozygous and homozygous Brattleboro rats. Neuroendocrinology 50:292–298

Shioya M, Tanaka J (1989) Inputs from the nucleus of the solitary tract to subfornical organ neurons projecting to the paraventricular nucleus in the rat. Brain Res 483:192–195

Simon-Oppermann Ch, Gray DA, Simon E (1986) Independent osmoregulatory control of central and systemic angiotensin II concentrations in dogs. Am J Physiol 250:R918–R925

Simonnet G, Rodriguez F, Fumoux F, Czernichow P, Vincent JD (1979) Vasopressin release and drinking induced by intracranial injection of angiotensin II in monkey. Am J Physiol 237:R20–R25

Simpson JB, Routtenberg A (1973) Subfornical organ: site of drinking elicitation by angiotensin II. Science 181:1172–1175

Sladek CD, Armstrong WE (1987) Effect of neurotransmitters and neuropeptides on vasopressin release. In: Gash DM, Boer GJ (eds) Vasopressin. Principles and properties. Plenum Press, New York, pp 275–333

Stanley BG, Leibowitz SF (1984) Neuropeptide Y: stimulation of feeding and drinking by injection into the paraventricular nucleus. Life Sci 35:2635–2642

Stricker M (1978) The renin–angiotensin system and thirst: some unanswered questions. Fed Proc 37: 2704–2710

Szczepanska-Sadowska E, Kozlowski S, Sobocinska J (1974) Blood antidiuretic hormone level and osmotic reactivity of thirst mechanism in dogs. Am J Physiol 227:766–770

Szczepanska-Sadowska E, Sobocinska J, Sadowski B (1982). Central dipsogenic effect of vasopressin. Am J Physiol 242:R372–R379

Szczepanska-Sadowska E, Simon-Oppermann Ch, Gray DA, Simon E (1984a) Control of central release of vasopressin. J Physiol (Paris) 79:432–439

Szczepanska-Sadowska E, Simon-Oppermann Ch, Gray DA, Simon E (1984b) Plasma and cerebrospinal fluid and osmolality in relation to thirst. Pflügers Arch 400:294–299

Szczepanska-Sadowska E, Sobocinska J, Kozlowski S (1987) Thirst impairment elicited by intraventricular administration of vasopressin antagonists. Peptides 8:1003–1009

Szczepanska-Sadowska E, Szmydynger-Chodobska J, Chodobski A (1988). Effect of vasopressin on cerebrospinal fluid formation and composition under anisoosmotic conditions. Eur J Clin Invest 18:A27, abstract 121

Tanaka J, Kaba H, Saito H, Seto K (1986a) Lateral hypothalamic area stimulation excites neurons in the region of the subfornical organ with efferent projections to the hypothalamic paraventricular nucleus in the rat. Brain Res 379:200–203

Tanaka J, Saito H, Seto K (1986b) Subfornical organ efferents influence the activity of median preoptic neurons projecting to the hypothalamic paraventricular nucleus in the rat. Exp Neurol 93:647–651

Tanaka J, Saito H, Kaba H (1987) Subfornical organ and hypothalamic paraventricular nucleus connections with median preoptic nucleus neurons: an electrophysiological study in the rat. Exp Brain Res 68: 579–585

Tarjan E, Denton DA, Ong F, Tregear G, Wade J (1987) Effect of icv administration of CRF and POMC peptides on the sodium and water metabolism of wild rabbits. Second World Congress of Neuroscience, Budapest, Abstract

Tarjan E, Denton DA, McBurnie MI (1988a) Water and sodium intake of sheep and rabbits during intracerebroventricular infusion of eledoisin. XI Congress of European Neuroscience Association, Zurich 1988, Abstract

Tarjan E, Denton DA, McBurnie MI, Weisinger RS (1988b) Water and sodium intake of wild and New Zealand rabbits following angiotensin. Peptides 9:677–679

Thomas WG, Sernia C (1985) Regulation of rat brain angiotensin II (AII) receptors by intravenous AII and low dietary Na⁺. Brain Res 345:54–61

Thornton SN, Fitzsimons JT (1989) ICV porcine relaxin stimulates water intake but not sodium intake in male and female rats. Appetite 12:242

Thrasher TN, Keil LC, Ramsay DJ (1982) Hemodynamic, hormonal and drinking responses to reduced venous return in the dog. Am J Physiol 243:R354–R362

Tribollet E, Barberise C, Jard S, Dubois-Dauphin M, Dreifuss JJ (1988) Localization and pharmacological characterization of high affinity binding sites for vasopressin and oxytocin in the rat brain by light microscopic autoradiography. Brain Res 442:105–118

Vijande M, Marin B, Brime J, Lopez-Sela P, Bernando R, Diaz F, Costales M (1989) Water drinking induced by insulin in humans. Appetite 12:243

Weindl A (1983) The blood–brain barrier and its role in the control of circulating hormone effects on the brain. In: Ganten D, Pfaff D (eds) Current topics in neuroendocrinology. Central cardiovascular control. Springer-Verlag, Berlin, Heidelberg, pp 151–186

Weisinger RS, Coghlan JP, Denton DA et al. (1980) ACTH-elicited sodium appetite in sheep. Am J Physiol 239:E45–E50.

Whitaker MD, McArthur RG, Corenblum B, Davidman M, Haslam RH (1979) Idiopathic, sustained, inappropriate secretion of ADH with associated hypertension and thirst. Am J Med 67:511–515

Wilson KM, Sumners C, Hathaway S, Fregly M (1986) Mineralocorticoids modulate central angiotensin II receptors in rats. Brain Res 382:87–96

Chapter 8

Mineral Appetite: An Overview

D. Denton

Overview of Mineral Appetite

It is now firmly established that large areas of the mountains and the interior of continents, have very low sodium concentrations in soil and plants, and sodium deficiency of animals does occur (Blair-West et al. 1968; Denton 1982). In the absence of geological sources, rain water is the source of sodium, and the sodium content declines with distance from sea coast and the marine aerosols. Similarly there are large areas of continents such as the veldt of Africa, southern Texas, and areas of Australia where phosphorus content of vegetation is very low, and physiological function in animals may be grossly impaired. As a result of a variety of circumstances, particularly attendant on lactation, calcium deficiency may develop and, similarly, magnesium deficiency may occur acutely. Given the high potassium content of cellular material, whether plant or animal in origin, the occurrence of potassium deficit in nature is probably unusual except perhaps as a result of bacterial infection of the gut, most likely in gregarious species, where copious diarrhoea and disturbance of extracellular chemistry can cause serious depletion of intracellular potassium, a phenomenon first fully documented in man in the US.

It follows from the above that there are circumstances in nature where powerful selection pressure would have favoured the emergence during evolution of brain mechanisms determining selective choice and intake of foods, licks or fluids with high content of minerals of which the animals were likely to become deficient or to be in borderline status. In this phylogenetic context, and given the crucial importance of these minerals in the functions of the circulatory milieu and in intracellular mechanisms, it is evident that there has been great survival advantage in the 'prewired' or innate systems which have developed in the brain to subserve the seeking, recognition, and quantitatively appropriate intake of specific minerals. These innate behaviour patterns, the dissection of which will be addressed with reference to sodium appetite, parallel other vegetative systems of high survival importance which are correspondingly subserved by integrated activity of hypothalamic, limbic, cortical and hormonal systems. These are, for example, hunger, thirst, temperature regulation, sexual and maternal behaviour, together with a gamut of species-specific territorial behaviours, often hormonally driven, and linked to certain of these innate vegetative behaviours.

Before proceeding to sodium, and for the purpose of showing the limitations of present knowledge and also something of the diversity of organization of these mineral appetites, the matter of phosphate appetite can be considered. This is seen in the wild and also in the pastoral industry. It occurs in large-framed animals, for example, giraffes, cattle, reindeer and deer. The classic behavioural manifestation is bone chewing, which with bird faeces, is about the only accidentally available source of phosphate in nature. Phosphate-deficient animals will chew bones avidly, and in extreme conditions, cattle will even attack rotting carcasses and may die of botulism. The work by Theiler et al. (1924) in South Africa established the relation of incidence of the behaviour to blood phosphate level. A detailed physiological analysis of the behaviour has been in progress at the Howard Florey Institute using cattle with parotid fistulae. They are fed a low phosphate diet but with a sodium supplement to cover the sodium, which together with phosphate, is lost in the saliva. The emergence of the bone appetite in the naive animal occurred as blood phosphate fell below 1 mmol/l (Denton 1982).

The bone appetite is eliminated by a rapid intravenous infusion of buffered sodium phosphate which raises blood phosphate to 2–3 mmol/l, but with a latent period of 45–60 min before the behavioural change. Intraventricular infusion of phosphate raising CSF concentration 2–8-fold is without effect on phosphate appetite. At present, the nature and whereabouts of the sensing system is unknown. This is also true of the cue which takes the phosphate-deficient but not the phosphate-replete animal to bone and initiates the chewing. Parenthetically, the cigar like chewing of long bones by deer in northern Europe and other areas can fashion a long bone into a remarkable pronged fashion which, as Sutcliffe (1977) of the British Museum has pointed out, led some archaeologists in Crete and Okinawa to postulate erroneously that such bones are artifacts produced by an osteokeratic civilization. Cafeteria experiments where trays of various salts – calcium phosphate, sodium phosphate, calcium sulphate, calcium carbonate, chopped new bone and old bone – have been offered to naive phosphate-deficient animals showed that the animals would proceed within a minute or two to identify and ingest the old bone. Incineration of the old bone at 500°C in an oven destroys the attractant property and a substantial series of differential extraction experiments on old bones indicates that the essential element is in the medullary fraction. Blood, fresh or old, and a large range of fatty acids are not effective and at this point the cueing factor(s) remain to be identified.

With calcium appetite, the experiments of Richter (Richter and Eckert 1937; Richter and Helfrick 1943) established both the augmented calcium appetite and phosphate aversion which followed parathyroidectomy in rats, and also the increased appetite for calcium and phosphorus seen in the lactation phase of the reproductive process. In wild rabbits, Denton and Nelson (1971) have shown an increase of calcium appetite in pregnancy and lactation. The work of Rodgers (1967) using a calcium-deficient diet and examining the evocation of appetite suggested the essential learned nature of calcium appetite, based on a retroactive sense of benefit. There was an analogy to the data of specific vitamin appetite of rats depleted of particular components of the B complex. However, it is possible to elicit an increased calcium intake by administration in appropriate sequence of a quintet of steroid and peptide hormones mimicking the secretion in pregnancy and lactation (Denton and Nelson 1978). The appetite reverts to basal by 48 hours after cessation of administration and thus would appear to suggest, as with Richter's data, the likelihood of innate physiological mechanisms activated by chemical and hormonal change of the milieu interieur similar to the appetite for sodium and bone. However, as with all such behaviour, learning components may be major elements in the final behaviour manifest.

Species Differences

With sodium appetite, it is clear that the main impact of environmental deficiency will fall on herbivores and vegans ingesting the very low sodium vegetation. Carnivores with the high sodium content of flesh and viscera are largely emancipated from the problem whatever the status of the environment, and the situation of omnivores may be intermediate. In this regard, in the Snowy Mountains of Australia where introduced and native herbivores are severely sodium deficient in early spring and summer it has been found that the foxes which are omnivores, may also be severely sodium depleted and have enlarged zona glomerulosa, high blood aldosterone, and extremely low urinary sodium content (Denton 1982).

As the number of animal species submitted to physiological analysis has been increased, it has emerged that there are large species differences in the organization of sodium appetite. To some extent, differences can be related to the differing impact of low sodium environment according to the eating behaviour of the species. This could well underlie the fact that a change of sodium concentration of cerebrospinal fluid (CSF) and brain extracellular fluid (ECF) is a direct major stimulus of sodium hunger in sheep and cattle and perhaps, also, in most of the grazing herbivores and ruminant species inhabiting the planet. These creatures bear the direct impact of a sodium-deficient environment which in the first instance may cause decrease in plasma and consequently CSF sodium concentration. Such a change in brain ECF sodium concentration does not appear to have a direct influence on sodium appetite in the rat, an omnivore, where the hormonal changes secondary to the sodium-deficient status are likely to be determinant. Further, in mice and wild rabbits reduction of CSF sodium concentration does not evoke sodium appetite in the sodium-replete animal.

Another consideration of selection pressure arises in the case of taste. There are specific or "salt best" taste fibres subserving salt taste. It is well attested under cafeteria conditions where the naive salt-deficient animal is offered, for example, sodium, potassium, calcium and magnesium chlorides that it will make a specific choice of sodium. However, if offered lithium chloride solution, for example, sheep or rats cannot distinguish it from sodium and may ingest it freely causing severe sickness, anorexia or death. Lithium salts do not occur freely in nature in lick form as do sodium. Thus, no selection pressure has existed to favour evolution of capacity to distinguish between the two ions.

A further general biological consideration in relation to sodium appetite lies in the severe demands the reproductive process places on mammals in a sodium-deficient environment. Sequestration of sodium in the tissues and fluids of the fetus, the uterine fluids and that consequent on the hypervolaemia of the dam followed by the loss of sodium in lactation can cause severe stress on sodium homeostasis. This will clearly be greatest in those species which have large litters and frequent pregnancies of short duration with rapid increase of body weight during gestation, in contrast to long-duration pregnancy with a single offspring. A spectacular sodium appetite is seen in pregnancy in mice (McBurnie et al. 1988), wild rabbits (Denton and Nelson 1971) and in rats (Richter 1956), whereas increased intake in pregnant and lactating sheep with a single lamb is small (Denton 1982).

Sodium Appetite is Innate

Independently of carefully controlled laboratory experiments which directly address the question, there is a large body of natural history and observation of behaviour of animals in the wild which points to salt hunger being innate. This in recounted in detail in Denton (1982), but attention is drawn here to illustrative instances. Wild rabbits in the Snowy Mountains of Australia become severely sodium deficient in spring when reproduction begins and grass sodium content is often only 1 mmol/kg dry weight. On the first occasion when sticks heavily impregnated with sodium, potassium, calcium, and magnesium salts are placed out on warrens, the sodium sticks may be gnawed away completely overnight without much attention to the other salts. Similarly wild rats (*Rattus fuscipes*) were trapped in the Snowy Mountains. When introduced into "drinkometer" intake metering cages with solutions of various chloride salts available, all solutions were usually tasted at the outset. By 30 minutes the rats had drunk large amounts of sodium solution and little of the other cations. As all are usually tasted, it follows that short-term sense of retroactive benefit, recognized as arising from one solution alone (i.e. sodium), is very unlikely to account for the large sodium intake in the first hour. Wild kangaroos in these mountains will selectively demolish filter paper pulp blocks impregnated with sodium salts without great interest in blocks impregnated with other salts. The moose of the Isle Royale in Canada do not have enough sodium available in browse food consumed on the island to account for sequestration involved in the annual increase of animal mass including reproduction. During summer, however, they graze knee deep in water and selectively ingest the submerged sodium accumulator plants (44–408 mmol sodium/kg) growing round the edge of the island (Jordan et al. 1973).

The initial observations on augmentation of appetite with adrenalectomy in rats (Richter 1956), followed by the data on the relation of extent of deficit and sodium intake (Epstein and Stellar 1955) pointed to sodium appetite being innate and not learned. Handal (1965) and Wolf (1969) systematically showed in naive rats with first experience of sodium deficiency that specific intake of an otherwise aversive concentration of sodium salt occurred immediately. Other experiments (Kriekhaus and Wolf 1968) showed how rats which had learned previously when without any prior experience of sodium deficiency, that lever pressing would deliver salt solution, continued to press the lever for salt when first made deficient though no salt was delivered.

Naive sheep of known history had permanent unilateral parotid fistulae made through which they lost 0.75–2.0 l/day (120–300 mmol of sodium) during the first postoperative week. During the control period the week preceding operation, a cafeteria of solutions of 300 mmol/l of sodium bicarbonate, sodium chloride, calcium chloride, potassium chloride and magnesium chloride were offered for either one hour per day at a regular time, or not at all. After operation all animals had access for one hour per day and behaviour was recorded in detail. Those animals which tasted solutions before operation chose specifically within 2–3 days to drink an amount of sodium commensurate with deficit and this also occurred within 3–6 days with animals which had no prior experience. The behaviour indicated an innate basis of choice of sodium (Denton 1982) though it was evident, as with many other parallels in the study of instinctive behaviour, that learning greatly facilitates the effectiveness of the behaviour. Indeed, the innate character of the behaviour probably embodies the propensity to learn choices pertinent to increased intake.

It has also been shown with Balb/c mice reared from birth on a diet with low but adequate sodium content (4–7 mmol/kg) that on the initial instance of sodium deficiency caused by frusemide, a selective appetite for sodium is immediately manifest in a cafeteria situation when given the same choice of salts as presented to sheep. This occurs whether or not the mice have had any prior experience of access to sodium salts (Denton et al. 1988).

There has been little work on sodium appetite in non-human primates. Hofmann et al. (1954) showed a poor survival capacity in adrenalectomized rhesus monkeys given isotonic saline to drink. Shulkin et al. (1984) showed rhesus monkeys responded to sodium depletion induced by frusemide by doubling intake of 9% sodium chloride solution. Barnwell et al. (1985) reported that baboons dislike salted food. Rowland and Fregly (1988) in their overview of salt appetite state that the behavioural contribution to sodium homeostasis in primates is unimpressive.

Preliminary experiments carried out in metabolism cages on baboons at the South West Medical Foundation of Texas have shown in naive animals that hedonic intake of 300 mm-sodium chloride is small (5–10 ml/day) (Shade, Eichberg and Denton, unpublished). Administration of frusemide by daily injection for three days which caused negative sodium balance resulted in a significant increase of sodium intake over the three day period. Then with a second occasion of a 3 day regime of frusemide administered a week later, a similar increased intake occurred. However on both occasions intake over the 3 days was less than deficit, and balance was repaired during the post period. On a third occasion, access to sodium solution was withheld during the 3 days of frusemide, and it was then presented. There was immediate intake of sodium solution, as much as half the audited deficit being replaced in 2 hours and intake during 24 h corrected the deficit. Thus the superimposition of a learning process on an innate appetite evoked by an initial experience of sodium deficiency was shown. With the higher apes such knowledge as exists resides in the observations in the wild made by Shaller (1963) on gorillas and by Goodall (1971) on chimpanzees. Interest in salt blocks and natural licks was observed. In man, the evocation of sodium appetite by sodium deficiency as caused by adrenal insufficiency is variable with about 15%–20% of patients presenting with salt craving (Thorn et al. 1942).

The Mechanism of Evocation of Sodium Appetite: General Considerations

Until about 10–15 years ago, experimental analysis of sodium appetite centred largely on two species, namely certain strains of laboratory rat and the sheep. The biological spectrum has been substantially widened since with studies on mice, cattle, wild and laboratory rabbits, hamsters, dogs, primates and further strains of rat, such as Fischer rats. It is now clear that substantial species differences exist in the chemical genesis of sodium appetite and in the characteristics of satiation – perhaps rather more than with thirst. There is an hedonic component of appetite, manifest in different degrees between species, and such elective intake, free of any metabolic need, has been postulated to have survival advantage (Bare 1949). Though sodium cannot be stored in the body to anticipate future deficit, the simple stimulus–response reaction to the taste may keep some measure of positive status in an environment of scarcity and also establishes a liking which will be an advantage when actual deficiency occurs. It will facilitate recall of location of sources of sodium, and the rapid repair of deficit. This behaviour, perhaps most evident in the laboratory rat as exemplified by the preference aversion curve for sodium solution (Richter 1956), reflects a pleasurable taste analogous to sweet, and in contrast to bitter. However, it is not entirely autonomous and independent of the body sodium status. It has been shown that intracerebroventricular (i.c.v.) infusion of 500 mmol sodium chloride artificial CSF beginning one hour before access to bar-press will reduce the need-free sodium intake of sodium-replete sheep by 50% (Weisinger et al. 1986b). The vector of influence of this change on the hedonic mechanisms is unknown. On the other hand, in mice, change from a high sodium diet (126 mmol sodium/kg) to a low sodium diet (7 mmol sodium/kg) did not change daily drinking of 200 mmol sodium chloride solution (Denton et al. 1988).

As well as the hedonic behaviour, it is clear that changes in the sodium concentration of brain ECF, and the actions of the hormonal secretions stimulated by sodium deficit will stimulate appetite. Furthermore, hormones of pregnancy and lactation and those evoked by stress, as well several neuropeptides may generate, or inhibit, appetite. The neurons and neuronal circuits which subserve innate sodium appetite are responsive to a number of chemical and hormonal influences. Correspondingly the neurochemical processes involved may be very complex.

Experiments to Determine the Effect of Changes in Sodium Concentration of Brain ECF and Intracellular Fluid on Sodium Appetite and Thirst

In sheep made deficient of sodium by salivary loss from a parotid fistula, i.c.v. infusion of 500 mmol sodium artificial CSF at 1 ml/h which raised CSF sodium concentration by 10–15 mmol/l reduced sodium intake by 75%. In the control experiment where the osmotic pressure of CSF was increased to a comparable degree by infusion of 0.7 M-mannitol artificial CSF, it was found that sodium appetite approximately doubled. The infusion reduced sodium concentration of CSF by 10–20 mmol/l (Weisinger et al. 1982). Appetite increase could be produced in the sodium-replete animal by the same infusion. Infusion of isotonic artificial CSF had no effect. It was also found that this effect occurred only with reduction of CSF sodium by those saccharides, e.g. 1-glucose, mannitol, sucrose, which did not enter cells or cross the blood–brain barrier. This indicated that the neuronal elements subserving sodium appetite and responsive to change in sodium concentration were deep within the neuropil. It required a front of fluid of reduced sodium concentration to move through the ependyma and penetrate the tortuous channels. If sugars such as D–glucose or 2-deoxyglucose were used this did not happen (Weisinger et al. 1985).

The result contrasted with effect on thirst of lowering CSF sodium. Thirst was engendered by brief intracarotid infusion of 4M-sodium chloride. Here, reduction of CSF sodium by any of several saccharides, independently of their propensity to enter cells or cross the blood–brain barrier, reduced water drinking (Park et al. 1989). This result is consistent with the proposal that sodium sensors which influence thirst are juxta ventricular (Olsson 1973; Andersson 1978).

Chiaraviglio and Lozada (1986) report that vanadate, an inhibitor of sodium, potassium adenosine triphosphatase activity, will reduce sodium appetite of sodium-deplete rats when infused into the CSF.

Essentially similar findings with regard to sodium appetite have been made in cattle (Blair-West et al. 1987). However, in wild rabbits (Denton et al. 1985) similar reduction of CSF sodium concentration over 2 days did not increase sodium appetite though the reduction of CSF sodium caused water diuresis. In rats both Epstein et al. (1984) and the Florey group (Denton et al. 1984; Osborne et al. 1987) showed no effect of change of CSF sodium on sodium appetite, and investigation of mice indicates reduction of CSF sodium of the sodium-replete mouse is without effect (Osborne et al. in press). The CSF infusion data in sheep and cattle are compatible with an early hypothesis (Denton and Sabine 1961; Denton 1966) that the specific stimulus to sodium appetite was reduction in intracellular sodium concentration of neurons subserving sodium appetite. As a corollary to this postulate, it was also proposed in the light of the basic discoveries of Jacob and Monod (1961), that the ionic changes may initiate transcriptional events leading to increased synthesis of transmitters and other elements modifying membrane characteristics and this was the basis of increased electrical excitation. In this regard it has been shown in sheep that intracarotid infusion of ouabain in the sodium-deplete animal aimed to contrive a concentration of 10^{-6}M in cerebral blood will cause a large reduction of sodium appetite but not water intake of the water-deplete animal whereas intravenous infusion at the same rate is without effect on sodium appetite (Denton et al. 1969). In a more refined approach, with a push–pull infusion of 10^{-5}M ouabain in anterodorsal and antero-ventral third ventricle, Tarjan et al. (1986) have shown that significant inhibition of sodium appetite occurs.

Conversely, it is known that in epithelial tissues phlorizin will inhibit sodium-coupled glucose transport. Phlorizin is a glycoside present in the barks of several species of Rosaceae and appears to act as a competitive antagonist at the receptor to prevent glucose entry to cells. Glick and Mayer (1968) showed that intracerebral infusion of phlorizin increased food appetite which suggested that it may act to decrease intracellular glucose in at least some neural tissues. It has been shown in sheep that intraventricular (i.v.t.) infusion of 2.3 mM-phlorizin will cause a 50%–100% increase in sodium appetite (Weisinger et al. 1985), a finding consistent with the postulate of reduced intracellular sodium concentration generating sodium appetite. An important concomitant finding in this regard was that phlorizin infusion, when coupled with 500 mmol sodium chloride artificial CSF which increased brain ECF sodium concentration, blocked the characteristic effect of high brain ECF sodium in reducing sodium appetite but did not influence the generation of thirst caused by high brain ECF sodium. It is proposed that phlorizin eliminated the decrease of sodium appetite characteristic with i.v.t. infusion of hypertonic sodium chloride by reducing glucose and thus sodium movement into cells. This prevented an increase in ICF sodium concentration which normally follows increased brain ECF sodium caused by hypertonic sodium chloride CSF infusion.

Thus the results obtained with both phlorizin and ouabain are consistent with the possibility that stimulation or inhibition of sodium appetite depends on changes occurring in the ICF sodium concentration of sodium sensor cells subserving sodium appetite. There is evidence to suggest that unlike D-glucose, phlorizin does not enter cells (Stirling 1967) and crosses the blood–brain barrier only slowly if at all (Brondsted 1970) and therefore it would penetrate deep into the neuropil.

Renin–Angiotensin System in the Brain

The weight of experimental evidence in the rat, and also accumulating in the mouse, suggests that intracerebral angiotensin is a major factor in the physiological control of sodium appetite. In sheep and cattle there is also evidence of involvement of the renin–angiotensin system in sodium appetite but the data are of a different character which may indicate a different quantitative contribution as well as different anatomical organization underlying the contribution of the circumventricular organs to physiological control.

Unger et al. (1986) summarized accumulated data on the presence of the components of the renin–angiotensin systems in the brain. Angiotensin (Ang) I, II and III have been extracted from brain of nephrectomized rats, rabbits and primates using HPLC with comparison to synthetic fragments. The amino acid sequence appeared to be the same as with plasma angiotensin, and, further to this, renin cleaved the same peptide from both brain and plasma angiotensinogen (Ganten et al. 1983). With immunocytochemical techniques a main location of Ang II is the hypothalamus in the median eminence, the nucleus paraventricularis, the supraoptic nucleus and the subfornical organ (SFO) (Fuxe et al. 1976). Within the paraventricular nucleus neurosecretory neurons which synthesize arginine vasopressin (AVP) in the magnocellular portion were stained with antisera to Ang II as were neurons throughout the parvocellular division (Lind et al. 1985).

Converting enzyme (CE) is widely distributed throughout the brain and has been localized by microdissection techniques, immunohistochemical studies and by autoradiographic visualization of CE with [^{3}H]captopril. The highest concentration occurs in the choroid plexus, subfornical organ, caudate nucleus, putamen, zona reticularis, substantia nigra, globus pallidus and the median eminence, and the cell culture evidence is that the enzyme is located intraneuronally as well as in vessels.

Campbell et al. (1984) using cell free translation of mRNA showed the same angiotensinogen molecule is synthesized in liver and brain. Dexamethasone administration and nephrectomy increased liver translation much more than brain indicating that although the two angiotensinogens appear to be the product of the same gene(s) the regulation of transcription is tissue specific. In contrast to immunocytochemical localization of Ang II in elements with neuronal morphology, Deschepper et al. (1986) localized angiotensinogen to astrocytes. Using hybridization histochemistry with several probes Lynch et al. (1987) determined the topographical distribution of angiotensinogen in the brain. Low levels of activity were detected throughout the brain while high levels were restricted to specific areas often corresponding to classical nuclear boundaries. Areas in which Ang II-like activity had been previously detected contained moderate to high immunocytochemical levels of angiotensinogen mRNA but some areas not identified on this basis were located. They note that the SFO was lightly labelled despite evidence for Ang II immunoreactive cells there, and its proposed importance in physiological reactions involving Ang II. With their methods where mRNA was dense they were not able to ascertain whether grain clusters originated from neurons or other cells. It is also interesting to note that there are areas of the brain where Ang II is shown to occur immunocytochemically but no CE is detected such as the bed nucleus of the stria terminalis and central nucleus of the amygdala. Mendelsohn et al. (1988) demonstrated high concentrations of Ang II receptors in specific areas of the brain in rat, sheep rabbit and man and, with regard to the hypothalamus, the anterior wall of the third ventricle including the circumventricular organs is particularly rich. There is binding also in the paraventricular, supraoptic and arcuate nuclei. Felix et al. (1986) using iontophoretic techniques have identified cells in the SFO which are specifically excitable by application of Ang II and do not respond to acetylcholine or substance P.

Schelling et al. (1980) stated that renin was immeasurable in CSF of rats and dogs. Unger et al. (1986) state renin isolated from brain will generate Ang I from angiotensinogen at neutral pH in contrast to acid proteases and its action is inhibited by specific renin antibodies. They state that ultimate proof of local synthesis of renin in the CNS will come from recombinant cDNA techniques as with angiotensinogen. They give evidence for demonstration of renin mRNA in mouse brain. At this stage localization of synthesis of renin in discrete areas of the brain by use of hybridization histochemistry techniques has not been forthcoming.

In rats it is now clearly established that i.c.v. infusion of Ang II will generate sodium appetite and that this may occur in the absence of preceding natriuresis (Bryant et al. 1980; Avrith and Fitzsimons 1980; Epstein 1986; Fitzsimons 1986). Furthermore, a striking synergy has been demonstrated between Ang II and mineralocorticoid (either DOCA or aldosterone). In combination there was a clear appetite-generating effect at a dosage which in either case alone was ineffective, and this was associated with positive sodium balance (Epstein 1986). Further, in the sodium-deficient animal Sakai et al. (1986) have shown that i.c.v. infusion of RU 28318, a potent aldosterone antagonist, greatly reduced sodium appetite. This synergistic effect is also reported to occur in the pigeon (Epstein and Massi 1987).

A further facet of the above data is the enhancement of salt appetite which follows on the first experience of sodium deficiency (Sakai et al. 1987). This involves an approximate doubling of intake on all subsequent episodes relative to the first experience, and is not dependent on whether or not the rat had access to salt on the first episode. However the phenomenon is not seen following the first episode if at that time the actions of the two hormones are blocked by i.c.v. RU 28318 and adequate dosage of captopril. The lifelong increase in avidity for sodium so produced appears to reflect a permanent change in the organization of the several mechanisms subserving sodium appetite in this species, and presumably is a component in the conspicuous overdrinking relative to deficit exhibited in rats in contrast to other species. The effect is not seen in mice or sheep.

The administration of low dose of captopril in the rat has been known for a long time to have the paradoxical effect of increasing both water and sodium intake. This increased sodium appetite is inhibited by i.c.v. administration of the angiotensin II antagonist saralasin, just as this peptide will block the increased sodium appetite caused by sodium depletion (Buggy and Jonklaas 1984). Intracerebroventricular infusion of captopril also greatly reduced sodium appetite (Weiss et al. 1986) and the combination of captropril with i.c.v. RU 28318 abolished salt appetite of sodium deficiency (Sakai et al. 1986). With the effect of low dose captopril enhancing sodium appetite, the present hypothesis is that systemic low doses block conversion of Ang I into Ang II peripherally, the absence of Ang II feedback enhances renin production and there is selective penetration of Ang I which is at high concentration in areas of the brain which converting enzyme I does not reach. This Ang I is converted there into a surfeit of Ang II which entrains sodium appetite. As Rowland and Fregly (1988) point out, whereas converting enzyme I infused intracerebroventricularly will inhibit the appetite produced by peripheral low dose captopril, the differential permeability has yet to be demonstrated. They note that there is no evidence that circulating Ang I or Ang II will access areas of the brain other than the circumventricular organs. This would seem a pertinent observation on limitation of present knowledge. However, one puzzle which might be added to the eventual analysis is the finding in sheep that intracarotid infusion of Ang II to give a high concentration is a powerful stimulus of thirst and that this effect is largely eliminated by beginning i.c.v. infusion of saralasin about 10 minutes beforehand. Conversely i.c.v. infusion of Ang II is a powerful immediate stimulus of thirst, but the effect is abolished by beginning intracarotid infusion of saralasin about 10 minutes beforehand. Given that saralasin is essentially a competitive antagonist it points to the fact that the octapeptides reach the same site whichever side of the blood–brain barrier they are delivered, though, as in other species, simple infusion of Ang II at high rates does not cause a rise of CSF Ang II (Abraham et al. 1976; Denton 1982).

The classic robust salt appetite of the adrenalectomized rat has also to be considered in relation to formal demonstration of the validity of the hypothesis that the synergism of angiotensin and aldosterone is the sole determinant of salt appetite of sodium deficiency in the rat.

The presumption is that excess generation or excess presence of Ang II presumably compensates for the absence of aldosterone. That is the i.c.v. infusion of Ang II supposedly mimes a situation where Ang II is generated in the brain in sodium deficiency, and thus in the adrenalectomized rat this endogenous intracerebral process will be greatly amplified. In the experiments on which the hypothesis is based (Epstein, 1986) only the highest dose of Ang II alone (100 pg/min) aroused sodium appetite compared with 10 pg/min which had no effect. Perhaps, therefore, a big difference would be expected to occur in brain Ang II content between the adrenally intact and the adrenalectomized sodium-deficient rat for equivalent sodium deficit. The point is that ultimately it will require measurement in vivo of the changes which the experimental infusions are presumed to mimic. Sodium deficiency could, of course, alter sensitivity of the systems to action of hormonal agents but this may be equal in the two situations for comparable sodium deficit. However, such sensitivity change could, perhaps, be caused by change of ionic composition of neurons representing another component in the genesis of appetite. Furthermore, it is interesting that Epstein (1986) reports that sodium appetite is suppressed with lower doses of captopril in the adrenalectomized sodium-deficient animal than in the adrenally intact sodium-deficient animal. Also Sakai et al. (1987) report, in relation to pretreatment of rats with angiotensin and aldosterone, a procedure which induced the permanent augmentation of sodium drinking with sodium deficiency, that either hormone alone is insufficient.

Clearly, quantitative analysis of brain Ang II in the sodium-deficient adrenal-ectomized rat relative to the sodium-deficient adrenal intact animal will be of great interest. The study by McEwan et al. (1986) on binding of aldosterone in the brain and the overlap with location of Ang II receptors and CE activity is generally consistent with the synergy hypothesis. The question remains whether sodium deficiency initiates a classical renin–angiotensin cascade in localized areas of the brain which generates the behaviour, or whether angiotensin is acting as a neurotransmitter or neuromodulator and is not necessarily formed by intraneuronal action of renin. The fact that captopril does not reduce the sodium appetite generated by deoxycorticosterone (DOC) (Weiss et al. 1986) possibly argues against angio-tensin acting as the major neurotransmitter in the circuits subserving sodium appetite.

In sheep there are considerable differences in the data, though Ang II would appear to have an important involvement. With intravenous (24 µg/h) or intraventricular (3.8 µg/h) infusion of Ang II, external sodium balance studies, coupled with experiments in which return of urine into the rumen was devised, showed that natriuresis caused by Ang II infusion was the predominant factor inducing sodium appetite (Weisinger et al. 1986a). There is some reservation about this in that short-term (3 h) intracerebroventricular infusion of Ang II may cause a small increase of sodium intake, an effect clearly anteceding any appetite caused by negative sodium balance. In the sodium-deficient sheep, whereas i.c.v. Ang II caused a large increase in water intake, no change in sodium intake resulted consistent with the fact that no natriuresis occurred in the sodium-deficient animal. Furthermore whereas the influence of reduction of brain ECF sodium concentration and phlorizin were additive in increasing sodium appetite, i.c.v. administration of angiotensin negated the influence of reduction of brain ECF sodium in increasing sodium appetite (Weisinger et al. 1987a). The explanation of this is not evident. Captopril, however, did not reduce the appetite-stimulating effect of reduction of brain ECF sodium concentration (Weisinger 1987b). Again this might argue against Ang II having a major neurotransmitter role in neuronal circuits subserving salt appetite.

The combination of infusion of Ang II with aldosterone at high physiological rate in both replete and deplete sheep had no influence on sodium intake over 2–4 days. However, systemic captopril infusion halved sodium appetite in the sodium-deplete animal, an effect which was reversed by intravenous infusion of Ang II at a very low physiological rate (3.8 μg/h). i.c.v. infusion of Ang II at 3.8 μg/h did not restore the appetite (Weisinger et al. 1987b). The same result was seen in sodium-deplete cattle in as far as systemic angiotensin infusion restored sodium appetite. Cattle differed from sheep in that in the sodium-deficient animal intravenous infusion of Ang II stimulated sodium appetite (Blair-West et al. 1988). In mice systemic Ang II infusion does not increase sodium appetite but i.c.v. infusion is a very powerful stimulus of both salt appetite and thirst. In the rabbit, i.c.v. Ang II infusion causes sodium appetite after 3–4 days, but the effect is preceded by natriuresis and this may be the sole cause (Tarjan et al. 1988a).

It would appear that angiotensin plays a dominant role in sodium appetite in some species with variation as to whether the locale is predominantly intracerebral (rat) or probably systemic via circumventricular organs (sheep and cattle). At present the evidence in rabbits is against either brain ECF sodium concentration or Ang II playing a major role. Our present knowledge would seem to fall far short of a coherent picture of a functioning organization in any species.

Richter (1956) first described the large effect of pregnancy and lactation on sodium appetite. A series of investigations (Denton 1982) showed that this large behavioural effect, involving turnover of total extracellular sodium content daily, could be reproduced by administration of progesterone and oestradiol followed by the peptides implicated in lactogenesis – ACTH, prolactin and oxytocin. As well as these hormonal effects, a large body of data in wild rabbits, mice, rats and sheep has shown that the adrenal steroid hormones evoked by ACTH release and stress also have a large salt appetite-stimulating effect.

With regard to neuropeptides, the possible wide influence of the cascade evoked by stress is underlined by the finding (Tarjan and Denton in press) that 24 h i.c.v. corticotropin releasing factor has a significant salt appetite-stimulating effect in the rabbit which is sustained for several days after cessation of infusion. Fitts et al. (1985) report that i.c.v. ANF reduces salt appetite in rats, and Tarjan et al. (1988b) found sodium appetite in response to sodium deficiency, and also water and food intake were reduced by i.c.v. ANF. De Caro (1986) and Tarjan et al. (1990) report that i.c.v. eledoisin reduced sodium intake in several species. Stricker and Verbalis (1987) have suggested that a central inhibitory control of sodium appetite is correlated with stimulation of oxytocin secretion as a result of activity in paraventricular and supraoptic nuclei.

Brain lesions have revealed some facts of great interest in the organization of salt appetite. In sheep (McKinley et al. 1986) destruction of the anterior wall of the third ventricle may render the animal temporarily or permanently adipsic, yet intake of hypertonic sodium bicarbonate solution in response to sodium deficit is unimpaired. The salt appetite induced by i.c.v. Ang II infusion in the sodium-replete animal is lost or impaired.

The data indicate a clear anatomical and functional separation of the mechanisms of thirst and sodium appetite. Discrete lesions of the ventral subcommisural portion of the nucleus medianus augment hedonic sodium intake in the rat but had no effect on depletion-induced intake (Gardiner et al. 1986). Nitabach et al. (1989) have shown that lesion of the medial amygdala abolished sodium appetite evoked by mineralocorticoids in rats, but did not affect sodium appetite in response to sodium depletion. Hedonic intake was greatly reduced. Weisinger et al. (in press) have shown that SFO lesion in the rat caused a large reduction of sodium intake of sodium-deplete rats, an effect reversed by low captopril dosage. However, Shulkin et al. (1983) did not find that SFO disconnection reduced sodium appetite. These and other data testify to a separation of the thirst and salt appetite function, and also show anatomic or functional separation of elements of the salt appetite system which subserve respectively hedonic, response to sodium deficiency, and mineralocorticoid-induced behaviours.

The overall picture of sodium appetite is that of a very complex system. One initial conjecture could be of the neuronal elements subserving the appetite being pluripotential, and, as such, responsive to many hormonal and some ionic environmental changes. This is analogous to the fashion a breast epithelial cell responds to many hormones and in appropriate sequence (Denton 1982). However, the dysjunction of effects of captopril on sodium appetite on the one hand, and not on mineralocorticoid-induced appetite on the other, would seem to speak against a common role of angiotensin II as a neurotransmitter or prime mover in all cells of the system. The different effect of lesions of the amygdala on mineralocorticoid-induced but not sodium deficit-induced appetite point to spatial separation of elements. Whereas significant neuronal elements of the system may be pluripotential in response characteristics, the system may embody functional integration of anatomically discrete populations with different reactivities which contrive the overall behaviour of the organism.

References

Abraham SF, Denton DA, McKinley MJ, Weisinger RS (1976) Effect of an angiotensin antagonist rat 1-ala 8-angiotensin II on physiological thirst. Pharmacol Biochem Behav 4:243

Andersson B (1978) Regulation of water intake. Physiol Rev 58:502–603

Avrith DB and Fitzsimons JT (1980) Increased sodium appetite in the rat induced by intracranial administration of components of the renin–angiotensin system. J Physiol (Lond) 310:349–364

Bare JK (1949) The specific hunger for sodium chloride in normal adrenalectomized white rats. J Comp Physiol Psychol 42:242

Barnwell GM, Dollahite J, Mitchell DS (1985) Salt taste preference in baboons. Physiol Behav 37: 279–284

Blair-West JR, Coghlan JP, Denton DA et al. (1968) Physiological, morphological and behavioural adaptation to a sodium deficient environment by wild native Australian and introduced species of animals. Nature 217:922–925

Blair-West JR, Denton DA, Gellately DR, McKinley MJ, Nelson JF, Weisinger RS (1987) Changes in sodium appetite in cattle induced by changes in CSF sodium concentration and osmolality. Physiol Behav 39:465–469

Blair-West JR, Denton DA, McKinley MJ, Weisinger RS (1988) Angiotensin - related sodium appetite and thirst in cattle. Am J Physiol 255:R205–R211

Brondsted H (1970) Cerebrospinal fluid glucose and phlorizin. Acta Neurol Scand 46:637–641

Bryant RW, Epstein AN, Fitzsimons JT, Fluharty SJ (1980) Arousal of specific and persistent sodium appetite in the rat with continuous intracerebroventricular infusion of angiotensin II. J Physiol (Lond) 301:365–382

Buggy J, Jonklaas J (1984) Sodium appetite decreased by central angiotensin blockade. Physiol Behav 32:737–742

Campbell DJ, Bouhnik J, Menard J, Corvol P (1984) Identity of angiotensinogen precursors of rat brain and liver. Nature 308:206–208

Chiaraviglio E, Lozada C (1986) Effect of intracerebroventricular vanadate administration on salt and water intake and excretion in the rat. Pharmacol Biochem Behav 24:1503–1508

de Caro G (1986) Effects of peptides of the "gut-brain-skin triangle" on drinking behaviour of rats and birds. In: de Caro G, Epstein AN, Massi M (eds) The physiology of thirst and sodium appetite. Plenum Press, New York, London, p 213

Denton DA (1966) Some theoretical considerations in relation to innate appetite for salt. Conditional Reflex 1:144

Denton DA (1982) The hunger for salt. Springer-Verlag, Berlin, Heidelberg

Denton DA, Nelson JF (1971) Effect of pregnancy and lactation on the mineral appetites of wild rabbits (*Oryctolagus cuniculus* L). Endocrinology 88:31

Denton DA, Nelson JF (1978) The control of salt appetite in wild rabbits during lactation. Endocrinology 103:1880

Denton DA, Sabine JR (1961) The selective appetite for sodium shown by sodium deficient sheep. J Physiol (Lond) 157:97

Denton DA, Kraintz F, Kraintz L (1969) The inhibition of salt appetite of sodium deficient sheep by intracarotid infusion of oubain. Commun Behav Biol 4:183–186

Denton DA, McKinley MJ, Nelson JF et al. (1984) Species differences in the effect of decreased CSF sodium concentration on salt appetite. J Physiol (Paris) 79:499–504

Denton DA, Nelson JF, Tarjan E (1985) Water and salt intake of wild rabbits (*Oryctolagus cuniculus*) following dipsogenic stimuli. J Physiol (Lond) 362:285–301

Denton DA, McBurnie M, Ong F, Osborne P, Tarjan E (1988) Sodium deficiency and other physiological influences on voluntary sodium intake of BALB/c mice. Am J Physiol 255:R1025–R1034

Deschepper GF, Bouhnik J, Ganong WF (1986) Co-localization of angiotensinogen and glial fibrillary acidic protein in astrocytes in rat brain. Brain Res 374:195–198

Epstein AN (1986) Hormonal synergy as the cause of salt appetite. In: de Caro G, Epstein AN, Massi M (eds) The physiology of thirst and sodium appetite. Plenum Press, New York, p 395

Epstein AN, Massi M (1987) Salt appetite in the pigeon in response to pharmacological treatments. J Physiol (Lond) 393:555–568

Epstein AN, Stellar E (1955) The control of salt preference in the adrenalectomized rat. J Comp Physiol Psychol 48:167

Epstein A, Zhang AD, Schultz J, Rosenberg M, Kupsha T, Stellar E (1984) The failure of ventricular sodium to control sodium appetite in the rat. Physiol Behav 32:683–686

Felix D, Gambino MC, Yong Y, Schelling P (1986) Angiotensin sensitive sites in the central nervous system. In: de Caro G, Epstein AN, Massi M (eds) The physiology of thirst and sodium appetite. Plenum Press, London, p135

Fitts DA, Thunhorst RL, Simpson JB (1985) Diuresis and reduction of salt appetite by lateral ventricular infusion of etriopeptin II. Brain Res 348:118–124

Fitzsimons JT (1986) Endogenous angiotensin and sodium appetite. In: de Caro G, Epstein AN, Massi M (eds) The physiology of thirst and sodium appetite. Plenum Press, New York, p 383

Fuxe K, Ganten D, Hockfelt, Bomme P (1976) Immunohystochemical evidence for the existence of angiotensin II containing nerve terminals in the brain and spinal cord of the rat. Neurosci Lett 2: 229–234

Ganten D, Hermann K, Bayer C, Unger TH, Lang RE (1983) Angiotensin synthesis in the brain and increased turnover in hypertensive rats. Science 221:869–871

Gardiner TW, Jolley JR, Vagnucci AH, Stricker EM (1986) Enhanced sodium appetite in rats with lesions centered on the nucleus medianus. Behav Neurosci 100:531–535

Glick Z, Mayer J (1968) Hyperphagia caused by cerebral ventricular infusion of phlorizin. Nature 219:1374

Goodall J, van Lawick (1971) In the shadow of man. Collins, London

Handal PJ (1965) Immediate acceptance of sodium salts by sodium deficient rats. Psychonomic Sci 3:315

Hofmann FG, Knobil E, Greep RO (1954) Effects of saline on the adrenalectomized monkey. Am J Physiol 178:361–366

Jacob F, Monod J (1961) Genetic regulatory mechanisms in the synthesis of proteins. J Mol Biol 3:318

Jordan PA, Botkin DB, Dominiski AS, Lowendorf HS, Belovsky GE (1973) Sodium as a critical nutrient for the moose of Isle Royale. In: Proceedings of the North American Moose Conference Workshop 9:13

Kriekhaus EE, Wolf G (1968) Acquisition of sodium by rats; interaction of innate mechanisms and latent learning. J Comp Physiol Psychol 65:197

Lind RW, Swanson LN, Bruhn TO, Ganten D (1985) The distribution of AII immunoreactive cells and fibres in the paraventriculo-hypophysial system of the rat. Brain Res 338:81–89

Lynch KR, Hawelu-Johnson CL, Guyenemt PG (1987) Localization of brain angiotensinogen mRNA by hybridization histochemistry. Brain Res 388:149–158

McBurnie M, Denton DA, Tarjan E (1988) Influence of pregnancy and lactation on sodium appetite of Balb/c mice. Am J Physiol 255:R1020–R1024

McEwen BS, Lambdin LT, Rainbow TC, De Nicola AF (1986) Aldosterone effects on salt appetite in adrenalectomized rats. Neuroendocrinology 43:38–43

McKinley MJ, Denton DA, Leventer M et al. (1986) Adipsia in sheep caused by cerebral lesions. In: de Caro G, Epstein AN, Massi M (eds) The physiology of thirst and sodium appetite. Plenum Press, New York, London, p 213

Mendelsohn FAO, Allen AM, Clevers J, Denton DA, Tarjan E, McKinley MJ (1988) Localization of angiotensin II receptor binding in rabbit brain by in vitro autoradiography. J Comp Neurol 270: 372–384

Nitabach MN, Shulkin J, Epstein AN (1989) The medial amygdala is part of a mineralocorticoid sensitive circuit controlling NaCl intake in the rat. Behav Brain Res 35:127–134

Olsson K (1973) Further evidence for the importance of CSF sodium concentration in central control of fluid balance. Acta Physiol Scand 88:183–188

Osborne PG, Denton DA, Weisinger RS (1987) Effect of variation of the composition of CSF in the rat upon drinking of water and hypertonic NaCl solution. Behav Neurosci 101:371–377

Osborne PG, Denton DA, McBurnie M, Tarjan E, Weisinger RS (in press) Decreased intracerebral sodium concentration and sodium appetite in Balb/c mice. Am J Physiol

Park R, Denton DA, McKinley MJ, Penington G, Weisinger RS (1989) Intracerebroventricular saccharide infusions inhibit thirst induced by systemic hypertonicity. Brain Research 493:123–128

Richter CP (1956) Salt appetite of mammals. Its dependence on instinct and metabolism. In: L'Instinct dans le comportment des animaux et de l'homme. Masson et cie, Paris, p 577

Richter CP, Eckert JF (1937) Increased calcium appetite of parathyroidectomized rats. Endocrinology 21:50–54

Richter CP, Helfrick S (1943) Decreased phosphorus appetite of parathyroidectomized rats. Endocrinology 33:349–352

Rodgers WL (1967) Specificity of specific hungers. J Comp Physiol Psychol 64:49–58

Rowland NE and Fregly MJ (1988) Sodium appetite; species and strain differences and role of renin–angiotensin–aldosterone system. Appetite 11:143–178

Sakai RR, Nicolaïdis S, Epstein AN (1986) Salt appetite is suppressed by interference with angiotensin II and aldosterone. Am J Physiol 251:R762–R768

Sakai RR, Fine WB, Epstein AN, Frankmann SP (1987) Salt appetite is enhanced by one prior episode of sodium depletion in the rat. Behav Neurosci 101:724–731

Schelling P, Ganten U, Sponer G, Ungar T, Ganten D (1980) Components of the renin angiotensin system in the cerebrospinal fluid of rats and dogs with special consideration of the origin and the fate of angiotensin II. Neuroendocrinology 31:297–308

Schulkin J, Eng R, Miselis RR (1983) The effects of disconnection of the subfornical organ on behavioral and physiological responses to alterations of body sodium. Brain Research 263:351–355

Schulkin J, Leibman D, Ehrman RN, Norton NW, Ternes JW (1984) Salt hunger in the rhesus monkey. Behav Neurosci 98:753–756

Shaller GB (1963) The mountain gorilla – ecology and behaviour. University of Chicago Press, Chicago

Stirling C (1967) High resolution radioautography of phlorizin-^{3}H in rings of hamster intestine. J Cell Biol 35:605–618

Stricker EM and Verbalis JG (1987) Central inhibitory control of sodium appetite in rats: correlation with pituitary oxytocin secretion. Behav Neurosci 101:560–567

Sutcliffe AJ (1977) Further notes on bones and antlers chewed by deer and other ungulates. Deer 4:73–80

Tarjan E, Denton DA (in press) Sodium/water intake of rabbits following administration of hormones of stress. Brain Res Bull

Tarjan E, Cox P, Denton DA, McKinley MJ, Weisinger RS (1986) The effect of local change of CSF sodium concentration in the anterior third ventricle on salt appetite. In: de Caro G, Epstein AN, Massi M (eds) The physiology of thirst and sodium appetite, Plenum Press, New York, London, pp 473–478

Tarjan E, Denton DA, McBurnie MI, Weisinger RS (1988a) Water and sodium intake of wild and New Zealand rabbits following angiotensin. Peptides 9:677–679

Tarjan E, Denton DA, Weisinger RS (1988b) Atrial natriuretic peptide inhibits water and sodium intake in rabbits. Regul Pept 23:63–75

Tarjan E, Blair-West JR, de Caro G et al. (1990) Sodium and water intake of sheep, rabbits and cattle during icv infusion of eledoisin. Pharmacol Biochem Behav 35:823–828

Theiler A, Green HH, du Toit TJ (1924) Phosphorus in the livestock industry. S Afr Dept Agric J 8:460

Thorn GW, Dorrance SS, Day E (1942) Addisons disease; evaluation of synthetic deoxycorticosterone acetate therapy in 158 patients. Ann Intern Med 16:1053

Unger T, Ganten D, Ludwig G, Lang RE (1986) The brain renin angiotensin system: update. In: de Caro G, Epstein AN, Massi M (eds) The physiology of thirst and sodium appetite, Plenum Press, New York, London, p 123

Weisinger RS, Considine P, Denton DA et al. (1982) Role of sodium concentration of the cerebrospinal fluid in the salt appetite of sheep. Am J Physiol 242:R51

Weisinger RS, Denton DA, McKinley MJ, Tarjan E (1985) Cerebrospinal fluid sodium concentration and salt appetite. Brain Res 326:95–105

Weisinger RS, Denton DA, McKinley MJ, Muller AF, Tarjan E (1986a) Angiotensin and sodium appetite of sheep. Am J Physiol 251:R690–R699

Weisinger RS, Denton DA, McKinley MJ, Simpson JB, Tarjan E (1986b) Cerebral sodium sensors and sodium appetite in sheep. In: de Caro G, Epstein AN, Massi M (eds) The physiology of thirst and sodium appetite. Plenum Press, New York, London, pp 485–490

Weisinger RS, Denton DA, McKinley MJ, Osborne PG, Tarjan E (1987a) Decrease of brain extracellular fluid [Na] and its interaction with other factors influencing sodium appetite in sheep. Brain Res 420:135–143

Weisinger RS, Denton DA, de Nicolantonio R, McKinley MJ, Muller AE, Tarjan E (1987b) Role of angiotensin in sodium appetite of sodium deplete sheep. Am J Physiol 253:R482–R488

Weisinger RS, Denton DA, de Nicolantonio R, Hards DK, McKinley MJ, Oldfield BJ, Osborne PG (in press) Subfornical organ lesion decreases sodium appetite in the sodium-depleted rat. Brain Res

Weiss ML, Moe KE, Epstein AN (1986) Interference with central actions of angiotensin II suppresses sodium appetite. Am J Physiol 250:R250–R259

Wolf G (1969) Innate mechanism for regulation of sodium intake. In: Pfaffmann C (ed) Proceedings of the third international symposium on olfaction and taste. Rockefeller University Press, New York, pp 548–553

Commentary

Szczepanska-Sadowska: Thus far, the studies concerning mineral appetite have focused mainly on sodium appetite. Much less attention has been paid to phosphate and calcium appetites although the latter phenomena can be manifested, at least in some species, as shown by spectacular experiments performed in the Howard Florey Institute. Much work remains to be done to explain the mechanism and significance of phosphate and calcium appetites. Specifically, it seems essential to elucidate the role of hormones involved in regulation of phosphate and calcium metabolism (parathyroid hormone (PTH), calcitonin, vitamin D_3 as well as cartilage and bone growth factors such as bone derived growth factor (BDGF), transforming growth factor B (TGFB), ILF1, platelet derived growth factor (PDGF), tumour necrosis factors (TNFs), interleukin 1 (IL1), fibroblast growth factors and others) in initiation and inhibition of phosphate and calcium intakes. The studies aimed at elucidating whether these appetites are also expressed by man should also be encouraged. Apart from anecdotal information there are no systematic studies concerning calcium and phosphate appetite in humans, though deficits in these two minerals occur in several clinical states and demands for them change during development of the organism. It has also been hypothesized that content of calcium and magnesium in water may play a role in regulation of blood pressure.

Regarding sodium intake there is abundant evidence of the regulatory role of CSF composition and hormonal inputs in initiation of this behaviour. In particular, the role of the renin–angiotensin system and mineralocorticoids has been carefully investigated. These studies are in progress and presumably will soon elucidate the nature of the synergism between angiotensin, mineralocorticoids and a decrease of sodium content. Since sodium depletion is associated with a decrease of extracellular fluid it would be pertinent to elucidate whether and how the cardiovascular reflexes contribute to regulation of salt appetite. It would also be interesting to know whether the rate of development of sodium depletion influences expression of sodium appetite, i.e., whether or not time-dependent adaptive changes occur.

Section III

Neural Pathways of Water Deficit Signals

Chapter 9

Central Projections of Osmotic and Hypovolaemic Signals in Homeostatic Thirst

A.K. Johnson and G.L. Edwards

Introduction

It has been more than 50 years since Lashley (1938) first called attention to the fact that drives like hunger and thirst are the result of the integrative action of the central nervous system. That is, a motivated behaviour such as drinking does not result from the activity of a single peripheral drive stimulus but is the central state arising from a summation of multiple neural and neurohumoral signals. As evidenced in many of the chapters of this volume, there has been significant progress since Lashley's influential review in the identification of the concomitants of hydrational status which function as "afferent" signals to the brain and which contribute to the generation of thirst in response to perturbations in body fluid balance and/or distribution. The study of dehydration-induced water intake (i.e., homeostatic drinking or homeostatic thirst) is unique in the investigation of motivated behaviours because there is a reasonable understanding of the nature of many of the adequate stimuli (i.e., changes in body fluid parameters; see Epstein 1973, 1982, for a discussion of the double depletion hypothesis) which are capable of effecting this complex behavioural response.

In addition to notable advances in developing a "psychophysiology" of homeostatic thirst, recent progress has also been made in identifying portions (or at least narrowing the regions) of the body and brain which are likely to house the receptors that are sensitive to hydration-related stimuli (see Chapters 5 and 6). Information about the location and nature of receptors can be combined with new developments in the neuroanatomical elucidation of the "visceral neuraxis" (Miselis 1986). This correlation of functional and morphological information provides the opportunity to develop concepts about the organization and nature of brain pathways involved in thirst-related information processing. This chapter will focus on the flow and handling of information derived from different types of body fluid-related receptors as it enters, progresses through, and ultimately interacts within the CNS.

Mechanisms of Afferent Signalling in the Control of Drinking

The Distribution of Receptors and Nature of Afferent Signalling in the Monitoring of Body Fluid Status

Because depletion of either the cellular or the extracellular compartment generates drinking and other complementary behavioural and internal responses, it is reasonable to assume that each compartment has one or more sets of sensory elements to detect volume changes. It has been proposed that cellular dehydration influences the control responses for the maintenance of fluid balance through the activation of osmoreceptors (Verney 1947; Wolf 1950) or sodium receptors (Andersson 1971) or both osmo- and sodium receptors (McKinley et al.1978) (see Chapter 5).

Since the initial osmoreceptor hypothesis of Verney (1947) and the experimental findings of Andersson (1953) demonstrating the induction of drinking by intracranial injection of hypertonic sodium chloride, the search for osmo-/sodium receptors has focused almost exclusively on the brain (see Chapter 5 for review). However, this centripetal preoccupation should not lead one to ignore the likelihood that afferent input from systemically located osmo-/sodium receptors may also play a participatory role and contribute to establishing the level of activation of central pathways in response to intracellular fluid volume changes. Manipulation of regional systemic osmolality or sodium concentration in the periphery have been shown to influence other control mechanisms involved in the maintenance of body fluid balance and distribution. For example, other fluid balance controls have been shown to be inhibited or facilitated in response to hypo- and hyperosmotic solutions delivered into the hepatic circulation (Haberich 1968; Baertschi and Vallet 1981; Blake and Lin 1978; Tordoff et al. 1986). Although selective hepatic vagotomy has been found not to reduce the drinking response to cellular dehydration (Jerome and Smith 1982), complete subdiaphragmatic vagotomy (Kraly et al 1975; Kraly 1978; Jerome and Smith 1982; Martin 1981; Martin and Novin 1981; see Smith 1986 for review) or deafferentation of carotid sinus and aortic arterial baroreceptors (Rettig and Johnson 1986) will abolish or significantly attenuate the drinking response to systemically administered hypertonic saline.

In contrast to cellular dehydration, which depends upon a receptor or receptors which are influenced by the status of extracellular humoral factor(s) (i.e., osmoles and/or sodium concentration), extracellular thirst is in all likelihood facilitated and inhibited by changes in the level of neural afferent inputs arising from mechano-receptors in the vasculature in addition to one or more humoral factors. Under hypovolaemia, elevated angiotensin II (Ang II; Fitzsimons 1969, 1979) acts through receptors in the subfornical organ (SFO; Simpson and Routtenberg 1973; Simpson et al. 1978; Simpson 1981) and complements the action of neurally conducted visceral input into the CNS. Such afferent neural activity reflects blood volume and arterial blood pressure and is derived from vascular receptors in the cardiopulmonary circulation and aortic arch/carotid sinus regions, respectively. The primary input from these volume/pressure receptors reaches the CNS via the ninth and tenth cranial nerves which terminate in the nucleus of the solitary tract (NTS), area postrema (AP), and the spinal trigeminal nucleus in the medulla and the first few segments of the spinal cord (Norgren 1981). It is also possible that other viscerally derived input from areas such as the mesenteric circulation (see Smith 1986, for review) and the kidneys (Ciriello and Calaresu 1980) may contribute to generating the central state, which accompanies thirst (see Johnson 1990, for a more detailed discussion of these afferent mechanisms in the control of fluid balance) but evidence for such a role is not yet available.

Although the humoral and multiple neural signals may appear parallel, and under some experimental conditions to be redundant, it is probable that in states of hypovolaemia/hypotension they act for thirst as they do for other controls of fluid balance, that is, in a collective, facilitatory manner, complementing the actions of one another. The relative contribution of any specific afferent, whether humoral or neural may vary as a function of severity or rapidity of induction of hypovolaemia (e.g., hypovolaemia induced with or without hypotension) (see Chapter 1). Understanding the nature of the interactions of humoral and various neural afferent inputs is critical to understanding the mechanisms of thirst generated under both natural and experimental conditions.

The Interaction of the Facilitatory and Inhibitory Influences in Deprivation-Induced and Experimental Thirsts

Material covered in chapters throughout this volume (see Chapters 1,2,6,7,11 and 19) that pertain specifically to thirst, as well as to other controls of fluid balance is consistent with the view that hyperosmolality, hypovolaemia, hypotension and elevated Ang II facilitate the activation of mechanisms to acquire and retain water (Table 9.1). Conversely, hyposmolality, hypervolaemia, hypertension and probably atrial natriuretic peptide (ANP) inhibit thirst and facilitate water loss. The interplay between multiple facilitatory and inhibitory components of thirst has significant implications for understanding the physiological mechanisms underlying the control of water intake and appreciating certain limitations of the methods commonly applied in the experimental analysis of thirst.

Table 9.1. Factors which may accompany thirst-inducing stimuli which increase or decrease the probability of drinking water

Experimental treatment	Facilitation present at the start of drinking				Inhibition present at the start of drinking			Inhibition that accrues as a result of water intake	
	Hyperos-molality	Elevated Ang II	Hypo-tension	Hypovol-aemia	Hypo-os-molality	Hyper-tension	Hypervol-aemia	Hypo-os-molality	Hypervol-aemia
Water deprivation	yes	yes	yes[a]	yes	no	no	no	yes	yes
Hypertonic saline (systemic)	yes	no	no	no	no	yes	yes	no	yes
Hypovolaemia	no	yes	yes[a]	yes	no	no	no	yes	no
Isoproterenol treatment	no	yes	yes	no	no	no	no[b]	yes	yes
Angiotensin II (systemic)	no	yes	no	no	no	yes	no	yes	yes

[a] Blood pressure will range between normotensive and hypotensive level depending upon the rapidity of onset and duration.
[b] Isoproterenol has been reported to increase central venous pressure in dog.

The interaction of facilitatory and inhibitory afferent mechanisms has significance in determining the onset and termination of drinking. The implications of inhibitory mechanisms on satiety are obvious (see Chapter 19). Facilitatory and inhibitory interactions can occur among inputs derived from either one or both of the two fluid compartments. For example, one of the best characterized intercompartmental facilitatory/inhibitory interactions is the cessation of water intake in the face of an existing extracellular hypovolaemia as a result of osmotic dilution (Stricker 1969; also see Chapter 11).

From a methodological standpoint it is important to appreciate that each of the experimental manipulations commonly employed to induce homeostatic drinking varies in the number and extent of each of the facilitatory and inhibitory components at the onset and throughout the course of drinking. For example, drinking induced by moderate water deprivation is probably accompanied by maximum facilitatory and minimal inhibitory components when compared with other types of experimental thirst. In contrast, drinking induced by a manipulation such as systemic delivery of Ang II appears in the face of several inhibitory inputs (e.g., arterial hypertension) and is probably the result of a single facilitatory component (Robinson and Evered 1987) (see Table 9.1). It is likely that the dipsogenic potency of Ang II has been underestimated, because of the failure by some investigators to appreciate the total constellation of both excitatory and inhibitory input (Johnson et al. 1986; Mann et al. 1987, for further review and discussion).

Mechanisms of Central Integration and Thirst

Introduction

The amount of information about sensory mechanisms and channels of afferent signalling for the generation of thirst is steadily accumulating, but there are many unresolved issues. Nevertheless, at the present time there is sufficient information to permit initial steps toward describing some of the central sites and mechanisms involved in the processing of fluid balance information that is specifically related to drinking. The regions of the brain which are targets for humoral input and of termination of first order visceral afferents are logical points of focus in this consideration.

Other chapters of this book (see Chapters 5, 6 and 7) discuss circumventricular organs, especially those associated with the lamina terminalis, as structures that sense changes in plasma osmolality (or sodium concentration) or angiotensin titres. The outputs of the SFO and organum vasculosum laminae terminalis (OVLT) pass into the periventricular nuclei of the anteroventral third ventricle (AV3V) and, collectively, the SFO, OVLT and AV3V function in the early processing of humorally derived information with neural inputs arising from other portions of the nervous system.

As noted above, much of the first order input from the viscera projects to the NTS and AP, the most caudal of the circumventricular organs. The AP and NTS are intimately interconnected with one another. In effect, the SFO–AV3V and the AP–NTS represent the two poles of the brain through which hydration-related information is funnelled into the neural network subserving the maintenance of fluid homeostasis. Due to its anatomical and neurochemical complexity, the entirety of the neural substrate subserving the controls of fluid balance and thirst is not yet defined. However, important information is available about the processing of input pertinent to hydromineral balance within the complexes comprising the SFO–AV3V and the AP–NTS as well as about anatomical and functional relationships between these two regions.

The Integrative Role of the SFO–AV3V

Studies Implicating the AV3V In Thirst and Body Fluid Regulation

Some of the studies which ultimately implicated the periventricular tissues of the AV3V in thirst and body fluid homeostasis were begun in the early 1970s. These investigations in rats were directed at determining the site and mechanisms of Ang II action as a dipsogen. In early experiments, it was observed that centrally injected angiotensin was most effective when the peptide gained entry into the cerebral ventricular system (Johnson and Epstein 1975; Johnson 1975). Based on this observation, it was proposed that this dipsogenic peptide stimulated drinking by acting on a periventricular site (Johnson and Epstein 1975).

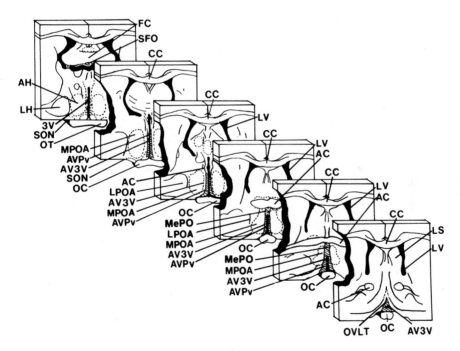

Fig. 9.1. Diagram of a portion of rat basal forebrain depicting the periventricular region of the AV3V. The diagonal lines depict the extent of a typical AV3V lesion. Abbreviations: AC, anterior commissure; AH, anterior hypothalamic nucleus; AV3V, anteroventral third ventricle; AVPv, anteroventral periventricular nucleus; CC, corpus callosum; FC, fornical commissure; LH, lateral hypothalamus; LPOA, lateral preoptic area; LS, lateral septal nucleus; LV, lateral ventricle; MePO, median preoptic nucleus; MPOA, medial preoptic area; OC, optic chiasm; OT, optic tract; OVLT, organum vasculosum of the laminae terminalis; SFO, subfornical organ; SON, supraoptic nucleus; 3V, third ventricle. (From Johnson (1990).)

Although the work of Simpson and Routtenberg (1973) had implicated the SFO as target tissue for Ang II, studies by Buggy et al. (1975) and Buggy and Fisher (1976) indicated that periventricular tissue in the brain, in addition to the SFO, must be sensitive to centrally injected peptide. In an attempt to define the locus of other periventricular sites responsive to Ang II, a series of brain lesions was made throughout the tissues surrounding the ventral portion of the third ventricle (Johnson and Buggy 1977). Destruction of the most anterior periventricular regions of the ventral third ventricle (i.e., an AV3V lesion) produces a complete cessation of water intake without other obvious signs of behavioural disruption. That is, the drinking of water ceases, but there are no indications of general arousal disorders or impaired motor function; food intake continues and, in fact, rats with the adipsia-inducing lesion will drink palatable sucrose or saccharine solutions even though water is refused and appears to be aversive when placed in the mouth.

The lesion which produces acute adipsia destroys the periventricular tissue between the optic chiasm and the anterior commissure (Fig. 9.1). Along the lamina terminalis, the lesion destroys the OVLT and the ventral portion of the median preoptic nucleus; extending caudal from the lamina terminalis, the preoptic periventricular nuclei (which includes the anteroventral periventricular nuclei described by Terasawa et al. 1980) and portions of the preoptic anterior hypothalamic nuclei are typically ablated. The medial preoptic and anterior hypothalamic nuclei receive only minor damage along their borders with the periventricular stratum of cells (Buggy and Johnson 1977a,b; Johnson and Buggy 1978; see Johnson 1982 1985a,b, 1990 for reviews). Although the total volume of tissue destroyed by the typical AV3V lesion is relatively small, experimental characterization of the functional deficits in the

Table 9.2. The acute and chronic effects of AV3V lesions

Function or effect	Result
Acute	
Drinking	Adipsia
Vasopressin release	Impaired
Hypernatraemia	Severe
Weight	Severe Loss
Extracellular fluid	Severe Loss
Chronic	
Drinking	24 h water intake recovers
	Impaired drinking responses to challenges
Vasopressin release	Impaired to humoral stimuli
	Intact to extracellular depletion
Hypernatraemia	Due to impaired thirst and vasopressin release
Natriuretic response	Impaired to volume expansion
Natriuretic factor(s)	Low circulating levels after volume load
Blood volume	Increased
Plasma renin	Increased
Basal arterial pressure	Normal
Pressor responses	Attenuated to centrally acting agents
Experimental hypertension	Blocked or attenuated in most models

systems maintaining fluid balance indicate a plethora of regulatory deficits in the postoperative animal.

During the period of immediate post-lesion-induced adipsia, the brain-damaged animals lose more weight and become markedly hypernatraemic and haemo-concentrated (Table 9.2). These indices of severe dehydration reflect not only the failure of the animals to drink but also disordered antidiuretic mechanisms. Circulating levels of vasopressin do not increase in spite of severe dehydration of both the intracellular and extracellular fluid compartments (Johnson 1985b). Morphological studies of the neurohypophysial system indicate that there is a deficit in the normal response to dehydration which increases the synthesis of vasopressin in magnocellular neurons (Carithers and Johnson 1988, for review).

If rats with typical AV3V lesions are not actively hydrated by the experimenter or provided with palatable solutions, they die of dehydration approximately one week after surgery. Animals that receive adequate hydration during the acute period usually recover "spontaneous" drinking within 10 days to two weeks and maintain themselves on water and normal laboratory chow and require no special attention within a benign laboratory environment. However, when the integrity of the systems mediating the behavioural and internal responses to dehydration are evaluated, there is evidence of chronic disruption (Tables 9.2 and 9.3).

When tested, rats with chronic AV3V lesions show impaired drinking responses to experimental manipulations, which induce or simulate dehydration of either the cellular or extracellular fluid compartments (Buggy and Johnson 1977a, b; Lind and Johnson 1983). Experimental drinking that is mediated solely by humoral factors (i.e., hyperosmolality or increased sodium concentration and Ang II) is more severely affected by chronic AV3V lesions. In contrast, thirst induced by manipulations which activate afferent neural input from systemic vascular receptors, in addition to humoral signalling, is less severely impaired by AV3V lesions. For example, water intake following subcutaneous polyethylene glycol treatment is attenuated but not abolished

Table 9.3. Chronic effects of AV3V lesions on experimental thirst

Stimuli or state producing water intake	Experimental treatment (thirst challenge)	Drinking response	Reference
Water deprivation			
Combined cellular and extracellular fluid losses	Water withheld for 24 h	Attenuated	Buggy and Johnson (1977a)
Cellular dehydration	Systemic injection of hypertonic saline	Abolished	Buggy and Johnson (1977a)
Extracellular dehydration			
Hormonal component	Systemic or central injection of angiotensin II	Abolished	Buggy and Johnson (1977a, 1978)
Hormonal and hypotensive components			
Beta-adrenoceptor agonist	Systemic isoproterenol injection	Attenuated	Lind and Johnson (1981)
Reduced cardiac return	Ligation of the inferior vena cava	Attenuated	Shrager and Johnson (1978)
Hormonal and hypovolaemic components without hypotension			
Sequestration of extracellular fluids	Systemic polyethylene glycol treatment	Attentuated	Lind and Johnson (1983)

Table 9.4. Parallels between patients with the syndrome of essential hypernatraemia and rats with lesions of the AV3V

	Patients with essential hypernatraemia	Animals with AV3V lesions
Hypothalamic damage	Yes	Yes
Chronically elevated serum sodium	Yes	Yes
Reduced or absent thirst	Yes	Yes
Elevated plasma renin	Yes	Yes
Elevated aldosterone	No[a]	No
Reduced blood volume	No	No

[a] 1 Patient

in animals with chronic AV3V lesions (Buggy and Johnson 1977a; Lind and Johnson 1983). This finding suggests that AV3V lesions may remove the Ang II-related component of extracellular thirst but that the contribution to the drinking response which is mediated by visceral afferent input (e.g., input from high pressure and cardiopulmonary baroreceptors) is still intact.

Chronic AV3V lesions in the rat produce a constellation of signs which appear to model the human clinical syndrome of essential or neurogenic hypernatraemia (Ross and Christie 1969, for review; see Table 9.4 for summary of parallels between rats with chronic AV3V lesions and humans with essential/neurogenic hypernatraemia). Essential hypernatraemia in humans occurs only when there are deficits in both thirst mechanisms and vasopressin secretion (Ross and Christie 1969). In human subjects with impaired vasopressin secretion (i.e., diabetes insipidus) and intact thirst mechanisms, plasma osmolality (and sodium) remains within the normal range. There are many clinical case studies of essential hypernatraemia wherein the patients report a lack of thirst sensation. Although the neural defect in patients with essential hypernatraemia may result from varied causes that involve major portions of the nervous system, almost invariably, when the damage is characterized, it is found to encroach upon the periventricular tissue surrounding the preoptic recess (i.e., comparable to the AV3V). Therefore, humans are comparable to other species such as dog (Witt et al. 1952), goat (Andersson et al. 1975), and sheep (McKinley et al.1982) in which extensive forebrain lesions which include the smaller, critical AV3V region have been shown to produce adipsia and disordered control of fluid homeostasis.

The Neural Connections Between the SFO and the AV3V: Functional Studies on the Control of Extracellular Fluid Depletion-Induced Drinking

The studies, which indicate that both the SFO (Simpson and Routtenberg 1973; Simpson et al. 1978) and the AV3V are angiotensin sensitive (Buggy et al. 1975; Buggy and Fisher 1976; Buggy and Johnson 1977a,b) and somehow collectively involved in the generation of thirst to exogenous octapeptide, provided a major impetus to investigate the neural mechanisms and organization of systems in structures along the lamina terminalis and around the ventricle of the rostral basal forebrain. Lesions directed at different structures along the lamina terminalis produce different effects on Ang II-induced drinking after systemic peptide administration in contrast to the drinking produced by intraventricular injections (Lind et al.1979, 1981, 1984;

Shrager 1981). Specifically, these studies carried out in the rat show that: (a) destruction of the OVLT does not produce drinking deficits to Ang II administered by either the systemic or central routes, (b) destruction of the SFO abolishes drinking to peripherally but not intraventricularly injected Ang II, and (c) lesions of the median preoptic nucleus abolish Ang II-induced drinking produced by either route of administration (see Johnson 1982; Lind and Johnson 1982b for reviews). The explanation of these observations derived from functional studies is more clearly understood in light of the emerging clarification of anatomical interconnectivity of the SFO and the structures of the AV3V.

Early in development, the rostral end of the neural tube closes to form the lamina terminalis. In later stages the telencephalon enlarges to encompass the rostral end of the neural tube and the tissues which ultimately become the rostral circumventricular organs, that is, the SFO and the OVLT, are separated by the anterior commissure. In spite of the spatial dissociation of the extra-blood–brain barrier tissues associated with the lamina terminalis, the elements along the rostral third ventricle remain intimately associated by neural (Hernesniemi et al.1972) and possibly vascular interconnections (Spoerri 1963). The potential of the functional significance of the "descending" connections of lamina terminalis structures was first emphasized in the studies by Miselis and colleagues (Miselis et al. 1979; Miselis 1981) which described the projection of the SFO efferents into the AV3V and into magnocellular nuclei (i.e., paraventricular and supraoptic nuclei). Tract tracing techniques, which facilitated the visualization of the course of SFO efferents (Lind et al. 1982), aided in the formulation of the hypothesis that information derived from the action of circulating Ang II on the SFO is carried into the AV3V, and in particular into the median preoptic nucleus (Lind and Johnson 1982a,b).

If the bundle of fibres which emerges from the SFO and descends into the AV3V is interrupted by the placement of a small knife cut at the level of the median preoptic nucleus, drinking to systemically administered Ang II is significantly impaired (Eng and Miselis 1981; Lind and Johnson 1982a, b). In contrast, drinking to intraventricularly injected Ang II, after a knife cut of the descending bundle of the lamina terminalis, is minimally affected (Lind and Johnson 1982a, b, c). Collectively, the functional studies using knife cuts and electrolytic lesions (described above) in conjunction with Ang II delivered systemically and centrally and the anatomical findings suggest a plausible model to account for the interaction of the SFO and AV3V in the processing of information, reflecting elevated circulating Ang II (Lind and Johnson 1982b). That is, it has been hypothesized that: (a) circulating Ang II acts on the SFO in the mode of a blood-borne hormone, (b) information is carried from the SFO through the descending bundle of the lamina terminalis into the AV3V, and (c) Ang II or an Ang II-like peptide contained in the descending axons in the SFO–AV3V projection is released into structures such as the median preoptic nucleus to act in the manner of a neurotransmitter/modulator to, in turn, activate a higher order projection to other CNS integrative regions involved in the expression of drinking behaviour.

When the model of Ang II action and interaction within the SFO–AV3V was formulated (Lind and Johnson 1982b), the proposed role for a descending Ang projection from the SFO to the AV3V was based on the functional evidence described above and the hypothesized existence of a brain renin–angiotensin system that was biochemically independent of the renal renin–angiotensin system, i.e., what might be considered the "classic" system (Chapter 10 and Ganten et al. 1976, 1982 for reviews). Although the brain and renal renin–angiotensin systems were conceived as possibly functionally interrelated, it is believed that the brain system is biochemically independent of circulating Ang II. That is, enzymes and substrates which comprise the metabolic cascade are proposed as being formed de novo in the brain. Since the proposal of an angiotensinergic connection between the SFO and the AV3V, several new lines of evidence have emerged that are consistent with the outlined principles of the model.

The presence of an angiotensinergic projection from the SFO to the AV3V received initial support with the identification of Ang II-like staining fibres within the descending bundle of the the lamina terminalis and the presence of Ang II-like cell bodies in the SFO (Gray et al. 1982). Subsequent work of Lind and Swanson and their colleagues demonstrated with combined retrograde tracing methods and immunocytochemistry that some of the Ang II-like staining in the descending SFO–AV3V projection originated in the SFO and probably terminated in the median preoptic nucleus (see Lind and Ganten 1990 for review). Other lines of converging histological evidence for the model derives from the demonstration of a high concentration of Ang II receptors in the structures along the lamina terminalis and throughout the AV3V region (Mendelsohn et al. 1984; Plunkett et al. 1987) and from the results of in situ hybridization studies indicating the presence of high concentrations of angiotensinogen mRNA throughout the AV3V (Lynch et al. 1987). In addition, in vitro and in vivo electrophysiological studies indicate that many neurons in structures along the lamina terminalis and throughout the AV3V are sensitive to exogenous Ang II (Gronan and York 1978; Nicolaïdis et al. 1983; Nelson and Johnson 1985; Tanaka et al. 1987). Thus, collectively, data have steadily accumulated to support the idea that both the SFO and AV3V are anatomically and functionally closely interconnected and that a portion of the information carried and processed within this region relies on an Ang II-like peptide.

Ascending Catecholamine Input into the AV3V: Evidence of an Interaction Between Angiotensin and Noradrenaline

Injection of 6-hydroxydopamine (6-OHDA), a neurotoxin which destroys catecholamine neurons, into the cerebral ventricles of the rat produces an animal with many of the same deficits in the control systems which maintain normal fluid homeostasis that are found following electrolytic destruction of the AV3V (Stricker and Zigmond 1974; Fitzsimons and Setler 1975; Gordon et al. 1979; Miller et al. 1979) (Table 9.5). For example, both treatments yield an animal: (a) with acute adipsia from which recovery of ad libitum drinking can occur if adequate hydration is provided over the postoperative period; and (b) with long-term drinking response deficits to many cellular and extracellular thirst challenges. The similarities between the effects of these two lesions led to the consideration of the possibility that the

Table 9.5. Parallels between rats with AV3V lesions and rats following 6-hydroxydopamine treatment

Temporary adipsia – must receive hydrational "therapy"
Permanent deficits to thirst challenges
Prevents the development of most forms of experimental hypertension
Pressor deficits to central angiotensin II and hypertonic saline injections
Chronic impairments in vasopressin release
Chronic reduction in body weight

two methods of destroying central neural tissue produce similar results due to the interruption of the catecholaminergic innervation of the AV3V (Brody and Johnson 1980).

Intraventricular injection (i.v.t) of 6-OHDA results in depletions of noradrenaline (NA) and dopamine (DA) throughout the brain and spinal cord. Consequently, experiments investigating the role of catecholamines in AV3V-related function have focused on, first, establishing which specific catecholamine is tied to thirst and, second, determining how the particular catecholamine is associated with the AV3V. By judiciously selecting a 6-OHDA treatment regime which takes into account the dose of 6-OHDA, frequency of administration, and whether or not the treatment is administered in the presence of a NA reuptake blocker, it is possible to achieve selective depletions of either DA or NA (Breese and Traylor 1970, 1971; Simmonds and Uretsky 1970; see Kostrzewa and Jacobowitz 1974 for review).

Experiments that investigated the relationship between the nature of catecholamine depletion and drinking deficits induced by centrally injected 6-OHDA indicated that central depletion of NA but not of DA resulted in a drinking deficit to centrally injected Ang II (Bellin et al. 1987a). In further studies, it was possible to locate the critical area of NA depletion in the region of the ventral lamina terminalis (i.e., median preoptic nucleus–OVLT; Bellin et al. 1987b, 1988; Cunningham and Johnson 1989). Further evidence that the presence of NA is critical in structures along the ventral lamina terminalis is provided in experiments where NA has been "replaced" after injection of 6-OHDA into the region of the median preoptic nucleus–OVLT.

Replacement studies in rats with depletion of catecholamines along the ventral lamina terminalis have taken two experimental directions. Repletion has been performed by administering exogenous NA by central infusion and by transplanting fetal NA-containing cells into the NA-depleted ventral lamina terminalis. Intraventricular infusion of NA starting 15 min before an intraventricular bolus injection of Ang II produced a significant increase in water intake in NA-depleted animals which did not respond to Ang II administered without NA (Cunningham and Johnson 1988). Although drinking increased significantly after the combined dose of Ang II and NA in comparison to Ang II alone, the drinking response was not fully restored (i.e., equivalent to that of intact rats). At this point, it is unclear whether the failure to obtain the complete recovery of the evoked drinking response with infused NA reflects an inappropriate dose of the catecholamine or whether the 6-OHDA treatment removed another neurohumoral factor(s). The transplantation of fetal noradrenergic cells into the NA-depleted lamina terminalis restored drinking in the 6-OHDA treated animals more effectively.

Neural transplantation studies have followed the procedure of placing a suspension of cells from the anlage of either locus coeruleus or of the hindbrain region containing the A1A2/C1C2 adrenergic/noradrenergic cell bodies from day 17 fetal rat brains into the parenchyma along the ventral lamina terminalis. The drinking response to Ang II was restored when the NA-depleted animals were tested, beginning two weeks after placement of the fetal cells and where there was anatomical evidence that the transplanted cells were viable (McRae-Degueurce et al.1986, 1987).

The results of the depletion/repletion studies indicate that at least a minimal amount of NA must be present in the tissue along the ventral lamina terminalis for normal drinking responses to exogenous Ang II. These experiments provide insight into the physiology of the AV3V and suggest that there may be an important interaction between peptidergic (Ang II) and catecholaminergic (NA) mechanisms in states where both neurotransmitter/neuromodulator systems are likely to be activated.

A Model for the Interaction of Angiotensinergic and Noradrenergic Systems in the AV3V

As discussed earlier (also see Chapter 6), there are multiple afferent signals which convey information to the brain regarding the status of vascular/extracellular fluid volume. Moderate hypovolaemia elevates circulating Ang II to dipsogenic levels (Johnson et al. 1981) and alters afferent nerve activity from cardiopulmonary receptors (Zimmerman et al. 1981; Moore-Gillon and Fitzsimons 1982; Kaufman 1984). It is likely that such afferent humoral and neural input interacts within common CNS sites to generate extracellular thirst.

There are several lines of experimental evidence which indicate that hypotension and/or reduced blood volume activates brainstem noradrenergic cell groups which project to the forebrain. Electrophysiological recording of single unit activity in the locus coeruleus (i.e., the A6 noradrenergic cell group) in anaesthetized rats indicates that cells in this region increase firing to haemorrhage and to hypotension induced with nitroprusside (Elam et al. 1985). In contrast, the hypertension that results from phenylephrine administration inhibits activity of these cells. Because the pressure/volume evoked changes in locus coeruleus activity were blocked by bilateral vagotomy, these results were interpreted as indicating that locus coeruleus activity is regulated tonically by vagal afferents from cardiac volume receptors and that ascending projections from the A6 region convey information about the status of the cardiovascular system to rostral regions of the neuraxis. In comparable studies, Moore and Guyenet (1983) recorded single units in the A2 catecholamine cell group (i.e., within the nucleus of the tractus solitarius) which were antidromically activated by electrical stimulation of the medial forebrain bundle. These catecholamine-containing cells were inhibited by increases in blood pressure produced by systemic administration of NA and vasopressin. These findings indicate that activity of the NA-containing cells of the dorsal caudal medulla is altered by peripheral input from vascular pressure receptors.

Using in vivo voltometry, Quintin et al. (1987) examined the changes in catecholamine metabolism in the A1 region of the ventrolateral medulla which accompany changes in blood volume and arterial pressure. The results of these experiments indicate that catecholamine metabolism is increased by hypotension produced by nitroprusside administration or by controlled haemorrhage. Conversely, catecholamine metabolism in this region decreased when blood pressure was elevated with phenylephrine. Deafferentation of high pressure baroreceptors abolished the changes in catecholamine metabolism in the A1 region that were normally produced by haemorrhage.

Information reflecting the status of vascular volume and arterial pressure projects rostrally (see Swanson and Sawchenko 1983; Ciriello et al. 1986 for reviews). A prime candidate as a neurotransmitter/neuromodulator involved in conveying such viscerally derived input from the hindbrain into the basal forebrain is noradrenaline (see Blessing and Willoughby 1988 for review). Among the anterior regions receiving input from hindbrain catecholamine cell groups are the structures contained within the AV3V (Lindvall and Bjorklund 1974). The findings of Saper et al. (1983) which indicate that the median preoptic nucleus receives afferents from the A1C1/A2C2 noradrenaline and adrenaline containing cells of the medulla are of particular significance.

The studies indicating the elevation of neural activity of brainstem noradrenergic neurons in conjunction with the forebrain anatomical projections of these neurons

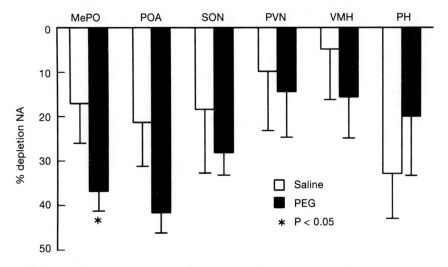

Fig. 9.2. The effect of polyethylene glycol (PEG)-induced hypovolaemia on noradrenaline (NA) turnover in several diencephalic regions. Hypovolaemia was induced by subcutaneous injection of 5 ml of 30% PEG. Percentage changes (+/- SEM; with respect to 0-time control) in NA content are shown 240 min after alpha-methyl-para-tyrosine (aMPT) treatment in saline (SAL) and PEG-treated groups. This index of NA turnover is related to the depletion of NA following aMPT administration as compared to NA levels in untreated control animals. The larger percentage depletions indicate higher NA turnover levels. *$P<0.05$ compared to SAL. Micropunches were taken from the median preoptic nucleus (MePO), preoptic area (POA), supraoptic nucleus (SON), paraventricular nucleus (PVN), ventromedial hypothalamus (VMH), and posterior hypothalamus (PH). Although several of the fluid related regions in the hypothalamus (e.g., MePO, SON, PVN) tended to show an increase in NA turnover, only the MePO was significant at the time interval tested. (From Wilkin et al. (1987).)

suggest that in states of hypotension and/or hypovolaemia there may be increased noradrenaline release in the rostral terminal regions. Recently, this hypothesis was tested by studying the effects of hypovolaemia induced by subcutaneous polyethylene glycol treatment on the decrease in forebrain NA levels in rats with NA synthesis blocked by alpha-methyl-para-tyrosine pretreatment (Wilkin et al. 1987). Of six forebrain areas studied, 240 min after polyethylene glycol treatment, the median preoptic region showed the greatest decline in NA (Fig. 9.2). This result suggests that more NA is released into the AV3V as a result of the gradually induced hypovolaemic state.

Fig. 9.3 is a model that describes the actions of Ang II within the SFO–AV3V, the interaction of angiotensin with NA within the AV3V, the effects of hypovolaemia on NA turnover within the median preoptic nucleus, and the relevance of these phenomena for the integrative control of extracellular thirst (Johnson 1985a, b, 1990; Johnson and Cunningham 1987; Johnson and Wilkin 1987). This model is an extension of the one initially proposed by Lind and Johnson (1982b; see above). It proposes that in states of hypovolaemia an angiotensinergic projection from the SFO is activated in proportion to the amount of circulating Ang II and that this "descending" peptidergic input into one or more structures of the AV3V (e.g., the median preoptic nucleus) interacts with "ascending" noradrenergic input which is increased in states of hypovolaemia/hypotension. In effect, angiotensin and NA within the AV3V "reinforce" one another and are mutually facilitatory. It is unclear whether angiotensin and NA are likely to function as "classic" neurotransmitters in this area or whether one or

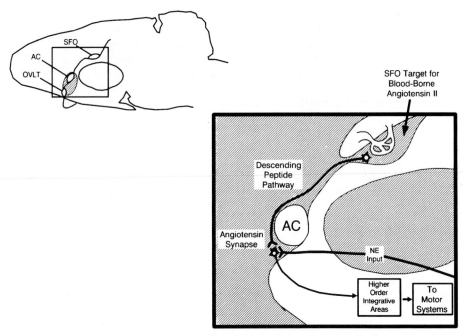

Fig. 9.3. A schematic representation of the interaction of angiotensinergic and noradrenergic inputs into the median preoptic (MePO) nucleus that are involved in the mediation of extracellular thirst. The model is based on one proposing a dual role for angiotensin (Ang) in thirst (Lind and Johnson 1982a, b). Blood-borne Ang II penetrates the fenestrated capillaries of the subfornical organ (SFO) and activates receptors. A midline "descending" pathway conducts the plasma Ang-related information down the lamina terminalis and into the AV3V (particularly the MePO). The "descending" pathway contains Ang and is proposed to use Ang as a peptidergic neurotransmitter. Ascending noradrenergic input into the MePO arises from the A1, A2 and A6 noradrenergic cell groups in the brainstem. It is proposed that noradrenaline (NA) availability in the MePO is inversely proportional to the systemic blood volume and/or blood pressure. Increased synaptic NA in the MePO modulates the action of increased Ang II to facilitate neural output to higher order neurons and regions involved in the control of water intake. (From Johnson (1990).)

both of these putative neurotransmitters serve(s) in the role of neuromodulator. However, an appealing hypothesis is that angiotensin acts in the more traditional mode of a neurotransmitter and NA serves the role of neuromodulator. NA has been demonstrated to modulate the informational throughput at various loci in the CNS, including synapses in "classic" sensory pathways by enhancing the signal to noise ratio (Woodward et al. 1979; Jahr and Nicoll 1982; Foote et al. 1983).

Hindbrain Mechanisms in the Control of Fluid and Electrolyte Balance

Early Functional Studies and Anatomical Substrates of the Visceral Neuraxis for Thirst and Fluid Homeostasis

Catecholaminergic cell bodies that innervate the AV3V are located in the caudal medulla. The anatomical location of these soma are in regions that receive primary afferent information from cardiovascular and gastrointestinal receptors; which suggests that these hindbrain areas might be involved in fluid balance. Indeed, the importance of the caudal medulla in the control of fluid balance has been known for many years. In the mid-nineteenth century, Claude Bernard noted that puncture of the floor of the fourth cerebral ventricle resulted in increased diuresis (see Cushny 1926). More recently, studies have focused on the region of the caudal rhomboid fossa near obex and the role of this region in sodium and water excretion. The results from these studies suggest that CNS structures near the obex may be important in excretion of sodium and water. Snyder and Sutin (1961) observed that lesions placed in the region of obex in rats resulted in an elevation of the sodium/potassium ratio in the urine during the 24 h period following surgery. This effect was not observed in dogs, however, as lesions placed near obex failed to produce marked alterations in water and electrolyte balance (Wise and Ganong 1960a). However, Wise and Ganong (1960b) did report that stimulation in and around the area postrema in the dorsal medulla caused increased excretion of sodium and chloride as well as increased urine volume and creatinine clearance. While these results suggested an important role for the dorsomedial medulla in the control of fluid homeostasis, the focus of the central neural control of fluid balance has remained the forebrain, particularly the hypothalamus.

The AP, a circumventricular organ, and the NTS, which receives primary afferent input from the digestive tract as well as the cardiovascular system, are located in the dorsomedial medulla near obex. The AP, like other circumventricular organs, lacks a blood–brain barrier. Thus, substances that normally do not gain access to the brain can readily penetrate the parenchyma of the AP. In fact, the AP is reported to contain binding sites for numerous circulating substances (Leslie 1986). Moreover, recent studies indicate that substances such as Ang II or vasopressin activate neurons in the AP (Carpenter et al. 1983). Thus, it is possible that the AP monitors changes in plasma concentrations of substances such as angiotensin and vasopressin that might influence water and electrolyte ingestion and excretion.

The AP and caudal NTS adjacent to the AP receive direct neural input from thoracic and abdominal viscera such as the heart, liver and kidneys (Ciriello et al. 1981; Kalia and Sullivan 1982; Wyss and Donovan 1984). Additionally, primary afferents from the carotid sinus and carotid bodies synapse in the caudal NTS with slight innervation of the AP (Kalia and Mesulam 1980; Panneton and Loewy 1980; Housley et al. 1987). Moreover, the AP and caudal NTS are interconnected both neurally and vascularly (Shapiro and Miselis 1985; Roth and Yamamoto 1968). These characteristics suggest this region is a likely site for initial integration of visceral afferent and humoral information pertaining to the status of water and electrolyte balance. This region may then act to modulate the behavioural response elicited by stimuli such as circulating Ang II that activate water and electrolyte ingestion.

There are three pathways by which output from the AP and adjacent NTS ascends the neuraxis to the forebrain and which seem to be of significance to the central control of fluid homeostasis. The first route is a direct projection from the dorsomedial medulla to forebrain nuclei. Many investigators have reported that neurons in the NTS project directly to forebrain nuclei (Ricardo and Koh 1978; Saper et al. 1983; Sawchenko and Swanson 1983; Zardetto-Smith and Gray 1987; Edwards et al. 1989; Ter Horst et al. 1989). It is also possible that the AP contributes to this direct projection, as one report suggests that cells in the AP project directly to the hypothalamus (Iovino et al. 1988). A second route by which output from the AP and NTS ascends to the forebrain involves a secondary relay nucleus, the lateral parabrachial nucleus (lPBN) of the pons (Shapiro and Miselis 1985; van der Kooy and Koda 1983). The lPBN, in turn, projects to many forebrain nuclei important in the control of fluid balance. Projections from the lPBN include the SFO, the AV3V, the hypothalamic paraventricular nucleus, the supraoptic nucleus and the amygdala (Saper and Loewy 1980; Fulwiler and Saper 1984). Finally, the AP and NTS send efferents to the caudal ventrolateral medulla which contains the A1 catecholamine cell group. The A1 catecholamine cell group provides catecholaminergic input to much of the forebrain (Sawchenko and Swanson 1983; Saper et al. 1983). Although many forebrain sites receive input via these projections, it is important to emphasize that several rostral areas reported to be critical to the control of fluid and electrolyte consumption and excretion are prominent among these structures (see above). Thus, afferent information received in the AP and caudal NTS can be processed and transmitted directly to the forebrain or the information may be transmitted to secondary relay nuclei in the medulla and pons which then convey the information to forebrain nuclei.

In addition to the three ascending pathways just described, it should be noted that fibres from the AP and NTS also project to hindbrain regions implicated in the control of efferent autonomic function. These efferent projection sites include the rostral ventrolateral medulla as well as the dorsal motor nucleus of the vagus nerve (Shapiro and Miselis 1985; van der Kooy and Koda 1983). Thus, the AP and caudal NTS can potentially modulate efferent autonomic outflow and visceral function. This is yet another likely mechanism for the AP and caudal NTS to influence the control of water and electrolyte ingestion.

Physiological and Behavioural Evidence Suggests a Role for the Dorsal Hindbrain in Fluid Balance

The AP and/or caudal NTS are reported to contain neural elements responsive to changes in body fluids and electrolytes. For example, the region of the caudal rhomboid fossa containing the AP is reported to be osmo- or sodium sensitive (Clemente et al. 1957; Adachi and Kobashi 1985). Additionally, afferent nerves from peripheral receptive elements such as high pressure baroreceptors in the aortic arch and carotid sinus, low pressure baroreceptors in the vena cava, pulmonary and hepatic vasculature and osmoreceptors in the liver alter neural firing in the caudal NTS (Adachi et al. 1976; Rogers et al. 1979; Kahrilas and Rogers 1984). Moreover, as mentioned above, the AP also contains binding sites for many circulating proteins and peptides that are considered important in regulation of fluid balance and cardiovascular function (Leslie 1986). Most importantly, substances known to be significant in the regulation of body fluids and electrolytes (Ang II, vasopressin and ANP) have binding sites in the AP (Harding et al. 1981; Nazarali et al. 1987; Gutkind et al. 1988; Tribollet et al. 1988; Saavedra 1987; Skofitsch and Jacobowitz 1988). In addition, neural elements within the AP are reported to be responsive to Ang II and vasopressin (Carpenter et al. 1983), suggesting that the binding sites for these substances are functional. Further support for a physiological role for Ang II and vasopressin binding sites in the caudal dorsomedial medulla is derived from the reported action of vasopressin or Ang II in the AP to modulate efferent function of the autonomic nervous system (Leslie 1986). The neural input from peripheral receptors that monitor volume and osmotic concentration and the sensitivity to circulating substances suggests that the AP and caudal NTS are sites involved in pathways that effect physiological responses to alterations in body water and electrolyte levels.

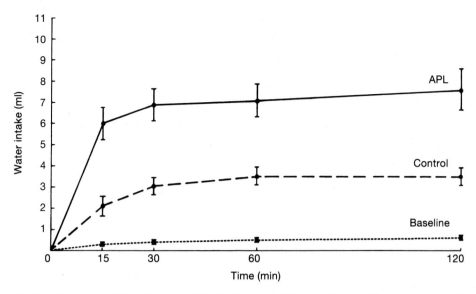

Fig. 9.4. Angiotensin II (Ang II) -induced drinking in AP/mNTS-lesion (APL, $n=21$) and control ($n=23$) rats. For the two groups the non-treatment baselines were not different ($P > 0.5$) . Therefore, the baseline intakes were pooled for the two groups. (From Edwards and Ritter (1982).)

One area of recent interest has been the role of the hindbrain in the control of water and sodium ingestion. Several investigators have found that lesions centred in the AP in the dorsomedial medulla result in profound changes in water and sodium intake (Contreras and Stetson 1981; Edwards and Ritter 1982; Hyde and Miselis 1984; Miselis et al. 1986; Ohman and Johnson 1989). It is important to indicate that although these lesions are centred in the AP, and are often referred to as AP-lesions, complete lesions of the AP include a portion of the immediately adjacent NTS. Thus, this lesion is referred to as an AP/mNTS-lesion to indicate the slight damage that occurs to the caudal NTS.

Although the effect of AP/mNTS-lesions on water intake remains controversial, there is emerging evidence that lesions of the AP/mNTS may produce a selective enhancement of drinking to stimuli that mimic depletion of the extracellular fluid compartment (Edwards and Ritter 1982; Ohman and Johnson 1989). In earlier studies, Miselis and co-workers reported that rats with AP/mNTS-lesions had a permanent polydipsia and drank more water than unlesioned controls when deprived of water for 24 h or challenged with hypertonic saline to produce cellular dehydration (Hyde and Miselis 1984). Other investigators found no difference in 24 h intake of water as well as no difference in 24 h water deprivation-induced drinking and cell dehydration-induced drinking (Contreras and Stetson 1981; Edwards and Ritter 1982). On the other hand, Edwards and Ritter (1982) found that rats with AP/mNTS lesions consume more water than unlesioned control rats after challenges that mimicked depletion of the extracellular fluid compartment. Injection of Ang II, or isoproterenol which directly stimulates the renin–angiotensin system, causes rats with AP/mNTS-lesions to consume significantly greater quantities of water than intact control rats (Fig. 9.4). Conversely, when these rats were subcutaneously injected with hypertonic saline to deplete the intracellular fluid space, there was no difference between the water intake of rats with AP/mNTS-lesions and unlesioned control rats. Thus, in

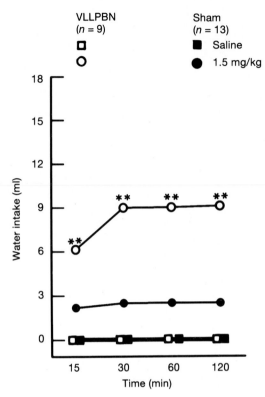

Fig. 9.5. Cumulative mean water intakes over 2 h after subcutaneous injections of angiotensin II (Ang II; 1.5 mg/kg) or vehicle (isotonic saline) in rats with sham lesions or with lesions of the ventrolateral region of the lateral parabrachial nucleus (VLLPBN) **=P<0.01. (From Ohman and Johnson (1989).)

these studies, AP/mNTS-lesions appeared selectively to accentuate drinking to depletion of the extracellular fluid compartment. More recently, Ohman and Johnson (1989) found that rats with lesions of the AP/mNTS consume more water after (i.c.v.) Ang II than unlesioned control rats. Additionally, Ohman and Johnson (1989) found that i.c.v. injection of carbachol did not produce increased drinking in the AP/mNTS-lesioned rats. Since carbachol is reported to activate central neural systems involved in cell dehydration-induced thirst (Block and Fisher 1975), these observations, as well as those of Edwards and Ritter, suggest that some factor or factors act through the AP and/or caudal NTS to modulate water intake when thirst is activated by depletion of the extracellular fluid compartment.

The role of one of the secondary projection sites of the AP and adjacent NTS in water intake has also been investigated. Ohman and Johnson (1986) found that bilateral electrolytic lesions of the lPBN resulted in a constellation of alterations in drinking behaviour that were remarkably similar to those reported by Edwards and Ritter after lesions of the AP/mNTS. These behavioural alterations included enhanced water ingestion after subcutaneous administration of Ang II or isoproterenol, dipsogenic stimuli that mimic depletion of the extracellular fluid compartment (Fig. 9.5). This observation is particularly interesting in light of the direct neural projection from the AP and caudal NTS to the lPBN (Shapiro and Miselis 1985). The existence of this pathway suggests that the AP and/or caudal NTS, and the lPBN are involved in a hindbrain system that acts to modulate drinking when drinking is initiated by depletion of the extracellular fluid compartment.

We have recently attempted to determine if neurons in the lPBN are involved in the behavioural alterations observed after lesions of the lPBN, or if fibres projecting through the lPBN are responsible for the enhanced drinking. These studies utilized the excitatory neurotoxin ibotenic acid to lesion cell bodies in the lPBN and leave fibres of passage intact. Lesion of the lPBN with ibotenic acid, resulted in the same pattern of altered drinking behaviour that was reported to follow electrolytic lesions of the lPBN (Edwards and Johnson 1987). That is, rats with ibotenic acid lesions of the lPBN consume significantly greater quantities of water after dipsogenic treatment that mimic depletion of the extracellular fluid compartment. Conversely, water intake after depletion of the cellular fluid compartment is normal in rats with ibotenic acid lesions of the lPBN. Thus, it appears that loss of cells within the lPBN is responsible for the enhanced water intake after depletion of the extracellular fluid compartment.

In summary, both functional and anatomical evidence suggests that the AP and adjacent NTS as well as the lPBN are components of a dorsal hindbrain system important in the regulation of water intake. This system seems selectively to inhibit drinking induced by manipulations which increase circulating Ang II or mimic the central state that accompanies hypovolaemia. Since elevated plasma Ang II levels do not decline rapidly after the ingestion of water (see Chapter 19), the AP/mNTS–lPBN pathway may normally play an important role in preventing overhydration due to the presence of a sustained peptidergic signal.

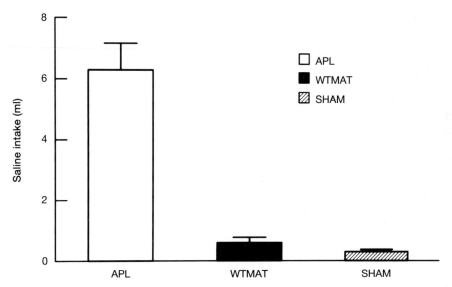

Fig. 9.6. Saline intake by AP/mNTS-lesioned (APL) and by weight-matched (WTMAT) and non-weight matched (SHAM) non-lesioned (i.e. control) rats during 3 h exposure to 3% NaCl. The weight-matched non-lesioned rats (WTMAT) were given reduced quantities of food so that the body weight loss after surgery was similar to the AP/mNTS-lesion group. Prior to the 3 h test all animals were maintained ad libitum on normal chow and water. Note that the 3% NaCl intake by the lesioned rats is significantly greater than for both groups of the unlesioned control rats.

In addition to the enhanced water intake after dipsogens that mimic depletion of the extracellular fluid compartment, recent studies have revealed a remarkable sodium appetite in rats with lesions of the AP and immediately adjacent NTS (Contreras and Stetson 1981; Watson 1985; Edwards et al. 1988). This increased consumption is observed both over 24 h period and in short 30–180 min tests (Fig. 9.6). Moreover, the consumption is noteworthy in that rats with AP/mNTS lesions will consume as much as twice their total exchangeable sodium in a short period (2–3 hr) when given access to concentrated saline solution. Although lPBN-lesions have similar effects on drinking behaviour, the effect of lesions of the lPBN on sodium intake is unclear at this time. However, preliminary studies indicate that rats with lesions of the lPBN do not overconsume concentrated saline solution as do rats with lesions of the AP/mNTS. Therefore, it is possible that the mechanism responsible for the overingestion of concentrated saline solution differs from that which produces increased water consumption after lesions of the AP and adjacent mNTS. On the other hand, the lPBN may be important in the expression of the overingestion of concentrated saline solutions exhibited by rats with lesions of the AP and adjacent NTS. In somewhat related studies, it is reported that lPBN lesions prevent or reverse the overconsumption of highly palatable foods exhibited by rats with lesions of the AP and adjacent NTS, whereas lPBN lesions alone have no apparent effect on the consumption of highly palatable food (Edwards and Ritter 1989). It is possible that lesions of the lPBN may also reverse the overingestion of concentrated saline solutions by rats with lesions of the AP and adjacent NTS.

Little is known of afferent neural or humoral fluid balance signals to the region of the dorsomedial medulla that AP/mNTS-lesions may interrupt. Recent studies have reported that distension of the great veins and pulmonary vessels acts to inhibit drinking (Kaufman 1984; Moore-Gillon 1980). Moreover, rats with lesions of the lPBN exhibit an attenuated inhibitory response to inflation of an atrial balloon which would be expected to activate low pressure baroreceptors (Ohman and Johnson 1987). This observation is consistent with the possibility that the AP, caudal NTS and lPBN are involved in a pathway that conveys low pressure cardiopulmonary information to the forebrain and acts to inhibit drinking behaviour. Recent anatomical evidence supports this hypothesis. When the anterograde label cholera toxin-conjugated horseradish peroxidase is injected around the cardiac atria, afferent terminals are observed in the caudal NTS adjacent to the AP (Bradd et al. 1989). It would be expected that some of the information conveyed via this projection would be from low pressure receptors in the atria and great veins. Interestingly, the region of the caudal NTS that receives afferent input from the cardiac atria also receives heavy input from the AP (Shapiro and Miselis 1985). Thus, it is possible the AP may act to modulate low pressure afferent activity in the caudal NTS.

The signal(s) that might activate neurons in the AP to inhibit fluid intake remain elusive. Certainly, either a circulating factor or afferent neural input, or both, would seem most logical given the anatomical and physiological characteristics of the AP. However, the precise nature of the factor(s) remains to be elucidated.

Summary and Conclusions

Over the past 50 years, several notable advances have increased our understanding of the mechanisms involved in the signalling and processing of thirst-related information. Among these are:

1. Recognition of the significance of cellular dehydration as an effective stimulus for drinking
2. Development of the concept of an osmoreceptor
3. Appreciation that output from the osmoreceptor can have both facilitatory and inhibitory influences
4. The discovery that humoral stimuli can act directly on the brain to generate thirst
5. The demonstration that pure extracellular hypovolaemia elicits water intake
6. Implication of the renal renin–angiotensin system in thirst
7. Implication that afferent input from cardio-pulmonary and high pressure baroreceptors has facilitatory and inhibitory influences on thirst
8. The discovery of an endogenous brain renin–angiotensin system
9. The elucidation of CNS noradrenergic mechanisms and pathways.

In this chapter, recent studies were reviewed which implicate the regions associated with the lamina terminalis and the AP/NTS as important complexes in the "early" processing of hydration-related input reflecting the status of extracellular fluid osmolality and volume. The SFO/AV3V region has been proposed to both receive and process facilitatory input from osmo-/sodium receptors and angiotensin receptors. Plasma levels of Ang II may have a central representation through the level of neuronal activity in a projection which descends from the SFO into the AV3V and which is angiotensinergic. Current evidence also suggests that facilitatory input from systemic vascular pressure/volume receptors arrives in the hindbrain and then ascends the neuraxis over adrenergic pathways to facilitate ongoing activity within the AV3V.

There is evidence that indicates the presence of a hindbrain system, which acts to protect against overexpansion of extracellular fluid volume. The AP/NTS and lPBN are two components of this proposed inhibitory system. It is very possible that ascending input from the AP/lPBN ascends the neuraxis to influence activity in the periventricular basal forebrain region that has been implicated in thirst.

The sites and mechanisms discussed in this review are located at sites high on the afferent limb. Although what lies between early processing areas of the medial basal forebrain and the mesencephalic regions involved in the generation of locomotor patterns has been a subject of creative and insightful hypothetical discussion (Swanson and Mogenson 1981), much work is required before the complexities involved in the neural organization of drinking will be comprehended. There is still much to be learned as research progresses toward the "final common path" of the motor act of ingesting water.

Acknowledgements. Studies from the authors' laboratories discussed in this chapter were supported in part by USPHS grants HLP 14388, HL 33447, HL 35600, and MH 00064 and by the Iowa Affiliate of the American Heart Association.

The authors are grateful for the constructive comments that were provided by Dr Robert Thunhorst and for the excellent editorial and secretarial assistance of Ms Norma Mottet.

References

Adachi A, Kobashi M (1985) Chemosensitive neurons within the area postrema of the rat. Neurosci Lett 55:137–140

Adachi A, Niijima A, Jacobs HL (1976) An hepatic osmoreceptor mechanism in the rat: electrophysiological and behavioral studies. Am J Physiol 231:1043–1049

Andersson B (1953) The effect of injections of hypertonic NaCl-solutions into different parts of the hypothalamus of goats. Acta Physiol Scand 28:188–201

Andersson B (1971) Thirst and brain control of water balance. Am Sci 59:408–415

Andersson B, Leksell LG, Lishajko F (1975) Perturbations in fluid balance induced by medially placed forebrain lesions. Brain Res 99:261–275

Baertschi AJ, Vallet PG (1981) Osmosensitivity of the hepatic portal vein area and vasopressin release in rats. J Physiol (Lond) 314:217–230

Bellin SI, Bhatnagar RK, Johnson AK (1987a) Periventricular noradrenergic systems are critical for angiotensin-induced drinking and blood pressure responses. Brain Res 403:105–112.

Bellin SI, Landas SK, Johnson AK (1987b) Localized injections of 6-hydroxydopamine into lamina terminalis-associated structures: effects on experimentally-induced drinking and pressor responses. Brain Res 456:9–16

Bellin SI, Landas SK, Johnson AK (1988) Selective catecholamine depletion of structures along the ventral lamina terminalis: effects on experimentally-induced drinking and pressor responses. Brain Res 456:9–16

Blake WD, Lin KK (1978) Hepatic portal vein infusion of glucose and sodium solutions on the control of saline drinking in the rat. J Physiol (Lond) 274:129–139

Blessing WW, Willoughby JO (1988) Adrenoreceptor agents and baroreceptor-initiated secretion of vasopressin. In: Cowley AW Jr, Liard J-F, Ausiello DA (eds) Vasopressin: cellular and integrative functions. Raven Press, New York, pp 349–353

Block ML, Fisher AE (1975) Cholinergic and dopaminergic blocking agents modulate water intake elicited by deprivation, hypovolemia, hypertonicity and isoproterenol. Pharmacol Biochem Behav 3:251–262

Bradd J, Dubin J, Due B et al. (1989) Mapping of carotid sinus inputs and vagal cardiac outputs in the rat. Soc Neurosci Abstr 15:593

Breese GR, Traylor T (1970) Effect of 6-hydroxydopamine on brain norepinephrine and dopamine: evidence for selective degeneration of catecholamine neurons. Pharmacol Exp Ther 174:413–420

Breese GR, Traylor T (1971) Depletion of brain noradrenaline and dopamine by 6-hydroxydopamine. Br J Pharmacol 42:88–99

Brody MJ, Johnson AK (1980) Role of the anteroventral third ventricle region in fluid and electrolyte balance, arterial pressure regulation and hypertension. In: Martini L, Ganong WF (eds) Frontiers in neuroendocrinology vol 6. Raven Press, New York, pp 249–292

Buggy J, Fisher AE (1976) Anteroventral third ventricle site of action for angiotensin induced thirst. Pharmacol Biochem Behav 4:651–660

Buggy J, Johnson AK (1977a) Preoptic-hypothalamic periventricular lesions: thirst deficits and hypernatremia. Am J Physiol 233:R44–R52

Buggy J, Johnson AK (1977b) Anteroventral third ventricle periventricular ablation: temporary adipsia and persisting thirst deficits. Neurosci Lett 5:177–182

Buggy J, Johnson AK (1978) Angiotensin-induced thirst: effects of third ventricular obstruction and periventricular ablation. Brain Res 149:117–128

Buggy J, Fisher AE, Hoffman W, Johnson AK, Phillips MI (1975) Ventricular obstruction: effect on drinking induced by intracranial injection of angiotensin. Science 190:72–74

Carithers JR, Johnson AK (1988) Fine structural studies of the effects of AV3V lesions on the hypothalamo-neurohypophyseal neurosecretory system. In: Cowley AW, Liard J-F, Ausiello DA (eds) Vasopressin: cellular and integrative functions. Raven Press, New York, pp 301–320

Carpenter DO, Briggs DB, Strominger N (1983) Responses of neurons of canine area postrema to neurotransmitters and peptides. Cell Mol Neurobiol 3:113–126

Ciriello J, Calaresu FR (1980) Hypothalamic projections of renal afferent nerves in the cat. Can J Physiol Pharmacol 58, 574–576

Ciriello J, Hrycyshyn AW, Calaresu FR (1981) Glossopharyngeal and vagal afferent projection to the brain stem of the cat: a horseradish peroxidase study. J Auton Nerv Syst 4:63–79

Ciriello J, Caverson MM, Polosa C (1986) Function of the ventrolateral medulla in the control of the circulation. Brain Res Rev 11:359–391

Clemente CD, Sutin J, Silverstone JT (1957) Changes in electrical activity of the medulla on the intravenous injection of hypertonic solutions. Am J Physiol 188:193–198

Contreras RJ, Stetson PW (1981) Changes in salt intake after lesions of the area postrema and the nucleus of the solitary tract in rats. Brain Res 211:355–366

Cunningham JT, Johnson AK (1988) Models for the integration of humoral and neural factors critical to body fluid homeostasis. In: Yoshida S, Share L (eds) Recent progress in posterior pituitary hormones. Elsevier, Amsterdam, pp 97–105

Cunningham JT, Johnson AK (1989) Decreased norepinephrine in the ventral terminalis region is associated with angiotensin II drinking response deficits following local 6-hydroxydopamine injections. Brain Res 480:54–71

Cushny AR (1926) The secretion of the urine. Longmans Green, Lond, pp 135–137

Edwards GL, Johnson AK (1987) Ibotenic acid lesions of the lateral parabrachial nucleus causes enhanced drinking to extracellular thirst challenges. Fed Proc 46:1235

Edwards GL, Ritter RC (1982) Area postrema lesions increase drinking to angiotensin and extracellular dehydration. Physiol Behav 29:943–947

Edwards GL, Ritter RC (1989) Lateral parabrachial lesions attenuate ingestive effects of area postrema lesions. Am J Physiol 256:R306–R312

Edwards GL, Beltz TG, Johnson AK (1988) Enhanced excretion of sodium by area postrema-lesioned rats after intragastric saline loads. Soc Neurosci Abstr 14:316

Edwards GL, Cunningham JT, Beltz TG, Johnson AK (1989) Neuropeptide Y immunoreactive cells in the caudal medulla project to the median preoptic nucleus. Neurosci Lett 105:19–26

Elam M, Svensson TH, Thoren P (1985) Differentiated cardiovascular afferent regulation of the locus coeruleus neurons and sympathetic nerves. Brain Res 358:77–84

Eng R, Miselis RR (1981) Polydipsia and abolition of angiotensin-induced drinking after transections of subfornical organ efferent projections in the rat. Brain Res 225:200–206

Epstein AN, (1973) Epilogue: retrospect and prognosis. In: Epstein AN, Kissileff HR, Stellar E (eds) The neuropsychology of thirst: new findings and advances in concepts. Wiley, New York, pp 315–332

Epstein AN (1982) The physiology of thirst. In: Pfaff DW (ed) The physiological mechanisms of motivation. Springer-Verlag, New York, Heidelberg, Berlin, pp 165–214

Fitzsimons JT (1969) The role of a renal thirst factor in drinking induced by extracellular stimuli. J Physiol (Lond) 201:349–368

Fitzsimons JT (1979) The physiology of thirst and sodium appetite. Cambridge University Press, Cambridge

Fitzsimons JT, Setler PE (1975) The relative importance of central nervous catecholaminergic and cholinergic mechanisms in drinking in response to angiotensin and other thirst stimuli. J Physiol (Lond) 250:613–631

Foote SK, Bloom FE, Aston-Jones G (1983) Nucleus locus coeruleus: new evidence of anatomical and physiological specificity. Physiol Rev 63:844–914

Fulwiler CE, Saper CB (1984) Subnuclear organization of the efferent connections of the parabrachial nucleus in the rat. Brain Res Rev 7:229–259

Ganten D, Hutchinson JS, Schelling P, Ganten U, Fischer H (1976) The iso-renin angiotensin systems in extrarenal tissue. Clin Exp Pharmacol Physiol 3:103–126

Ganten S, Printz M, Phillips MI, Scholkens BA (eds) (1982) The renin angiotensin system in the brain. Springer, Heidelberg

Gordon FG, Brody MJ, Finsk GP, Buggy J, Johnson AK (1979) Role of central catecholamines in the control of blood pressure and drinking behaviour. Brain Res 178:161–173

Gray TS, Cassell MD, Williams TH, Lind RW, Johnson AK (1982) The subfornical organ-median preoptic pathway: evidence suggesting that an AII-like substance may be a neurotransmitter in this projection. Neurosci Abstr 8:102

Gronan RJ, York DH (1978) Effects of angiotensin II and acetylcholine on neurons in the preoptic area. Brain Res 154:172–177

Gutkind JS, Kurihara M, Castren E, Saavedra JM (1988) Increased concentration of angiotensin II binding sites in selected brain areas of spontaneously hypertensive rats. J Hypertens 6:79–84

Haberich FJ (1968) Osmoreception in the portal circulation. Fed Proc 27:1137–1141

Harding JW, Stone LP, Wright JW (1981) The distribution of angiotensin II binding sites in rodent brain. Brain Res 205:265–274

Hernesniemi J, Kawana E, Bruppacher H, Sandri C (1972) Afferent connections of the subfornical organ and of the supraoptic crest. Acta Anat 81:321–336

Housley GD, Martin-Body RL, Dawson NJ, Sinclair JD (1987) Brain stem projections of the glossopharyngeal nerve and its carotid sinus branch in the rat. Neuroscience 22:237–250

Hyde TM, Miselis RR (1984) Area postrema and adjacent nucleus of the solitary tract in water and sodium balance. Am J Physiol 247:R173–R182

Iovino M, Papa M, Monteleone P, Steardo L (1988) Neuroanatomical and biochemical evidence for the involvement of the area postrema in the regulation of vasopressin release in rats. Brain Res 447: 178–182

Jahr E, Nicoll RA (1982) Noradrenergic modulation of dendrodenritic inhibition in the olfactory bulb. Nature (Lond) 297:227–229

Jerome C, Smith GP (1982) Gastric vagotomy inhibits drinking after hypertonic saline. Physiol Behav 28:371–374

Johnson AK (1975) The role of the cerebral ventricular system in angiotensin-induced thirst. In: Peters F, Fitzsimons JT, Peters-Haefeli L (eds) Control mechanisms of drinking. Springer, Berlin Heidelberg New York, pp 117–122

Johnson AK (1982) Neurobiology of the periventricular tissue surrounding the anteroventral third ventricle (AV3V) and its role in behavior, fluid balance, and cardiovascular control. In: Smith OA, Galosy RA, Weiss SM (eds) Circulation, neurobiology and behavior. Elsevier, New York, pp 277–295

Johnson AK (1985a) The periventricular anteroventral third ventricle (AV3V): its relationship with the subfornical organ and neural systems involved in maintaining body fluid homeostasis. Brain Res Bull 15:595–601

Johnson AK (1985b) Role of the periventricular tissue surrounding the anteroventral third ventricle (AV3V) in the regulation of body fluid homeostasis. In: Schrier RW (ed) Vasopressin. Raven Press, New York, pp 319–382

Johnson AK (1990) Brain mechanisms and the control of body fluid homeostasis. In: Gisolfi CV, Land DR (eds) Perspectives in exercise. science and sports medicine: fluid homeostasis during exercise. Benchmark Press, Indianapolis, pp 347–424

Johnson AK, Buggy J (1977) A critical analysis of the site of action for the dipsogenic effect of angiotensin II. In: Buckley JP, Ferrario C, Lokhandwale MF (eds) Central action of angiotensin and related hormones. Pergamon Press, Elmsford, NY, pp 357–386

Johnson AK, Buggy J (1978) Periventricular preoptic-hypothalamus is vital for thirst and normal water economy. Am J Physiol 234:R122–R129

Johnson AK, Cunningham JT (1987) Brain mechanisms and drinking: the role of lamina terminalis-associated systems and extracellular thirst. Kidney Int 32:S35–S42

Johnson AK, Epstein AN (1975) The cerebral ventricles as the avenue for the dipsogenic action of intracranial angiotensin. Brain Res 86:399–418

Johnson AK, Wilkin LD (1987) The integrative role of neural systems of the lamina terminalis in the regulation of body fluid homeostasis. In: Gross P (ed) Circumventricular organs and body fluid homeostasis. CRC Press, Boca Raton, FT, pp 125–141

Johnson AK, Mann JFE, Rascher W, Johnson JK, Ganten D (1981) Plasma angiotensin II concentrations and experimentally induced thirst. Am J Physiol 240:R229–R234

Johnson AK, Robinson MM, Mann JFE (1986) The role of the renal renin–angiotensin system in thirst. In: de Caro G, Epstein AN, Massi M (eds) The physiology of thirst and sodium appetite. Plenum Press, New York, pp 161–180

Kahrilas PJ, Rogers RC (1984) Rat brainstem neurons responsive to changes in portal blood sodium concentration. Am J Physiol 247:R792–R799

Kalia M, Mesulam MM (1980) Brain stem projections of afferent and efferent fibers of the vagus nerve in the cat. II. Laryngeal, tracheobronchial, pulmonary, cardiac and gastrointestinal branches. J Comp Neurol 193:523–553

Kalia M, Sullivan JM (1982) Brainstem projections of sensory and motor components of the vagus nerve in the rat. J Comp Neurol 211:248–264

Kaufman S (1984) Role of right atrial receptors in the control of drinking in the rat. J Physiol (Lond) 349:389–396

Kostrzewa RM, Jacobowitz DM (1974) Pharmacological actions of 6-hydroxydopamine. Pharmacol Rev 26:199–288

Kraly FS (1978) Abdominal vagotomy inhibits osmotically induced drinking in the rat. J Comp Physiol 92:999–1013

Kraly FS, Gibbs J, Smith GP (1975) Disordered drinking after abdominal vagotomy in rats. Nature 257:226–228

Lashley KS (1938) Experimental analysis of instinctive behavior. Psychol Rev 45:445–471

Leslie RA (1986) Comparative aspects of the area postrema: fine-structural considerations help to determine its function. Cell Mol Neurobiol 6:95–120

Lind W, Ganten D (1990) Angiotensin. In: Bjorklund A, Hokfelt T, Kuhar MJ (ed) Neuropeptides in the CNS (Handbook of Chemical Neuroanatomy, vol 9, part II) Elsevier, Amsterdam (in press)

Lind RW, Johnson AK (1981) Periventricular preoptic-hypothalamic lesions: effects on isoproterenol induced thirst. Pharmacol Biochem Behav 15:563–565

Lind RW, Johnson AK (1982a) Subfornical organ-median preoptic connections and drinking and pressor responses to angiotensin II. J Neurosci 2:1043–1051

Lind RW, Johnson AK (1982b) Central and peripheral mechanisms mediating angiotensin-induced thirst. In: Ganten D, Printz M, Phillips MI, Scholkens BA (eds) The renin angiotensin system in the brain. Exp Brain Res (Suppl. 4). Springer, Berlin, Heidelberg, New York, pp 353–364

Lind RW, Johnson AK (1982c) On the separation of functions mediated by the AV3V region. Peptides 3:495–499

Lind RW, Johnson AK (1983) A further characterization of the effects of AV3V lesions on ingestive behavior. Am J Physiol 245:R83–R90

Lind RW, Shrager EE, Bealer SL, Johnson AK (1979) Critical tissues within the periventricular region of the antero ventral third ventricle (AV3V) associated with specific thirst deficits. Soc Neurosci Abstr 5:220

Lind RW, Hartle DK, Brody MJ, Johnson AK (1981) Separation of the functional deficits induced by lesions of the AV3V. Fed Proc 40:390

Lind RW, Van Hoesen GW, Johnson AK (1982) An HRP study of the connections of the subfornical organ of the rat. J Comp Neurol 210:265–277

Lind RW, Thunhorst RL, Johnson AK (1984) The subfornical organ and the integration of multiple factors in thirst. Physiol Behav 32:69–74

Lindvall O, Bjorklund A (1974) The organization of the ascending catecholamine neuron systems in the rat brain as revealed by the glyoxylic acid fluorescence method. Acta Physiol Scand 412:1–48

Lynch KR, Hawelu-Johnson CL, Guyenet PG (1987) Localization of brain angiotensinogen mRNA by hybridization histochemistry. Mol Brain Res 2:149–158

Mann JFE, Johnson AK, Ganten D, Ritz E (1987) Thirst and the renin–angiotensin system. Kidney Int 32:S27–S34

Martin JR (1981) Effects of partial and complete vagal denervation on spontaneous ingestion and drinking induced with volemic and osmotic regulatory challenges. J Neurosci Rev 6:243–250

Martin JR, Novin D (1981) Response to dipsogenic stimuli after abdominal vagotomy in rats. Physiol Psychol 9:181–186

McKinley MJ, Denton DA, Weisinger RS (1978) Sensors for antidiuresis and thirst-osmoreceptors or CSF sodium detectors? Brain Res 141:89–103

McKinley MJ, Denton DA, Leksell LG et al. (1982) Osmoregulatory thirst in sheep is disrupted by ablation of the anterior wall of the optic recess. Brain Res 236:210–215

McRae-Degueurce A, Bellin SI, Landas SK, Johnson AK (1986) Fetal noradrenergic transplants into amine-depleted basal forebrain nuclei restore drinking to angiotensin. Brain Res 374:162–166

McRae-Degueurce A, Cunningham JT, Bellin S, Landas S, Wilkin L, Johnson AK (1987) Fetal noradrenergic cell suspensions transplanted into amine-depleted nuclei of adult rats. Ann NY Acad Sci 495: 757–759

Mendelsohn FAO, Quirion R, Saavedra JM, Aguilera G, Catt KJ (1984) Autoradiographic localization of angiotensin II receptors in rat brain. Proc Natl Acad Sci USA 81:1575–1579

Miller TR, Handelman WA, Arnold PE, McDonald KM, Molinoff PB, Schrier RW (1979) Effect of central catecholamine depletion on osmotic and nonosmotic stimulation of vasopressin (antidiuretic hormone) in the rat. J Clin Invest 64:1599–1607

Miselis RR (1981) The efferent projections of the subfornical organ of the rat: a circumventricular organ within a neural network subserving water balance. Brain Res 230:1–23

Miselis RR (1986) The visceral neuraxis in thirst and renal function. In: de Caro G, Epstein AN, Massi M (eds) The physiology of thirst and sodium appetite. Plenum Press, New York, pp 345–354

Miselis RR, Shapiro RE, Hand PJ (1979) Subfornical organ efferents to neural systems for control of body water. Science 205:1022–1025

Miselis RR, Hyde TM, Shapiro EE (1986) Disturbances in water balance controls following lesions to the area postrema and adjacent solitary nucleus. In: de Caro G, Epstein AN, Massi M (eds) The physiology of thirst and sodium appetite. Plenum Press, New York, London, pp 287–297

Moore SD, Guyenet PG (1983) Effect of blood pressure and clonidine on the activity of A2 noradrenergic neurons. Soc Neurosci Abstr 9:549

Moore-Gillon MJ (1980) Effects of vagotomy on drinking in the rat. J Physiol (Lond) 308:417–426

Moore-Gillon MJ, Fitzsimons JT (1982) Pulmonary vein-atrial junction stretch receptors and the inhibition of drinking. Am J Physiol 11:R452–R457

Nazarali AJ, Gutkind JS, Saavedra JM (1987) Regulation of angiotensin II binding sites in the subfornical organ and other rat brain nuclei after water deprivation. Cell Mol Neurobiol 7:447–455

Nelson DO, Johnson AK (1985) Subfornical organ projections to nucleus medianus: electrophysiological evidence for angiotensin II synapses. Fed Proc 44:1010

Nicolaïdis S, Ishibaski S, Gueguen B, Thornton SW, Beaurepaire R (1983) Iontophoretic investigation of identified SFO angiotensin in responsive neurons firing in relation to blood pressure changes. Brain Res Bull 10:357–363

Norgren R (1981) The central organization of the gustatory and visceral afferent systems in the nucleus of the solitary tract. In: Katsuki Y, Norgren R, Sato M (eds) Brain mechanisms of sensation. Wiley, Chichester, New York, pp 143–160

Ohman LE, Johnson AK (1986) Lesions in lateral parabrachial nucleus enhance drinking to angiotensin II and isoproterenol. Am J Physiol 251:R504–R509

Ohman LE, Johnson AK (1987) Brainstem mechanisms and the inhibition of drinking. Fed Proc 46:1434

Ohman LE, Johnson AK (1989) Brain stem mechanisms and the inhibition of angiotensin-induced drinking. Am J Physiol 256:R264–R269

Panneton WM, Loewy AD (1980) Projections of the carotid sinus nerve to the nucleus of the solitary tract in the cat. Brain Res 191:239–244

Plunkett LM, Shigematsu K, Kurihara M, Saavedra JM (1987) Localization of angiotensin II receptors along the anteroventral third ventricle area of the rat brain. Brain Res 405:205–212

Quintin L, Gillon J-Y, Ghignone M, Renaud B, Pujol J-F (1987) Baroreceptor-linked variations of catecholamine metabolism in the caudal ventrolateral medulla: an in vivo electrochemical study. Brain Res 425:319–326

Rettig R, Johnson AK (1986) Aortic baroreceptor deafferentation diminishes saline-induced drinking in rats. Brain Res 370:29–37

Ricardo JA, Koh ET (1978) Anatomical evidence of direct projections from the nucleus of the solitary tract to the hypothalamus, amygdala, and other forebrain structures in the rat. Brain Res 153:1–26

Robinson MM, Evered MD (1987) Pressor action of intravenous angiotensin II reduces drinking response in rats. Am J Physiol 252:R754–R759

Rogers RC, Novin D, Butcher LL (1979) Electrophysiological and neuroanatomical studies of hepatic portal osmo- and sodium-receptive afferent projections within the brain. J Auton Nerv Syst 1: 183–202

Ross EJ, Christie SBM (1969) Hypernatremia. Medicine 48:441–473

Roth GI, Yamamoto WS (1968) The microcirculation of the area postrema in the rat. J Comp Neurol 133:329–340

Saavedra JM (1987) Regulation of atrial natriuretic peptide receptors in the rat brain. Cell Mol Neurobiol 7:151–153

Saper CB, Loewy AD (1980) Efferent connections of the parabrachial nucleus in the rat. Brain Res 197:291–317

Saper CB, Reis DJ, Joh T (1983) Medullary catecholamine inputs to the anteroventral third ventricular cardiovascular regulatory region in the rat. Neurosci Lett 42:285–291

Sawchenko PE, Swanson LW (1983) The organization of noradrenergic pathways from the brainstem to the paraventricular and supraoptic nuclei in the rat. Brain Res Rev 4:275–291

Shapiro RE, Miselis RR (1985) The central neural connections of the area postrema of the rat. J Comp Neurol 234:344–364

Shrager EE (1981) The contribution of periventricular structures of the lamina terminalis to the control of thirst. Doctoral dissertation. University of Iowa, Iowa City

Shrager EE, Johnson AK (1978) Ablation of periventricular tissue surrounding the anteroventral third ventricle (AV3V) blocks drinking to caval ligation but not renin release. Soc Neurosci Abstr 4:180

Simmonds MA, Uretsky NJ (1970) Central effects of 6-hydroxydopamine on the body temperature of the rat. Br J Pharmacol 40:630–638

Simpson JB (1981) The circumventricular organs and the central actions of angiotensin. Neuroendocrinology 32:248–256

Simpson JB, Routtenberg A (1973) Subfornical organ: site of drinking elicited by angiotensin II. Science 181:1172–1175

Simpson JB, Epstein AN, Camardo Jr JS (1978) Localization of receptors for the dipsogenic action of angiotensin II in the subfornical organ of rat. J Comp Physiol Psychol 92:581–601

Skofitsch G, Jacobowitz DM (1988) Atrial natriuretic peptide in the central nervous system of the rat. Cell Mol Neurobiol 8:339–391

Smith GP (1986) Peripheral mechanisms for the maintenance and termination of drinking in the rat. In: de Caro G, Epstein AN, Massi M (eds) The physiology of thirst and sodium appetite. Plenum Press, New York, pp 265–277

Snyder RL Jr, Sutin J (1961) Effect of lesions of the medulla oblongata on electrolyte and water metabolism in the rat. Exp Neurol 4:424–435

Spoerri V (1963) Uber die Gefassversorgung des Subfornikalorgans der Ratte. Acta Anat (Basel) 54:333

Stricker EM (1969) Osmoregulation and volume regulation in rats: inhibition of hypovolemic thirst by water. Am J Physiol 217:98–105

Stricker EM, Zigmond MJ (1974) Effects on homeostasis of intraventricular injections of 6-hydroxydopamine in rats. J Comp Physiol Psychol 86:973–994

Swanson LW, Mogenson GJ (1981) Neural mechanisms for the functional coupling of autonomic endocrine and somatomotor responses in adaptive behaviour. Brain Res Rev 3:1–34

Swanson L, Sawchenko PE (1983) Hypothalamic integration: organization of the periventricular and supraoptic nuclei. Ann Rev Neurosci 6:269–324

Tanaka J, Saito H, Kaba H, Seto K (1987) Subfornical organ neurons act to enhance the activity of paraventricular vasopressin neurons in response to intravenous angiotensin II. Neurosci Res 4: 424–427

Terasawa E, Wiegand SJ, Bridson WE (1980) A role for medial preoptic nucleus on afternoon of proestrus in female rats. Am J Physiol 238:E533–E539

Ter Horst GJ, deBoer P, Luiten PGM, van Willigan JD (1989) Ascending projections from the solitary tract nucleus to the hypothalamus. A *Phaseolus vulgaris* lectin tracing study in the rat. Neuroscience 31:785–797

Tordoff MG, Schulkin J, Friedman MI (1986) Hepatic contribution to satiation of salt appetite in rats. Am J Physiol 251:R1095–R1102

Tribollet E, Barberris C, Jard S, Dubois-Dauphin M, Dreifuss JJ (1988) Localization and pharmacological characterization of high affinity binding sites for vasopressin and oxytocin in the rat brain by light microscopic autoradiography. Brain Res 442:105–118

van der Kooy D, Koda LY (1983) Organization of the projections of a circumventricular organ: the area postrema in the rat. J Comp Neurol 219:328–338

Verney EB (1947) The antidiuretic hormone and the factors which determine its release. Proc R Soc 135B:25–106

Watson WE (1985) The effect of removing area postrema on the sodium and potassium balances and consumptions in the rat. Brain Res 359:224–232

Wilkin LD, Patel KP, Schmid PG, Johnson AK (1987) Increased norepinephrine turnover in median preoptic nucleus following reduced extracellular fluid volume. Brain Res 423:369–372

Wise BL, Ganong WF (1960a) The effect of ablation of the area postrema on water and electrolyte metabolism in dogs. Acta Neuroveg (Vienna) 22:14–32

Wise BL, Ganong WF (1960b) Effect of brain stem stimulation on renal function. Am J Physiol 198: 1291–1295

Witt DM, Keller AD, Batsel HL, Lynch JR (1952) Absence of thirst and resultant syndrome associated with anterior hypothalamectomy in the dog. Am J Physiol 171:780

Wolf AV (1950) Osmometric analysis of thirst in man and dog. Am J Physiol 161:75–86

Woodward DJ, Moises HC, Waterhouse BD, Hoffer BJ, Freedman R (1979) Modulating actions of norepinephrine in the central nervous system. Fed Proc 38:2109–2116

Wyss JM, Donovan MK (1984) A direct projection from the kidney to the brainstem. Brain Res 298: 130–134

Zardetto-Smith AM, Gray TS (1987) A direct neural projection from the nucleus of the solitary tract to the subfornical organ in the rat. Neurosci Lett 80:163–166

Zimmerman MB, Blaine EH, Stricker EM (1981) Water intake in hypovolemic sheep: effects of crushing the left atrial appendage: Science 211:489–490

Chapter 10

Neurochemistry of the Circuitry Subserving Thirst

B.J. Oldfield

Introduction

Many regions of the brain including parts of the limbic system, lateral hypothalamus and preoptic area have been nominated as the focus of generation of thirst. However, a recurrent theme of many of the chapters of this book is that the lamina terminalis, the neural tissue forming the anterior wall of the third cerebral ventricle, is the crucial area for the maintenance of fluid and electrolyte homeostasis in the body. This region of the brain described recently as a "complex unity" (Lind 1988a) because of its common functional and morphological features, is in fact composed of three discrete regions or nuclei, the subfornical organ (SFO) and the organum vasculosum lamina terminalis (OVLT) which both lack a blood–brain barrier, and an intervening tissue, the median preoptic nucleus (MnPO) which like the brain at large, is protected from the haemal milieu by non-fenestrated blood vessels. The lack of a blood–brain barrier in at least part of the lamina terminalis confers upon it the ability to sample the haemal environment. Furthermore, the extensive neural interconnections of its component nuclei and its afferent and efferent projections to other parts of the brain (Fig. 10.1a) give it the potential to integrate both humoral and neural information.

Two observations have focused attention on the lamina terminalis as a site for the genesis of drinking behaviour. One was the finding that the subfornical organ and the ventral lamina terminalis together with its surrounding periventricular structures (AV3V region) mediate, with remarkable sensitivity, the dipsogenic response to elevated levels of the hormone angiotensin II (Ang II) introduced into the circulation (Simpson and Routtenberg 1975) or into the cerebrospinal fluid (CSF) (Buggy and Fisher 1976). The other is that in a wide range of species, ablation of the AV3V region produces an adipsia, albeit sometimes transient (Buggy and Johnson 1977). Although this focus has been maintained, it has become increasingly apparent that the lamina terminalis cannot be viewed as solely important, but rather it should be seen as one level of the integration of osmotic, humoral and neural information relating to total body fluid homeostasis. This integration can be appreciated from examination of lesions of the ventral lamina terminalis which not only produce an adipsic response but, in the face of the ensuing dehydration, produce an inappropriate loss of fluid via the kidneys (Johnson and Buggy 1976). Therefore, the pathways subserving thirst are inextricably involved with those controlling arginine vasopressin (AVP) release from the pituitary. Furthermore, it has been suggested that during hypovolaemia, in addition to a direct action on the lamina terminalis through Ang II, information conveyed from low pressure/volume receptors via catecholaminergic pathways from the brainstem to the median preoptic nucleus may be responsible for

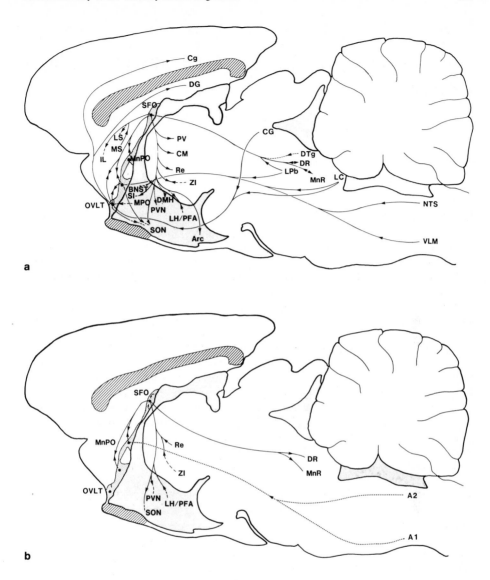

Fig. 10.1 a. A midsagittal section of the rat brain showing most of the afferent and efferent projections of the lamina terminalis (lightly striped area). The direction of pathways is indicated by arrows and reciprocal projections are denoted by double arrows. **b** Angiotensinergic (solid lines) and catechol-aminergic (dotted lines) pathways projecting to and from the lamina terminalis. Note the presence of intrinsic angiotensingeric neurons in the SFO. Abbreviations: Arc, arcuate nucleus; BNST, bed nucleus stria terminalis; Cg, cingulate cortex; CG, central grey; CM, central medial thalamic nucleus; DG, dentate gyrus; DMH, dorsal medial hypothalamus; DR, dorsal raphe; DTg, dorsal tegmentum; IL, infralimbic cortex; LC, locus coeruleus; LH/PFA, lateral hypothalamus perifornical area; LPb, lateral parabrachial nucleus; LS lateral septum; MnPO, median preoptic nucleus; MnR, median raphe; MPO, medial preoptic area; MS, medial septum; NTS, nucleus solitary tract; OVLT, organum vasculosum lamina terminalis; PV, paraventricular nucleus of thalamus; PVN, paraventricular nucleus hypothalamus; Re, reuniens nucleus; SI, substantia innominata; SON, supraoptic nucleus; VLM, ventrolateral medulla; ZI, zona incerta.

augmenting drinking responses (Johnson and Cunningham 1987). Far less is known of the interaction with higher centres which must provide on one hand a framework of memory and experience on which the drinking response is based, and on the other, the somatomotor pathways responsible for its activation. Neurochemists working in the field have attempted to assign chemical specificity to what we know of the circuitry of thirst. The remainder of this chapter is based on these attempts and is underpinned by two important facts: (a) the lamina terminalis is heavily favoured as the site of initiation of thirst, and therefore, at least some of its first order connections are good candidates for the circuits of thirst, and (b) angiotensin II is the predominant neurochemical feature of many of these circuits.

The Role of Angiotensin II in the Initiation of Thirst

The motivational state of thirst, and its behavioural expression as drinking may be initiated by the action of Ang II on the brain in either of two ways. One is via the circulation, where it has been reasoned that the elevation of Ang II in the blood which accompanies hypovolaemia may trigger thirst, and the other is via Ang II-containing pathways intrinsic to the central nervous system (CNS).

Circulating Ang II may access the SFO and OVLT by virtue of their deficient blood–brain barrier. Most experimental evidence involving lesions and selective knife cuts supports a direct dipsogenic action of circulating Ang II on the SFO (Simpson and Routtenberg 1973; Lind and Johnson 1982a,b), although a role for the OVLT has not been discounted (Phillips and Hoffman 1976; Phillips 1987). Centrally administered Ang II on the other hand produces its potent dipsogenic effect by acting at the "AV3V region" (Buggy and Fisher 1976). The AV3V is a collective term loosely describing the ventral median preoptic nucleus, OVLT, anterior periventricular nucleus and adjacent medial preoptic area as well as the anteroventral periventricular nucleus. The site of action of centrally administered Ang II has not always been so clearly defined. In early controversies the relative importance of the SFO and OVLT in the mediation of drinking produced by central Ang II was debated (see Lind 1988a for review). A close examination of the potential barriers to diffusion of Ang II puts the median preoptic nucleus, or a related area behind the blood–brain barrier, among the most likely candidates (Fig. 10.2). As noted above the two circumventricular organs (CVOs) forming the dorsal and ventral poles of the lamina terminalis contain plexuses of fenestrated blood vessels whereas the intervening median preoptic nucleus does not. As a consequence of the fact that Ang II does not readily cross the blood–brain barrier, the SFO and OVLT are the only sites in the lamina terminalis accessible to blood-borne Ang II. A peculiarity which is not as well appreciated is that the CVOs are protected from the CSF at their ventricular surface by tight junctions between overlying ependymal cells. These junctions are not present in other parts of the ventricles. Furthermore, the CVOs are isolated from the CSF of the interstitial spaces of the brain by a complex array of tanycytic processes, sealed by tight junctions, which encircle the vessels of the neuropil adjoining the CVOs. These junctions are effective in precluding the free passage of hormones such as LHRH and vasopressin between the CVOs and the adjacent brain parenchyma (Krisch et al. 1987).

In view of these barriers, the action of centrally administered Ang II on the AV3V must exclude the OVLT, although Landas et al. (1980) have suggested that the processes of tanycytes form a conduit to the OVLT from the ventricular CSF

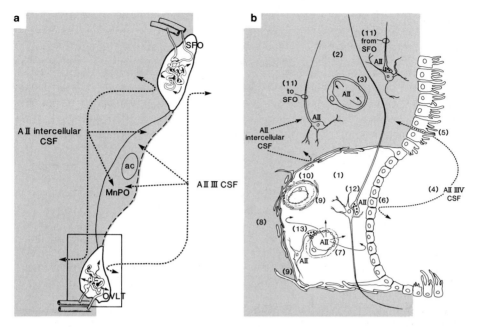

Fig. 10.2. a A schematic sagittal diagram of the lamina terminalis showing the hypothetical barriers within the CNS to angiotensin II (AII) in various intracranial compartments. **b** A more detailed view of the OVLT (1) and overlying ventral median preoptic nucleus (2). The features depicted within the OVLT apply to other CVOs including the SFO. The blood–brain barrier characteristic of most brain regions (3) prevents access of AII to the parenchyma. AII within the ventricular CSF (4) can permeate periventricular regions via the intercellular clefts of the ventricular ependymal lining (5), but cannot penetrate the CVOs because of tight junctions (6) between the overlying specialized ependyma. Fenestrations within capillaries of the SFO and OVLT (7) allow free access of AII to the internal milieu of the CVOs, but further penetration into the adjacent brain parenchyma (8) is restricted by a complex gial boundary. The barrier formed by tight junctions between tanycytic processes (9) is too extensive and convoluted to be accurately represented in a simple two-dimensional sketch. The true form of the CVO–brain barrier more closely approximates the arrangement of processes around individual blood vessels (10) (see Krisch and Leonhardt 1978; Krisch et al. 1987) than the "corral" arrangement drawn around the OVLT. As outlined in the text, AII-containing neurons project reciprocally between the SFO and the MnPO (11) and angiotensingergic terminals in the OVLT are of both synaptic (12) and secretory (13) types (see Oldfield et al. 1989). As yet there is no direct evidence to suggest that secretory Ang II neurons are intrinsic to the OVLT as drawn, although this appears to be the case in the SFO. (Redrawn and expanded from Phillips (1987).)

thus enabling the octapeptide to penetrate the glial boundary. Unless this route can be further substantiated it seems most likely that Ang II introduced into the CNS acts to induce thirst at a point in the AV3V behind the blood–brain barrier such as the median preoptic nucleus.

Central Angiotensinergic Pathways

Central Production of Angiotensin II

The significance to thirst of angiotensin-containing neural pathways in the brain is derived from two facts described above: (a) that Ang II introduced into the brain causes thirst and (b) that circulating Ang II does not cross the blood–brain barrier. Given these points the concept arises that neuronally released Ang II may be implicated in the neurocircuitry of thirst. Before discussion of Ang II-containing pathways, it seems pertinent to review very briefly some of the observations which have led to the acceptance of a brain renin–angiotensin system (for reviews see Ganong 1981; Printz 1988). Twenty years ago, two groups (Ganten et al. 1971; Fischer-Ferraro et al. 1971) independently isolated a renin-like enzyme from the brains of rats and dogs. These findings, together with the presence in the brain of renin substrate, and the enzyme responsible for conversion of Ang I to Ang II, angiotensin converting enzyme (Roth et al. 1969), strengthened the claim for a brain renin–angiotensin system independent of that operating in peripheral, non-neural tissues. The dynamic nature of central Ang II production was demonstrated by the intracerebroventricular administration of an inhibitor of angiotensin converting enzyme which caused a reduction of local Ang II (Ganten et al. 1983). However, acceptance of some of these concepts has not been universal. The presence of brain isorenin was vigorously questioned by Day and Reid (1976) who claimed that the enzyme properties attributed to renin were more likely those of the cytosomal acid protease cathepsin D. It was not until Hirose et al. (1978) and others claimed to separate renin-like activity from other proteases in the brain, that scepticism concerning brain renin was largely removed. Despite the fact that renin has been identified immunocytochemically by a number of groups (Inagami et al. 1980; Fuxe et al. 1980; Calza et al. 1982; Healy and Printz 1984), the mRNA for renin, the critical indicator of local production, has not been consistently identified in the brain until recently (Field et al. 1984; Dzau et al. 1984). The most recent studies utilizing sensitive solution hybridization techniques and large amounts of input RNA have identified renin mRNA in the brains of mice and rats (Suzuki et al. 1988; Miller et al. 1989), but only at the limits of resolution of these procedures. Hybridization histochemistry which would give an indication of the tissue location of renin mRNA has as yet been unsuccessful.

Much of the impetus in the investigation of the brain renin–angiotensin system over the last 10 years has been directed at the colocalization within single neurons of different components of the renin–angiotensin cascade. If this co-compartmentalization were identified, the concept of an intracellular biosynthetic pathway for the generation of Ang II could be easily supported. In the tradition of study of the brain renin–angiotensin system, the findings in this regard are still marred by inconsistency; however, it is safe to say that colocalization has proven incomplete in all cases. In regions such as the paraventricular nucleus (PVN) for example, where the colocalization of all components seems tantalizingly close, angiotensin II immunoreactivity is reportedly found in magnocellular neurons (Healy and Printz 1984; Lind et al. 1985a; Aronsson et al. 1988) as is the mRNA for angiotensinogen in some (Aronsson et al. 1988) but not all studies (Lynch et al. 1987), renin is present in a different population of magnocellular neurons (Healy and Printz 1984; Aronsson et al. 1988) and Ang I is likely to be present in the parvocellular PVN (Healy and Printz 1984). In addition, angiotensinogen mRNA and its peptide product is variously reported as being solely in neurons (Aronsson et al. 1988), solely in glia (Deschepper et al. 1986; Lynch et al. 1987; Stornetta et al. 1988) or in both neurons and glia (Richoux et al. 1988; Thomas and Sernia et al. 1988). Converting enzyme, identified by in vitro autoradiography, has been described in the PVN (Chai et al. 1987; Strittmatter and Snyder 1987) but in immunocytochemical studies it has either not been detected at this site or only with insufficient resolution to determine colocalization with other components of the renin–angiotensin system (Strittmatter and Snyder 1987). Our recent ultrastructural studies of the distribution of converting enzyme in the circumventricular organs show an association with glia (Oldfield et al. 1988). The weight of evidence therefore, is shifting away from a complete intra-neuronal biosynthetic pathway toward a far more novel mode of peptide production involving interaction of neurons and glia. In one such hypothetical scenario, glia may be the source of angiotensinogen which is secreted into the extracellular space and may be the starting point of a cascade resulting in the production of Ang II which is subsequently incorporated into specific neurons.

Neural Circuits Containing Angiotensin II

Irrespective of its mode of production, it is clear from immunocytochemical studies that Ang II-containing neuronal cell bodies, fibres and terminals can be identified throughout the CNS (Lind et al. 1985b). A complete description of the extensive network of Ang II-containing neurons is beyond the scope of this chapter but these essentially overlap with regions involved with homeostatic mechanisms, either in relation to the putative pathways of thirst described above, or associated with networks important in cardiovascular control. Cell bodies are conspicuously present in the SFO and to a lesser extent in the MnPO and the OVLT. They are also abundant in the perifornical area of the hypothalamus, PVN, supraoptic nucleus (SON), zona incerta, parts of the thalamus, the central nucleus of the amygdala, and the nucleus of the solitary tract in the brainstem. A number of lines of experimental evidence considered together, support the concept that Ang II may be acting as a neurotransmitter or neuromodulator in these pathways (Fuxe et al. 1976). They include the close correlation of Ang II terminal fields with Ang II receptors (see below), the somatofugal transport of the peptide within neurons (Lind et al. 1984b), the responsiveness of neurons in the vicinity of Ang II terminals to iontophoretically applied Ang II (Nicholl and Barker 1971; Okuya et al. 1987), and the identification of Ang II within typical synaptic vesicles in terminals forming synaptic associations (Oldfield et al. 1989).

In the CVOs, at the light microscope level of resolution, large varicose Ang II-containing "terminals" in the vicinity of fenestrated capillaries have been described and it is proposed that these are acting in a neurosecretory rather than in a neurotransmitter role (Lind 1988a). Lind further described these as a peculiarity of the Brattleboro strain of rat. Our ultrastructural studies in the CVOs of normal Sprague–Dawley rats indicated that such terminals do have the classical configuration of neurosecretory terminals seen in the external layer of the median eminence, in that they lack synaptic specialization and are clustered around the perivascular spaces of fenestrated vessels. That they and their parent cell bodies are intrinsic to the CNS, and indeed to the SFO, is shown by their persistence after nephrectomy (Oldfield et al. 1989) and surgical isolation of the SFO (Lind 1988a). The possible neurosecretory function of Ang II in each of the CVOs is discussed by Oldfield et al. (1989).

It has been noted that the dipsogenic response to circulating Ang II is mediated by the SFO and that lesions of the SFO effectively remove this response. In addition, knife cuts interrupting efferent pathways of the SFO either just below the nucleus or further ventrally as they enter the MnPO are equally effective in eliminating drinking (Lind and Johnson 1982a,b). Therefore a probable thirst function has been assigned to the SFO to MnPO projection that has been well established by tracing studies (Miselis et al. 1979; Miselis 1981; Lind et al. 1982; McKinley et at 1989). It would be possible better to define the observed drinking response to centrally administered

Ang II acting in the vicinity of the MnPO, if this pathway were shown to utilize Ang II as a neurotransmitter. Some of the neuropharmacological and ligand binding experiments outlined below certainly support this scheme, and double-labelling studies employing the retrograde tracing of dyes and immunocytochemical localization of Ang II, have shown this to be a very likely prediction (Lind et al. 1984a).

Angiotensin-containing projections from the SFO have also been identified which are directed to the PVN (Lind et al. 1984b) and SON (Jhamandas et al. 1989) as well as to midbrain structures (Lind 1986). Conversely, Ang II-positive inputs to the SFO are derived from the perifornical region of the hypothalamus, the zona incerta, the thalamus (Lind et al. 1984a) and from the median preoptic nucleus (Lind 1988a). It is suspected, but not yet proven in all cases, that Ang II projections are reciprocal with the SFO from each of these areas. Angiotensin-containing pathways associated with the lamina terminalis are depicted in Fig. 10.1b.

As alluded to earlier, there is a commonality of function associated with the lamina terminalis in terms of both water ingestion and excretion. This is evidenced by electrophysiological studies which have elucidated the dual action of Ang II in the SFO on drinking (Gutman et al. 1988) and control of the magnocellular neurosecretory system (Ferguson and Renaud 1986; Tanaka et al. 1987b). Activation of the SFO by Ang II applied either directly or into the circulation, changes the excitability of putative vasopressin neurons (Ferguson and Renaud 1986; Tanaka et al. 1986, 1987a, Tanaka 1989) and this activation can be interrupted by pretreatment of the MnPO with an angiotensin II antagonist, saralasin (Tanaka et al. 1986, 1987a; Tanaka 1989). In conjunction with double-labelling studies at the ultrastructural level which have provided evidence for a direct projection from the MnPO to vasopressin neurons in the SON (Oldfield et al. 1986; McKinley et al. 1989), these data indicate that at least part of the input from the lamina terminalis to the magnocellular system is mediated via the MnPO. Projections directly from the SFO to the PVN and SON have also been well established, and some of these synapse with vasopressin-containing neurons in the rat (unpublished observations).

Angiotensin Receptors and Their Involvement in the Circuitry of Thirst

The distribution of Ang II receptors, localized with radiolabelled analogues, correlates remarkably well with brain regions involved in the control of fluid and electrolyte balance (McKinley et al. 1989). This is most prominent in the lamina terminalis, where Ang II receptors in the CVOs, the SFO and OVLT, by virtue of their deficient blood–brain barrier, are the targets of circulating Ang II (Van Houten et al. 1980; Simpson 1980) but are also likely to be receptive to neuronally released Ang II from afferent fibres (Fig. 10.1b). In addition, it is conceivable given the high concentration of converting enzyme present in the SFO and OVLT (Chai et al. 1987) that Ang II is formed locally from circulating Ang I. A high density of Ang II receptors in the median preoptic nucleus completes a continuum of binding in the lamina terminalis that is conserved across all the species studied (Fig. 10.3; Mendelsohn et al. 1984; Speth et al. 1985; Gehlert et al. 1986; McKinley et al. 1986; Mendelsohn et al. 1988) including man (McKinley et al. 1987) and forms part of the basis for the functional unity described by Lind (1988a). Other discrete areas of the brain containing high densities of Ang II receptors include the remaining CVOs, the median eminence, the anterior pituitary and area postrema, as well as other structures implicated in the pathways of thirst such as the paraventricular nucleus. Ang II receptor binding in

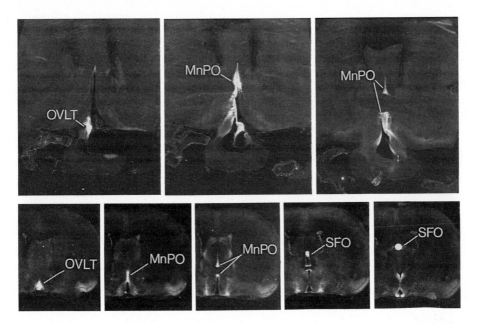

Fig. 10.3. Binding of [125]I-[Sar1 Ile8]-Ang II in the lamina terminalis of the human (upper panel) and the rat (lower panel) lamina terminalis. Sections have been cut in a coronal plane and are arranged in order from left to right in a rostrocaudal direction. (Data kindly provided by A.M. Allen, M.J. McKinley, and F.A.O. Mendelsohn (McKinley et al. 1987; Allen et al. 1988).)

parts of the lamina terminalis can be modulated by the hydrational state of the animal (Mendelsohn et al. 1983; Nazarali et al. 1987) but as yet changes in related forebrain nuclei have not been detected in response to dehydration.

In many studies of Ang II as a putative neurotransmitter it has been assumed that its action is postsynaptic; however, in some sensory systems such as the nodose ganglia or spinal dorsal root ganglia, receptors are transported in a form recognizable by the radiolabelled antagonist [125]I-[Sar1 Ile8] -Ang II from cell body to axon terminal. The somatofugal transport of the receptor to a presynaptic site has been shown by the ligation of the vagus nerve and the subsequent accumulation of label proximal to the ligation. There is also a reduction of binding in the nucleus of the solitary tract, the central site of termination of sensory fibres travelling in the vagus (Lewis et al. 1986). It is unlikely that presynaptic Ang II receptors are peculiar to sensory systems. Structures such as the median eminence which lack neuronal cell bodies or processes which might form suitable postsynaptic sites are still labelled with radioactive Ang II analogues.

There are other characteristics of the Ang II receptor that remain unresolved or controversial. There are even some suggestions that Ang III may be the more biologically important form of the peptide in the brain (Wright et al. 1985; Felix and Harding 1986; Harding and Felix 1987), a concept that draws some support from the observation that Ang II and Ang III compete with radioiodinated Ang II with equal affinity for binding sites in membranes isolated from rat brain (Printz et al. 1987). It remains to be seen, perhaps through advances in the understanding of the molecular biology of the angiotensin receptor, whether there is in fact a family of Ang II receptors in the brain. Recently it has been demonstrated that the *mas* oncogene may code for the angiotensin receptor (Jackson et al. 1988); however, the concentration of the *mas* transcript largely in the cerebral cortex does not correspond well with our knowledge of the receptor studied by in vitro autoradiography or the actions of Ang II assessed by other means.

Correlation of Angiotensin Receptors with Angiotensin-Containing Pathways

In general, the correspondence of Ang II receptors to Ang II immunoreactive terminals is extremely good (Lind 1988b), adding credibility to the argument that Ang II acts as a neurotransmitter. There are mismatches however, and perhaps the most outstanding of these are in the supraoptic nucleus and central nucleus of the amygdala of the rat. Here moderate densities of Ang II-positive terminals in the case of the SON and extremely high densities in the central nucleus of the amygdala correspond with very low levels of Ang II receptors identified by in vitro autoradiography. The absence of binding in areas such as the amygdala, where immunoreactive Ang II is present in large 100 nm vesicles within synaptic terminals (Fig. 10.4) is difficult to resolve but may indicate that Ang II is responsible for the modulation of release of a more conventional neurotransmitter at these sites.

Other Peptides and Monoamines Involved in Thirst

Catecholamines – Lateral Hypothalamus

The disruption of drinking behaviour resulting from lateral hypothalamic lesions has been variously attributed to cell bodies within the hypothalamus itself and to catecholaminergic fibres of passage, particularly those ascending in the dopaminergic nigrostriatal pathway. In early work, the lateral hypothalamus was the focus of attention. In addition to the gross deficits in drinking and eating produced by lesions in the lateral hypothalamus (Teitelbaum and Epstein 1962), drinking could be elicited by electrical stimulation of this region (Andersson and McCann 1955; see also Rolls 1979). In regard to the specificity of the latter, a well-cited cautionary note was sounded by the observation that stimulus-bound ingestive behaviour could be initiated by such a general and presumably unrelated stimulus as the application of a paper clip to a rat's tail (Antelman et al. 1976).

An understanding of the trajectory of catecholaminergic fibres through the lateral hypothalamus (Ungerstedt 1971a) and a suspicion that these were recruited in lateral hypothalamic lesions, led to attempts to eliminate catecholaminergic fibres using local or intracisternal injections of 6-hydroxydopamine (6-OHDA), a neurotoxin which acts on both noradrenaline- and dopamine-containing neurons. These chemical lesions produced adipsia and aphagia and therefore mimicked the deficits produced by non-specific lateral hypothalamic damage (Ungerstedt 1971b; Breese et al. 1973; Stricker and Zigmond 1974). Whether the neurotoxin was infused into the ventricular CSF or directly into the nigrostriatal bundle, rats failed to drink appropriately to challenges including hypertonic saline or Ang II either in the CSF or the circulation (Fibiger et al. 1973; Gordon et al. 1979). Even though these early data underline the importance of fibres in transit through the hypothalamus, it is unlikely that their recruitment in lateral hypothalamic lesions accounts for the entire behavioural deficit.

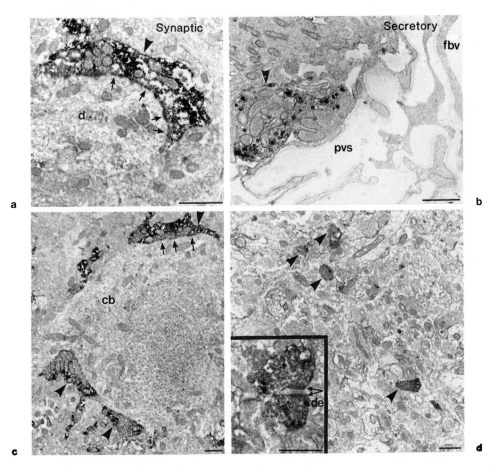

Fig. 10.4. Electron micrographs of Ang II within nerve terminals in the CNS. **a** and **b** contrast the appearance of Ang II-positive nerve terminals (arrowheads) in synaptic and secretory configurations, the latter being typical of the CVOs. In **c** and **d**, Ang II-positive terminals form synapses with cell bodies (cb) and dendrites (de) within the central nucleus of the amygdala (**c**) and the median preoptic nucleus (**d**). Despite the fact that in both regions, immunoreactive Ang II is contained in large (100 nm) vesicles and synaptic specializations (arrows) are present, Ang II receptors, detectable using autoradiographic techniques, are present in the median preoptic nucleus but not in the central nucleus of the amygdala. Bar = 1 μm.

Injections of kainic acid into the hypothalamus which destroy cell bodies but leave fibres of passage intact, cause rats to drink far less than controls in response to the cellular dehydration produced by infusions of hypertonic saline (Stricker et al. 1978).

There are other regions and fibre pathways which have been implicated in the catecholaminergic modulation of drinking. Given the common features of the deficient drinking in response to elevated levels of Ang II in the CSF, in both animals treated with 6-OHDA and those with AV3V lesions, it has been suggested that the ventral lamina terminalis is one such region (Gordon et al. 1979; Johnson and Cunningham 1987). Here it seems that noradrenaline rather than dopamine is important. If desmethylimiprimine (DMI), an agent which prevents uptake of 6-OHDA specifically in noradrenergic fibres, is injected into the region of the MnPO the action of the neurotoxin is virtually nullified. More localized injections of 6-OHDA throughout the lamina terminalis indicate that when an area encompassing the ventral median preoptic nucleus and OVLT is involved, drinking in response to intracerebroventricular Ang II is attenuated, but the dipsogenic effects of the cholinergic agonist carbachol or of hypertonic saline are unaffected (Bellin et al. 1988). These results indicate first, that there is regional specificity in the noradrenergic innervation of the lamina terminalis, and also that there is functional specificity in that the effects of 6-OHDA are most pronounced in relation to Ang II-induced drinking. The loss of noradrenergic fibres from the ventral MnPO has been verified histochemically. Furthermore, it has been shown that the depletion of noradrenergic fibres, caused by 6-OHDA, can be circumvented, along with the associated behavioural deficits, if the region is pretreated with DMI (Bellin et al. 1988). Noradrenergic terminals in the MnPO arise, at least in part, from A2 and A1 noradrenergic cell groups of the brainstem (Saper et al. 1983). These and other data have been incorporated into a model describing a means by which sensory information from blood volume/pressure receptors, relayed via noradrenergic pathways, may be integrated with the thirst initiated in the lamina terminalis following hypovolaemia (see Chapter 9).

Acetylcholine

The identification of acetylcholine as a possible chemical substrate involved in the initiation of drinking in the CNS has, to some extent, paralleled that of Ang II. Both substances were initially described as having widespread action throughout the diencephalon. In the case of acetylcholine, or more often the cholinergic agonist carbachol, these sites extended throughout the lateral hypothalamus (Grossman 1962) and in some studies involved a range of limbic structures (Fisher and Coury 1962). The involvement of the lateral hypothalamus, preoptic area, hippocampus, septum, diagonal band, cingulate cortex, mammillary body and thalamic reuniens nucleus led the authors to point to the similarities with the so-called Papez-circuit which had earlier been assigned a role in the expression of emotion in man and animals (Papez 1937). Subsequent studies using lower doses of carbachol to reduce diffusion, and allow more precise anatomical localization, indicated that drinking was produced most effectively by activation of the region centred around the anterior aspect of the third ventricle (Swanson and Sharpe 1973). The trend toward a more circumscribed site of action was continued and the SFO became the focus of attention. Here direct application of carbachol caused the greatest drinking with shortest latency. Furthermore, lesion of the SFO reduced the dipsogenic response to carbachol into the third ventricle (Simpson and Routtenberg 1972). At the time, this was seen as consistent with the SFO being the site of action of intracerebroventricularly administered carbachol; however, these data are not so easily aligned with contemporary knowledge of the boundary between the CVOs and substances circulating in the CSF (see Fig. 10.2). On the other hand, cholinergic fibres which originate in the septum and project to the SFO have been identified by histochemical staining of acetylcholinesterase in the rat (Lewis and Shute 1967). Further evidence in support of a role for endogenous acetylcholine in drinking comes from the prevention of breakdown of the neurotransmitter with agents such as physostigmine, which results in the elevation of thirst (Winson and Miller 1970).

Drinking modulated by a cholinergic mechanism has however, failed to maintain the focus enjoyed by angiotensin. This is, to some extent, because the notion has not been supported by current neurochemistry and receptor studies. The availability of specific monoclonal and polyclonal antibodies to choline acetyl transferase (Chat) has made possible the detailed mapping of cholinergic pathways (Armstrong et al. 1983; Mesulum et al. 1983; Kasa 1986). Unfortunately, some of these have only demonstrated Chat-positive neuronal cell bodies (Mesulum et al. 1983; Armstrong et al. 1983). Others have detected terminal fields, but these have not been described in areas such as the SFO, periventricular structures, or OVLT considered by earlier workers to be possible sites of action of acetylcholine in the genesis of drinking (Simpson and Routtenberg 1972; Swanson and Sharpe 1973; Buggy 1978). Furthermore, in vitro autoradiographic analyses of the distribution of cholinergic muscarinic receptors, thought to mediate the drinking response to exogenous acetylcholine or carbachol (Stein and Seifter 1962), do not mention the areas noted as critical for cholinergic drinking.

Prostaglandins

Prostaglandins of the E series (E_1 and E_2) have an antidipsogenic action on the thirst created by a range of dipsogenic stimuli, but have the most profound effect on the drinking initiated by elevated levels of Ang II either in the circulation or in the brain (Kenny 1986). From this it has been proposed that endogenous prostaglandins may act, perhaps in conjunction with other naturally occurring substances, as a satiety factor responsible for the termination of bouts of drinking (Kenny and Moe 1981). When inhibitors of prostaglandin synthesis are administered into the cerebrospinal fluid (Phillips and Hoffman 1976) or into the peritoneum (Perez Guaita and Chiaraviglio 1980), the thirst arising from intracerebroventricular Ang II seems to be prolonged. Other authors have not been able to reproduce this effect (Fluharty 1981; Kenny and Moe 1981) and Fluharty (1981) has suggested that the antidipsogenic effect of prostaglandins is a pharmacological response perhaps secondary to an increase in core temperature. The potential of endogenous prostaglandins as antidipsogens may be revealed by more complete understanding of the distribution of their binding sites in the brain. In this regard, Malet et al. (1982) have described prostaglandin E_2 binding in membrane fractions of the hypothalamus, amygdala and septum; however, a greater level of resolution is necessary for the better definition of the function of the prostaglandins in the neurocircuitry of thirst.

Tachykinins

The tachykinins isolated in the mammalian CNS to date include substance P, neurokinin A, neurokinin B, and neuropeptide K (Arai and Emson 1986). Consistent with the presence of multiple tachykinins in the mammalian brain, three corresponding subtypes of receptor have been proposed (Quirion 1985; Lee et al. 1987). Eledoisin, a member of the tachykinin group isolated from a Mediterranean cephalopod and claimed to be immunoreactive in the rat brain (Saria et al. 1986), is thought to have a non-selective agonist action on all subtypes of receptor but a particular affinity for the NK3 receptor (Bergstrom et al. 1987). Eledoisin, when injected into the brain, causes water-replete pigeons to drink water with a short latency, similar to that associated with Ang II. The related peptides, substance P and physalaemin are less effective in initiating drinking in the pigeon (Evered et al. 1977). Conversely in the rat, eledoisin and substance P have the opposite effect when administered intracerebroventricularly, causing a dose-dependent inhibition of the drinking caused by a range of dipsogenic stimuli (de Caro et al. 1977; Fitzsimons and Evered 1978). However, recent studies of the wild rabbit and sheep, have shown a dose-dependent dipsogenic response to eledoisin injected into the CSF (Tarjan et al. 1990). The variable action of eledoisin in different species may be resolved when data on the relative distribution of tachykinin receptor subtypes in different species become available.

Angiotensin Fragments

The catabolic products of angiotensin, the heptapeptides A(2–8) and A(1–7) have been variously described as being mildly dipsogenic, acutely dipsogenic or without effect on drinking. Most comparisons of Ang II and Ang III (A2–8) have indicated that Ang II is by far the more potent form of the peptide with regard to drinking induced by intracerebroventricular injection (Fitzsimons 1971; Fitzsimons and Kucharczyk 1978; Evered and Fitzsimons 1981); however, these results have been challenged on the basis of their failure to take into account the greater lability of Ang III in the CSF (Wright et al. 1985). The half life of Ang III in the CSF of the lateral ventricle is 6.5 s compared with 22.5 s for Ang II. It has been suggested that if the relative degradation rate is considered and appropriate doses used, the two forms of the peptide are equiactive when injected into the brain (Wright et al. 1985). Furthermore, in electrophysiological investigations, iontophoretically applied Ang III has been shown to be more potent than Ang II in initiating a response from neurons in the paraventricular nucleus (Harding et al. 1986). In these studies, use of the specific receptor antagonist Sar1 Ile8-Ang II, and the manipulation of aminopeptidases indicates that Ang III and Ang II are likely to occupy the same receptor and that neuronal responses in the paraventricular nucleus are largely due to the action of Ang III converted from Ang II (Felix and Harding 1986). Taken together the recent studies by Harding, Felix and Wright have provided evidence, supported to some extent by binding studies (see Printz 1988), which suggests that Ang II may have to be converted into Ang III before becoming active.

The literature in relation to A(1–7) is far less extensive, and while many authors have been prepared to discount the peptide as an inactive product of the catabolism of Ang II produced by the cleavage of the phenylalanine residue from its C terminus (Green et al. 1982), others (Chappel et al. 1989) have recently suggested a more active role. Although the lack of dipsogenic activity reported by Evered and Fitzsimons (1981) has not been challenged, it has been suggested that A(1–7) may have a role in modulating release of vasopressin (Schiavone et al. 1988) and a similar action to Ang II when injected into the brainstem (Campagnole-Santos et al. 1989).

Summary

The lateral hypothalamus, septum, preoptic area and amygdala have all been implicated in the modulation of drinking behaviour, but the most widely studied and enduring area of interest has been the lamina terminalis. In this region, studies of anatomical connections, neurochemical content and receptor localization have provided a framework for many of the physiological and behavioural observations on thirst. It seems likely that angiotensin II plays a key role in the circuitry of thirst, and the emphasis given to this hormone/neurotransmitter in the present chapter reflects the importance attached to it in the literature during the last 20 years. However, it should be appreciated that the circuits containing Ang II are likely to be involved in the regulation of other modalities in addition to thirst. Elevated levels of Ang II in the peripheral circulation act at circumventricular organs, most notably at the SFO, to initiate thirst. It has been proposed that information is then relayed to the median preoptic nucleus by pathways utilizing Ang II as a neurotransmitter/neuromodulator. It is conceivable, given the demonstration of noradrenergic pathways from the brainstem to the MnPO, and projections from the MnPO directly to magnocellular AVP neurons, that the MnPO integrates sensory information pertinent to hydrational status which is subsequently manifest in changes in water intake and excretion.

In regions other than the lamina terminalis, neurochemical information is fragmentary by comparison. The current challenge to neurochemists is to apply the rigorous attention concentrated on the circuitry and neurochemistry of the lamina terminalis to other parts of the relay implicated in the generation and expression of thirst. The development of ultrastructural techniques to identify, individually and in combination, neuronal tracers and immunocytochemically labelled transmitters will make possible the description of the circuitry and synaptology of thirst with a greater degree of precision.

Acknowledgement. The author is supported by the National Health and Medical Research Council of Australia.

References

Allen AM, McKinley MJ, Mendelsohn FAO (1988) Comparative neuroanatomy of angiotensin receptor localisation in the mammalian hypothalamus. Clin Exp Pharmacol Physiol 15:137–145

Andersson B, McCann SM (1955) A further study of polydipsia evoked by hypothalamic stimulation in the goat. Acta Physiol Scand 33:333–346

Antelman SM, Rowland NE, Fisher AE (1976) Stimulation bound ingestive behavior: a view from the tail. Physiol Behav 17:743–748

Arai H, Emson PC (1986) Regional distribution of neuropeptide K and other tachikinins (neurokinin A, neurokinin B and substance P) in rat central nervous system. Brain Res 399:240–249

Armstrong DM, Saper CB, Levey AI, Wainer BW, Terry RD (1983) Distribution of cholinergic neurons in rat brain: demonstrated by the immunocytochemical localisation of choline acetyl transferase. J Comp Neurol 216:53–68

Aronsson M, Almason K, Fuxe K et al. (1988) Evidence for the existence of angiotensinogen mRNA in magnocellular paraventricular hypothalamic neurons. Acta Physiol Scand 132:585–586

Bellin SI, Landas SK, Johnson AK (1988) Selective catecholamine depletion of structures along the ventral lamina terminalis: effects on experimentally-induced drinking and pressor responses. Brain Res 456:9–16

Bergstrom L, Torrens V, Saffroy M et al. (1987). [^{3}H] Neurokinin B and ^{125}I-Bolton-Hunter eledoisin label identical tachikinin binding sites in the rat brain. J Neurochem 48:125–133

Breese GK, Smith RM, Cooper BR, Grant LD (1973) Alteration in consummatory behavior following intracisternal injection of 6-hydroxydopamine. Pharmacol Biochem Behav 1:319–328

Buggy J (1978) Block of cholinergic-induced thirst after obstruction of anterior ventral third ventricle or periventricular preoptic ablation. Soc Neurosci Abstr 4:172

Buggy J, Fisher AG (1976) Anteroventral third ventricle site of action for angiotensin induced thirst. Pharmacol Biochem Behav 4:651–660

Buggy J, Johnson AK (1977) Preoptic hypothalamic periventricular lesions: thirst deficits and hypernatremia. Am J Physiol 233:44–52

Calza L, Fuxe K, Agnati LF et al. (1982) Presence of renin-like immunoreactivity in oxytocin immunoreactive nerve cells of the paraventricular and supraoptic nuclei in the rat hypothalamus. Acta Physiol Scand 116:313–316

Campagnole-Santos MJ, Diz DI, Santos RAS, Khosla MC, Brosnihan KB, Ferrario CM (1989) Cardiovascular effects of angiotensin (1–7) injected into the dorsal medulla of rats. Am J Physiol 257:H324–H329

Chai SR, Mendelsohn FAO, Paxinos E (1987) Angiotensin converting enzyme in rat brain visualised by quantitative in vitro autoradiography. Neuroscience 20:615–627

Chappel MC, Brosnihan KB, Diz DI, Ferrario CM (1989) Identification of angiotensin (1–7) in rat brain: evidence for differential processing of angiotensin peptides. J Biol Chem 264:16518–16523

Day RP, Reid IA (1976) Renin in dog brain: enzymological similarity to cathepsin D. Endocrinology 99:93–100

de Caro G, Micossi LG, Piccinin G (1977) Antidipsogenic effect of intraventricular administration of eledoisin to rats. Pharmacol Res Commun 9:488–493

Deschepper CF, Crumrine DA, Ganong WF (1986) Evidence that the gonadotrophs are the site of production of angiotensin II in the anterior pituitary of the rat. Endocrinology 119:36–43

Dzau VJ, Ellison K, McGowan D, Gross KW, Ouellette A (1984) Hybridization studies with a renin cDNA probe: evidence for widespread expression of renin in the mouse. J Hypertens 2 (Suppl 3): 235–237

Evered MD, Fitzsimons JT (1981) Drinking and changes in response to precursors, fragments and analogues of angiotensin II in the pigeon, *Columba livia*. J Physiol (Lond) 310:353–366

Evered MD, Fitzsimons JT, de Caro G (1977) Drinking behavior induced by intracranial injections of eledoisin and substance P in the pigeon. Nature 208:332–333

Felix D, Harding JW (1986) Manipulation of aminopeptidase activities: differential effects of iontophoretically applied angiotensins in rat brain. J Hypertens 4:S398–S401

Ferguson AV, Renaud LP (1986) Systemic angiotensin acts at subfornical organ to facilitate activity of neurohypophysial neurons. Am J Physiol 251:R712–R717

Fibiger HC, Zis AP, McGeer G (1973) Feeding and drinking deficits after 6-hydroxydopamine administration in the rat. Similarities to the lateral hypothalamic syndrome. Brain Res 55:135–148

Field LJ, McGowan RA, Dickinson DP, Gross KW (1984) Tissue and gene specificity of mouse renin expression. Hypertension 6:597–603

Fischer-Ferraro C, Nahmod VE, Goldstein DJ, Finkielman S (1971) Angiotensin and renin in rat and dog brain. J Exp Med 133:353–361

Fisher AE, Coury JN (1962) Cholinergic tracing of a central neural circuit underlying the thirst drive. Science 138:691–693

Fitzsimons JT (1971) The effect on drinking of peptide precursors and of shorter brain peptide fragments of angiotensin II injected into rat diencephalon. J Physiol (Lond) 214:295–303

Fitzsimons JT, Evered MD (1978) Eledoisin, substance P and related peptides, intracranial dipsogens in the pigeon and anti dipsogens in the rat. Brain Res 150:533–542

Fitzsimons JT, Kucharczyk J (1978) Drinking and haemodynamic changes induced in dog by intracranial injection of components of the renin–angiotensin system. J Physiol (Lond) 276:419–434

Fluharty SJ (1981) Cerebral prostaglandin biosynthesis and angiotensin-induced drinking in rats. J Comp Physiol Psychol 95:915

Fuxe K, Ganten D, Hokfelt T, Bolme P (1976) Immunohistochemical evidence for the existence of angiotensin II-containing nerve terminals in the brain and spinal cord of the rat. Neurosci Lett 2: 229–234

Fuxe K, Ganten D, Hokfelt T (1980) Renin-like immunocytochemical activity in the rat and mouse brain. Neurosci Lett 18:245–250

Ganong WF (1981) The brain and the renin–angiotensin system. In: Buckley JP, Ferrario CM (eds) Central nervous system mechanisms in hypertension. Raven Press, New York, pp 283–292

Ganten D, Boucher R, Ernest J (1971) Renal activity in brain tissue of puppies and adult dogs. Brain Res 33:557–559

Ganten D, Hermann K, Bayer C, Unger TH, Lang RE (1983) Angiotensin synthesis in the brain and increased turnover in hypertensive rats. Science 221:896–871

Gehlert DR, Speth RC, Wamsley JK (1986) Distribution of [^{125}I]-angiotensin II binding sites in the rat brain: a quantitative autoradiography study. Neuroscience 18:837–856

Gordon FJ, Brody MJ, Fink GD, Buggy J, Johnson AK (1979) Role of central catecholamines in the control of blood pressure and drinking behavior. Brain Res 178:161–163

Green LJ, Spadaro ACC, Martins AR, Perussi De Jesus WD, Camargo ACM (1982) Brain endo-oligopeptidase B: a post proline cleaving enzyme that inactivates angiotensin I and angiotensin II. Hypertension 4:178–184

Grossman SP (1962) Direct adrenergic and cholinergic tracing of a central neural circuit underlying the thirst drive. Science 138:691–693

Gutman MB, Jones DL, Ciriello J (1988) Effect of paraventricular nucleus lesion on drinking and pressor responses to Ang II. Am J Physiol 255:R882–R887

Harding JW, Felix D (1987) Angiotensin-sensitive neurones in the rat paraventricular nucleus: relative potencies of angiotensin II and angiotensin III. Brain Res 410:130–134

Harding JW, Imboden H, Felix D (1986) Is angiotensin III the centrally active form of angiotensin? Experientia 42:706

Healy DP, Printz MP (1984) Distribution of immunoreactive angiotensin II, angiotensin I and renin in the central nervous system of intact and nephrectomised rats. Hypertension 6(Suppl 1):1130–1136

Hirose S, Yokasawa H, Inagami T (1978) Immunocytochemical identification of renin in rat brain and distinction from acid proteases. Nature 274:392–393

Inagami T, Clemens DL, Celio MR et al. (1980) Immunohistochemical localisation of renin in mouse brain. Neurosci Lett 18:91–98

Jackson TR, Blair LAC, Marshall J, Goedert M, Hanley MR (1988) The *mas* oncogene encodes an angiotensin receptor. Nature 335:437–440

Jhamandas JH, Lind RN, Renaud LP (1989) Possible transmitter role for angiotensin II in a subfornical organ-supraoptic nucleus pathway: electrophysiological and anatomical evidence in the rat. Brain Res 487:52–61

Johnson AK, Buggy J (1976) A critical analysis of the site of action for the dipsogenic effect of angiotensin II. In: Buckley JP, Ferrario CM (eds) Central actions of angiotensin and related hormones. Pergamon Press, New York, pp 357–386

Johnson AK, Cunningham JT (1987) Brain mechanisms and drinking. The role of the lamina terminalis associated systems in extracellular thirst. Kidney Int 32(Suppl 21):S35–S42

Kasa P (1986) The cholinergic systems in brain and spinal cord. Progr Neurobiol 26:211–272

Kenny NJ (1986) Suppression of water intake by the E prostaglandins. In: de Caro G, Epstein AN, Massi M (eds) The physiology of thirst and sodium appetite. Plenum Press, New York, pp 227–238

Kenny NJ, Moe KE (1981) The role of endogenous prostaglandin E in angiotensin-II-induced drinking in rats. J Comp Physiol Psychol 95:383

Krisch B, Leonhardt H (1978) The functional and structural border between the CSF-and blood-milieu in the circumventricular organs (organum vasculosum lamina terminalis, subfornical organ, area postrema) of the rat. Cell Tissue Res 195:485–497

Krisch B, Leonhardt H, Oksche H (1987) Compartments in the organum vasculosum lamina terminalis of the rat and their delineation against the outer cerebrospinal fluid containing space. Cell Tissue Res 250:331–347

Landas S, Phillips MI, Stamler JF, Raizada MK (1980) Visualisation of specific angiotensin II binding sites in the brain by fluorescent microscopy. Science 210:791–793

Lee CM, Campbell NJ, Williams BJ, Iversen LL (1987) Multiple tachykinin binding sites in peripheral tissue and brain. Eur J Pharmacol 130:209–218

Lewis PR, Shute CC (1967) The cholinergic limbic-system: projections to hippocampus formations, medial cortex, nuclei of the ascending cholinergic reticular system, and the subfornical organ and the supraoptic crest. Brain 90:521–550

Lewis SJ, Allen AM, Verberne AJM, Figdor R, Jarrot B, Mendelsohn FAO, (1986) Angiotensin II receptor binding in the rat nucleus solitarius is reduced after unilateral nodose ganglionectomy or vagotomy. Eur J Pharmacol 125:305–307

Lind RW (1986) Bidirectional, chemically-specified neural connections between the subfornical organ and the midbrain raphe system. Brain Res 384:250–261

Lind RW (1988a) Angiotensin and the lamina terminalis: illustrations of a complex unity. Clin Exp Hypertens [A] (Suppl 1):79–105

Lind RW (1988b): Sites of action of angiotensin in the brain. In: Harding JW, Wright JW, Speth RC, Barnes CD (eds) Angiotensin and blood pressure regulation. Academic Press, New York, pp 135–163

Lind RW, Johnson AK (1982a) Subfornical organ-median preoptic connections and drinking and pressor responses to angiotensin II. J Neurosci 2:1043–1051

Lind RW, Johnson AK (1982b) Central and peripheral mechanisms mediating angiotensin-induced thirst. Exp Brain Res Suppl 4:353–364

Lind RW, Van Hoesen GW, Johnson AK (1982) An HRP study of the connections of the subfornical organ of the rat. J Comp Neurol 210:265–277

Lind RW, Swanson LW, Ganten D (1984a) Angiotensin II immunoreactivity in the neural afferents and efferents of the subfornical organ of the rat. Brain Res 321:209–215

Lind RW, Swanson LW, Ganten D (1984b) Angiotensin II immunoreactive pathways in the central nervous system of the rat. Evidence for a projection from the subfornical organ to the paraventricular nucleus of the hypothalamus. Clin Exp Hypertens [A]6:1915–1920

Lind RW, Swanson LW, Brunn TO, Ganten D (1985a) The distribution of angiotensin immunoreactive cells and fibres in the paraventricular–hypophysial system of the rat. Brain Res 338:81–89

Lind RW, Swanson LW, Ganten D (1985b) Organization of angiotensin immunoreactive cells and fibres in the rat central nervous system: an immunohistochemical study. Neuroendocrinology 40:2–24

Lynch KR, Hawelu-Johnson CL, Guyenent PG (1987) Localization of brain angiotensin mRNA by hybridisation histochemistry. Brain Res 388:149–158

Malet C, Scherrer H, Saavedra JM, Dray F (1982) Specific binding of [^{3}H] prostaglandin E_2 to rat brain membranes and synaptosomes. Brain Res 236:227

McKinley MJ, Allen AM, Clevers J, Denton DA, Mendelsohn FAO (1986) Autoradiographic localisation of angiotensin receptors in the sheep brain. Brain Res 375:373–376

McKinley MJ, Allen AM, Clevers J, Paxinos G, Mendelsohn FAO (1987) Angiotensin receptor binding in human hypothalamus: autoradiographic localisation. Brain Res 4210:375–379

McKinley MJ, Allen AM, Chai SR, Hards DK, Mendelsohn FAO, Oldfield BJ (1989) The lamina terminalis and its neural connections: neural circuitry involved in angiotensin action and fluid and electrolyte homeostasis. Acta Physiol Scand 136 (Suppl 583):113–118

Mendelsohn FAO, Aguilera G, Saavedra JM, Quirion R, Catt KJ (1983) Characteristics and regulation of angiotensin II receptors in pituitary, circumventricular organs and kidney. Clin Exp Hypertens [A]5:1081–1097

Mendelsohn FAO, Quirion R, Saavedra JM, Aguilera G, Catt KJ (1984) Autoradiographic localisation of angiotensin II receptors in rat brain. Proc Natl Acad Sci USA 81:1575–1579

Mendelsohn FAO, Allen AM, Clevers J, Denton DA, Tarjan E, McKinley MJ (1988) Localisation of angiotensin receptor binding in rabbit brain by in vitro autoradiography. J Comp Neurol 270:372–384

Mesulum MM, Mufson EJ, Wainer BH, Levey AI (1983) Central cholinergic pathways in the rat: an overview based on an alternative nomenclature Ch1-Ch6. Neuroscience 10:1185–1201

Miller CCJ, Canter AT, Brooks JI, Lovell-Badge RH, Brammer WJ (1989) Differential extra-renal expression of the mouse renin genes. Nucleic Acids Res 17:3117–3129

Miselis RR (1981) The efferent projections of the subfornical organ of the rat: a circumventricular organ within a neural network subserving water balance. Brain Res 230:1–23

Miselis RR, Shapiro RE, Hand PJ (1979) Subfornical organ efferents to neural systems for control of body water. Science 205:1022–1025

Nazarali AJ, Gutkind JS, Saavedra JM (1987) Regulation of angiotensin binding sites in the subfornical organ and other rat brain nuclei after water deprivation. Cell Mol Neurobiol 7:447–455

Nicholl RA, Barker JL (1971) Excitation of supraoptic neurosecretory cells by angiotensin II. Nature 233:172–174

Okuya S, Inenaga K, Kaneko T, Yamashita H (1987) Angiotensin II sensitive neurons in the supraoptic nucleus, subfornical organ, anteroventral third ventricle of the rat in vitro. Brain Res 402:58–67

Oldfield BJ, Clevers J, McKinley MJ (1986) A light and electron microscopic study of the projections of the nucleus medianus with special reference to inputs to vasopressin neurons. Soc Neurosci Abst 12:445

Oldfield BJ, Mendelsohn FAO, McKinley MJ (1988) Ultrastructural distribution of angiotensin converting enzyme in rat brain. Neurosci Lett [Suppl] 30:106

Oldfield BJ, Ganten D, McKinley MJ (1989) An ultrastructural analysis of the distribution of angiotensin II in the rat brain: J Neuroendocrinol 1:121–128

Papez JW (1937) A proposed mechanism of emotions. Arch Neurol Psychiatry 38:725–743

Perez Guaita MF, Chiaraviglio E (1980) Effect of prostaglandin E1 and its biosynthesis inhibitor indomethacin on drinking in the rat. Pharmacol Biochem Behav 13:787

Phillips MI (1987) Brain angiotensin. In: Gross PM (ed) Circumventricular organs and body fluids Vol III. CRC Press, Boca Raton, pp 163–182

Phillips MI, Hoffman WE (1976) Sensitive sites in the brain for blood pressure and drinking responses to angiotensin II. In: Buckley JP, Ferrario C (eds) Central actions of angiotensin and related hormones. Pergamon Press, New York

Printz MP (1988) Regulation of the brain angiotensin system: a thesis of multicellular involvement. Clin Exp Hypertens [A] 10:17–35

Printz MP, Chen FM, Slivka S, Maclejewski AR (1987) Comparison of neural and peripheral angiotensin receptors. In: Buckley JP, Ferarrio CM (eds) Brain peptides and catecholamines in cardiovascular regulation. Raven Press, New York, pp 233–243

Quirion R (1985) Multiple tachikinin receptors. Trends Neurosci 8:183–815

Richoux JP, Bouhnik J, Clauser E, Corvol P (1988) The renin angiotensin system in the rat brain. Immunocytochemical localisation of angiotensinogen in glial cells and neurons. Histochemistry 89: 323–331

Rolls ET (1979) Effects of electrical stimulation of the brain on behavior. In Connolly K (ed) Psychology surveys, vol 2. George Allen and Unwin, London, pp 151–169

Roth M, Weitzman AF, Piquillord Y (1969) Converting enzyme content of different tissues of the rat.

Experientia 25:1247

Saper CB, Reis DJ, Joh T (1983) Medullary catecholamine inputs to anteroventral third ventricular cardiovascular regulatory region in the rat. Neurosci Lett 42:285–291

Saria A, Garnse R, Petermann J, Fisher JA, Theodorson-Norheim H, Lundberg J (1986) Simultaneous release of several tachikinins and calcitonin gene-related peptide from rat spinal cord slices. Neurosci Lett 63:310–314

Schiavone MT, Santos RAS, Brosnihan KB, Khosla MC, Ferrario CM (1988) Release of vasopressin from the rat hypothalamoneurohypophysial system angiotensin 1–7 heptapeptide. Proc Natl Acad Sci USA 85:4095–4098

Simpson JB (1980) The circumventricular organs and the central action of angiotensin. Neuroendocrinology 32:248–256

Simpson JB, Routtenberg A (1972) The subfornical organ and carbachol-induced drinking. Brain Res 45:135–152

Simpson JB, Routtenberg A (1973) Subfornical organ: site of drinking elicitation by angiotensin II. Science 181:1172–1175

Simpson JB, Routtenberg A (1975) Subfornical organ lesions reduce intravenous angiotensin induced drinking. Brain Res 88:154–161

Speth RC, Wamsley JK, Gehlert DR, Chernicky CL, Barnes KL, Ferrario CM (1985) Angiotensin II receptor localisation in the canine CNS. Brain Res 326:137–143

Stein L, Seifter J (1962) Muscurinic synapses in the hypothalamus. Am J Physiol 202:751–756

Stornetta RL, Hawelu-Johnson CL, Guyenet PG, Lynch KR (1988) Astrocytes synthesise angiotensinogen in brain. Science 242:1444–1446

Stricker EM, Zigmond MJ (1974) Effects on homeostasis of intraventricular injection of 6-hydroxydopamine in rats. J Comp Physiol Psychol 86:973–994

Stricker EM, Swerdloff AF, Zigmond MJ (1978) Intrahypothalamic injections of kainic acid produce feeding and drinking deficits in rats. Brain Res 158:470–473

Strittmatter SM, Snyder SH (1987) Angiotensin converting enzyme immunohistochemistry in rat brain and pituitary gland: correlation of isoenzyme type with cellular localisation. Neuroscience 21:407–420

Suzuki F, Ludwig G, Hellmann W (1988) Renin gene expression in rat tissue: a new quantitative assay method for rat renin mRNA using synthetic cRNA. Clin Exp Hypertens [A]10:345–359

Swanson L, Sharpe LG (1973) Centrally induced drinking: comparison of angiotensin II- and carbachol-sensitive sites in rats. Am J Physiol 225:566–573

Tanaka J (1989) Involvement of the median preoptic nucleus in the regulation of paraventricular vasopressin neurons by the subfornical organ in the rat. Exp Brain Res 76:47–54

Tanaka J, Kaba H, Saito H, Seto K (1986) Subfornical organ efferents influence the activity of median preoptic neurones projecting to the hypothalamic paraventricular nucleus in the rat. Exp Neurol 93:647–651

Tanaka J, Saito H, Kaba H (1987a) Subfornical organ and hypothalamic paraventricular nucleus connections with median preoptic neurons: an electrophysiological study in the rat. Exp Brain Res 68:579–585

Tanaka J, Saito H, Seto K (1987b) Subfornical organ neurons act to enhance the activity of paraventricular vasopressin neurons in response to intravenous angiotensin II. Neurosci Res 4:424–427

Tarjan E, Blair-West JR, de Caro G et al. (1990) Sodium and water intake of sheep, rabbits and cattle during ICV infusion of eledoisin. Pharmacol Biochem Behav 35:823–828

Teitelbaum P, Epstein AN (1962) The lateral hypothalamic syndrome: recovery of feeding and drinking after lateral hypothalamic lesions. Psychol Rev 69:74–90

Thomas WA, Sernia C (1988) Immunocytochemical localization of angiotensinogen in the rat brain. Neuroscience 25:319–341

Ungerstedt U (1971a) Stereotaxic mapping of the monoamine pathways in the rat brain. Acta Physiol Scand Suppl 367:1–48

Ungerstedt U (1971b) Adipsia and aphagia after 6-hydroxydopamine induced degeneration of the nigro-striatal dopamine system. Acta Physiol Scand Suppl 367:95–122

Van Houten M, Schiffrin EL, Mann JFE, Posner B, Boucher R (1980) Radioautographic localisation of specific binding sites for blood borne angiotensin in the rat brain. Brain Res 186:480–485

Winson J, Miller NE (1970) Comparison of drinking elicited by eserine or DFP injected into preoptic area of rat brain. J Comp Physiol Psychol 73:233–237

Wright JW, Morseth SL, Abhold RH, Harding JW (1985) Pressor action and dipsogenicity induced by angiotensin II and III in rats. Am J Physiol 249:R514–R521

Chapter 11

Central Control of Water and Sodium Chloride Intake in Rats During Hypovolaemia

E.M. Stricker

Introduction

Considerations of body fluid homeostasis inevitably focus on water and sodium homeostasis. It has long been recognized that animals do not monitor body water or sodium content per se but become aware of dehydration or sodium loss when appropriate sensors detect specific consequences of each deficiency. There are two dimensions in which such deficiencies become manifest and can be detected, related to the osmolality and volume of blood. The receptors responsive to changes in plasma osmolality appear to be located in the anterior hypothalamic area, and stimulation by osmotic dehydration elicits thirst and secretion of the antidiuretic hormone, arginine vasopressin (AVP). The regulation of blood volume is considerably more complex than osmoregulation. The same considerations of thirst and AVP secretion associated with osmoregulation are also relevant to volume regulation, but additionally there are parallel issues involved in the maintenance of sodium balance, namely, sodium chloride intake motivated by sodium appetite and renal sodium conservation mediated in large part by the adrenocortical hormone, aldosterone. Reduction in plasma volume (hypovolaemia) elicits each of these compensatory responses, which are stimulated by neural signals from vascular baroreceptors and/or by the renin–angiotensin system (see Chapter 6).

This chapter focuses exclusively on the behavioural contributions to volume regulation motivated by thirst and sodium appetite in rats, and the central mechanisms by which water and sodium chloride consumption may be controlled and integrated.

Hypovolaemia

Studies of the rapid physiological responses of animals to hypovolaemia, involving the secretion of AVP, aldosterone, and various pressor hormones, have traditionally used episodes of controlled haemorrhage to provoke those responses. However, studies of the more gradual behavioural responses to hypovolaemia require that animals remain capable of moving about and drinking for prolonged periods, and therefore preclude procedures like haemorrhage that remove red blood cells or precipitate profound arterial hypotension. Thus, in investigating the biological bases of thirst and sodium appetite, a long-lasting and substantial hypovolaemia has been produced in laboratory rats by injecting subcutaneously small amounts of a hyperoncotic colloidal solution (Stricker 1966). The colloid gradually draws isosmotic plasma fluid across the capillary membrane to the interstitium in amounts proportional to the volume and concentration of the injected solution. Fluid accumulation increases progressively over 12–16 hours and its magnitude can be quite considerable: 5 ml of a 30% solution of polyethylene glycol (PEG) injected into an adult rat causes up to 30% reduction of plasma volume, and twice that dose can produce so much haemoconcentration that the kidneys fail. The induced plasma volume deficits linger for at least 24 hours, until the lymphatic system drains the injected colloid and sequestered fluid from the subcutaneous tissue.

After PEG treatment, rats begin to drink water within an hour or two and continue to drink for several hours in frequent but brief episodes. The cumulative amount of water that they consume is proportional to the induced deficit in plasma volume over a wide range of depletions (Fitzsimons 1961; Stricker 1968). However, most of the ingested fluid does not remain in the intravascular compartment but is distributed intracellularly by osmosis. Nevertheless, PEG-treated rats do not continue to ingest water but are inhibited from doing so by the osmotic dilution of body fluids that results from the renal retention of ingested water (Stricker 1969). What the animals need to restore body fluid osmolality to normal, and to repair their plasma volume deficit, is sodium chloride.

Appropriate to their need for a dilute saline solution isosmotic with plasma, PEG-treated rats increase their intakes of both water and 0.5 M sodium chloride solution (Stricker and Wolf 1966; Stricker and Jalowiec 1970). At first, the PEG-treated rats drink water exclusively while excreting small amounts of sodium-poor urine. After 5 hours, the rats begin to drink the concentrated saline solution in rapid alternation with water and thereby steadily increase the cumulative intake of each fluid. Nevertheless, the renal conservation of water and sodium persists because most of the ingested fluid is also sequestered in the oedema at the injection site. However, ultimately tissue turgor limits further fluid accumulation at that site, and thereafter ingested fluid remains largely in the circulation and serves to repair plasma volume. Renal conservation of sodium ends at this time, approximately 12–18 hours after injection of the colloid. It is noteworthy that the PEG-treated rats do not stop drinking saline when the hypovolaemia is abolished, but continue to consume both sodium chloride solution and water throughout the 24-hour test period.

Rats maintained on a sodium-deficient diet instead of standard laboratory chow for several days before PEG treatment have generally similar responses: they increase water and sodium chloride intakes, conserve water and sodium in urine, and thereby repair their plasma volume deficits (Stricker 1981). Of interest, however, is the fact that the rats drink saline solution within 30 minutes after PEG treatment, well before any thirst is evident. These observations suggest that the delay in sodium appetite normally seen after PEG treatment is an effect of the common practice of feeding rats sodium-rich laboratory chow.

Teleologically, the consumption of water and sodium chloride by hypovolaemic rats makes sense; the animals consume exactly what they need to repair the induced deficit. Nevertheless, the mechanisms that mediate these appropriate drinking behaviours have remained uncertain. A full consideration of those issues has been presented elsewhere (Stricker and Verbalis 1990). This chapter briefly summarizes the conclusions reached and then considers their implications for the central control of water and sodium chloride intake. Before beginning, some general principles regarding the control of ingestive behaviour will be mentioned.

Interaction of Neural and Endocrine Signals

Ingestive behaviours in mammals are controlled by both excitatory and inhibitory stimuli. With regard to thirst during hypovolaemia, the plasma volume deficits appear to provide two separate stimuli: one is a neural signal that arises from low-pressure baroreceptors (e.g., Zimmerman et al. 1981;Kaufman 1984), whereas the other is an endocrine signal, involving the action of angiotensin in the brain, that arises when hypovolaemia elicits release of renin from the kidneys (e.g., Fitzsimons 1969; Leenen and Stricker 1974). It seems clear that these two stimuli of drinking can act independently of one another; exogenous angiotensin elicits drinking in the absence of hypovolaemia (Fitzsimons and Simons 1969; Mann et al. 1980), whereas hypovolaemia elicits normal water consumption in nephrectomized rats (Fitzsimons 1961; Stricker 1973). It is also evident that these separate neural and endocrine excitatory signals are each inhibited by osmotic dilution resulting from renal retention

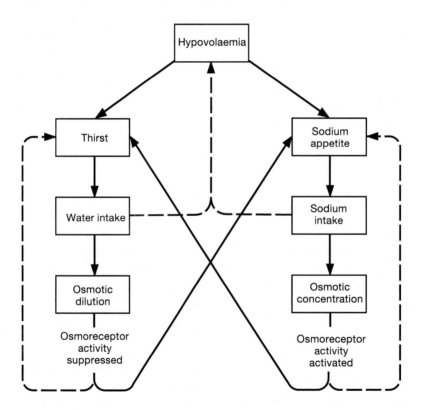

Fig.11.1. Schematic representation of the physiological mechanisms controlling thirst and sodium appetite during hypovolaemia. Solid lines indicate stimulation, broken lines indicate inhibition. Hypovolaemic rats alternately drink water and concentrated NaCl solution, depending on the current plasma osmolality, and ultimately consume sufficient fluid at an isotonic concentration to repair the volume deficit. However, the water and NaCl intakes are limited when animals have access to only one of the drinking fluids, due to activation of the respective inhibitory osmoregulatory pathways. Conversely, neither inhibitory pathway is activated when the rats drink isotonic NaCl solution, and consequently intakes proceed unabated in response to the hypovolaemic stimulus. (Adapted from Stricker and Verbalis, 1988.)

of ingested water (Stricker 1969, 1971a), which presumably is detected by the same cerebral osmoreceptors that stimulate thirst during osmotic dehydration (Fig. 11.1).

The control of sodium appetite is similar; indeed, the excitatory stimuli of sodium appetite probably involve the same combination of neural input from vascular baroreceptors plus angiotensin that also stimulates thirst, hence the inclination of rats to consume both water and saline during hypovolaemia (Fig. 11.1). Whether water or saline is consumed, therefore, depends not on the presence of differentiating excitatory stimuli but on the presence of differentiating inhibitory stimuli. More specifically, inhibition of thirst results from osmotic dilution of body fluids, as mentioned, whereas inhibition of sodium appetite results in part from the osmotic concentration of body fluids that follows the consumption of salt. However, even before such increases in plasma osmolality dehydrate cells sufficiently to stimulate thirst, they remove the inhibition of water consumption created by osmotic dilution and thereby allow hypovolaemic thirst to reappear; subsequent water intake then continues until the diluted plasma osmolality again inhibits thirst and disinhibits sodium appetite, thus allowing a new cycle of behaviour. This scheme explains nicely the alternating pattern of water and saline intakes observed in PEG-treated rats, by which the hypovolaemic animals ultimately consume sufficient amounts of both fluids to restore plasma volume.

When rats have been fed sodium-deficient diet before PEG treatment, hyponatraemia results from dietary sodium deprivation and therefore an initial episode of water consumption is not required to produce osmotic dilution of body fluids. Thus, thirst is inhibited then and the animals begin to drink sodium chloride solution immediately. Thereafter, the controls of drinking are as described above and summarized in Fig. 11.1.

Integrative Function of Paraventricular Nucleus

The following central mechanism to account theoretically for these and related observations is described fully elsewhere (Stricker and Verbalis 1990). The excitatory and inhibitory stimuli for drinking appear to influence a brain system that controls inhibition of sodium appetite, as if they activated a "switch" in the direction of water intake rather than sodium chloride intake. We have presumed that this switching system is located in some subset of oxytocinergic neurons in the paraventricular nucleus (PVN) of the hypothalamus because the presence of sodium appetite in rats has been found to vary inversely with plasma levels of the peptide hormone, oxytocin (OT).

For example, osmotic dilution, which inhibits water intake and facilitates sodium chloride intake, also inhibits neurohypophyseal secretion of OT (and AVP; Stricker and Verbalis 1986). Sodium appetite similarly occurs during other circumstances in which plasma levels of OT (but not AVP) are low, such as in PEG-treated rats after volume repletion had occurred or when prefed a sodium-deficient diet, or in adrenalectomized rats after a period of imposed sodium deficiency, or in rats treated daily with the mineralocorticoid hormone desoxycorticosterone (DOC) (Stricker and Verbalis 1987; Stricker et al. 1987). Indeed, low plasma levels of OT represent the only factor observed consistently in diverse conditions in which sodium appetite is seen in rats, and sodium appetite has not been observed in rats at times when plasma OT levels are elevated concurrently.

Despite such correlational evidence, other experiments demonstrated that circulating OT does not directly affect the intake of sodium chloride solution in rats. Continuous administration of synthetic OT from an implanted osmotic minipump, raising plasma OT to physiological levels and beyond, did not inhibit sodium appetite in PEG-treated rats, nor did systemic administration of an OT receptor antagonist enhance sodium chloride intake in PEG-treated rats (Stricker and Verbalis 1987). These results, therefore, indicate that increased circulating levels of OT do not, by themselves, provide an inhibitory stimulus of sodium appetite. Instead, plasma levels of OT may best be viewed as a peripheral marker of the parallel activity of some system of brain neurons that are involved in the inhibitory control of sodium appetite.

OT secreted from the neurohypophysis is synthesized in magnocellular neurons in the PVN and supraoptic nucleus (SON) of the hypothalamus. Adjacent to the magnocellular neurons in the PVN, but not the SON, are parvocellular OT-containing neurons that project widely throughout the brain to sites including the limbic system and the dorsal motor nucleus of the vagus in the brainstem (Swanson and Sawchenko 1983). Thus, central oxytocinergic projections appear to be advantageously located to influence ingestive behaviour and associated autonomic functions. Perhaps these centrally projecting hypothalamic nuclei are stimulated simultaneously with the magnocellular neurons projecting to the pituitary so that the induced activity somewhere in the brain serves to inhibit sodium appetite; conversely, sodium appetite may develop when activity is suppressed in this subset of PVN neurons. Of course, the PVN is an extraordinarily complex structure containing many different neuropeptidergic cells projecting widely throughout the brain, any of which might be activated concurrently with magnocellular OT neurons, and thus the switching system obviously could be located in non-OT-containing cells or even in some other area in the brain whose function was also inversely related to that of magnocellular OT neurons in the PVN.

A putative role of PVN neurons in the inhibitory control of sodium appetite would appear to be inconsistent with the sodium appetite seen in lactating rats (Richter and Barelare 1938; Thiels et al. 1990), inasmuch as suckling is an established stimulus of magnocellular PVN activity and pituitary OT secretion. However, experiments have indicated that parvocellular PVN activity may not increase during suckling (Lawrence and Pittman 1985; Helmreich et al. 1988). Thus, suckling would not be expected to provide a stimulus that directly affected the proposed central control of sodium chloride consumption. Even if it did, parvocellular PVN neurons still might mediate inhibition of sodium appetite because lactating animals are inactive when nursing, and therefore suckling-induced OT secretion and saline ingestion necessarily occur at different times.

This hypothesis has been evaluated in a variety of ways. These experimental observations are grouped here according to the potential input to oxytocinergic neurons in the PVN that various treatments might have provided.

Osmotic Dehydration

Neural signals from osmoreceptors in the basal forebrain to the PVN are known to increase secretion of OT and AVP in rats (Carithers and Johnson 1985; Stricker and Verbalis 1986; see Chapters 5 and 9). AVP is well recognized for its role in conserving water in urine, whereas more recently it has become clear that OT serves the useful function of promoting renal sodium excretion after sodium chloride loads (Balment et al. 1980; Verbalis et al. 1988). Complementing these physiological responses are water intake and inhibition of sodium appetite, respectively, which are also observed after sodium chloride loads.

When hypovolaemic or sodium-deficient rats consume concentrated sodium chloride solution, the normal negative feedback from ingested salt inhibits further sodium chloride consumption temporarily while stimulating thirst. With injections of hypertonic sodium chloride solution, the same temporary inhibition of sodium appetite and stimulation of thirst also has been obtained, both in hypovolaemic rats and in adrenalectomized rats (Stricker 1983; Stricker and Verbalis 1987). The inhibitory effect of sodium chloride loads on sodium appetite is presumably mediated by the same osmoreceptors that stimulate secretion of OT and AVP. Thus, destruction of the ventral portion of nucleus medianus (vNM), a small midline periventricular structure within this forebrain region, blunts the secretion of OT and AVP in rats in response to the administration of hypertonic sodium chloride solution but not in response to hypovolaemia or arterial hypotension (Mangiapane et al. 1983; Gardiner et al. 1985). Moreover, rats with lesions of vNM showed a marked spontaneous appetite for concentrated sodium chloride solution and drank 15–35 ml of 0.5 M sodium chloride solution daily for at least several months postsurgery (Gardiner et al. 1986). Examination of these animals indicated that they did not become sodium deficient after the brain lesions, nor was the renin–angiotensin system activated. Instead, we presume that such lesions blocked the normal inhibitory effects on sodium appetite that occur when rats consume concentrated sodium chloride solution, thus permitting the ingestion of saline to persist unchecked.

Hypovolaemia

Hypovolaemia and angiotensin both increase pituitary OT secretion (e.g. Lang et al. 1981; Stricker and Verbalis 1986) and therefore presumably increase activity of the subset of PVN neurons involved in the central inhibition of sodium appetite. Thus, when volume depletion exceeds 20%–25% and/or blood levels of angiotensin are supraphysiological, the consumption of water, but not sodium chloride, is elicited (Stricker 1971b, 1973, 1981; Hosutt and Stricker 1981). Conversely, both water and saline are consumed when magnocellular (and perhaps parvocellular) PVN activity is suppressed before the central injection of angiotensin, as by prefeeding rats a sodium-deficient diet (Buggy and Fisher 1974) or by treating them with DOC (Fluharty and Epstein 1983).

In addition to input from vascular baroreceptors to the PVN, there are known anatomical connections between the subfornical organ (SFO) and the PVN (Lind et al. 1982; Miselis 1982), which appear to be importantly involved in mediating the stimulating effects of angiotensin on drinking and neurohypophyseal hormone secretion (Miselis et al. 1979; Ferguson and Kasting 1988). Drinking is elicited by the administration of remarkably small amounts of angiotensin into the SFO (Simpson and Routtenberg 1973; Simpson et al. 1978), whereas administration of angiotensin receptor antagonists into the SFO blocks drinking induced by systemic angiotensin (Simpson et al. 1978). Although electrolytic lesions destroying this tissue have also been reported to abolish or severely attenuate the drinking response of rats to angiotensin administered into the systemic circulation (Simpson and Routtenberg 1975; Simpson et al. 1978), they have much less effect on drinking elicited by PEG-induced hypovolaemia, as might be expected because neural signals from vascular baroreceptors are still present (Hosutt et al. 1981).

Stress

In addition to these effects of osmo- and volume regulatory stimuli, there are other important influences on the putative switching system in the PVN that have nothing to do with sodium or water homeostasis. Some non-specific factors seem to involve certain kinds of traumatic stressors, especially those accompanied by anuria. Such conditions always set the "switch" towards water intake and away from sodium chloride intake, and are so influential that sodium appetite will not occur then despite the presence of frank hypovolaemia and/or sodium deficiency. This result occurs whether anuria is produced in rats by bilateral nephrectomy or ureter ligation, or by puncturing the bladder (Fitzsimons and Stricker 1971; Stricker 1971b, 1973), and invariably is associated with elevated levels of plasma OT (Stricker et al. 1987).

A comparable effect on sodium appetite in rats is produced by the injection of chemical agents causing nausea. Thus, lithium chloride was found to increase pituitary OT secretion in rats (Verbalis et al. 1986) and also to decrease consumption of sodium chloride but not water in PEG-treated rats (Stricker and Verbalis 1987). These effects reflect activation of the magnocellular PVN neurons, and suggest that parvocellular oxytocinergic neurons mediating inhibition of sodium appetite might also have been stimulated (McCann et al. 1989).

Other Elements in Central Neural Control

There are two other factors that have been demonstrated to influence fluid ingestion by rats during hypovolaemia and/or sodium deficiency. One is a circadian influence on drinking, presumably mediated in part by the suprachiasmatic nucleus. Thus, both adrenalectomized rats and PEG-treated rats are more inclined to drink sodium chloride at night than by day (Rowland et al. 1985). Moreover, in rats with vNM lesions, the pronounced increase in daily consumption of sodium chloride solution occurs exclusively by night (Gardiner et al. 1986; see also Gardiner and Stricker 1985a), and the observed impairments of water drinking responses to osmotic dehydration and hypovolaemia occur by day but not by night (Gardiner and Stricker 1985b). Thus, the apparent elimination of thirst in these animals is not likely to reflect damage to osmoreceptors and a failure to detect plasma volume deficits, respectively, but an inability to translate the messages into appropriate behaviour during the day.

Another factor of known significance in the control of fluid ingestion is the central system that mediates behavioural activation, the non-specific drive that characterizes all voluntary responses to stimulation. In this regard, it has long been known that bilateral lesions of the lateral hypothalamic area disrupt water and sodium chloride intakes in rats during hypovolaemia and/or sodium deficiency (Stricker and Wolf 1967; Wolf and Quartermain 1967). These impairments are particularly promi-nent when the testing period is brief rather than prolonged (Stricker 1976), apparently because the brain-damaged animals are generally more behaviourally responsive at night than by day (Rowland 1976), when studies typically are conducted. These and other results (e.g. Stricker et al. 1979; Snyder et al. 1985) indicate the importance of dopamine-containing systems, which pass through the lateral hypothalamic area, in the mediation of motivated ingestive behaviours as well as other voluntary activities (see Chapter 14).

It should be noted that an impairment of behavioural activation may also contribute to the dysfunctions in the drinking responses to acute dehydration that were observed in rats after lesions of vNM or SFO. In this regard, systemic administration of the stimulant agent, caffeine, was often able to restore water consumption in the brain-damaged animals (Hosutt et al. 1981; Gardiner and Stricker 1985b). Thus, the drug may have promoted ingestion because it increased the readiness of animals to respond, and thereby compensated for focal impairments in the activational component of drinking behaviour.

Summary and Conclusions

During hypovolaemia, neural signals from vascular baroreceptors act together with angiotensin to provide excitatory signals both for thirst and for sodium appetite in rats. The differential stimulation of water or sodium chloride intake depends in large part on the presence of inhibitory stimuli. During hypovolaemia, osmotic dilution inhibits thirst but disinhibits sodium appetite, whereas osmotic concentration inhibits sodium appetite but disinhibits thirst. This symmetrical arrangement permits water and sodium chloride intakes to be stimulated alternately in the hypovolaemic rat and thereby insures volume repletion.

Various stimuli inhibit sodium appetite other than osmotic dehydration. These include severe hypovolaemia with or without accompanying arterial hypotension, anuria, and nausea. In each case, inhibition of sodium appetite in rats is associated with increased plasma levels of OT, indicating increased activity in magnocellular neurons in the PVN. Conversely, various treatments that potentiate sodium appetite blunt OT secretion, including osmotic dilution, sodium deficiency, and daily treatment with DOC.

A subset of parvocellular neurons in the PVN may play a critical role in mediating inhibition of sodium appetite and thereby determine whether thirst or sodium appetite is stimulated during hypovolaemia. In this regard, there is known neural input into oxytocinergic neurons in PVN from vascular baroreceptors, from cerebral osmoreceptors, from angiotensin receptors in SFO, and from brainstem chemoreceptors that detect circulating toxins.

Orthogonal to these considerations are endogenous rhythms that influence consumption of water and sodium chloride. In general, during hypovolaemia and/or sodium deficiency rats are much more likely to behave promptly and appropriately by night than by day. Also relevant to these considerations are activational systems in the brain that actually permit behaviour to occur. These systems are not centred in the lateral hypothalamus, as had been conceptualized initially, but appear to involve ascending catecholaminergic neurons coursing through the diencephalon.

References

Balment RJ, Brimble MJ, Forsling ML (1980) Release of oxytocin induced by salt loading and its influence on renal excretion in the male rat. J Physiol (Lond) 308:439–449

Buggy J, Fisher AE (1974) Evidence for a dual role for angiotensin in water and sodium intake. Nature 250:733–735

Carithers J, Johnson AK (1985) Lesions of the tissue surrounding the preoptic recess (AV3V region) affect neurosecretory cells in the paraventricular nuclei in the rat. Brain Res 337:233–243

Ferguson AV, Kasting NW (1988) Angiotensin acts at the subfornical organ to increase plasma oxytocin concentrations in the rat. Regul Pept 23:343–352

Fitzsimons JT (1961) Drinking by rats depleted of body fluid without increase in osmotic pressure. J Physiol (Lond) 159:297–309

Fitzsimons JT (1969) The role of a renal thirst factor in drinking induced by extracellular stimuli. J Physiol (Lond) 201:349–368

Fitzsimons JT, Simons BJ (1969) The effect on drinking in the rat of intravenous infusion of angiotensin, given alone or in combination with other stimuli of thirst. J Physiol (Lond) 203:45–57

Fitzsimons JT, Stricker EM (1971) Sodium appetite and the renin–angiotensin system. Nature, New Biol 231:58–60

Fluharty SJ, Epstein AN (1983) Sodium appetite elicited by intracerebroventricular infusion of angiotensin II in the rat: II. Synergistic interaction with systemic mineralocorticoids. Behav Neurosci 97:746–758

Gardiner TW, Stricker EM (1985a) Hyperdipsia in rats after electrolytic lesions of nucleus medianus. Am J Physiol 248:R214–R223

Gardiner TW, Stricker EM (1985b) Impaired drinking responses of rats with lesions of nucleus medianus: circadian dependence. Am J Physiol 248:R224–R230

Gardiner TW, Verbalis JG, Stricker EM (1985) Impaired secretion of vasopressin and oxytocin in rats after lesions of nucleus medianus. Am J Physiol 249:R681–R688

Gardiner TW, Jolley JR, Vagnucci AH, Stricker EM (1986) Enhanced sodium appetite in rats with lesions centered upon nucleus medianus. Behav Neurosci 100:531–535

Helmreich DL, Thiels E, Verbalis JG, Stricker EM (1988) Suckling does not affect gastric motility in rats. Soc Neurosci Abstr 14:629

Hosutt JA, Stricker EM (1981) Hypotension and thirst after phentolamine treatment. Physiol Behav 27:463–468

Hosutt JA, Rowland N, Stricker EM (1981) Impaired drinking responses of rats with lesions of the subfornical organ. J Comp Physiol Psychol 95:104–113

Kaufman S (1984) Role of right atrial receptors in the control of drinking in the rat. J Physiol (Lond) 349:389–396

Lang RE, Rascher W, Heil J, Unger T, Wiedmann G, Ganten D (1981) Angiotensin stimulates oxytocin release. Life Sci 29:1425–1428

Lawrence D, Pittman QJ (1985) Response of rat paraventricular neurones with central projections to suckling, haemorrhage or osmotic stimuli. Brain Res 341:176–183

Leenen FH, Stricker EM (1974) Plasma renin activity and thirst following hypovolemia or caval ligation in rats. Am J Physiol 226:1238–1242

Lind RW, Van Hoeson GW, Johnson AK (1982) An HRP study of the connections of the subfornical organ in the rat. J Comp Neurol 210:265–277

Mangiapane ML, Thrasher TN, Keil LC, Simpson JB, Ganong WF (1983) Deficits in drinking and vasopressin secretion after lesions of nucleus medianus. Neuroendocrinology 37:73–77

Mann JFE, Johnson AK, Ganten D (1980) Plasma angiotensin II: dipsogenic levels and angiotensin-generating capacity of renin. Am J Physiol 238:R372–R377

McCann MJ, Verbalis JG, Stricker EM (1989) LiCl and CCK inhibit gastric emptying and feeding and stimulate OT secretion in rats. Am J Physiol 256:R463–R468

Miselis RR (1982) Recent advances in subfornical organ morphology. In: Rodriguez EM, vanWimersma Greidanus TB (eds) Frontiers in hormone research, vol 9. Karger, Basel, pp 79–87

Miselis RR, Shapiro RE, Hand PJ (1979) Subfornical organ efferents to neural systems for control of body water. Science 205:1022–1025

Richter CP, Barelare B Jr (1938) Nutritional requirements of pregnant and lactating rats studied by the self-selection method. Endocrinology 23:15–24

Rowland N (1976) Circadian rhythms and partial recovery of regulatory drinking in rats after lateral hypothalamic lesions. J Comp Physiol Psychol 90:382–393

Rowland NE, Bellush LL, Fregly MJ (1985) Nycthemeral rhythms and sodium chloride appetite in rats. Am J Physiol 249:R375–R378

Simpson JB, Routtenberg A (1973) Subfornical organ: site of drinking elicitation by angiotensin II. Science 181:1172–1175

Simpson JB, Routtenberg A (1975) Subfornical organ lesions reduce intravenous angiotensin-induced drinking. Brain Res 88:154–161

Simpson JB, Epstein AN, Camardo JS Jr (1978) Localization of receptors for the dipsogenic action of angiotensin II in the subfornical organ of rat. J Comp Physiol Psychol 92:581–608

Snyder AM, Stricker EM, Zigmond MJ (1985) Stress-induced neurological impairments in an animal model of Parkinsonism. Ann Neurol 18:544–551

Stricker EM (1966) Extracellular fluid volume and thirst. Am J Physiol 211:232–238

Stricker EM (1968) Some physiological and motivational properties of the hypovolemic stimulus for thirst. Physiol Behav 3:379–385

Stricker EM (1969) Osmoregulation and volume regulation in rats: inhibition of hypovolemic thirst by water. Am J Physiol 217:98–105

Stricker EM (1971a) Inhibiton of thirst in rats following hypovolemia and/or caval ligation. Physiol Behav 6:293–298

Stricker EM (1971b) Effects of hypovolemia and/or caval ligation on water and NaCl solution drinking by rats. Physiol Behav 6:299–305

Stricker EM (1973) Thirst, sodium appetite, and complementary physiological contributions to the regulation of intravascular fluid volume. In: Epstein AN, Kissileff HR, Stellar E (eds), The neuropsychology of thirst. H.V. Winston, New York, pp 73–98

Stricker EM (1976) Drinking by rats after lateral hypothalamic lesions: a new look at the lateral hypothalamic syndrome. J Comp Physiol Psychol 90:127–143

Stricker EM (1981) Thirst and sodium appetite after colloid treatment in rats. J Comp Physiol Psychol 95:1–25

Stricker EM (1983) Thirst and sodium appetite after colloid treatment in rats: role of the renin–angiotensin–aldosterone system. Behav Neurosci 97:725–737

Stricker EM, Jalowiec JE (1970) Restoration of intravascular fluid volume following acute hypovolemia in rats. Am J Physiol 218:191–196

Stricker EM, Verbalis JG (1986) Interaction of osmotic and volume stimuli in regulation of neurohypophyseal secretion in rats. Am J Physiol 250:R267–R275

Stricker EM, Verbalis JG (1987) Central inhibitory control of sodium appetite in rats: correlation with pituitary oxytocin secretion. Behav Neurosci 101:560–567

Stricker EM, Verbalis JG (1988) Hormones and behavior: the biological bases of thirst and sodium appetite. Am Sci 76:261–267

Stricker EM, Verbalis JG (1990) Sodium appetite. In: Stricker E (ed) Handbook of behavioral neurobiology, vol 10. Plenum Press, New York, pp 387–419

Stricker EM, Wolf G (1966) Blood volume and tonicity in relation to sodium appetite. J Comp Physiol Psychol 62:275–279

Stricker EM, Wolf G (1967) The effects of hypovolemia on drinking in rats with lateral hypothalamic damage. Proc Soc Exp Biol Med 124:816–820

Stricker EM, Cooper PH, Marshall JF, Zigmond MJ (1979) Acute homeostatic imbalances reinstate sensorimotor dysfunctions in rats with lateral hypothalamic lesions. J Comp Physiol Psychol 93:512–521

Stricker EM, Hosutt JA, Verbalis JG (1987) Neurohypophyseal secretion in hypovolemic rats: inverse relation to sodium appetite. Am J Physiol 252:R889–R896

Swanson LN, Sawchenko PE (1983) Hypothalamic integration: organization of the paraventricular and supraoptic nuclei. Ann Rev Neurosci 6:269–324

Thiels E, Verbalis JG, Stricker EM (1990) Sodium appetite in lactating rats. Behav Neurosci 104:742–750

Verbalis JG, McHale CM, Gardiner TW, Stricker EM (1986) Oxytocin and vasopressin secretion in response to stimuli producing learned taste aversions in rats. Behav Neurosci 100:466–475

Verbalis JG, Mangione M, Stricker EM (1988) Oxytocin is natriuretic at physiological plasma concentrations. Soc Neurosci Abstr 14:628

Wolf G, Quartermain D (1967) Sodium chloride intake of adrenalectomized rats with lateral hypothalamic lesions. Am J Physiol 212:113–118

Zimmerman MB, Blaine EH, Stricker EM (1981) Water intake in hypovolemic sheep: effects of crushing the left atrial appendage. Science 211:489–491

Commentary

Epstein: Stricker and his colleagues have been developing the interesting idea that sodium chloride intake is inhibited by an oxytocinergic "switch" that is located in the parvocellular PVN. The idea has the virtue of specificity. It predicts, first, that ablation of the parvocellular PVN will exaggerate sodium chloride intake in the sodium-deficient animal, second, that intracerebral injection of oxytocin will suppress the behaviour especially if the injections are made into the parvocellular PVN, and third, that oxytocin blockers will have the opposite effect. These experiments would be crucial to acceptance of this idea.

Stricker: We have proposed that parvocellular neurons in the PVN may play an important role in the inhibition of sodium appetite in rats. This hypothesis may be evaluated by determining sodium appetite in rats after those neurons have been damaged, as you suggest. However, in the event that such data when collected do not support the proposal and ultimately it is rejected, we should not neglect the data that led to the proposal because they would still be unexplained. That is, the fact would still remain that in a great number of conditions – indeed, in all studies of sodium appetite in rats in which appropriate measurements have been made – sodium appetite is present only when plasma OT levels are low and it is inhibited when plasma OT levels are elevated, even when animals are sodium-deficient and/or hypovolaemic. A new, more satisfactory proposal would be needed to account for those findings.

Our present hypothesis that centrally projecting neurons in PVN mediate inhibition of sodium appetite would not be evaluated by the two other experiments you suggest. Central administration of OT agonist or antagonist would have to be made into the relevant terminal fields of the parvocellular projections, not into the cell bodies in PVN. Until we have some idea as to where those critical brain sites are, I believe this issue can be approached only by administering OT agonist and antagonist into the cerebrospinal fluid and hoping that it diffuses to the critical target area. I do not want to discourage investigators from doing those experiments, but I do want to encourage caution in interpreting the results.

Fregly: A mechanism by which aldosterone–angiotensin synergy might work involves the upregulation of angiotensin II receptors in the diencephalon of the rat by aldosterone (Wilson et al. 1986; Sumners and Fregly 1989). Under these conditions, the drinking response to acutely administered (either centrally or peripherally) angiotensin II is enhanced (Fregly et al. 1979; Wilson et al. 1986).

We have also observed upregulation of angiotensin II receptors on neurons in culture when either aldosterone or deoxycorticosterone acetate was added to the medium bathing the neurons (Sumners and Fregly 1989). This suggests a direct effect of this steroid on receptors for angiotensin II.

References

Fregly MJ, Katovich MJ, Barney CC (1979) Effect of chronic treatment with deoxycorticosterone on the dipsogenic response of rats to isoproterenol and angiotensin. Pharmacology 19:165–172

Sumners C, Fregly MJ (1989) Modulation of angiotensin II binding sites in neuronal cultures by mineralocorticoids. Am J Physiol 256:C121–C129

Wilson KM, Sumners C, Hathaway S, Fregly MJ (1986) Mineralocorticoids modulate central angiotensin II receptors in rats. Brain Res 382:87–96

Stricker: These interesting findings are consistent with the angiotensin–aldosterone synergy hypothesis, although it has not been demonstrated that an upregulation of angiotensin receptors in the brain makes the rat more likely to drink concentrated sodium chloride solution in large amounts.

Epstein: It is puzzling that Stricker and his colleagues were unable to demonstrate increased sodium chloride intake during lactation. The phenomenon has been demonstrated in at least four laboratories (Richter and Barelare 1938; Denton and Nelson 1978; McBurnie, et al. 1988; Schleifer and Woodside 1990; Frankmann et al. submitted) in three species (rat, mouse, rabbit). Richter's finding presents a problem for the idea that high central oxytocin activity suppresses salt intake.

References

Denton DA, Nelson JF (1978) The control of salt appetite in wild rabbits during lactation. Endocrinology 103:1880–1884

Frankmann SP, Ulrich P, Epstein AN (submitted) Transient and lasting effects of reproductive episodes on NaCl intake of the female rat.

McBurnie M, Denton DA, Tarjan E (1988) Influence of pregnancy and lactation on sodium appetite of BALB/c mice. Am J Physiol 225:R1020–R1024

Richter CP, Barelare B Jr (1938) Nutritional requirements of pregnant and lactating rats studied by the self-selection method. Endocrinology 23:15–24

Schleifer LA, Woodside BC (1990) The effect of reproductive status on sodium chloride intake and preference curves. Proc Eastern Psychol Assoc 61:35

Stricker: It is true that we have been unable to find substantial sodium appetite in lactating rats during standard maintenance conditions (Thiels et al. 1990). In these experiments lactating rats were maintained on a sodium-deficient diet and given water and 0.5 M sodium chloride solution to drink. They drank only 2–5 ml of saline each day, amounts that could easily replace the estimated daily loss of 1–2 mEq sodium in milk. These findings actually are quite comparable to the results reported by Richter and Barelare. A more marked sodium appetite would not be expected in rats under these circumstances because their sodium losses are so modest. Indeed, when animals consume standard sodium-rich chow, their intake of sodium chloride in food far exceeds the loss of sodium in milk.

On the other hand, we do find a marked sodium appetite in lactating rats after a period of dietary sodium deprivation. It is conceivable that hormones associated with lactation stimulate sodium appetite directly, as was suggested by studies in rabbits (Denton and Nelson 1978), but there is no evidence to date that a comparable phenomenon occurs in rats independently of sodium deficiency. Those studies of rabbits (and perhaps of mice) may indicate different mechanisms by which sodium appetite is controlled than are present in rats. Similarly, your study of multiparous lactating rats appears to involve different influences on sodium appetite than exist in primiparous rats.

Note that pituitary OT secretion reflects the activity of magnocellular neurons in PVN, whereas we are proposing that it is a marker of parallel activity in parvocellular neurons that might be part of that central inhibitory system. There is no evidence to date that suckling increases activity in the parvocellular PVN neurons. Thus, the well-known secretion of pituitary OT by lactating rats during nursing, taken together with the marked intake of saline by lactating rats after a period of imposed dietary sodium deprivation, does not present a problem for the proposal that elevated parvocellular PVN activity may be incompatible with sodium appetite in rats. Besides, the release of milk into the mammary ducts requires that the dam is in an inactive, sleep-like state, and consequently it does not co-occur with ingestive behaviour. Rather, nutrient ingestion invariably occurs between bouts of nursing and associated episodes of pituitary OT secretion.

References

Denton DA, Nelson JF (1978) The control of salt appetite in wild rabbits during lactation. Endocrinology 103:1880–1884
Thiels E, Verbalis JG, Stricker EM (1990) Sodium appetite in lactating rats. Behav Neurosci 104: 742–750

Chapter 12

Rostro-Sagittal Brain: Site of Integration of Hydrational Signals in Body Fluid Regulation and Drinking

S. Nicolaïdis, M. El Ghissassi and S.N. Thornton

Introduction

Evolution of Ideas

The evolution of ideas on the anatomical localization of structures related to drinking, and body fluid regulation in general, seems to have followed a caudorostral course. Following this rule, the course may be reaching its end because the structure we shall describe in the last part of this chapter is the most anterior point in the midline brain since it occupies the entire rostral limitation of the medial diencephalon and telencephalon. This structure is the organum cavum prelamina terminalis (OCPLT) (Nicolaïdis and El Ghissassi 1989, 1990), and evidence will be presented showing that it may play a crucial role in the control of drinking.

Thanks to the thorough study of the history of thirst by Fitzsimons (1979) we know that the control of drinking has been related to specific areas in the brain since the nineteenth century. Nothnagel (1881) made the remarkable observation of a patient who had fallen from a horse and who complained of acute thirst before any diuresis could have accounted for a primary diuresis and subsequent secondary thirst. This led to the suggestion that the location which was implicated in this symptom should be somewhere in the base of the brain. This idea was also expressed by Longet (1850, 1861) from other clinical observations.

There followed a long interval before a more circumscribed area, the anterior hypothalamus, was found to house the "centres" of thirst as well as antidiuresis. The evidence that the hypothalamoneurohypophyseal system was responsible for the control of diuresis rapidly drew attention to the anterior portion of the diencephalon. The observation by French neurosurgeons and neurologists of a dramatic complaint of thirst in response to the puncture of a suprachiasmatic cyst oriented attention more precisely towards this rostral area (Kourilski et al. 1942). A more detailed anatomical localization was made possible when stereotaxic lesions in the anterolateral hypothalamus resulted in adipsia and aphagia (Anand and Brobeck 1951; Andersson and McCann 1955). The pivotal experiment, however, was performed by Andersson (1952) who induced copious drinking in the goat in response to intrahypothalamic application of hypertonic sodium chloride, or by electrical stimulation (Andersson 1953). The critical area was the anterior hypothalamus between the fornical columns, the mammillothalamic tracts and lateral preoptic area. This latter area was also found to trigger the best drinking responses in rats and rabbits to microinjections of hypertonic solutions of osmotically active solutes (Blass and Epstein 1971; Peck and Novin 1971).

The voloreceptive network for drinking was explored later when neurons sensitive to volaemic changes were found to be located within more sagittal areas posterior and anterior to the organum vasculosum of the lamina terminalis (OVLT) (Nicolaïdis 1970; Thornton et al. 1984). Neurons in these areas respond to hypovolaemia and may complement the neural and humoral atrial blood volume related messages.

An important step in our understanding of the way thirst is controlled was made when it appeared that, besides the biophysical (osmo- and volumetric) signals of dehydration, thirst was also modulated by neurochemical messages of dehydration and among them by the renin–angiotensin system (Epstein et al. 1970; Fitzsimons 1979). That is, another parameter seems to be taken into account in order to tune the final state of thirst.

All these investigations, therefore, have established that the signals of the hydrational states (intra- and extracellular, haemodynamic and peptidergic) reach the anterior hypothalamus and contribute to the sensing up of the hydrational state of the body. Obviously, however, all the concurrent signals of dehydration, in order to co-operate in the onset and offset of drinking, have to converge towards some common integrative neurons.

Integration of Neural and Humoral Hydrational Signals

The ability of neurons to respond to signals of both intracellular and extracellular dehydration and to trigger regulatory responses was first reported by Nicolaïdis (1970). This report was also the first to locate the integrative neurons along the most rostral and medial structures of the brain which included the diagonal band of Broca, the dorsal chiasmatic nucleus and populations of neurons found in the anterior wall of the third cerebral ventricle. Later, iontophoretic investigations showed that the same area contained angiotensin-sensitive neurons (Ishibashi et al. 1985; Thornton et al. 1985). Besides these iontophoretic extensions of the initial electrophysiological findings, the role of this region in the regulation of water and electrolyte balance was subsequently confirmed by lesion and stimulation techniques (Montemurro and Stevenson 1957; Buggy et al. 1975; McKinley et al. 1982; Thrasher and Keil 1987; Johnson and Buggy 1978). In addition, the anatomy of this circuitry has also been extensively studied (Miselis et al. 1987).

The subfornical organ (SFO) was first recognized as a site of action of the extracellular dehydration by Palkovits et al. (1968). When circulating angiotensin was shown to act upon brain structures in order to affect drinking, the role of the fenestrated circumventricular organs in giving access of such messages to thirst-related structures was shown at the level of the SFO (Simpson and Routenberg 1973) and subsequently extensively investigated. The area around the OVLT (or the suprachiasmatic recess or the anteroventral third ventricle) was then shown to be particularly sensitive to the dipsogenic action of angiotensin II (Nicolaïdis and Fitzsimons 1975; Buggy et al. 1975). Subsequently, in an attempt to improve the localisation of the crucial area for drinking two methods were used: (a) metabolic, using the 2-deoxyglucose technique of Sokoloff (Nicolaïdis et al. 1981) and (b) lesions (Johnson 1982). The results of both studies mentioned the "nucleus medianus" as a possible crucial structure. This nucleus lines the dorsal rostral and ventral aspect of the anterior commissure and its anterior limit reaches the OCPLT, a point that will be discussed later.

It is noteworthy that it was electrophysiological investigations which revealed the interest of the midline rostral structures in water and electrolyte balance. They also revealed the voloreceptive and integrative properties of neurons located in this region. After discussing the main findings of these electrophysiological studies their role in highlighting interesting properties in an area which delineates an as yet undescribed structure of interest, the OCPLT, will be presented.

Electrophysiological Findings on the Integrative Properties of Neurons

A neuron can be considered as an elementary integrative system if it is capable of receiving more than one relevant input and responding in a way that takes into account all the afferent information. In the case of water and electrolyte balance the relevant information concerns osmometric, volumetric and cardiovascular parameters, their neurochemical consequences and afferents from the orogastrointestinal tract.

Thirst indeed depends on the sum of osmometric and volumetric signals (Fitzsimons and Oatley 1968) and is quenched by orally administered water and aggravated by oral salt. Are there neurons capable of adding algebraically thirst-related information?

The next section shows that such information converges towards the rostral wall of the third ventricle.

Convergent Systemic and Gustatory Projection of Sodium Chloride Versus Water Stimulations

The first single unit recordings in response to sodium chloride or water carotid injections made in the supraoptic and paraventricular nucleus confirmed that Verney's (1947) idea about specialized neurons, the so-called osmoreceptors, was feasible (Cross and Green 1959). A wider exploration along the thirst-related area revealed that sodium chloride-sensitive neurons were more widely distributed in the anterior hypothalamus (Nicolaïdis 1968). Besides the supraoptic nucleus they were found in the anterior medial forebrain bundle, the anterior parafornical area, the pallidofugal bundle (Nicolaïdis 1968), the diagonal band of Broca and the suprachiasmatic and arcuate nuclei (Nicolaïdis 1970).

Neurons in the medial forebrain bundle and the pallidofugal bundle respond in an opposite way to hypertonic sodium chloride solutions and to water whether applied in the carotid artery and/or directly onto the tongue (Nicolaïdis 1968, 1969). This was the first example ever reported of integration of information, which in this particular case was converging from the inner and from the external milieu.

Integration of Volaemic, Osmotic and Blood Pressure Signals

The juxtaventricular area extending from the suprachiasmatic nucleus to the diagonal band of Broca, possesses neurons which respond to volaemic as well as blood pressure and osmotic changes (Nicolaïdis 1970; Thornton et al. 1984). However, the number of volosensitive neurons is much smaller than the neurons responding to osmotic and blood pressure changes. Also, the volosensitive neurons were found within a more restricted area than the neurons responding to osmotic and blood pressure changes. Finally, the volosensitive neurons found in the OVLT were also responsive to the iontophoretic application of angiotensin II (Thornton et al. 1984).

Angiotensin II was also found to be able to modulate the firing of barosensitive neurons. Such neurons, particularly in the diagonal band of Broca showed an enhancement of their activity in response to both angiotensin II and to lowering of blood pressure (Thornton et al. 1984; Ishibashi et al. 1985).

Neuronal Response to Angiotensin II Iontophoretic Application

Neuronal responsiveness to angiotensin II is widespread. Most iontophoretic investigations using angiotensin II or III have been oriented towards structures suspected to play some role on drinking or vasopressin secretion (Nicoll and Barker 1971; Wayner et al. 1973; Felix and Akert 1974; Buranarugsa and Hubbard 1979; Simonnet et al. 1980). In a more systematic study, responsiveness to angiotensin II and to blood pressure changes showed the large distribution of angiotensin-responsive neurons, although they were mainly concentrated in the barosensitive rostrodiencephalic region including the SFO and the diagonal band of Broca and the neurons surrounding the OVLT (Thornton et al. 1984; Ishibashi et al. 1985).

Co-action of Angiotensin and Vasopressin on Neuronal Activity

Accumulating evidence indicates that the effect of multiple regulatory hormones is not accounted for by the sum of the individual effects of each of the hormones (Ramsay et al. 1983). This may be true as far as the neurohormones are concerned, particularly since the coexistence of several hormones in the same neuron is now universally admitted. Thus, the idea that one single unit could respond to more than one peptide in an additive or non-additive way was explored. In contrast to angiotensin,

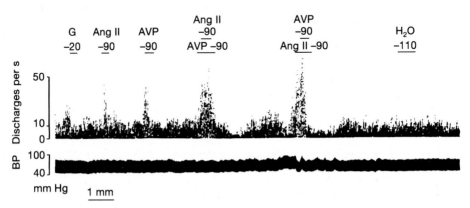

Fig. 12.1. One-second itegrated unit discharges from one glutamate (G) responsive neuron located in the medial septal nucleus. This neuron responded to either or both angiotensin II (Ang II) and AVP. When the elctrophoresis of one of these peptides was superimposed on the other the response was additive (double or less). Equal or stronger application of vehicle (H_2O) was ineffective.

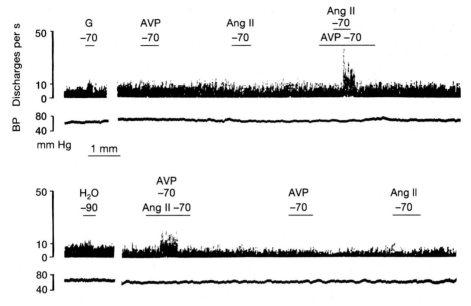

Fig. 12.2. As in Fig. 12.1. Recording of an SFO neuron unresponsive to either Ang II or AVP and also to various currents through the vehicle. Only when AVP and Ang II were applied together was a response observed. This response was decreased as soon as the second molecule application was discontinued. Blood pressure readings were added in this case to show that accidental fluctuations did not affect the neuronal activity. This is not a barosensitive neuron.

the role of vasopressin in drinking is unclear (Fitzsimons 1979; Nicolaïdis 1985). Both of these neurohormones are activated by dehydration, and, therefore, they might be suspected to act synergistically on integrative neurons which would have their activity modulated by hormonal derivatives of the primitive events, namely intracellular and extracellular dehydration.

Hormonal co-action was investigated using single-neuron recordings while angiotensin II and arginine vasopressin (AVP) were applied iontophoretically on the surface of the neuronal membrane (Nicolaïdis and Jeulin 1984). This investigation was restricted within the structures that surround the rostral limit of the third ventricle. A total of 49 rats was used and 218 neurons were explored for their responsiveness to either angiotensin or AVP or to both of these peptides.

The most often encountered co-action was additive (Fig. 12.1). Additive responses were found at various levels of the midline structures but were scarce when the exploration was more than 0.5 mm lateral. The responsive areas included the medial preoptic area, medial corticohypothalamic nucleus and paraventricular thalamic nucleus.

Another interesting co-action was found only in the SFO and periventricular thalamic nucleus and consisted in a type of permissive action. Response to either angiotensin II or to AVP was obtained only if both peptides were applied concomitantly. Repeated application of one of these two peptides was ineffective as if one of them was negating the other one (Fig. 12.2). Three histologically identified neurons in the OVLT responded to angiotensin only but not to AVP or to the simultaneous application of AVP and angiotensin II.

Parallel research established that all the neurons tested in this area responded to the V1 (pressor response related) agonist of AVP but not to the V2 (antidiuresis related) agonist. Furthermore, neural responses to AVP were blocked by V1 but not V2 antagonists.

Interestingly, a number of the neurons responding to angiotensin and to AVP also respond to blood pressure changes (osmosensitivity was not explored in these iontophoretically tested neurons).

Neuronal Response to Angiotensin and Na$^+$ Iontophoretic Application: Peculiarity of Neurons Lining the OCPLT

During the above iontophoretic investigations the possibility that some neurons, particularly close to the ependymal wall, could show an enhancement of their activity when angiotensin was co-applied with Na$^+$ was systematically tested. The reason for this attention was related to Andersson's (1953) hypothesis that angiotensin acquires dipsogenic properties by altering the sodium permeability of some periventricular specific neurons. Practically all of the neurons tested in this way failed to show a synergistic response to angiotensin and Na$^+$. Only two exceptions were observed which happened to line the cavity which is visible in the midline between the corpus callosum and the septum, one in the septohippocampal nucleus and the other in the area of taenia tecta. A systematic investigation of the properties of the neurons lining the above mentioned cavity revealed two additional neurons lining the same cavity, one in the vertical limb of the diagonal band of Broca, in front of the nucleus medianus, and the other in the septohippocampal nucleus.

Further experiments were performed (Nicolaïdis and El Ghissassi 1989, 1990) to understand the anatomofunctional meaning of the cavity which was ignored in practically all the atlases and, simultaneously, seemed to be lined by such peculiar neurons.

Evidence for an Undescribed Hollow System, the OCPLT

The neurons with the peculiar properties described in the previous experiment were found to line a series of hollow structures that communicate with each other. In the rat, these communicating hollow structures extend from the velum (membranous roof of the third ventricle in the posterior septum) to the genu of the corpus callosum (horizontal horn) and along the diagonal bands of Broca, down to the anterior aspect of the OVLT (vertical horn).

Only the portion of these cavities between the genu of the corpus callosum and the septal region above the SFO corresponds to what was first described in the mid-seventeenth century by Sylvius de le Boe (1680) as the fifth ventricle because it was believed to belong to the cerebroventricular system. This known part of the whole system was subsequently given the name of cavum septi pellucidi (Kappers et al. 1936) and occasionally dealt with as nothing more than a more or less mysterious subarachnoid-like space invaginating the brain tissue (Oteruelo 1986). The rest of the system was probably considered as a fixation artifact and was just ignored.

The following experiments were designed to determine whether the system of cavities corresponded to a communicating coherent anatomical entity and to characterize its possible properties and role in the regulation of hydromineral balance.

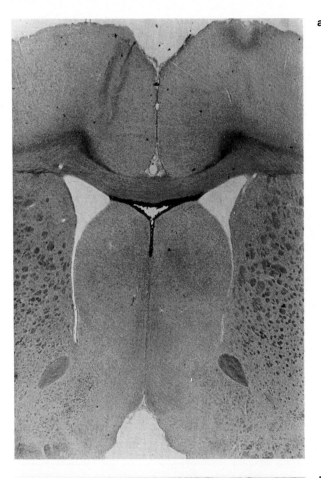

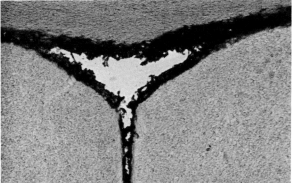

Fig. 12.3. Coronal slice (**a**) and its magnification (**b**) of the horizontal horn (cavum septi pellucidi) of the OCPLT at a level close to the genu of the corpus callosum (CC) and at a diagonal band of Broca. A solution (1 µl) of pontamine sky blue was injected via the injector into a rat responsive to angiotensin II before death. **b,** The entire extent of the space under the CC is shown together with the delimitation of the OCPLT from the cerebroventricular fluid, in particular by the ependymal epithelium on the ventricular side. Both the subcallosal and the midseptal wings were stained at this relatively anterior level.

Delineation of the OCPLT: the Extent of the Hollow Organ

To visualize the full extent of the hollow structures pontamine blue or India ink was injected stereotaxically in one part of the system (usually below the genu of the corpus callosum). Dyes were also used to assess a possible communication between the cavities of the OCPLT and the vascular or the subarachnoid space.

Fig. 12.3 shows a frontal section of the OCPLT at the level of the genu of the corpus callosum where the cavity reaches its maximum size. This is also the level where the cavity (belonging still to the horizontal horn near the shoulder) is T-shaped. The lateral edges of this T-shaped cavity reach the border of the lateral ventricles but they are separated from them by a tight wall made in part by the ependymal epithelium lining the ventricles.

Fig. 12.4 shows a schematic representation of the entire OCPLT. It is composed of two components (or horns), one horizontal and one vertical. The horizontal horn runs along the septal midline under the corpus callosum and, at the level of the genu of the corpus callosum, reaches its largest dimensions before following a dorsoventral direction (vertical horn). In its vertical trajectory the OCPLT runs within the point of separation of the midline brain in its two lateral hemispheres. The rostral edge of the vertical horn is located in the interhemispheric space from which it is separated by a cellular membrane which also separates the OCPLT from the anterior cerebral

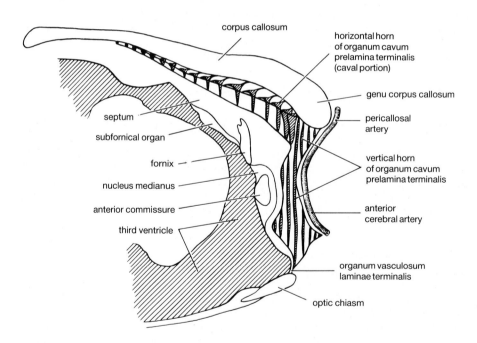

Fig. 12.4. Schematic view of the OCPLT in successive sections showing the sagittal flat irregular pouch of the vertical horn, widening under the genu of the corpus callosum (CC) and becoming T-shaped (also see Fig. 12.3), then a tubular cavity. The latter makes its way between the CC and the velum transversum. The bottom of the vertical horn ends on the outer wall of the lamina terminalis at a level corresponding to the vestigial neuroporic recess. There is no communication with either the pericallosal cistern, the cisterns of the basis and the longitudinal cerebral fissure or the space surrounding the anterior cerebral artery.

artery and its perivascular adventitia. As a result of this membranous border, whenever a dye was injected in the OCPLT it never reached the interhemispheric subarachnoid space, and the converse, a dye injected in the interhemispheric space never reached the OCPLT.

The continuum of cavities composing the OCPLT is lined by a number of structures including the fornices, the septohippocampal nucleus, the taenia tecta, the rostral extension of the nucleus medianus, the vertical limbs of the diagonal band of Broca and the midline area where the midline brain splits in its hemispheric prolongations and, finally, the anterior aspect of the lamina terminalis.

Other Anatomical Features of the OCPLT

Besides being a closed system the cavity (or the cavities) running from the mid-septal area down to the lamina terminalis seems to be filled with a fluid because the stain resulting from dye injection was always faded as compared to the stain resulting from dye injection within a plain brain tissue as if, within the OCPLT, the dye was diluted by an intracavital fluid.

Today it is difficult to know where the intra-OCPLT fluid comes from. One possibility is that the fluid originates from a highly vascular dorsoventral network lining the vertical horn of the OCPLT. The haemodynamic characteristics and the blood–brain barrier of this interesting vascular network are currently under investigation in the laboratory. Since it can be shown (see below) that angiotensin is very active as a dipsogen when microinjected in the OCPLT, it is tempting to suggest that this peptide (and perhaps other peptides) reaches the cavity via the abovementioned vascular network.

Drinking and Renal Effects of Angiotensin II into the OCPLT

Because electrophysiological investigations have shown that neurons lining various parts of the OCPLT were activated by iontophoretic applications of angiotensin II, the possibility was investigated that microinjections in the this hollow system could be particularly efficient in triggering drinking responses in the free-moving rat. A particular attention was directed to the possibility that the dipsogenic responses could be elicited wherever in the hollow system the microinjection was made. Microinjections of angiotensin were also made in adjacent to OCPLT brain structures to assess the specificity of responsiveness of the periluminal structures as compared to neighbouring plain structures. Post-mortem verification was always performed and the tip position was assessed by an examiner who was unaware of the behavioural results. More than 120 rats were used in the various stages of the experiments.

There was a drinking response wherever the angiotensin was injected, provided the fluid was able to reach either the horizontal or the vertical horn of the OCPLT. The response to 0.2 ng of angiotensin II was significant and was maximal above 5 ng. The only extra-OCPLT tissue that was sensitive to angiotensin II was the SFO but all of the canulae aimed at this circumventricular organ were transfixing one or another part of the horizontal horn of the OCPLT.

It was observed that, whenever the injection missed the cavity of the OCPLT there was no drinking response.

The exquisite sensitivity of microinjections of angiotensin II made in the OCPLT may account for some observations of "unexpected" drinking in response to tissue microinjection of angiotensin II in the septum (Booth 1968). It is possible that the leakage along the track of the injector brought about an injection within the cavity of the OCPLT, necessarily transfixed by the injector, and that the drinking response came from this leakage.

In a more recent study it has been shown that injection of dipsogenic doses of angiotensin II bring about antidiuretic and antinatriuretic responses after a latency of 5–10 minutes. Furthermore, injection of small doses of antinatriuretic factor (ANF) in the OCPLT is followed by diuresis and natriuresis after a latency similar to angiotensin. ANF injections in the OCPLT are also able to inhibit drinking (Nicolaïdis and El Ghissassi, unpublished).

Discussion

The data presented here show that the region which derives from the embryonic lamina terminalis is equipped with neuronal populations sensitive to signals relevant to water and electrolyte balance. Osmoreceptive, voloreceptive and baroreceptive neurons are found in the same areas. In addition, peptides known to modulate thirst and water balance are also active on receptive neurons of the same area. These neurons, in addition to their peptide-receptive property, have their activity modulated by both angiotensin and vasopressin. Co-action of these peptides shows the best example of additive or multiplicative integration of different messages.

The pathways used by the different messages to reach the integrative neurons are of interest and are discussed in Chapters 9 and 13. Gustatory afferents ascend via the solitarius-parabrachial pathway and project onto neurons, the topography of which has not been completely explored. It is possible that voloreceptive information projects also from the peripheral periatrial receptors to voloresponsive neurons.

Opinions vary on the pathway followed by peptides which carry information on the hydrational state of the body. As in the case of the non-chemical messages the angiotensin II message may be generated in situ or may affect the integrative area from the periphery. Blood-borne peptides are believed to interact with the neural tissue via the fenestrated vessels of circumventricular organs like the SFO before they act upon specific neurons around the nucleus medianus and OVLT areas. Another peptide-integrating system may be suggested by the studies reported in this chapter. Systemic and/or regional-borne peptides may be released in the lumen of the OCPLT from the bordering vascular network and from this sort of peptidic sink they may affect a vast neural network projecting to the nuclei that control drinking and autonomous regulatory responses. The organum cavum prelamina terminalis seems to function as the place of convergence of chemicals which, in turn, will modulate the responsiveness of neurons in the vicinity which, themselves, function as elementary integrators of neural information on hydrational state.

Acknowledgement. The research presented in this chapter was supported by a grant from the National Institute of Mental Health (1 PC 1 MH-43787).

References

Anand BK, Brobeck JR (1951) Hypothalamic control of food intake in rats and cats. Yale J Biol Med 24:123–140

Andersson B (1952) Polydipsia caused by intrahypothalamic injection of hypertonic NaCl solution. Experientia 8:157–158

Andersson B (1953) The effect of injections of hypertonic NaCl solutions into different parts of the hypothalamus of goats. Acta Physiol Scand 28:188–201

Andersson B, McCann SM (1955) The effect of hypothalamic lesions on water intake of the dog. Acta Physiol Scand 35:312–320

Blass EM, Epstein AN (1971) A lateral preoptic osmosensitive zone for thirst in the rat. J Comp Physiol Psychol 76:378–394

Booth DA (1968) Mechanism of action of norepinephrine in eliciting an eating response on injection into the rat hypothalamus. J Pharmacol Exp Ther 160:336–348

Buggy J, Fisher AE, Hoffman WE, Johnson AK, Phillips MI (1975) Ventricular obstruction effect on drinking induced by intracranial injection of angiotensin. Science 190:72

Buranarugsa P, Hubbard JI (1979) The neuronal organization of the rat subfornical organ in vitro and a test of the osmo- and morphine-receptor hypothesis. J Physiol (Lond) 291:101–116

Cross BA, Green JD (1956) Activity of single neurons in the hypothalamus effect of osmotic and other stimuli. J Physiol (Lond) 148:554–569

Epstein AN, Fitzsimons JT, Rolls BJ (1970) Drinking induced by injection of angiotensin into the brain of the rat. J Physiol (Lond) 210:457–474

Felix D, Akert K (1974) The effect of angiotensin II on neurones in the cat subfornical organ. Brain Res 76:350–353

Fitzsimons JT (1979) The physiology of thirst and sodium appetite. Monographs of the Physiological Society no 35, Cambridge University Press, Cambridge

Fitzsimons JT, Oatley K (1968) Additivity of stimuli for drinking in rats. J Comp Physiol Psychol 67:273–283

Ishibashi S, Oomura Y, Gueguen B, Nicolaïdis S (1985) Neuronal responses in subfornical organ and other regions of angiotensin II applied by various routes. Brain Res Bull 14:307–313

Johnson AK (1982) Neurobiology of the periventricular tissue surrounding the anteroventral third ventricle (AV3V) and its role in behaviour, fluid balance, and cardiovascular control. In: Smith OA, Galosy RA, Weiss SM (eds) Circulation neurobiology and behaviour. Elsevier, New York.

Johnson AK, Buggy J (1978) Periventricular preoptic-hypothalamus is vital for thirst and normal water economy. Am J Physiol 234:R122–R129

Kappers CUA; Uber GC, Crosby EC (1936) The comparative anatomy of the nervous system of vertebrates, including man. Vol III. Hafner, New York, pp 1409–1410

Kourilski R, David M, Sicard J, Galey JJ (1942) Diabète insipide post-traumatique; cessation subite de la soif au cours de l'ouverture d'un kyste arachnoïdien de la région optochiasmatique-guérison. Rev Neurol 74:264–280

Longet FA (1850) Traité de Physiologie, vol 2. Paris, Masson

Longet FA (1861) Traité de Physiologie, vol 1. Paris, Masson

McKinley MJ, Denton DA, Leksell LG et al. (1982) Osmoregulatory thirst in sheep is disrupted by ablation of the anterior wall of the optic recess. Brain Res 236:210–215

Miselis RR, Weiss ML, Shapiro RE (1987) Modulation of the visceral neuraxis. In: Gross PM (ed) Circumventricular organs and body fluids. CRC Press, Boca Raton, vol. 3, pp 143–162

Montemurro DG, Stevenson JAF (1957) Adipsia produced by hypothalamic lesions in the rat. Can J Biochem Physiol 35:31–37

Nicolaïdis S (1968) Réponses des unités osmosensibles hypothalamiques aux stimulations salines et aqueuses de la langue. C R Acad Sci Paris 267:2352–2355

Nicolaïdis S (1969) Early systemic responses to orogastric stimulation in the regulation of food and water balance. Functional and electrophysiological data. Ann N Y Acad Sci 157:1176–1203

Nicolaïdis S (1970) Mise en évidence de neurones basosensibles hypothalamiques antérieur et médian chez le chat. J Physiol (Paris) 62:199–200

Nicolaïdis S (1985) Thirst mechanisms and antidiuretic hormone in diabetes insipidus in man. Front Hormone Res 13:69–88

Nicolaïdis S, El Ghissassi M (1989) A new brain organ and its role on drinking. Appetite 12:227

Nicolaïdis S, El Ghissassi M (1990) A physiological role for a new brain organ in the rat: the organum cavum pre-lamina terminalis. Am J Physiol (in press)

Nicolaïdis S, Fitzsimons JT (1975) La dépendance de la prise d'eau induite par l'angiotensine II envers la fonction vasomotrice cérébrale locale chez le rat. C R Acad Sci 281D, pp 1417–1420

Nicolaïdis S, Jeulin AC (1984) Propriétés des neurones diencéphaliques antérieurs impliqués dans la régulation hydrominérale. J Physiol (Paris) 79:406–415

Nicolaïdis S, Le Poncin-Lafitte M, Danguir J, Grosdemouge C, Rapin JR (1981) Specific behaviour-bound brain cartography of the glucose uptake rate. Eur Neurol 20:180–182

Nicoll RA, Barker JL (1971) Excitation of supraoptic neurosecretory neurons by angiotensin II. Nature New Biol 233:172–174

Nothnagel H (1881) Durst und Polydipsie. Archiv Pathol Anat Physiol 86:435–437

Oteruelo FT (1986) On the cavum septum pellucidi and the cavum vergae. Anat Anz 162:271–278

Palkovits M, Zaborsky L, Magyar P (1968) Volume receptors in the diencephalon. Acta Morph Hung 16:391–401

Peck JW, Novin D (1971) Evidence that osmoreceptors mediating drinking in rabbits are in the lateral preoptic area. J Comp Physiol Psychol 74:134–147

Ramsay DJ, Thrasher TN, Keil LC (1983) The organum vasculosum laminae terminalis: a critical area for osmoreception: the neurohypophysis: structure, function and control. Prog Brain Res 60:91–98

Simonnet G, Bioulac B, Rodriguez F, Vincent JD (1980) Evidence of a direct action of angiotensin II on neurones in the septum and in the medial preoptic area. Pharmacol Biochem Behav 13:359–363

Simpson JB, Routenberg A (1973) Subfornical organ: site of drinking elicited by angiotensin II. Science 181:1172–1175

Sylvius F de le Boe (1680) Opera Medica, Tam Hactenus Insdita, Quam Varilis Locis et Formis Edita; Apud Danielem Elsevirum et Abrahamum Wolfgang. Amsterdam (A reprint of the 1679 edition)

Thornton SN, De Beaurepaire R, Nicolaïdis S (1984) Electrophysiological investigation of cells in the region of the anterior hypothalamus firing in relation to blood pressure changes. Brain Res 299:1–7

Thornton SN, Jeulin AC, De Beaurepaire R, Nicolaïdis S (1985) Iontophoretic application of angiotensin II, vasopressin and oxytocin in the region of the anterior hypothalamus in the rat. Brain Res Bull 14:211–215

Thrasher TN, Keil LC (1987) Regulation of drinking and vasopressin secretion: role of organum vasculosum laminae terminalis. Am J Physiol 253:R108–R120

Verney EB (1947) The antidiuretic hormone and the factors which determine its release. Proc R Soc (Lond) 135B:26–106

Wayner MJ, Ono T, Nolley D (1973) Effects of angiotensin applied electrophoretically on lateral hypothalamic neurons. Pharmacol Biochem Behav 1:223–226

Section IV

Neural Organization of Drinking
Behaviour

Chapter 13

Sensory Detection of Water

R. Norgren

Introduction

Although some species derive water exclusively from their food, most terrestrial animals seek out and consume water. This behaviour has three phases – location, discrimination and monitoring – and each phase may utilize distinct sensory receptors. The problem of locating water sources in the environment varies considerably with species and climate. Given that a creature knows its home range, however, locating water is probably more a cognitive than a sensory problem. Once located, assessing the quality of water is a sensory problem, presumably involving oral receptors. Monitoring the amount of fluid ingested also apparently uses sensory information, but the receptors may be different from those required for discrimination. This last assertion is based on complementary behavioural observations. At least in some species, water intake closely matches an imposed deficit, even when drinking ceases before significant absorption has taken place. If provided with an open gastric cannula, the same thirsty animal may drink to great excess (Blass and Hall 1976). Abundant other behavioural data attest to a role for both oropharyngeal and gastrointestinal afferent activity in regulating water intake (see Smith 1986 for a recent review). Direct electrophysiological evidence also exists, but it derives from far fewer experiments that can be over-interpreted.

Neurons throughout the hypothalamic–preoptic continuum respond to fluid intake, but only a small percentage of these responses can be attributed unambiguously to oral water stimulation or the physiological control of fluid balance (Nicolaïdis 1969; Norgren 1970). For many cells, water and saline were equally effective. In other cases, neurons that proved to be osmosensitive, i.e. their activity patterns were influenced by intracarotid infusions of hypertonic saline, also responded to intraoral fluids (Vincent et al. 1972; Sessler and Salhi 1983). Much of the neural response to intraoral fluids, however, was inhibitory and inhibition occurs commonly in the ventral forebrain during ingestion regardless of whether the behaviour is prompted by thirst or hunger (Edinger and Pfaffmann 1971; Hamburg 1971). In fact, some of these neurons might alter their responsiveness to peripheral stimuli depending on the animal's deprivation condition or learning experience (Kendrick and Baldwin 1989; Nakamura et al. 1989).

The most convincing evidence for a role of peripheral water receptors directly influencing the central neural mechanisms controlling fluid balance comes from vasopressinergic neurons with axons that reach the neurohypophysis. These neurons have an established role in the control of fluid balance and more of them responded to oral water stimulation than did their oxytocinergic counterparts. Even here the specificity of the response might be questioned. In the experiments with monkeys (Arnauld and du Pont 1982), the response latencies of supraoptic nucleus neurons were long enough (5–20 s) that gastric receptors could not be ruled out. In acute rat preparations, the inhibitory latency for paraventricular vasopressinergic neurons was not noted, but the neural activity reached a minimum 30 s after water was applied (Sakaguchi et al. 1989). In this experiment, the stimulus volume never exceeded 100 μl and was confined to the pharyngolaryngeal region.

Oral Sensory Cues

Touch

One reason that electrophysiological evidence is less than definitive is the inability to specify which sensory receptors of the oral cavity and upper gastrointestinal system may contribute to the identification and monitoring of water during ingestive behaviour. Without this knowledge, the central pathways that convey peripheral water sensibility cannot be determined with accuracy and thus the neural mechanisms controlling fluid intake lack an important sensory definition. In the oral cavity, potential water receptors present an embarrassment of riches; in the gut, few if any have been documented conclusively. In the mouth, both somatosensory and gustatory receptors provide sensory information about fluids. Tactile sensitivity alone cannot distinguish between water and aqueous solutions, unless the solute changes the viscosity of the fluid. In fact, it remains an assumption that intraoral touch and pressure receptors provide the sensory activity needed to assess viscosity or the 'mouth-feel' touted by the food industry. In my experience, central neurons that respond to light tactile stimulation in the anterior, trigeminally innervated part of the oral cavity often fail to respond to fluids. In the posterior oral cavity, in which the somatosensory innervation is provided by the glossopharyngeal and superior laryngeal nerves, so-called flow responses are the bane of gustatory research, because chemical sensitivity can be obscured by neural activity elicited by the application of fluids (Norgren and Pfaffmann 1975; Boudreau et al. 1987). Although it may not be sufficient to distinguish water from other fluids, the backdrop of neural activity provided by tactile sensitivity may be necessary to allow the discrimination by more specific receptors. Zeigler et al. (1985) have demonstrated that somatosensory denervation of the oral cavity eliminates or greatly reduces both food and water intake.

Temperature

The role of tactile sensibility in monitoring water intake cannot be assessed independently, because both denervation and mechanically bypassing the oral cavity eliminate the effects of thermal stimulation as well. On the other hand, temperature can be manipulated independently of touch and, when it is, plasma vasopressin (AVP) drops and water intake changes. In water-deprived human subjects, sucking on ice chips lowered plasma AVP levels within 10 minutes, but gargling for 2 minutes failed to change the same measure (Seckl et al. 1986; Salata et al. 1987). In water-deprived rodents, merely cooling the tongue with dry air or cold metal supports robust licking (Hendry and Rasche 1961; Mendelson and Chillag 1970).

However, cool water actually reduces intake (Kapatos and Gold 1972b; Gold 1973). The latter effect was explained by the observation that cool water empties more slowly from the stomach than does warm water, thus producing more gastric distension (Deaux 1973).

This rationale does not account for air-licking. Both air-licking and the satiating effects of cool water make sense if oral thermal receptors provide neural activity useful in the detection of water. The import of this neural activity might be associative. Normally water is cooler than the oral cavity and, thus, orolingual cooling might be associated with water, particularly in the context of deprivation. The apparent lack of extinction during air-licking, however, suggests some inherent value in coolness for detecting water. Perhaps, in the absence of other sensory cues, cooling is sufficient to signal water to a thirsty animal. An analogous situation holds for hungry rats, which respond to sucrose solutions as food in the absence of other stimuli, but tend to ignore sweet tastes when complex food cues are available (Mook 1974).

Taste

Water Rinse Responses

The gustatory system clearly contributes to evaluating the chemical composition of fluids and perhaps also to monitoring intake volume. The mechanisms through which taste buds distinguish water from sapid chemicals dissolved in water remain in dispute. The taste attributed to water may not arise from a water taste per se, i.e. receptors that respond better to pure water than to water solutions. In humans, the taste of water is determined primarily by the chemical composition of the fluids in the oral environment prior to the application of water. Normally, this would be saliva, which is a hypotonic solution of electrolytes, chiefly Na^+ and Cl^-. After adaptation at sodium chloride concentrations often found in saliva, water tastes bitter–sour (Bartoshuk et al. 1964). As Bartoshuk (1977) emphasizes, with the appropriate adapting solutions, water can be made to elicit each of the four basic tastes. In animals, an analogous contingent water response also appears at the electrophysiological level, both in whole nerve and single fibre preparations. Depending on the species and the chemical sensitivity of the nerve or fibre, a water rinse often evokes a substantial excitatory response above the tonic activity maintained by the sapid chemical. This "water response" usually is short lived and the ongoing activity returns to its prestimulus spontaneous levels with continued rinsing (see Bartoshuk 1977 for an excellent review).

Perceptually, the contingent water response is a powerful effect, but as a potential source of information for water detection it has several awkward attributes. First, it is a form of contrast or even opponent-process and thus the psychophysical effect is not unique to water. After adaptation, water elicits a perceived taste, but the same taste can be produced directly by a sapid chemical. Second, electrophysiologically at least, the contingent water response is brief relative to a drinking bout and is followed by sustained inhibition. The time course of the response perhaps suits it for the recognition of water, but not for the maintenance of drinking or for the monitoring of volume. Finally, the sapid adapting response required for eliciting a contingent water response is also excitatory, and usually occurs in a similar form, i.e. a relatively large amplitude, rapid phasic component followed by a lower, sustained level of activity.

Pharyngolaryngeal Water Responses

Most of the electrophysiological demonstrations of contingent water responses come from the chorda tympani nerve, which innervates the anterior tongue gustatory receptors. Similar effects can be observed in the glossopharyngeal nerve but, as mentioned above, the tactile responsiveness often exhibited by the gustatory axons can confuse the issue. Axons in the superior laryngeal nerve, which innervate gustatory receptors of the larynx, respond to both mechanical and chemical stimuli as well, but as many as half these fibres also respond to water with sustained increases in activity (Harding et al. 1978; Shingai 1980; Stedman et al. 1980; Dickman and Smith 1988). These gustatory axons are similar to those innervating the tongue, in that they typically respond to more than one sapid chemical. They differ, in that they often respond to water, but seldom to sodium chloride. In fact, water responses are common enough and robust enough, that the interstimulus rinse in these preparations is typically a saline solution.

The water responses evident in the superior laryngeal nerve overcome some of the disadvantages of the contingent water responses observed in the other gustatory nerves, and thus might provide the neural information necessary to maintain and monitor water intake. In a normal salivary environment, water would elicit a sustained, excitatory response and rinsing with saliva would inhibit that activity. This is the obverse of the pattern observed in a typical chorda tympani fibre that responds well to weak saline solutions. In one report, successive drops of water sustained the axonal discharge of a superior laryngeal fibre "for a few minutes," while liquid paraffin applied to the same receptive field had no effect (Shingai 1980).

Physiological and behavioural evidence also implicates the pharyngolaryngeal region in oral water responses. Based on the review of the 19th century literature and his own observations Cannon (1918) asserted not only that thirst was a sensation resulting from a dry mouth, but also that this sensation arose primarily from "the pharynx, where the respiratory tract crosses the digestive tract ..." (p 194). In a number of species, including humans, water is often the most effective fluid stimulus for eliciting swallowing, at least when applied to the larynx (Storey 1968; Shingai and Shimada 1976; Shingai et al. 1989). Similarly, water, but not saline, applied to the pharyngolaryngeal region induces sustained diuresis in both animals and humans (Shingai et al. 1988; Akaishi et al. 1989). In rats, this diuresis is abolished by local anaesthetization of the pharyngolaryngeal region or by cutting the superior laryngeal nerves. Local anaesthesia applied to the pharyngolaryngeal region also reduces spontaneous water intake for 48 hours but, when applied to the hard palate, it has no effect (Miyaoka et al. 1987a). As mentioned above, similar water stimulation also inhibits paraventricular vasopressinergic neurons, even in non-deprived animals, thus providing a direct neurophysiological link between the oral water stimulus and the renal diuresis (Sakaguchi et al. 1989). The same group of investigators has observed that laryngectomized humans report different thirst sensations and different drinking preferences when compared with normal controls (Miyaoka et al. 1987b,c).

Behavioural, physiological, and neurophysiological evidence implicates the pharyngolaryngeal region in the maintenance and monitoring of water intake. It must be emphasized, however, that much of the evidence is circumstantial and it must contend with contravening facts and interpretations. Foremost of these is the location of the gustatory receptors. The vast majority of the taste buds innervated by the superior laryngeal nerve are on the airway side of the larynx and epiglottis. In the normal course of ingestion, these receptors would not be exposed to swallowed fluids (see Norgren 1984, 1990 for short reviews and additional references). In one experiment, out of 57 superior laryngeal axons tested, only two responded maximally when fluids flowed through the oesophagus. The remainder were activated better by fluids in the trachea (Dickman and Smith 1988). Thus, even if the pharyngolaryngeal region contributes sensory information important for monitoring water intake, the receptors involved might not be taste buds innervated by the superior laryngeal nerve. There are extralingual taste buds elsewhere in the oropharyngeal area. The vast majority of these are on the soft palate and are innervated by the greater superficial petrosal branch of the facial nerve (Miller et al. 1978). Little is known about their chemical sensitivity, but water was not effective in one small sample of central gustatory neurons that responded to palatal sapid stimulation (Travers et al. 1986). There are scattered taste buds in the upper oesophagus, on the pharyngeal wall, and near the tonsils. A few, particularly on the posterior pharyngeal wall, are probably innervated by the superior laryngeal nerve; most rely on the glossopharyngeal nerve. A role for these gustatory receptors in monitoring water intake can be inferred from the behavioural and physiological data, but direct neurophysiological evidence is lacking.

Gastrointestinal Sensory Responses

Even without neurophysiological data, the primacy of oropharyngeal sensory receptors for detecting water and maintaining fluid ingestion can be established with behavioural evidence (see Smith 1986). Once satiation is considered, however, the presence of water in the gut plays a critical role. In the gastrointestinal tract, theoretically water can be detected both preabsorptively, by receptors in the gastric and intestinal mucosa, and postabsorptively, by receptors in the hepatic portal system and in the general circulation (Chapter 19). The mucosa of the oesophagus, stomach and intestines are endowed with a variety of tactile, thermal and chemical receptors served by the vagus and splanchnic nerves, but little or none of the electrophysiological evidence deals specifically with water (Mei 1983). The functional import of gastric afferent activity in terminating water intake is questionable. When water is prevented from reaching the intestine with a pyloric loop, drinking ceases "probably... only when a pathological degree of distention is achieved" (Hall and Blass 1977). When allowed to enter the duodenum, ingested water elicits normal satiety. Once in the intestine, however, absorption occurs rapidly enough that distinguishing between pre- and postabsorptive afferent influence becomes more difficult.

In the absence of specific water receptors, the best candidate for monitoring water in the gut would be osmoreceptors. Specific chemoreceptors in the lumen of the gut respond to carbohydrates (glucose) or amino acids. Osmoreceptors have not been reported in the gastric mucosa. Those identified in the small intestine either respond differentially to different molecules at the same osmotic pressure or they are frankly multimodal, responding to thermal and tactile stimuli as well as to molecules (Mei 1983; Andrews 1986). The only true osmoreceptors appear to be within the hepatic portal system, rather than on the luminal side of the intestine (Niijima 1969; Andrews and Orbach 1975; Adachi et al. 1976). As with oropharyngeal stimulation, water introduced into the stomach or duodenum elicits prompt diuresis that, at least for the gastric load, probably precedes any postabsorptive effects (Nicolaïdis 1980; Mei 1985).

In sum, the relationship between gastrointestinal sensory receptors and the monitoring of water intake is even less certain than for the oropharyngeal receptors. Ample behavioural evidence exists implicating the gastrointestinal system in the termination of water intake. Equally convincing electrophysiological observations have identified tactile, thermal, chemical, and osmotic receptors in the viscera that could provide the required sensory information. Which receptors actually play such a role remains to be determined. Cutting the subdiaphragmatic vagus or its branches produces complex effects on water intake that provide little hint of the specific sensory mechanisms involved (Smith and Jerome 1983). As far as is known, the effects of severing the visceral afferent axons in the sympathetic trunks have yet to be investigated.

Initiating diuresis and assessing the volume of water ingested may require quite distinct sensory information. At present, the physiological evidence indicates that preabsorptive sensory activity supports the former response, but similar evidence for the latter function is equivocal. Animals that ingest water rapidly seem to be able to judge volume from oropharyngeal cues alone. Those that drink more slowly may require both oropharyngeal and postabsorptive cues to maintain the same regulation. The participation of sensory mechanisms in the gut lumen remains to be demonstrated.

Central Neural Pathways

Although the receptors critical for detecting water and monitoring its intake cannot be specified with certainty, the nerves that convey the information to the CNS are known, as are their first central relays. The oral cavity is innervated by the trigeminal, facial, and glossopharyngeal nerves, and by the superior laryngeal branch of the vagus nerve. The trigeminal nerves, of course, terminate in the principal and spinal trigeminal nuclei. The trigeminal branches that innervate the oral cavity, however, also send axons into the nucleus of the solitary tract (NST). Conversely, the facial, glossopharyngeal and superior laryngeal nerves terminate primarily in the NST, but also make varying contributions to the spinal trigeminal nuclei. Of the three, only the glossopharyngeal projection to the trigeminal system is substantial (Beckstead and Norgren 1979; Contreras et al. 1982; Hamilton and Norgren 1984; Altschuler et al. 1989). The subdiaphragmatic vagus nerve projects only to caudal NST (Norgren and Smith 1988). The central pathways of the splanchnic visceral afferent system are less clearly delineated, but hepatic osmoreceptor information does reach the hypothalamus via the spinal cord (Schmidt 1973).

The nucleus of the solitary tract receives both gustatory and oral somatosensory afferent axons that probably transmit sensory information about water. As far as is known, the trigeminal system serves only tactile and thermal sensations. For this reason, the NST may be more important in the sensory detection of water. The NST receives not only primary intraoral tactile and thermal afferent axons from the trigeminal, glossopharyngeal, and superior laryngeal nerves, but also second order projections from the spinal trigeminal nuclei. Because it also receives chemical sensory information, the NST processes all the afferent activity that might aid in discriminating water from other fluids. In addition, at least in rodents, neurons at all levels of the NST project to the pontine parabrachial nuclei which, in turn, project to much of the ventral forebrain (see Norgren et al. 1989 for a brief review and references).

Nevertheless, incidental observations of animals with lesions in the rostral or gustatory areas of the NST revealed no obvious deficits in fluid balance. Similarly, lesions confined to the parabrachial nuclei do not result in persistent changes in spontaneous water intake (Norgren et al. 1985; Spector, Grill and Norgren, unpublished observations). About 20% of neurons in the gustatory zone of the parabrachial nuclei of behaving rats respond to water, but none do so differentially. In other words, the cells respond equally well to distilled water and sapid stimuli (Nishijo and Norgren 1990). Lesions of the area postrema and adjacent NST or destruction of the lateral parabrachial nucleus increase water intake in response to some, but not all dipsogens. The subdiaphragmatic vagus terminates in the NST near the area postrema and neurons in those zones project to the lateral parabrachial nuclei. Ohman and Johnson (1989) speculate that these systems are part of "an inhibitory neural network involved in the integration of circulatory and hydromineral information that minimizes overhydration." (p R269). The monitoring of ingested fluid volume would be an important aspect of any such function. Admittedly, this is thin evidence on which to judge the importance of the NST or the parabrachial nuclei in the sensory detection of water. At least in the lamb, neurons in the NST are much more sensitive to distilled water applied to the epiglottis than to 0.5 M sodium chloride (Sweazey and Bradley 1988). Unfortunately, at present, research that deals with pharyngolaryngeal afferent activity in the NST is interpreted almost exclusively in terms of the neural mechanisms of respiration and swallowing (Jean 1984).

Lesions of either peripheral trigeminal nerves or in the principal trigeminal nucleus, on the other hand, do produce aphagia and adipsia (Zeigler et al. 1985). Unlike the parabrachial nuclei, the trigeminal system does not distribute widely in the ventral forebrain. Apart from the traditional thalamocortical projections, however, brainstem trigeminal axons also terminate in the zona incerta (Smith 1973; Erzurumlu and Killackey 1980). Lesion, stimulation, and electrophysiological experiments have all implicated the zona incerta in the control of drinking behaviour, particularly to acute cellular dehydration (Huang and Mogenson 1972; Walsh and Grossman 1976; Brown and Grossman 1980; Mok and Mogenson 1986, 1987). The functional relationship between the zona incerta and the limbic forebrain systems implicated in the control of hydromineral balance remains uncertain. Only a few cells in the hypothalamus and the central nucleus of the amygdala project to the zona incerta (Roger and Cadusseau 1985; Shammah-Lagnado et al. 1985). Similarly, except for a few axons in the lateral hypothalamus, efferents from the zona incerta avoid the limbic system (Ricardo 1981).

The central neural systems implicated in the sensory evaluation of water compound the conundrum that arises on the periphery. In the oral cavity, several sensory modalities appear to provide information about water, but the evidence provides little basis for dividing that responsibility between them. In the brain, the nucleus of the solitary tract receives all the relevant oral sensory information and projects to most of the ventral forebrain areas implicated in the control of fluid balance. At present, however, the NST and the parabrachial nuclei appear to have only minor influence on drinking behaviour. Peripheral trigeminal neurons generate some, perhaps ambiguous, sensory messages about water and the central trigeminal system reaches only one area implicated in the response to specific fluid balance challenges. Nevertheless, damage in the trigeminal system produces substantial deficits in the initiation of ingestive behaviour, including drinking.

Acknowledgement. The author's research was supported by grants from the NIH (DC 00240) and ADAMHA (MH 43787). The chapter was written during the term of a Research Career Development Award, Level II, also from ADAMHA (MH 00653).

References

Adachi A, Niijima A, Jacobs HL (1976) An hepatic osmoreceptor mechanism in the rat: electrophysiological and behavioral studies. Am J Physiol 231:1043–1049

Akaishi T, Shingai T, Miyaoka Y, Homma S (1989) Hypotonic diuresis following oropharyngeal stimulation with water in humans. Neurosci Lett 107:70–74

Altschuler SM, Bao Y, Beiger D, Hopkins DA, Miselis RR (1989) Viscerotopic representation of the upper alimentary tract in the rat: sensory ganglia and nuclei of the solitary and spinal trigeminal tracts. J Comp Neurol 283:248–268

Andrews PL (1986) Vagal afferent innervation of the gastrointestinal tract. Prog Brain Res 67:65–86

Andrews WH, Orbach J (1975) Effects of osmotic pressure on spontaneous afferent discharges in the nerves of the perfused rabbit liver. Pflügers Arch 361:89–94

Arnauld E and Du Pont J (1982) Vasopressin release and firing of supraoptic neurosecretory neurones during drinking in the dehydrated monkey. Pflügers Arch 394:195–201

Bartoshuk LM (1977) Water taste in mammals. In: Weijnen JA, Mendelson J (eds) Drinking Behavior. Plenum Press, New York, pp 317–339

Bartoshuk LM, MacBurney DH, Pfaffmann C (1964) Taste of sodium chloride solutions after adaptation to sodium chloride: implications for the "water taste". Science 143:967–968

Beckstead RM, Norgren R (1979) Central distribution of the trigeminal, facial, glossopharyngeal, and vagus nerves in the monkey. J Comp Neurol 184:455–472

Blass EM, Hall WG (1976) Drinking termination: interactions among hydrational, orogastric, and behavioral controls in rats. Psychol Rev 83:356–374

Boudreau JC, Do LT, Sivakumar I, Oravec J, Rodriguez CA (1987) Taste systems of the petrosal ganglion of the rat glossopharyngeal nerve. Chem Senses 12:437–458

Brown JC, Grossman SP (1980) Evidence that nerve cell bodies in the zona incerta influence ingestive behavior. Brain Res Bull 5:593–597

Cannon W (1918) The physiological basis of thirst. Proc R Soc Lond [Biol] 90:283–301

Contreras R, Beckstead R, Norgren R (1982) An autoradiographic examination of the central distribution of the trigeminal, facial, glossopharyngeal and vagus nerves in the rat. J Auton Nerv Syst 6:303–322

Deaux E (1973) Thirst satiation and the temperature of ingested water. Science 181:1166–1167

Dickman JD, Smith DV (1988) Response properties of fibers in the hamster superior laryngeal nerve. Brain Res 450:25–38

Edinger H and Pfaffmann C (1971) Single unit activity during drinking. Fed Proc 30:376A

Erzurumlu RS, Killackey HP (1980) Diencephalic projections of the subnucleus interpolaris of the brainstem trigeminal complex in the rat. Neuroscience 5:1891–1901

Gold R (1973) Cool water suppression of water intake: one day does not a winter make. Bull Psychon Soc 1:385–386

Hall WG, Blass EM (1977) Orogastric determinants of drinking in rats: interaction between absorptive and peripheral controls. J Comp Physiol Psychol 91:2:365–373

Hamburg M (1971) Hypothalamic unit activity and eating behavior. Am J Physiol 220:980–985

Hamilton R, Norgren R (1984) Central projections of gustatory nerves in the rat. J Comp Neurol 222:560–577

Harding R, Johnson P, McClelland ME (1978) Liquid-sensitive laryngeal receptors in the developing sheep, cat and monkey. J Physiol (Lond) 277:409–422

Hendry D, Rasche R (1961) Analysis of a new non-nutritive positive reinforcer based on thirst. J Comp Physiol Psychol 54:477–483

Huang YH, Mogenson GJ (1972) Neural pathways mediating drinking and feeding in rats. Exp Neurol 37:269–286

Jean A (1984) Brainstem organization of the swallowing network. Brain Behav Evol 25:109–116

Kapatos G, Gold RM (1972a) Tongue cooling during drinking: a regulator of water intake in rats. Science 176:685–686

Kapatos G, Gold RM (1972b) Rats drink less cool water: a change in the taste of water? Science 178:1121

Kendrick KM, Baldwin BA (1989) The effects of sodium appetite on the responses of cells in the zona incerta to the sight or ingestion of food, salt and water in sheep. Brain Res 492:211–218

Mei N (1983) Sensory structures in the viscera. In: Ottoson D (ed) Progress in sensory physiology vol 4. Springer-Verlag, New York, pp 1–43

Mei N (1985) Intestinal chemosensitivity. Physiol Rev 65:211–237

Mendelson J, Chillag D (1970) Tongue cooling: a new reward for thirsty rodents. Science 170: 1418–1421

Miller I Jr, Gomez MM, Lubarsky EH (1978) Distribution of the facial nerve to taste receptors in the rat. Chem Senses Flav 3:397–411

Miyaoka Y, Sakaguchi T, Yamazaki M, Shingai T (1987a) Changes in water intake following pharyngolaryngeal deafferentation in the rat. Physiol Behav 40:369–371

Miyaoka Y, Sawada M, Sakaguchi T, Hasegawa A, Shingai T (1987b) Differences in drinking behavior between normal and laryngectomized man. Percept Mot Skills 64:1088–1090

Miyaoka Y, Sawada M, Sakaguchi T, Shingai T (1987c) Sensation of thirst in normal and laryngectomized man. Percept Mot Skills 64:239–242

Mok D, Mogenson GJ (1986) Contribution of zona incerta to osmotically induced drinking in rats. Am J Physiol 251:R823–R832

Mok D, Mogenson GJ (1987) Convergence of signals in the zona incerta for angiotensin-mediated and osmotic thirst. Brain Res 407:332–340

Mook DG (1974) Saccharin preference in the rat: some unpalatable findings. Psychol Rev 81:475–490

Nakamura Y, Ono T, Tamura R, Indo M, Takashima Y, Kawasaki M (1989) Characteristics of rat lateral hypothalamic neuron responses to smell and taste in emotional behavior. Brain Res 491:15–32

Nicolaïdis S (1969) Early systemic responses to oro-gastric stimulation in the regulation of food and water balance. Functional and electrophysiological data. Ann NY Acad Sci 157:1176–1203

Nicolaïdis S (1980) Hypothalamic convergence of external and internal stimulation leading to early ingestive and metabolic responses. Brain Res Bull 5 Suppl 4:97–101

Niijima A (1969) Afferent discharges from osmoreceptors in the liver of the guinea pig. Science 166: 1519–1520

Nishijo H, Norgren R (1990) Gustatory neural activity in the parabrachial nuclei of awake rats. J Neurophysiol 63:707–724

Norgren R (1970) Gustatory responses in the hypothalamus. Brain Res 21:63–77

Norgren R (1984) Central neural mechanisms of taste. In: Darien-Smith I (ed), Brookhart J, Mountcastle V (section eds) Handbook of physiology – the nervous system 111, Sensory processes. American Physiology Society Washington, DC, pp 1087–1128

Norgren R (1990) The gustatory system. In: Paxinos G (ed) The human nervous system. Academic Press, San Diego, pp 845–861

Norgren R, Pfaffmann C (1975) The pontine taste area in the rat. Brain Res 91:99–117

Norgren R, Smith GP (1988) Central distribution of subdiaphragmatic vagal branches in the rat. J Comp Neurol 273:207–223

Norgren R, Flynn FW, Grill HJ, Schwartz G (1985) Central gustatory lesions. I. Intake and taste reactivity tests. Soc Neurosci Absts 11:1259

Norgren R, Nishijo H, Travers SP (1989) Taste responses from the entire gustatory apparatus. Ann NY Acad Sci 575:246–264

Ohman L, Johnson AK (1989) Brain stem mechanisms and the inhibition of angiotensin-induced drinking. Am J Physiol 256:R264–R269

Ricardo JA (1981) Efferent connections of the subthalamic region in the rat. II. Zona incerta. Brain Res 214:43–60

Roger M, Cadusseau J (1985) Afferents to the zona incerta in the rat: a combined retrograde and anterograde study. J Comp Neurol 241:480–492

Sakaguchi T, Tamaki M, Akaishi T, Miyaoka Y (1989) Responses in discharge of vasopressinergic neurons in the hypothalamic paraventricular nucleus elicited by water application to the pharyngolaryngeal regions in the rat. Chem Senses 14:327–333

Salata RA, Verbalis JG, Robinson AG (1987) Cold water stimulation of oropharyngeal receptors in man inhibits release of vasopressin. J Clin Endocrinol Metab 65:561–567.

Schmidt M (1973) Influences of hepatic portal receptors on hypothalamic feeding and satiety centers. Am J Physiol 225:1089–1095

Seckl JR, Williams TD, Lightman SL (1986) Oral hypertonic saline causes transient fall of vasopressin in humans. Am J Physiol 251:R214–R217

Sessler FM, Salhi MD (1983) Convergences on lateral preoptic neurons of internal and external stimuli related to drinking. Neurosci Lett 36:151–155

Shammah-Lagnado SJ, Negrao N, Ricardo JA (1985) Afferent connections of the zona incerta: a horseradish peroxidase study in the rat. Neuroscience 15:109–134

Shingai T (1980) Water fibers in the superior laryngeal nerve of the rat. Jpn J Physiol 30:305–307

Shingai T, Shimada K (1976) Reflex swallowing elicited by water and chemical substances applied in the oral cavity, pharynx, and larynx of the rabbit. Jpn J Physiol 26:455–469

Shingai T, Miyaoka Y, Shimada K (1988) Diuresis mediated by the superior laryngeal nerve in rats. Physiol Behav 44:431–433

Shingai T, Miyaoka Y, Ikarashi R, Shimada K (1989) Swallowing reflex elicited by water and taste solutions in humans. Am J Physiol 256:R822–R826

Smith GP (1986) Peripheral mechanisms for the maintenance and termination of drinking in the rat. In: de Caro G, Epstein AN, Massi M (eds) The physiology of thirst and sodium appetite. Plenum Press, New York, pp 265–277

Smith GP, Jerome S (1983) Effects of total and selective abdominal vagotomies on water intake in rats. J Auton Nerv Syst 9:259–271

Smith RL (1973) The ascending fiber projections from the principal sensory trigeminal nucleus in the rat. J Comp Neurol 148:423–446.

Stedman HM, Bradley RM, Mistretta CH, Bradley BE (1980) Chemosensitive responses from the cat epiglottis. Chem Senses 5:233–245

Storey AT (1968) Laryngeal initiation of swallowing. Exp Neurol 20:359–365

Sweazey RD, Bradley RM (1988) Responses of lamb nucleus of the solitary tract neurons to chemical stimulation of the epiglottis. Brain Res 439:195–210

Travers SP, Pfaffmann C, Norgren R (1986) Convergence of lingual and palatal gustatory neural activity in the nucleus of the solitary tract. Brain Res 365:305–320

Vincent JD, Arnauld E, Bioulac B (1972) Activity of osmosensitive single cells in the hypothalamus of the behaving monkey during drinking. Brain Res 44:371–384

Walsh LL, Grossman SP (1976) Zona incerta lesions impair osmotic but not hypovolemic thirst: Physiol Behav 16:211–215

Zeigler HP, Jacquin M, Miller M (1985) Trigeminal orosensation and ingestive behavior in the rat. Prog Psychobiol Physiol Psychol 11:63–196

Commentary

Booth: One would have to agree that sensory evaluation of oral texture has told us remarkably little about tactile psychophysics – but also, so far, the other way round. Nevertheless, there is progress, particularly for fluids. What people describe as the thickness of a homogeneous fluid food, for example, is very strongly related to its viscosity measured as shear rates in the range the tongue moves fluid against the palate. A neat result for tactile (or kinaesthetic) patterns from the fat in milk has been obtained by using the multidimensional causal analysis expounded in Chapter 4. Creaminess ratings and judgements of fat content are often well correlated with each other. They are also rather poor at discriminating actual fat content. Viscosity is an essential part of the stimulus, but it has now been shown that the distribution of fat particle sizes in the emulsion is also crucial (Richardson, Birkett, Booth, in preparation). Furthermore, when we covary out of the sense of fat content the ratings of how like cream the test milk is, all discrimination of fat content is lost. It is, therefore, unlikely that prodding and stroking the mouth will tell us much about the neurophysiology of texture in fluids.

Saltmarsh: Viscosity is surely very important in detecting the "wateriness" of fluids, besides any negative taste effects. Beverages low in solute tend to be characterized as feeling "thin" as well as less tasty. Viscosity also relates to the perceived "thirst-quenching" character of beverages which relates inversely to viscosity.

Rolls: As water is ingested, the rated pleasantness of its taste declines. This may be related to body fluid changes, but also occurs if water is swilled in the mouth but not swallowed. Thus, sensory-specific satiety occurs to water (Rolls and Rolls 1982).

References:

Rolls BJ, Rolls ET (1982) Thirst: Cambridge University Press, Cambridge

Chapter 14

Neostriatal Mechanisms Affecting Drinking

T. Schallert

Introduction

As a thirsty animal drinks water in the wild, whether it be prey or predator, the temporal pattern of its drinking behaviour may be influenced not only by factors relating to water balance but also by reactivity to potentially distractive sensory events (Schallert and Hsiao 1979). From moment to moment a vigilant animal may disengage from drinking and orient to a wide range of stimuli, from salient to subtle, or it may continue to drink while covertly attending to, but ignoring behaviourally, the sensory stimulation. The neostriatum and related forebrain areas may be involved in the modulation of fluid intake as part of a unique mechanism for switching between ingestive and orienting movements.

Background

For 20 years it has been known that dopamine-specific (DA) lesions of the meso-telencephalic fibres that pass through the lateral hypothalamic area cause adipsia, as well as aphagia and sensory neglect (Ungerstedt 1971), which previously had been observed after focal lesions of the lateral hypothalamic area (Anand and Brobeck, 1951; Teitelbaum and Epstein 1962; Marshall et al. 1971; Schallert and Whishaw 1978; Schallert and Wilcox 1985; Grossman 1990). Ungerstedt (1971) suggested that general arousal was impaired by DA depletion and that aphagia, adipsia and sensory neglect were noticed simply because they are readily quantifiable and are disastrous to the animal. Later studies confirmed that both striatal DA depletion and large non-specific posterior-lateral hypothalamic damage affected movement initiation and sensory responsiveness to such a degree that the probability of consumatory behaviour was greatly decreased (Robinson and Whishaw 1974; Levitt and Teitelbaum 1975; Balagura et al 1969; Stricker and Zigmond 1976), particularly if mesocortico-limbic neurons were also involved (Nadaud et al 1984). Axon-sparing lesions (Grossman et al. 1978) or more anteriorly placed electrolytic lesions of the lateral hypothalamus (Schallert and Whishaw 1978; White 1986; Berridge 1989) caused active aphagia and adipsia without neglect or motor impairment.

Preoperative history of ingestive experience can importantly influence the outcome of the brain damage (Schallert 1989). For example, if the lesions were not too large, a preoperative regimen of restricted daily food intake attenuated or prevented aphagia (Powley and Keesey 1970; Grijalva et al. 1976; Schallert and Whishaw 1978), and a preoperative regimen of restricted daily watering attenuated or prevented adipsia (Schallert 1982). Neither regimen improved motor deficits, although some sensory impairments were ameliorated (Schallert and Whishaw 1978; Schallert 1982).

A role for central trigeminal structures in feeding, drinking and sensory deficits caused by lateral hypothalamic lesions was suggested by Zeigler and Karten (1974; see also Nadaud et al. 1984). However, the role must be indirect because trigeminal deafferentation does not yield active aversion to water or food (Berridge and Fentress 1985). Moreover, vibrissae removal, which compromises far less sensory processing than does trigeminal deafferentation, may substantially deplete the striatum of DA (Huston et al. 1988).

For a DA-depleting neurotoxin to induce adipsia, aphagia or sensory neglect that fails to recover, 95% of the DA content of the neostriatum must be depleted (Stricker and Zigmond 1976; Robinson et al. 1990). Microdialysis studies indicate that DA depletions less than 95% are associated with normal levels of extracellular (presumably synaptic) DA, possibly because DA uptake is reduced (Robinson et al. 1990). Unless the animal is stressed (e.g., by an osmotic or hypovolaemic challenge, Stricker and Zigmond 1976), these levels of DA appear to be sufficient for adequate ingestive and orienting behaviours to take place. However, as Teitelbaum and Epstein (1962) prescribed for lateral hypothalamic lesions, recovery may require an extensive period of intragastric feeding and hydration.

Severe impairment of drinking behaviour also occurs as a result of damage to other extrahypothalamic regions in the forebrain, including neocortex (Sorenson and Ellison 1970; Braun 1975; Kolb et al. 1977; Vanderwolf et al. 1978) and globus pallidus (Morgane 1961, 1975; Levine and Schwartzbaum 1973; Berridge and Cromwell 1990). Cortical damage causes permanent tongue extension deficits (Castro 1972; Whishaw et al. 1981), but does not cause active adipsia and only briefly affects water ingestion (in non-complex environments, such as the home cage). However, the animal's ability to drink from a standard drinking spout may be chronically impaired unless special training procedures are instigated which permit an effective spout-biting reaction. With training, spout drinking can emerge after one postoperative week (Whishaw et al. 1981). Before that, a number of behavioural abnormalities may interact to yield adipsia. The duration of training can be shortened to as few as one or two days if it is initiated after about one postoperative week (Whishaw et al. 1981). Thus, decortication does not yield a form of adipsia that is comparable to that observed after the lesions of basal ganglia or hypothalamus. The decorticate animal is more likely than control animals to drink water in an unfamiliar environment (Whishaw et al. 1981). In contrast, animals with lesions restricted to the orbital frontal area drink in the home cage soon after surgery but chronically fail to drink in an open field, even when severely deprived of water (Kolb et al. 1977).

Unlike cortical lesions or DA depletion, lesions in the region of the globus pallidus cause active aversion to water and food (White 1986; Berridge and Cromwell 1990), which may be more similar to the effects of lateral hypothalamic lesions, at least those lesions that are not localized too far posteriorly (Anand and Brobeck 1951; Teitelbaum and Epstein 1962; Schallert and Whishaw 1978; Schallert, 1985). DA depletion caused by intraventricular infusions (e.g., Stricker and Zigmond 1976) may cause non-specific striatopallidal damage, which yields active adipsia and aphagia in some animals (Whishaw et al. 1978).

Unilateral DA Deficiency and the Disengage Deficit

In the usual laboratory situation, normal animals readily disengage from drinking or eating behaviour in response to phasic somatosensory stimulation, such as that provided by thin wooden probes inserted through holes in the sides of the cage and brushed lightly against the vibrissae. Following recovery from sensory neglect caused by unilateral depletion of neostriatal DA, a residual sensorimotor impairment remains indefinitely. This impairment is a deficit in responding to sensory stimulation, but the deficit appears to be linked specifically to ingestive behaviour (Schallert and Hall 1988). When eating or drinking, these animals completely fail to orient to tactile stimulation of the body contralateral to the DA depletion, whereas ipsilateral sensory stimulation is never ignored (visual and auditory stimuli have not been investigated systematically).

During non-ingestive behaviour (at least that which has been examined so far) contralateral orienting is highly reliable, even when such behaviour involves intense oral stimulation. Bracelets fashioned from pipe cleaners were attached to the wrists of each forelimb so that they were extremely difficult for the animal to remove. Despite vigorous and persistent biting and licking of the bracelets, the animals instantly interrupted such behaviour and oriented to somatosensory probes that were brushed against the vibrissae on either side of the midline (Schallert and Hall 1988).

The disengage deficit is extremely robust and readily occurs in the absence of the well-established symptoms of marked DA deficiency (Hall and Schallert 1988). It is not ameliorated by otherwise functional intrastriatal mesencephalic fetal grafts that reverse sensory inattention and/or amphetamine-induced rotational behaviour (Mandel et al. 1989, 1990). Mandel et al. suggest that the capacity to divert responding from ingestive to orienting behaviour requires not only sufficient striatal levels of DA, but also the complex nigrostriatal coupling that exists in the fully intact brain. Although severe neonatal DA depletion spares the disengage function, a 1%–2% further DA depletion during adulthood causes the deficit to appear (Schallert et al. 1989).

Hemidecorticate animals do not show the disengage deficit, nor do animals with extensive damage to thalamus, hippocampus or the DA terminals in the nucleus accumbens. Although a complete anatomical–behaviour mapping analysis has not been carried out (notably, DA-specific lesions to frontal cortex have not been examined), based on the present set of observations it was speculated that a specialized mechanism within the neostriatum may serve to redirect attention from ingestive behaviour to external stimulation, and perhaps vice versa (Hall and Schallert 1988 and unpublished).

Striatal Unit Activity in Relation to Behaviour

Evidence from electrophysiological and behaviour–pharmacological analyses indicates that "automatic" (so-called type 2) behaviours such as lapping or chewing share a common neural substrate in the forebrain that is distinct from that associated with "voluntary" (so-called type 1) behaviours such as head movements or locomotion (Vanderwolf 1989). This polarization of behavioural organization (see also Craig 1918) would require a specialized gating mechanism. Striatal unit recording studies, in particular, may be useful in interpreting the failure to respond to sensory stimulation during ongoing ingestive behaviour. Remarkably few neurons in the neostriatum respond unconditionally to independent sensory, motor or ingestive events, but instead respond typically to particular combinations of events (Rolls and Williams 1987; Schneider and Lidsky 1987). Moreover, the striatum is linked anatomically not only to brain regions controlling the motor programmes for water intake (e.g., entopeduncular nucleus, substantia nigra pars reticulata, and other structures in the brain stem; (Grill and Norgren 1978; Joseph et al. 1985; and Chapters 15 and 16), but also to areas controlling lateral head movements involved in orienting reactions (e.g., substantia nigra pars reticulata and, in turn, the deep layers of superior colliculus; Chevalier and Deniau 1987; Barth and Schallert 1987; and Chapter 16).

Duncan et al. (1989) found that 48 hours of water deprivation causes metabolic activity to be greatly increased, relative to most other brain areas, not only in hypothalamic areas concerned with water balance, but also in the somatosensory cortex. Major striatal efferent targets, including entopeduncular nucleus and substantia nigra pars reticulata, project to thalamic areas (e.g., ventromedial and ventrolateral) having connections to the orofacial area of the somatosensory neocortex where a range of behaviourally significant information can be integrated (Joseph et al. 1985; Chevalier and Deniau 1987; Mogenson 1987).

Neurons Having a Disengage Function?

The neostriatum in the cat contains a population of "complex sensory neurons" that appear to act in a manner consistent with the disengage operation (Lidsky et al. 1985; Schneider et al. 1985; Schneider and Lidsky 1987). These cells respond dramatically when the vibrissae contralateral to the recording site are stroked while an animal is lapping milk, but not while an animal laps in the absence of vibrissae stimulation nor when vibrissae stimulation occurs while an animal is not lapping. It is possible that these neurons serve to disengage ingestive movements in response to environmentally significant events, and that adequate dopaminergic input to the neostriatum is essential for the normal operation of these neurons (Schallert and Hall 1988).

Neurons Having an "Ignore" Function?

The switch from ingestive behaviour to orienting in response to sensory stimulation is a discrete event, and yet these complex striatal neurons respond continuously while the vibrissae are stroked during fluid intake. This characteristic would appear to make them unsuitable for the disengage operation.

An alternative view may be advanced based on more careful consideration of behaviour in relation to unit activity and emphasizing phasic cortical and tonic nigral modulation of intrinsic neostriatal cells. In the original interpretation (Schallert and Hall 1988), it was imagined that the animals in the unit recording studies (e.g., Schneider and Lidsky 1987) would have disengaged from lapping movements and oriented to the vibrissae stimulation if head movements were permitted (the animals were immobilized), just as rats typically do when the vibrissae are stroked during ingestive behaviour. However, this response should not have seemed likely because, in fact, the animals in the studies by Schneider and Lidsky continued to lap despite ongoing vibrissae stimulation. The cats were apparently ignoring the vibrissae stimulation, presumably because they were well deprived and sufficiently tame. In other words, even had they been freely moving, they probably would not have interrupted their ingestive behaviour in response to somatosensory stimulation.

Therefore, the enhanced activity of the "complex sensory neurons" recorded exclusively during the combination of vibrissae stimulation and lapping behaviour may represent an explicit ignore function rather than a disengage function. It can be shown that a tame dog that ordinarily responds by orienting vigorously when petted, while eating or drinking may ignore (overtly, at least) even salient sensory stimulation. As soon as it finishes drinking, it responds instantly with its normal array of playful orienting behaviours.

If it is assumed that future research confirms the existence of a class of neurons in the neostriatum which have an ignore function, and that these cells serve that function by becoming active when distractive sensory stimulation occurs during ingestive behaviour, it potentially means that a loss of inhibitory (dopaminergic) input could result in tonic hyperactivity (disinhibition) of these cells. In the DA-depleted animal, this event may explain the failure to disengage from ongoing ingestive behaviour in situations where orienting would normally take place (Schallert and Hall 1988).

Evidence Supporting the "Ignore" Interpretation

It follows that damage to the intrinsic neurons of the neostriatum should not yield the disengage deficit. Indeed, if they are truly "ignore" cells, their loss should reverse the disengage deficit in DA-depleted animals. To test this hypothesis, Hall and Schallert (unpublished) infused the axon-sparing neurotoxin ibotenic acid into the neostriatum of one hemisphere. Consistent with the hypothesis, no disengage deficit was found in these animals (histological analysis confirmed the absence of intrinsic striatal neurons). Vibrissae stimulation on either side of the body elicited normal orienting responses regardless of ongoing behaviour. Moreover, in an additional experiment (n=8), damage to the intrinsic cells of the striatum prior to DA depletion (i.e., unilateral neostriatal ibotenic acid infusion followed one week later by ipsilateral infusion of 6-hydroxydopamine, in the same animals) protected against the disengage deficit expected from DA depletion.

Preliminary Neural Model of Ignore and Disengage Functions

A revised working model was, therefore, needed that could reflect the relationship among the neurons in the neostriatum and connected structures potentially involved in the selection of orienting vs ingestive behaviour. The neostriatum would contain a subgroup of neurons that serve an "ignore" function when activated. These cells could be inhibited by nigrostriatal dopaminergic projections, or excited by neocortical glutaminergic input which heavily innervates the striatal region. In this scheme, the disengage mechanism would be independent of the "ignore" neurons and might not be housed within the neostriatum itself, though its location is unknown. Ignore or disengage functions could be expressed via striato-nigro-fugal or striato-pallido-fugal circuits (Chevalier and Deniau 1987); however, in whatever extrastriatal region(s) the disengage mechanism may reside, it would receive a direct or indirect inhibitory influence from the neostriatal "ignore" cells.

Except for a limited set of behavioural circumstances, the intrinsic neurons involved in the ignore function would remain inactive, which would permit normal orienting to occur in response to sensory stimulation. Sensory stimulation would activate alternative neuronal elements that do not require the "complex sensory" (ignore) cells of the neostriatum. However, under conditions in which the animal must behaviourally ignore sensory stimulation, an excitatory input onto these cells could be activated. Theoretically this input could originate from neocortex, which presumably integrates information acquired from various subcortical regions regarding the physiological state (e.g., water balance), the nature of the sensory input (threatening or non-threatening; novel or familiar), baseline emotional status (e.g., tame vs cautious), taste, the environmental context and experience. As a secondary modulatory mechanism, partial inhibition of ascending nigrostriatal neurons could also increase the activity of the "ignore" cells, an effect which could provide a filtering process that tonically raises the threshold for reacting to sensory events.

In contrast, increased activation of ascending nigrostriatal neurons could override corticostriatal regulation, which would permit the disengage function to occur under a broader set of conditions. In a pathological counterpart to the DA-depleted animal, which failed to react to sensory stimulation during ingestive behaviour, overactivation of the nigrostriatal DA cells might produce an animal having impaired ability to ignore sensory stimulation. At the water hole, such an animal might seem "nervous" in that it would interrupt its drinking behaviour in response to the slightest sensory event.

Conclusion

Most neural studies of thirst have been concerned primarily with the physiological factors that contribute to the initiation of fluid intake. In contrast, this chapter addresses the potential role of brain mechanisms that are probably several synapses removed from neurons directly involved in water balance. The focus has been on cells in the striatum and allied systems that may comprise a complex interface between orienting and ingestive processes. However, this line of investigation is still in an early stage. Indeed, the organization of forebrain mechanisms involved in the modulation of fluid intake in general remains largely a mystery. Hopefully, the observations discussed above will provide a novel vantage point for investigating the hierarchical control of drinking.

Acknowledgement. This work was supported by NIH grant NS 23964.

References

Anand BK, Brobeck JR (1951) Hypothalamic control of food intake in rats and cats. Yale J Biol Med 24:123–140
Balagura S, Wilcox RH, Coscina DV (1969) The effect of diencephalic lesions on food intake and motor activity. Physiol Behav 4:629–633
Barth TM, Schallert T (1987) Somatosensorimotor function of the superior colliculus, somatosensory cortex, and lateral hypothalamus in the rat. Exp Neurol 95:661–678
Berridge KC (1989) Psychological routes of different neural aphagias. Appetite 12:199
Berridge KC, Cromwell HC (1990) Motivational/sensorimotor interaction controls aphagia and exaggerated treading after striatopallidal lesions. Behav Neurosci 104:778–795
Berridge KC, Fentress JC (1985) Trigeminal-taste interaction in palatability processing. Science 228: 747–750
Berridge KC, Venier IL, Robinson TE (1989) Taste reactivity analysis of 6-hydroxydopamine-induced aphagia: implications for arousal and anhedonia hypotheses of dopamine function. Behav Neurosci 103:36–45

Braun JJ (1975) Neocortex and feeding behavior in the rat. J Comp Physiol Psychol 89:507–522

Castro AJ (1972) The effects of cortical ablations on tongue usage in the rat. Brain Res 45:251–253

Chevalier G, Deniau JM (1987) Functional significance of a double GABAergic inhibitory link in the striato-nigro-fugal pathways. In: Sangler M, Feuerstein C, Scatton B (eds) Neurotransmitter interactions in the basal ganglia. Raven Press, New York

Craig W (1918) Appetites and aversions as constituents of instincts. Biol Bull Woods Hole 34:91–107

Duncan GE, Oglesby SA, Greenwood RS, Meeker RB, Hayward JN, Stumpf WE (1989) Metabolic mapping of functional activity in rat brain and pituitary after water deprivation. Neuroendocrinology 49:489–495

Grijalva CV, Lindholm E, Schallert T, Bicknell E (1976) Gastric pathology and aphagia following lateral hypothalamic lesions in rats: effects of preoperative weight reduction. J Comp Physiol Psychol 90: 505–519

Grill HJ, Norgren R (1978) The taste reactivity test. II Mimetic responses to gustatory stimuli in chronic thalamic and chronic decerebrate rats. Brain Res 143:281–297

Grossman SP (1990) Thirst and salt appetite. Academic Press, New York (in press)

Grossman SP, Dacey D, Halaris AE, Collier T, Routtenberg A (1978) Aphagia and adipsia after preferential destruction of nerve cell bodies in hypothalamus. Science 202:537–539

Hall S, Schallert T (1988) Striatal dopamine and the interface between orienting and ingestive functions. Physiol Behav 44:469–471

Huston JP, Steiner H, Schwarting RKW, Morgan S (1988) Parallels in behavioral and neural plasticity induced by unilateral vibrissae removal and unilateral lesion of the substantia nigra. In: H Flohr (ed) Post-lesion neural plasticity. Springer, Berlin, pp 537–551

Joseph JP, Boussaoud D, Biguer B (1985) Activity of neurons in the cat substantia nigra pars reticulata during drinking. Exp Brain Res 60:375–379

Kolb B, Whishaw IQ, Schallert T (1977) Aphagia, behavior sequencing, and body weight set point following orbital frontal lesions in rats. Physiol Behav 19:93–103

Levine MS, Schwartzbaum JS (1973) Sensorimotor functions of the striatopallidal system and lateral hypothalamus and consummatory behavior in rats. J Comp Physiol Psychol 85:615–635

Levitt DR, Teitelbaum P (1975) Somnolence, akinesia, and sensory activation of motivated behavior in the lateral hypothalamic syndrome. Proc Natl Acad Sci USA 72:2819–2823

Lidsky TI, Manetto C, Schneider JS (1985) A consideration of sensory factors involved in motor functions of the basal ganglia. Brain Res Rev 9:133–146

Mandel RJ, Norrman A, Haapaniemi C, Brundin P, Björklund A (1989) Deficits in disengage behavior are not ameliorated by intrastriatal mesencephalic grafts that reverse amphetamine-induced rotational behavior. Soc Neurosci Abs 15:1355

Mandel RJ, Brundin P, Björklund A (1990) The importance of graft placement and task complexity for transplant-induced recovery of simple and complex sensorimotor deficits in dopamine denervated rats. Eur J Neurosci (in press)

Marshall JF, Turner BH, Teitelbaum P (1971) Sensory neglect produced by lateral hypothalamic damage. Science 174:523–525

Mogenson GJ (1987) Limbic-motor integration. Prog Psychobiol Physiol Psychol 12:117–158

Morgane PJ (1961) Alterations in feeding and drinking behavior of rats with lesions in globi pallidi. Am J Physiol 201:420–428

Morgane PJ (1975) Anatomical and neurobiochemical bases of the central nervous control of physiological regulations and behavior. In: Mogenson GJ, Calaresu F (eds) Neural integration of physiological mechanisms and behavior. University of Toronto Press, Toronto, Canada

Nadaud D, Simon H, Herman JP, Le Moal M (1984) Contributions of the mesencephalic dopaminergic system and the trigeminal sensory pathway to the ventral tegmental aphagia syndrome in rats. Physiol Behav 33:879–887

Powley TL, Keesey RE (1970) Relationship of body weight to the lateral hypothalamic feeding syndrome. J Comp Physiol Psychol 70:25–36

Robinson TE, Whishaw IQ (1974) Effects of posterior hypothalamic lesions on voluntary behavior and hippocampal electroencephalograms in the rat. J Comp Physiol Psychol 86:768–786

Robinson TE, Castañeda E, Whishaw IQ (1990) Compensatory changes in striatal dopamine neurons following recovery from injury induced by 6-OHDA or methamphetamine: a review of evidence from microdialysis studies. Can J Psychol 44:253–275

Rolls ET, Williams GV (1987) Sensory and movement-related neuronal activity in different regions of the primate striatum. In: Schneider JS, Lidsky TI (eds) Basal ganglia and behavior. Hans Huber, Lewiston, NY, pp 37–59

Schallert T (1982) Adipsia produced by lateral hypothalamic lesions: facilitation of recovery by preoperative restriction of water intake. J Comp Physiol Psychol 96:604–614

Schallert T (1985) Behavioral correlates of neurochemistry. Appetite 6:169–170

Schallert T (1989) Preoperative intermittent feeding or drinking regimens enhance post-lesion sensorimotor function. In: Schulkin J (ed) Preoperative events: their effects on behavior following brain damage. Lawrence Erlbaum Association, NJ, pp 1–20

Schallert T, Hall S (1988) 'Disengage' sensorimotor deficit following apparent recovery from unilateral dopamine depletion. Behav Brain Res 30:15–24

Schallert T, Hsiao S (1979) Homeostasis and life. Behav Brain Sci 2:118

Schallert T, Whishaw IQ (1978) Two types of aphagia and two types of sensorimotor impairment after lateral hypothalamic lesions: observations in normal weight, dieted, and fattened rats. J Comp Physiol Psychol 92:720–741

Schallert T, Wilcox RE (1985) Neurotransmitter-selective brain lesions. In: Boulton AA, Baker GB (eds) Neuromethods (Series 1: Neurochemistry). General neurochemical techniques. Humana Press, Clifton, NJ, pp 343–387

Schallert T, Petrie BF, Whishaw IQ (1989) Neonatal dopamine depletion: spared and unspared sensorimotor and attentional disorders and effects of further depletion in adulthood. Psychobiology 17:386–398

Schneider JS (1987) Basal ganglia-motor influences: role of sensory gating. In: Schneider JS, Lidsky TI (eds) Basal ganglia and behavior. Hans Huber, Lewiston, NY, pp 103–121

Schneider JS, Lidsky TI (eds) (1987) Basal ganglia and behavior. Hans Huber, Lewiston, NY.

Schneider JS, Levine MS, Hull CD, Buchwald NA (1985) Development of somatosensory responsiveness in the basal ganglia in awake cats. Neurophysiology 54:143–154

Sorenson CA, Ellison GD (1970) Striatal organization of feeding behavior in the decorticate rat. Exp Neurol 29:162–174

Stricker EM, Zigmond MJ (1976) Recovery of function after damage to central catecholamine-containing neurons: a neurochemical model for the lateral hypothalamic syndrome. Prog Psychobiol Physiol Psychol 6:121–187

Teitelbaum P, Epstein AN (1962) The lateral hypothalamic syndrome: recovery of feeding and drinking after lateral hypothalamic lesions. Psychol Rev 69:74–90

Ungerstedt U (1971) Adipsia and aphagia after 6-hydroxydopamine induced degeneration of the nigro-striatal dopamine system. Acta Physiol Scand Suppl 367:95–122

Vanderwolf CS (1989) Cerebral activity and behavior: control by central cholinergic and serotonergic systems. Int Rev Neurobiol 30:225–340

Vanderwolf CH, Kolb B, Cooley RK (1978) Behavior of the rat after removal of the neocortex and hippocampal formation. J Comp Physiol Psychol 92:156–175

Whishaw IQ, Robinson TE, Schallert T, De Ryck M, Ramirez VD (1978) Electrical activity of the hippocampus and neocortex in rats depleted of forebrain dopamine and norepinephrine: relations to behavior and effects of atropine. Exp Neurol 65:748–767

Whishaw IQ, Schallert T, Kolb B (1981) An analysis of feeding and sensorimotor abilities of rats after decortication. J Comp Physiol Psychol 95:85–103

White NM (1986) Control of sensorimotor function by dopaminergic nigrostriatal neurons: influences on eating and drinking. Neurosci Biobehav Rev 10:15–36

Zeigler HP, Karten HP (1974) Central trigeminal structures and the lateral hypothalamic syndrome in the rat. Science 186:636–638

Commentary

Greenleaf: How influential can distractive sensory stimulation be in inhibiting drinking behaviour in normal animals?

Schallert: The drinking behaviour of an animal potentially can be enormously affected by distractive sensory input. The amount of water ingested would, of course, depend foremost on the degree of water deprivation and the salience of the sensory stimulation, but would also depend on other variables suggested in this chapter. A water-deprived rat can be kept from drinking for many hours. Severely deprived cattle can be driven away from water by people on horseback. Water-deprived animals are ever alert for predators at the water hole, and in response to signs of them an animal readily ceases drinking. The animal may avoid water for extremely long periods if predators reliably appear before or at the onset of drinking. However, if the predator is visible but at a safe distance, the animal may continue to drink, though it certainly does not neglect the predator.

These sorts of observations about fluid ingestion reflect the concept of homeostasis as conceived by Cannon (1939), who invented the term. Unlike Richter (1943), Cannon intentionally separated physiological from behavioural events associated with the maintenance of water balance in his use of the term homeostasis, although he discussed at length their close interrelationship. He emphasized that behaviour is allowed to vary considerably contingent on the situation, whereas water conservation mechanisms automatically provide relatively stable conditions sufficient to maintain the integrity of the cells in the internal environment. The advantages associated with terrestrial life would be lost if drinking behaviour were rigidly faithful to home-cage target values of homeostasis. We would be no more free to venture from the pond than were our amphibian ancestors. Cannon regarded homeostatic mechanisms as a means to free an animal to engage in an array of adaptive behaviours without the interruptions of moment to moment regulatory behaviour. In fact, because drinking behaviour is not the slave of homeostasis, it is in the long run better able to serve it (Hsiao and Schallert 1979; Schallert 1988; see also Chapter 31).

References

Cannon WB (1939) The wisdom of the body. WW Norton, New York
Richter CP (1943) Total self regulatory functions in animals and human beings. Harvey Lect 38:63–103
Schallert T (1988) Animal models of eating disorders: hypothalamic function. In: Blinder BJ, Chaitin BF, Goldstein R (eds) The eating disorders. PMA Publishing Corp, pp 39–47
Schallert T, Hsaio S (1979) Homeostatis and life. Behav Brain Sci 2:118

Szczepanska-Sadowska: It is regrettable that there is no information in this chapter on the influence of potentially distractive sensory events on human drinking behaviour. Experimental evidence points to the importance of integrity of the nigrostratial system for orienting and ingestive processes. In this context it would be important to know whether patients with parkinsonism also manifest the disengage deficit.

Schallert: This has not been investigated. However, one of the most characteristic features of Parkinson's disease is the patient's inability, while engaged in one task, to respond to requests to begin a second task (Marsden, 1986; see also Cools et al. 1984).

References

Cools AR, van de Bercken JHL, Horstink MWI, Berger HJC (1984) Cognitive and motor shifting aptitude disorder in Parkinson's disease. J Neurol Neurosurg Psychiatry 47:443–453
Marsden CD (1980) The enigma of the basal ganglia. Trends in Neurosci 284–287

McKinley: Allen and others will be reporting in the *Journal of Comparative Neurology* that they have found receptors for angiotensin II in human striatum and substantia nigra, and they note that these receptors may be absent in Parkinson's disease.

Chapter 15

Drinking in Mammals: Functional Morphology, Orosensory Modulation and Motor Control

H.P. Zeigler

Introduction

Drinking involves the acquisition of an aliquot of liquid, its transport from the front of the oral cavity to the caudal oropharyngeal region and its swallowing. In addition to water intake, it also encompasses behaviours such as non-nutritive drinking (the ingestion of water to which a non-nutritive tastant has been added) and nutritive drinking (the ingestion of liquid food or sugar solutions) which may engage ingestive systems related either to energy balance or water balance (Mook and Kenney 1977).

While both eating and drinking are mediated by the same (oromotor) effector apparatus, there are distinct differences in the functional morphology of the two behaviours (Thexton 1984). These arise primarily from differences in the physical characteristics of solids and liquids as reflected in their sensory (stimulative) properties. For example, because of its mass and physical structure, food must be transported through the oral cavity against varying degrees of resistance, whereas liquids, because they flow, offer minimal resistance to movement. Similarly, the transport of solids and liquids requires different amounts of clearance between the tongue and palate, since liquids form a thin film on the tongue whereas solids do not (Thexton and Crompton 1989). Furthermore, in mammals, eating requires a stage of mechanical reduction (comminution), both to facilitate the digestion of solids and to convert the food items into particles suitable for swallowing. This process of reduction is the function of mastication (chewing) and this aspect of ingestion is lacking in drinking behaviour. (In the guinea pig, for example, chewing is associated with both vertical and horizontal movements of the mandible (Byrd 1981) but during drinking the horizontal movements are absent (Gerstner and Goldberg 1989).) Finally, food items vary in their size, shape, texture, hardness, etc., generating complex patterns of sensory feedback which will change during the course of eating as the food items are processed by mastication. The increased complexity of eating behaviour reflects the addition of mastication and manipulation phases, required by the processes of mechanical reduction and modulated by the sensory properties of the food (e.g. Weijs and Dantuma 1975; Thexton et al. 1980; Thexton and McGarrick 1988; Thexton and Crompton 1989) In contrast, except for temperature, the sensory properties of water are likely to be relatively invariant. (However, since the largest proportion of fluid intake in humans consists of flavoured beverages, the sensory properties of these fluids are likely to be critical determinants of their ingestion (see, e.g, Chapters 4, 22 and 24).) The first part of this chapter reviews the structure of the oromotor system and its functional organization in the generation of drinking responses. The second part focuses on the role of orosensory factors in the initiation of drinking and the modulation of drinking response topography.

Anatomy of the Oromotor Apparatus

The oromotor apparatus is located at the most rostral portion of the digestive system. Some of its component structures, like the maxilla, teeth and hard palate, are fixed; others like the tongue, mandible, soft palate, pharynx and hyoid, possess varying degrees of mobility. (Because the hyoid functions both as a link bone in the chain of jaw opener muscles and as a lingual bone forming the base of the tongue it is involved in both jaw and tongue movements (Thexton 1984).)

As shown in Fig. 5.1, the mouth is divisible into two large compartments: the oral cavity and the pharyngeal cavity. The oral cavity extends from the lips back to the fauces, the passage between the two compartments. Above this passage lies the soft palate. It is flanked by the pillars of the fauces, which mark its lateral boundaries as well as the border between the oral and pharyngeal cavities. In all mammals the pharynx is divisible into an anterior (vallecula) and a posterior compartment. The

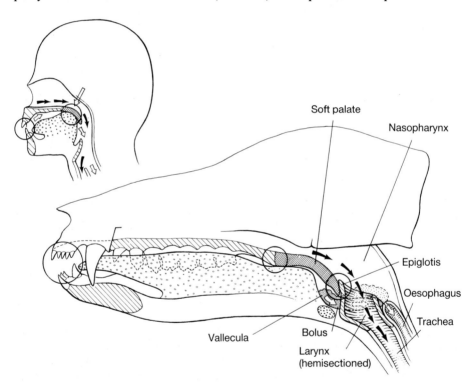

Fig. 15.1. Hemisections through the oral cavity and oropharynx of a human (top) and opossum (bottom) showing the main structures and the locations of the three oral seals. The anterior oral seal is shown at the lips. The middle oral seal is shown as a contact between the soft palate and the tongue and coincides with the posterior boundary of the oral cavity. The posterior oral seal (absent in humans) is represented by the contact point between the soft palate and the epiglottis. The space between the middle and posterior oral seals is part of the oropharynx while the space between the posterior part of the tongue and the anterior surface of the epiglottis forms the vallecular cavity. (From Hiiemae and Crompton (1985), with permission.)

floor of the vallecular compartment is formed by the posterior surface of the tongue, its ceiling by the soft palate and its posterior boundary by the epiglottis. The posterior compartment extends to the posterior wall of the pharynx.

The schematic drawings in Fig. 15.1 illustrate an important feature of the construction of the mouth. By appropriate placement of the lips, tongue and soft palate, specific segments of the oral passageway may be temporarily sealed off during ingestion. The most anterior of these temporary seals is formed by the lips, an intermediate one by the tongue and the hard and soft palates, and a posterior one by the soft palate and epiglottis. (The soft palate/epiglottal seal is lacking in adult humans.) The compartments formed by these seals can provide temporary storage for aliquots of either liquids or solids during transport through the mouth (Hiiemae and Crompton 1985).

Functional Morphology of Drinking Behaviour

In all mammals, including man, drinking involves ingestion, transport and swallowing (deglutition) components. However, the ingestion and transport components in the human and anthropoids have some unique features, correlated with the structure of the ingestive apparatus (see below). Different mammalian groups may be distinguished with respect to their mode of ingestion. For liquids, two mechanisms of ingestion are available: licking and/or lapping, and sucking. Licking/lapping are characteristic of the laboratory rat, the hamster, the opossum and the domestic cat and dog (Hiiemae and Ardran 1968; Hiiemae and Crompton 1971, 1985; Marowitz and Halpern 1973; Weijs and Dantuma 1975; Thexton and McGarrick 1988; see Halpern 1977, Table 4). Although lapping involves intake from more extensive liquid sources and contact only with the liquid (see Marowitz and Halpern 1973, Fig. 2), the two behaviours are conjoined here because both involve the repetitive collection of relatively small volumes of liquid at relatively high rates and its intraoral transport by the tongue.

Both behaviours begin with a downward (slow opening) movement of the lower jaw creating an enlargement of the slight pre-existing gape. During this movement the tongue is protracted so that its tip makes contact with the liquid (see Figs. 15.2 and 15.4a, below). Jaw closing and tongue retraction are closely coupled and the liquid captured at the tongue tip is brought into the oral cavity. In both behaviours the aliquot of liquid is transported through the oral cavity to the vallecula on the surface of the tongue which, because it is retracting, functions somewhat like a conveyor belt. Figure 15.2a, b illustrates tongue movements during licking from a sipper tube in the rat and lapping in the opposum.

In sucking, by contrast, a relatively large volume of fluid is ingested more slowly and the intraoral movement of the liquid involves the use of a suction/pressure mechanism, rather than lingual transport. The process has been studied in some detail in the miniature pig (Hering and Scapino 1973) and in a number of herbivores (Crompton 1989, personal communication). It begins with the snout in the water. In the first phase of transport, a seal is created between the tongue and the anterior part of the soft palate. The anterior portion of the tongue is then lowered within the oral cavity to suction liquid into its anterior portion. This compartment is then sealed by pressing the tongue tip against the tip of the palate. In the second phase, the seal between the tongue and the anterior part of the soft palate is broken, caudal to the compartment in which the water is trapped. The anterior part of the tongue is elevated, forcing the liquid back through the fauces while the back part of the tongue moves

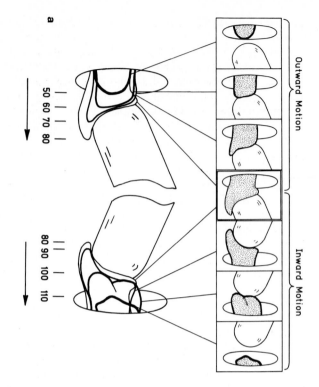

a

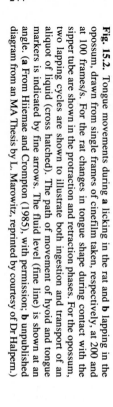

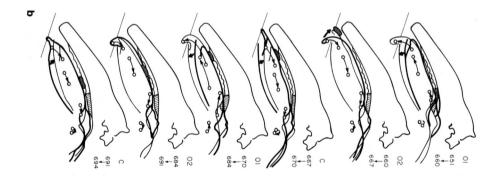

b

Fig. 15.2. Tongue movements during **a** licking in the rat and **b** lapping in the opossum, drawn from single frames of cinefilm taken, respectively, at 200 and at 100 frames/s. For the rat changes in tongue shape during contact with the sipper tube are shown in the protraction and retraction phases. For the opossum, two lapping cycles are shown to illustrate both ingestion and transport of an aliquot of liquid (cross hatched). The path of movement of hyoid and tongue markers is indicated by fine arrows. The fluid level (fine line) is shown at an angle. (**a** From Hiiemae and Crompton (1985), with permission; **b** unpublished diagram from an MA Thesis by L. Marowitz, reprinted by courtesy of Dr Halpern.)

forward, reducing the pressure within the vallecula and producing the second instance of suction.

(In humans, water is also drawn into the oral cavity by a suction mechanism. The tongue surface contacts the anterior part of the soft palate anterior to the liquid. Transport is achieved by rolling that (tongue) contact point backwards, somewhat like rolling up a toothpaste tube. However, because humans lack a soft palate/epiglottal seal, liquid is not stored in the vallecula, but swallowed from anterior to the fauces.)

In all mammals, the aliquot of liquid is transported to the anterior portion of the pharynx (the vallecula) and, except in humans, stored there as a bolus. Swallowing, the final stage of drinking, serves to move the aliquot from the oral cavity through the oesophagus. Swallowing has been intensively studied in man and in anaesthetized animals (reviews in Jean 1984; Miller 1982) and its relation to intraoral transport differs between licking/lapping and sucking. In sucking, the ingestion and swallowing of an aliquot of liquid takes place within the same cycle. In licking and lapping, swallowing seems to necessitate an interruption of the rhythmic cycle of jaw and tongue movements which mediate collection and transport of the aliquot. It thus occurs as an activity interpolated into the slow opening phase, with some ratio of orolingual/oropharyngeal activities. Swallowing has generally been conceptualized as a reflex, mediated by a central pattern generator and initiated when the stimulation of an oropharyngeal "trigger zone" by the bolus reaches some critical level (volumetric hypothesis). This model suggests that, regardless of the rate of intraoral transport (i.e., the rate of bolus accumulation), modulation of the swallowing pattern will involve primarily adjustments in the frequency of swallowing, while the volume per swallow should remain relatively constant.

The relation between intraoral transport and swallowing has been examined during water licking (Weijnen et al. 1984) and during intraoral infusion of a sucrose solution (Kaplan and Grill 1989), using electromyographic criteria to identify the occurrence of a swallow. In both drinking paradigms, swallows occurred at regular intervals (i.e. n licks/swallow), they were interpolated into bursts of oromotor activity and the occurrence of a swallow was highly correlated with transient interruption of that activity, as reflected in an extension of about 20–40 ms duration in the associated interlick interval. The volume per swallow proved not to be constant but, like the frequency of swallows, varied with the flow rate of the liquid, a finding inconsistent with a volumetric hypothesis. Although these results are not incompatible with a reflex model of swallowing (see Weijnen et al. 1984) they suggest that the notion of its moment-to-moment control by the tactile properties of the bolus is overly simplistic. They imply some separation of mechanisms timing the initiation of the swallow from those co-ordinating the sequential activation of the swallowing musculature (see, e.g. Turvey et al. 1989). An alternate model proposed by Kaplan and Grill (1989) postulates dynamic modulation of the central timing mechanism by inputs related to, for example the rate of intraoral transport. (For a discussion of the relation between sensory inputs and "central" pattern generation see, e.g., Brooks 1986; Pearson 1985).

Efferent control of the ingestive apparatus is vested in cranial (V, VII, IX, X, XI, XII) and upper cervical spinal motor nerves (C1, C2) which control the muscles of the jaw, tongue, hyoids and pharynx (Halpern 1977; Luschei and Goldberg 1981). The patterned activation of these muscles, in specific groups, mediates the various phases of drinking (Fig. 15.3). During the ingestion by licking/lapping (or infusion) of water or sucrose (at low concentrations) the pattern of muscle activation is highly stereotyped (Yamamoto et al. 1982; Travers and Norgren 1986; Thexton and

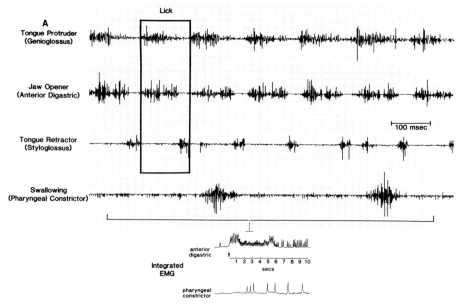

Fig. 15.3. Electromyographic (EMG) activity from jaw, tongue and pharyngeal muscles of a rat during the licking of 0.3 M sucrose solution from a water bottle spout. A single lick cycle is outlined. (From Travers and Norgren (1986), with permission.)

McGarrick 1988). The electromyographic sequence begins with synchronous activation of tongue protruder and jaw opener muscles which alternate their activity with that of the tongue retractors and jaw closers. Swallowing is interpolated into these licking cycles and is preceded by an extra contraction of the tongue retractor. This ingestion "lick" pattern is clearly distinguishable from the EMG pattern seen during rejection of a noxious liquid, which is correlated with a "gape" response (Grill and Norgren 1978).

Similarities and Differences in the Functional Morphology of Eating and Drinking

Thexton has pointed out that "although the biomechanics ultimately dictate a fixed set of relationships between the movements of the different anatomical structures, the relative magnitude and timing of the different components of action (jaw, hyoid, tongue) . . . are flexible" (1984, p. 1018). The interaction of these structures is related to their function. For example, during eating, the tongue functions in the selection, positioning and transport of food particles; during drinking it has only a transport function.

Analyses of eating and drinking in mammals (cat, opossum) have shown that intraoral transport during eating involves two distinct stages (Hiiemae and Crompton 1985). In stage I the unchewed bolus is transported from the front of mouth (incisors) to the cheek teeth; during stage II transport, the chewed bolus is moved from the cheek teeth to the posterior tongue and pharynx. In the interim, there is a period of oromotor activity whose function is not transport of the food but its local recirculation during mastication. In lapping, stage I and stage II transport appear to merge. However, in both behaviours, transport is accomplished by rhythmically reciprocating cycles of jaw and tongue movements which are out of phase with each other. Differences in the topography of eating and drinking reflect differences in the relation between jaw movement and tongue movement cycles (Thexton and McGarrick 1988, 1989; Thexton and Crompton 1989). For both behaviours, an initial slow jaw-opening movement is associated with tongue protraction and followed by tongue retraction. In drinking, this tongue retraction is associated with jaw-closing; in eating it is followed by a fast jaw-opening movement. (The difference between the movements may be functionally related to the intraoral clearance required for the posterior transport of solids but not liquids, which form a thin film on the tongue.) However, both eating and drinking contain a component in which protraction of the tongue is linked to the slow-open phase of jaw movement. In drinking, this tongue movement may "squeeze" liquid against the palate producing posterior transport. In eating, the movement presses the tongue against the palatal rugae, stripping off food particles as it advances. Thus, eating and drinking appear to share a common movement pattern which, although it serves different functions in the two behaviours, may reflect patterns of neuromuscular co-ordination "basic" to the two behaviours (Thexton 1984).

Orosensory Systems and the Initiation of Drinking

Orosensory inputs arise from an array of receptors, including gustatory (chemosensory) receptors on the tongue, pharynx, epiglottis and hard and soft palates, somatosensory (temperature and touch) receptors in the tongue and oral mucosa, and proprioceptors (spindles, tendon organs) in muscles and joints. (For reviews see Dubner et al. 1978; Miller and Spangler 1982; Norgren 1984; Zeigler et al. 1985.) The adequate stimulus for the elicitation of licking in the rat appears to be somatosensory, rather than gustatory, and to involve some combination of tactile and thermal properties. (That it is neither fluidity nor tactile stimulation per se is evident from the fact that licking is not continuously evoked by saliva or by intraoral contact of the palate by the tongue (Justenson 1977).) Evidence for this conclusion comes from descriptive accounts of drinking topography, from an analysis of the stimulus properties of water and from studies of air- and cold-licking behaviour. Analysis of videotape records (Zeigler et al. 1984) indicates that the rat's drinking sequence may be divided into approach, perioral contact and mouth opening phases. Mouth opening and tongue protrusion do not occur during the approach phase of eating or drinking but always follow a relatively prolonged period of perioral contact with the food or water source (Fig. 15.4a,b). Moreover, "short but consistent bouts of rhythmic oral activity are produced by intraoral infusions of water" (Travers and Norgren 1986). These observations suggest that, under normal conditions, the mouth opening and tongue protrusion phase of drinking in the rat is not elicited by distal stimuli (i.e., vision, olfaction) but by somatosensory stimulation of the oral and perioral region.

It has been observed, for example, that the number of licks and the intraburst lick rate elicited by a single drop are unrelated to its gustatory stimulus dimensions but directly proportional to its volume, i.e., to the amount of somatosensory stimulation (Hulse and Suter 1968, 1970). A role for temperature is suggested by the report that the rat's water intake is sensitive to variations in water temperature (Gold and LaForge 1977) and that water-deprived hamsters, guinea pigs and rats will "drink" (i.e. orally

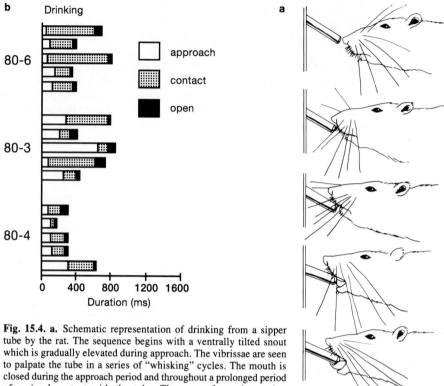

Fig. 15.4. a. Schematic representation of drinking from a sipper tube by the rat. The sequence begins with a ventrally tilted snout which is gradually elevated during approach. The vibrissae are seen to palpate the tube in a series of "whisking" cycles. The mouth is closed during the approach period and throughout a prolonged period of perioral contact with the tube. The start of mouth opening coincides with a forward movement of the snout that brings the tube into contact with the upper lip line. b. Temporal relations among the three phases of drinking for three normal rats: open bars, approach; dotted bars, perioral contact; closed bars, mouth opening. c. Duration of components of the ingestive behaviour sequence in trigeminally deafferented rats. Note the prolonged duration of perioral contact (dotted bars) and the absence of the mouth-opening/tongue-protrusion phase. After Zeigler et al. (1985), with permission.

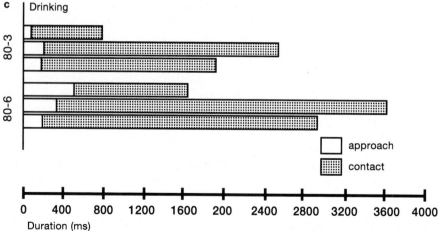

palpate) a cold dry metal tube (Mendelson 1977). The ingestive movement pattern used by each species in cold drinking is similar to that used in its water drinking. Thus, rats, which ingest water by licking, also lick the tube, whereas the guinea pig, a suction drinker, "ingests water ... by taking the tube into its mouth and then sucking/swishing its tongue around the orifice" (Mendelson 1977, p. 192).

Linking these observations on the role of temperature and touch is a body of work on air-licking. In addition to sharing motor patterns, air-licking resembles drinking in several respects. It is induced by manipulations that induce drinking (including deprivation and hypertonic saline), has a satiating effect upon the subsequent ingestion of water and may be used to reinforce instrumental behaviours (Mendelson 1977). Air-licking behaviour has been attributed to the reinforcing effects of evaporative cooling of the oral region. However, while rats lick water with their tongues, air-licking is maintained only when the airstream stimulates the interior of the mouth. Rats will twist their heads almost upside down to produce such stimulation and air-licking ceases in its absence. Taken together, the data on air-licking and cold licking suggest that these behaviours, and by implication, water drinking, are elicited and sustained by somatosensory inputs from extralingual oral tissue.

This conclusion is supported by studies of orosensory deafferentation. Gustatory denervation of the tongue does not produce a decrement in air-licking (Mendelson 1977) and has only a minimal effect on water intake (Jacquin 1983). In contrast, trigeminal (somatosensory) denervation of the oral cavity (exclusive of the tongue and sparing taste nerves and the motor innervation of the jaws and tongue) produces a massive disruption of drinking (and eating) behaviour in the rat (reviewed by Zeigler et al. 1985).

In the immediate postoperative period ... approaches to water in a dish or sipper tube were infrequent and hesitant. The initial orientation of the snout and lower lip was normal, yet neither mouth opening nor tongue extension occurred (Jacquin and Zeigler 1983, p. 69)

While trigeminal deafferentation does not appear to affect either the form or the duration of the approach phase, the perioral contact phase is markedly extended (compare Figs. 15.4b and c). Even after prolonged contact with the pellet, the jaw does not open and the tongue does not protrude to bridge the gap between the end of the sipper tube and the inside of the mouth. Under these conditions the animals behave as if they are "fixed" in the contact phase, continually rubbing the snout against the tube, but eventually aborting the ingestive sequence. This pattern of disruption is correlated with a decreased probability of mouth opening and tongue protrusion to perioral stimuli, although vigorous oromotor activity may be elicited by intraoral (cotton swab, food, water) or extraoral (tail pinch, paw pinch) stimuli.

Trigeminal orosensory deafferentation in the rat is followed by periods of adipsia/hypodipsia persisting for as long as three weeks. Fig. 15.5 compares the water intake of deafferented subjects with that of a yoked (food intake) control group. The deafferented subjects decrease their intake abruptly, rather than gradually, and are still adipsic at a time when the controls have begun to ingest substantial amounts of water.

A comparison of deafferented subjects with oromotor controls indicates that the reduced intake of the deafferented rat is not due to its sensorimotor impairments but reflects a reduction in responsiveness to the water source, as measured by the time spent in contact with the sipper tube. Rats with section of the trigeminal motor or hypoglossal nerves, which could not move the jaw or tongue, continued to respond vigorously and persistently to the water source, increasing their contact time from

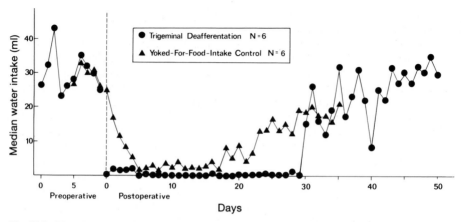

Fig. 15.5. Water intake in deafferented and yoked controls, maintained on the same type and amounts of food. Note the differences in the rate of decline of intake and the time course of recovery in the two groups. (From Jacquin and Zeigler (1983), with permission.)

preoperative levels of about 20 min/day to almost 120 min/day. In contrast, deafferented subjects reduced their contact time to less than 5 min/day (Zeigler et al. 1985, Figs. 31, 32). Trigeminal orosensory deafferentation also profoundly depresses performance of instrumental responses reinforced with water, including lever pressing, magazine approach and dipper contact (Jacquin and Zeigler 1984). Thus, trigeminal orosensory inputs play a critical role in eliciting and maintaining the drinking response.

Topography of Drinking Behaviour and the Invariance Hypothesis

In the present context, the term topography refers to the distinguishing features, both spatial and temporal, which characterize a pattern of behaviour. Although the morphology of the mammalian oromotor apparatus and the sensory properties of water obviously generate constraints on that pattern, the temporal organization of mammalian drinking behaviour has some degrees of freedom.

The parameters of that organization arise from the fact that drinking has both a macrostructure and a microstructure. After water deprivation, for example, animals may meet their water requirements by drinking to satiety in a single continuous bout or spread out their intake over a larger number of bouts, varying in duration and separated by intervals of non-responding (Rolls and Rolls 1982). The organization of ingestion into these "bursts" and "pauses" represents a mechanism of intake control at the macrobehavioural level. However, at the microbehavioural level, bursts of drinking are made up of individual movement elements (licks, laps, sucks). Thus intake may also be controlled by variations in the local lick rate, i.e. the interlick interval (ILI), measured over an uninterrupted burst of licks, or the global lick rate, i.e. licks/min, measured over a longer period which may include pauses (Justenson 1977). Finally, the ILI is itself the sum of two movement components: the duration of contact between the tongue and the water source (on-on time) and the time to the next contact (off-on time). As Allison (1971) pointed out "a constant rate of licking bursts may therefore conceal large systematic microbehavioral adjustments" (p. 409) which could also be used to control intake.

In one report a contact sensor (drinkometer) was used to monitor licking behaviour (Stellar and Hill 1952). The device functions by making the animal part of the monitoring circuit. With one of the leads from the apparatus attached to the cage floor and another to the water source; each tongue contact with the water completes the electrical circuit and activates a counter (Weijnen 1977, 1989).

Stellar and Hill reported a (measured) lick rate of 6–7/s and a (calculated) intake rate of 0.03 ml/s in rats deprived of water for periods ranging from 6 hours to 7 days. They noted individual differences in lick rate and commented on the effects of environmental constraints (Stellar and Hill 1952, p. 98). The main feature of drinking topography was found to be its division into bursts and pauses and a slight falling off in rate after the initial burst was noted (p. 100). Nevertheless, despite their caveats, these authors, unlike Smith and Smith (1939) did not characterize the rat's rate of licking as "relatively constant". Instead, they concluded that,

Whenever the rat drinks, it drinks at a constant rate regardless of its level of thirst. At the outset of drinking or at the end of a 2-hr drinking period, after a mild deprivation or after a severe one, the rat laps water at a rate of six to seven tongue laps per second or it does not drink at all. (Stellar and Hill 1952, p. 98).

Thus the basic message of their paper was that the rat's local rate of drinking (i.e. the averaged rate of an uninterrupted burst) is invariant.

For some psychologists, the invariance hypothesis was a welcome anchor point amidst the complexities of ingestive behaviour. If drinking rate was indeed invariant then the amount of intake was a relatively simple function of burst length x burst frequency. The invariance hypothesis was congruent with the then ubiquitous concept of the fixed action pattern, simplified the task of the motivational theorist (Premack 1965, cited in Cone 1974) and provided a stable behavioural baseline for studies of neurosensory and neuromotor mechanisms (Halpern 1985; Wiesenfeld et al. 1977). It also represented an empirical challenge and provided the impetus for analytical studies of variables controlling lick rate in a variety of mammals.

The initial conclusions of these studies supported the invariance hypothesis. Constancy of lick rates was reported in cats, gerbils, hamsters, rabbits and rats (both feral and domestic) as was lick invariance in the face of variations in deprivation and taste characteristics (e.g. saccharin, sucrose, sodium chloride; review in Cone 1974; Justenson 1977). Moreover, attempts at modulating lick rate using operant conditioning procedure have had only limited success (Hulse 1967; Welzl 1976; Hernandez-Mesa et al. 1985). Such findings led to the hypothesis that licking is produced by a central pattern/rhythm generator, influenced only minimally by peripheral sensory inputs (see, e.g., Juch et al. 1985; Hernandez-Mesa et al. 1988).

However, as Cone (1974) pointed out, many of the early studies used only descriptive statistics in their analyses. When the original data were reanalysed and in new studies with appropriate statistical analyses, significant variations in lick rate were found as a function of sex, age, time of day, deprivation level, phase of the test session, environmental constraints, experience with the fluid, saccharin or sucrose concentration, viscosity and drop volume (reviews in Cone 1974; Halpern 1977; Justenson 1977). In a recent study, for example, rats decreased their ILI by about 6% (i.e., increased the lick rate) when access to a saccharin solution was constrained to 30-s intervals (Mook and Wagner 1989). In all these studies, the extent of reported variation is small (e.g., 0.5–2.5 licks/s: Weijnen, personal communication). However, the consistency of these departures from invariance and their systematic relation to controlling variables suggests that they are not simply some type of biological noise. Moreover, Allison (1971) described significant differences in the microstructure of nutritive and non-nutritive drinking in rats.

Thus the available evidence is consistent with both "fixed" and "flexible" licking response topographies (cf. Malmo et al. 1986). The relative invariance of the behaviour, like that of so many other putative fixed action patterns (e.g., Zeigler et al. 1980) may reflect both the relative invariance of environmental factors during drinking in the laboratory (Hulse and Suter 1970, p. 314; Marowitz and Halpern 1973, p. 263) and the constraints imposed by the oromotor apparatus. Nevertheless, within the limits imposed by those constraints, the ingestive mechanism appears to possess some degree of adaptive flexibility. The extent of this flexibility is best appreciated by considering its role in the orosensory modulation of drinking and the regulation of intake.

Orosensory Modulation of Drinking Response Topography

The control of fluid intake may involve variations in both the macrostructure and microstructure of drinking behaviour. For example, after water deprivation dogs meet their water requirements by drinking to satiety in a single continuous bout whereas rats distribute their intake over a larger number of bouts whose duration and frequency varies with the level of deprivation (Rolls and Rolls 1982). Variations in the relation between the frequency and/or duration of bursts and pauses also characterizes the control of intake during non-nutritive and nutritive drinking. Thus, for many solutions (e.g., saccharin, sucrose), the duration of drinking bouts increases with the concentration of the tastant up to some level in a manner that corresponds to the classical preference–aversion curve for the intake of these solutions (see, e.g., Spector and Smith 1984; Travers and Norgren 1986). Conversely, under conditions of intermittent (30 s) access to saccharin solutions, rats decrease the frequency and the duration of pauses and increase slightly the local (licks/s) rates (Mook and Wagner 1988, 1989). They may also show variations in global rates (licks/min) related to the concentration of the tastant (e.g., Krimm et al. 1987; Davis 1989 and see Fig. 15.6a).

Unfortunately, the analysis of orosensory modulation is complicated by the fact that the ingestion of fluids may have both facilitatory and inhibitory consequences, related, respectively, to their orosensory and postingestional properties. Davis has suggested that "… the initial rate of ingestion reflects the stimulating effectiveness of the substance being consumed, that is, the excitatory influence on the ingestive mechanism, and… the rate of decline of ingestion is a measure of the effectiveness of a negative feedback inhibitory signal …" (Davis 1989). Increasing the concentration of a tastant will increase its stimulatory effectiveness but may simultaneously increase its postingestional inhibitory effect (Fig. 15.6a). Thus, a study of sucrose drinking patterns, suggested that "postingestional satiety mechanisms seem to be limiting bout volume with the high sucrose concentrations… whereas orosensory factors seem to be limiting bout volume with the low concentrations" (Spector and Smith 1984, p. 133).

Alternatively, the rate of decline of ingestion may also be related to the gustatory properties of the solution, e.g., the hypothesized "dual" (sweet/bitter) taste of saccharin (Smith et al. 1987). Thus, although the rat may ingest the same volumes of two solutions, that intake may be mediated by very different topographies (Fig. 15.6b).

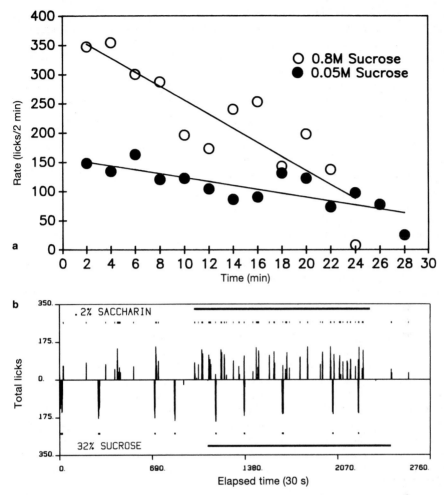

Fig. 15.6. a. Rate functions for (non-food-deprived) rats drinking two different solutions of sucrose. The initial licking rate is greater for the more concentrated solution but its rate of decline is steeper. There was no significant difference in the intake of the two solutions in a 30 min period. (From Davis 1989, Fig. 2, reprinted with permission.) **b.** Topography of saccharin (*top*) and sucrose (*bottom*) drinking over a 23-h period in non-deprived rats. Although there was no significant difference in the amounts ingested, bout frequencies, durations and interbout intervals differed significantly for the two solutions. Solid horizontal line indicates dark period; vertical lines indicate drinking bouts. (From Smith et al. (1987), with permission.)

Finally, Mook has summarized evidence that orosensory (non-gustatory) proprioceptive or central correlates of performance contribute to the termination ("oral satiety") of drinking. He suggests (a) that the rat monitors some performance-related consequence of the amount of licking which is correlated with volume intake and (b) that termination of drinking occurs when that feedback signal matches some preset reference value specifying the volume to be drunk (Mook 1990).

The studies reviewed above indicate that the act of drinking is accompanied by several types of orosensory (gustatory, somatosensory) and postingestive (osmotic and chemospecific) stimuli. Although the differential effects of orosensory and postingestional processes may be experimentally dissociated (Davis and Levine 1977; Mook and Kenney 1977; Mook 1990), it is the moment-to-moment interaction of these processes which generates the topography of the normal drinking sequence.

Conclusions

In recent years, there has been considerable progress in elucidating the functional morphology of mammalian drinking. While there are species differences in the ingestive phase of drinking (e.g., licking, lapping, sucking), the intraoral transport phase is characterized by a pattern of hyoid, jaw and tongue movements which is highly stereotyped and may be a common component of both eating and drinking in mammals (Thexton and McGarrick 1988; Thexton and Crompton 1989). Furthermore, recent studies have identified the sensory processes which elicit the drinking response and control its topography. These studies suggest that somatosensory stimuli elicit the drinking response, contribute to its maintenance and possibly, to its termination. In view of the systematic relation between the duration of a drinking bout and the concentration of the tastant, "gustatory stimuli appear to either prolong this behavior (preferred stimulus) or abort it (noxious stimuli)" (Travers and Norgren 1986, p. 554; see also Halpern 1985).

Studies of the orosensory and postingestive modulation of drinking have clarified the manner in which these factors interact to generate drinking topography. However, it is clear that volume measurements alone provide only limited information as to causal factors mediating intake. Microbehavioural studies of drinking topography can be of considerable utility in identifying such factors and in clarifying the deficits produced by brain lesions (Davis 1989; Evered and Mogenson 1976; Krimm et al. 1987). However, the use of such methodologies in the study of central drinking mechanisms is still the exception rather than the rule.

For example, it is generally assumed that all types of thirst engage the oromotor apparatus in the same way. It might thus be predicted that the increased intake produced by different dipsogens (e.g., water deprivation, polyethylene glycol, saline or angiotensin) would be mediated by a common effect on drinking topography. So far as I am aware, this hypothesis has never been systematically tested. Indeed, a major unfinished task in the study of drinking is to establish anatomical and neurobehavioural links between central processes controlling water balance and the peripheral oromotor ("final common path") mechanisms mediating water intake. Instead, studies of the neural control of drinking typically demonstrate that some brain structure mediates a construct called "thirst" whose activation somehow initiates a movement pattern called drinking. Until closer links are forged between the construct and the movement pattern, "thirst" will remain a Cartesian ghost in the neural machinery of drinking.

Acknowledgements: Preparation of this chapter was supported by Research Scientist Award MH-00320. I am indebted to the following individuals for a critical reading of an earlier draft of the MS: Dr Alfred W. Crompton, Museum of Comparative Zoology, Harvard University; Dr John D. Davis, Dept. of Psychology, University of Illinois, Chicago; Drs Harvey Grill and Joel Kaplan, Dept. of Psychology, University of Pennsylvania; Dr Bruce Halpern, Section on Neurobiology and Behavior, Cornell University; Dr Douglas Mook, Dept. of Psychology, University of Virginia; Dr James C. Smith, Dept. of Psychology, Florida State University; Dr A. J. Thexton,

Dept of Physiology, Guy's Hospital, London, UK; Dr J. A. W. M. Weijnen, Tilburg University, Netherlands; Professor Garth Zweers, University of Leiden, Netherlands.

References

Allison J (1971) Microbehavioral features of nutritive and nonnutritive drinking in rats. J Comp Physiol Psychol 76:408–415

Brooks VB (1986) The neural basis of motor control. Oxford University Press, New York

Byrd KE (1981) Mandibular movements and muscle activity during mastication in the guinea pig. J Morphol 170:147–169

Cone D (1974) Do mammals lick at a constant rate? A critical review of the literature. Psychol Rec 24:353–364

Davis JD (1989) The microstructure of ingestive behavior. In: Schneider LH, Cooper SJ (eds) The psychobiology of eating disorders: preclinical and clinical perspectives. Annals of the New York Academy of Sciences, vol. 575

Davis JD, Levine MN (1977) A model for the control of ingestion. Psychol Rev 84:379–412

Dubner R, Sessle BJ, Storey AT (1978) The neural basis of oral and facial function. Plenum Press, New York

Evered MD, Mogenson GJ (1976) Regulatory and secondary water intake in rats with lesions of the zona incerta. Am J Physiol 230:1049–1057

Gerstner GE, Goldberg LJ (1989) An analysis of mandibular movement trajectories and masticatory muscle EMG activity during drinking in the guinea pig. Brain Res 479:6–15

Gold RM, Laforge RG (1977) Temperature of ingested fluids: preference and satiation effects. In: Weijnen JAWM, Mendelson J (eds) Drinking behavior: oral stimulation, reinforcement and preference. Plenum Press, New York, pp 247–274

Grill H, Norgren R (1978) The taste reactivity test. I. Mimetic responses to gustatory stimuli in neurologically normal rats. Brain Res 143:263–279

Halpern BP (1977) Functional anatomy of the tongue and mouth of mammals. In: Weijnen JAWM, Mendelson J (eds) Drinking behavior: oral stimulation, reinforcement and preference. Plenum Press, New York, pp 1–92

Halpern BP (1985) Time as a factor in gustation: temporal patterns of taste stimulation and response. In: Pfaff DW (ed) Taste, olfaction and the central nervous system: a festschrift in honor of Carl Pfaffman. Rockefeller University Press, New York, pp 181–209

Hering SW, Scapino RP (1973) The physiology of feeding in miniature pigs. J Morphol 141:427–460

Hernandez-Mesa N, Mamedov Z, Bures J (1985) Operant control of the pattern of licking in rats. Exp Brain Res 58:117–124

Hernandez-Mesa N, Mamedov Z, Bures J (1988) Licking during forced spout alteration in rats: resetting the pacemaker or disconnecting the motor output. Exp Brain Res 70:561–568

Hiiemae KM, Ardran GM (1968) A cinefluorographic study of mandibular movement during feeding in the rat (Rattus norvegicus). J Zool (Lond) 154:139–154

Hiiemae KM, Crompton AW (1971) A cinefluorographic study of feeding in the American opossum. In: Dahlberg AA (ed) Dental morphology and evolution. University of Chicago Press, Chicago, pp 299–334

Hiiemae KM, Crompton AW (1985) Mastication, food transport and swallowing. In: Hildebrand M, Bramble DM, Liem KF, Wake DB (eds) Functional vertebrate morphology. Harvard University Press, Cambridge, pp 262–290

Hulse SH (1967) Licking behavior in rats in relation to saccharin concentration and shifts in fixed-ratio reinforcement. J Comp Physiol Psychol 64:478–484

Hulse SH, Suter S (1968) One drop licking in rats. J Comp Physiol Psychol 66:536–539

Hulse SH, Suter S (1970) Emitted and elicited behavior: an analysis of some learning mechanisms associated with fluid intake in rats. Learn Motiv 1:304–315

Jacquin M (1983) Gustation and ingestive behavior in the rat. Behav Neurosci 97:98–109

Jacquin MF, Zeigler HP (1983) Trigeminal orosensation and ingestive behavior in the rat. Behav Neurosci 97:62–97

Jacquin MF, Zeigler HP (1984) Trigeminal denervation and operant behavior in the rat. Behav Neurosci 98:1004–1022

Jean A (1984) Brainstem organization of the swallowing network. Brain Behav Evol 25:109–116

Juch PJW, Van Willigen JD, Broekhuijsen ML, Ballantijn CM (1985) Peripheral influences on the central pattern–rhythm generator for tongue movements in the rat. Arch Oral Biol 30:415–421

Justenson DR (1977) Classical and instrumental conditioning of licking: a review of methodology and data. In: Weijnen JAWM, Mendelson J (eds) Drinking behavior: oral stimulation, reinforcement and preference. Plenum Press, New York, pp 115–156

Kaplan J, Grill HJ (1989) Swallowing during ongoing fluid ingestion in the rat: Brain Res 499:63–80

Krimm RF, Mohssen SN, Smith JC, Miller IJ Jr, Beidler L (1987) The effect of bilateral sectioning of the chorda tympani and greater superficial petrosal nerves on the sweet taste in rats. Physiol Behav 41:495–501

Luschei ES, Goldberg LJ (1981) Neural mechanisms of mandibular control: mastication and voluntary biting. In: Brooks V (ed) Handbook of physiology: motor control, vol 2. American Physiological Society, Bethesda, Maryland, pp 1237–1274

Malmo HP, Malmo RB, Weijnen JAWM (1986) Individual consistency and modifiability of lapping rates in rats: a new look at the variance–invariance question. Int J Psychophysiol 4:111–119

Marowitz LA, Halpern BP (1973) The effects of environmental constraints upon licking patterns. Physiol Behav 11:259–263

Mendelson J (1977) Airlicking and cold licking in rodents. In: Weijen JAWM, Mendelson J (eds) Drinking behavior: oral stimulation, reinforcement and preference. Plenum Press, New York, pp 157–197

Miller AJ (1982) Deglutition. Physiol Rev 62:129–184

Miller IJ, Spangler KM (1982) Taste bud distribution and innervation of the palate of the rat. Chem Senses 7:99–108

Mook DG (1990) Satiety, specifications and stop-rules: feeding as voluntary action. In: Sprague J, Epstein AN (eds) Progress in psychobiology and physiological psychology, vol 16. Academic Press, New York

Mook DG, Kenney NJ (1977) Taste modulation of fluid intake. In: Weijnen JAWM, Mendelson J (eds) Drinking behavior: oral stimulation, reinforcement and preference. Plenum Press, New York, pp 275–315

Mook DG, Wagner S (1988) Adjustment to intermittent access in rats drinking saccharin: I. Pauses in drinking. Physiol Behav 44:21–26

Mook DG, Wagner S (1989) Adjustment to intermittent access in rats drinking saccharin: II. Adjustment of lapping rate. Physiol Behav 45:299–305

Norgren R (1984) Central neural mechanisms of taste. In: Darian-Smith I (ed) Handbook of physiology– the nervous system, vol 3. American Physiological Society, Bethesda, MD, pp 1087–1128

Pearson KG (1985) Are there central pattern generators for walking and flight in insects? In: Barnes WJP, Gladden MH (eds) Feedback and motor control in invertebrates and vertebrates. Croom Helm, London

Rolls BJ, Rolls ET (1982) Thirst. Cambridge University Press, Cambridge

Smith JC, Wilson L, Krimm R, Merryday D (1987) A moment-to-moment comparison of sucrose and saccharin drinking in the rat. Chem Senses 12:99–112

Smith MF, Smith KU (1939) Thirst motivated activity and its extinction in the cat. J Gen Psychol 21: 89–98

Spector AC, Smith JC (1984) A detailed analysis of sucrose drinking in the rat. Physiol Behav 33: 127–136

Stellar E, Hill HH (1952) The rat's rate of drinking as a function of water deprivation. J Comp Physiol Psychol 45:96–102

Thexton AJ (1984) Jaw, tongue and hyoid movement: a question of synchrony. J Res Soc Med 77: 1010–1019

Thexton AJ, Crompton AW (1989) Effect of sensory input from the tongue on jaw movement in normal feeding in the opossum. J Exp Zool 250:233–243

Thexton AJ, McGarrick JD (1988) Tongue movement in the cat during lapping. Arch Oral Biol 33: 331–339

Thexton AJ, McGarrick JD (1989) Tongue movement in the cat during the intake of solid food. Arch Oral Biol 34:239–248

Thexton AJ, Hiiemae KM, Crompton AW (1980) Food consistency and bite size as regulators of jaw movement during feeding in the cat. J Neurophysiol 44:456–474

Travers JB, Norgren R (1986) Electromyographic analysis of the ingestion and rejection of sapid stimuli in the rat. Behav Neurosci 100:544–555

Turvey MT, Schmidt RC, Rosenblum LD (1989) 'Clock' and 'motor' components in absolute coordination of rhythmic movements. Neuroscience 33:1–10

Weijnen JAWM (1977) The recording of licking behavior. In: Weijnen JAWM, Mendelson J (eds) Drinking behavior: oral stimulation, reinforcement and preference. Plenum Press, New York, pp 93–114

Weijnen JAWM (1989) Lick sensors as tools in behavioral and neuroscience research. Physiol Behav 46:923–928

Weijnen JAWM, Wouters J, vanHest JMHH (1984) Interaction between licking and swallowing in the drinking rat. Brain Behav Evol 25:117–127

Weijs WA, Dantuma R (1975) Electromyography and mechanics of mastication in the albino rat. J Morphol 146:1–34

Wiesenfeld Z, Halpern BP, Tapper DN (1977) Licking behavior. Evidence for a hypoglossal oscillator. Science 196:1122–1124

Welzl H (1976) Attempt to modify rate and duration of licking in rats by operant conditioning. Behav Proc 1:319–326

Yamamoto T, Matsuo R, Fujiwara T, Kawamura Y (1982) EMG activities of masticatory muscles during licking in rats. Physiol Behav 29:149–154

Zeigler HP, Levitt P, Levine RR (1980) Eating in the pigeon (*Columba livia*): response topography, stereotypy and stimulus control. J Comp Physiol Psychol 94:783–794

Zeigler HP, Semba K, Jacquin MF (1984) Trigeminal reflexes and ingestive behavior in the rat. Behav Neurosci 98:1023–1038

Zeigler HP, Jacquin MF, Miller MG (1985) Trigeminal orosensation and ingestive behavior in the rat. In: Sprague JM, Epstein AN (eds) Progress in psychobiology and physiological psychology. Academic Press, New York, pp 65–196

Chapter 16

Drinking: Hindbrain Sensorimotor Neural Organization

J.B. Travers

Introduction

Numerous studies suggest that the patterned motor responses of drinking and swallowing are organized at the brainstem level (reviewed in Berntson and Micco 1976). Decerebration experiments further indicate that tactile and gustatory receptors can modify fluid intake using brainstem mechanisms, but that drinking to correct a water imbalance and spontaneous drinking both require an intact forebrain (Grill and Miselis 1981). Because only a fraction of an animal's behaviour is evident when the forebrain and hindbrain are disconnected, decerebrate preparations provide an experimental method for establishing a gross anatomical location for function. Careful behavioural testing reveals those sensory and motor capacities that are intact, thereby allowing anatomical and physiological studies to be interpreted within the context of a limited, but specific, behavioural repertoire. The first section of this chapter reviews the sensorimotor capacity of the decerebrate preparation with respect to ingestive behaviour. This is followed by several sections discussing brainstem pathways mediating these responses.

Ingestive Responses in the Decerebrate

The oromotor components of the ingestive consummatory response, including mastication, suckling, licking and swallowing can all be elicited in the decerebrate and, therefore, are organized in the brainstem (e.g. Berntson and Micco 1976). This consummatory response consists of a continuous sequence of rhythmic and ballistic oropharyngeal movements that propel the food or fluid through the mouth to the stomach (see Chapter 15). Experimental studies further indicate that each stage of the sequence, substrate identification, intraoral transport and swallowing, is initiated and guided by specific perioral and intraoral sensory receptors (Chapter 13 and 15). Behavioural testing of decerebrate preparations further indicates that some of the sensory systems that influence the ingestive consummatory response do so through brainstem pathways.

In particular, intraoral stimulation is effective in producing the appropriate ingestive response in decerebrate preparations. Essentially normal oromotor components of ingestion and rejection can be elicited in the decerebrate rat (and anencephalic human) using the appropriate gustatory stimuli, and rhythmic masticatory movements can be elicited in the rat by tactile stimulation of the hard palate (Van Willigen and

Weijs-Boot 1984). Ingestive responses that involve interoceptors can also be elicited in decerebrate preparations. Decerebrate rats are less active following intragastric infusions of either mash or glucose and reject orally presented sucrose following intragastric infusions of sucrose (Grill and Norgren 1978; Junquera et al. 1984). Behavioural indices of satiety involving both oromotor and spinal-motor systems are thus influenced by interoceptor pathways in the absence of forebrain connections.

Certainly not all the interoceptor and intraoral influences on ingestion can be ascribed to the brainstem. Substrates for forming a conditioned taste aversion are lacking in the hindbrain (Grill and Norgren 1978) and, more important for present purposes, so also is the ability of the decerebrate rat to respond to dehydration or electrolyte imbalance with increased water or sodium intake (Grill and Miselis 1981; Grill et al. 1986). Thus, a rat with only a hindbrain will drink if water is placed directly into its mouth, but will neither seek out water nor demonstrate an increased preference for intraoral infusions of water if dehydrated.

It is unclear from experimental studies in decerebrate rats whether perioral stimulation leads to the acquisition of food or water. Although trigeminal denervation studies clearly demonstrate the importance of an intact perioral region to initiate both drinking and eating in several experimental animals (reviewed in Zeigler et al. 1985), the effectiveness of perioral stimulation to produce appropriate ingestive responses in the decerebrate is both species and age dependent, and there are contradictory descriptions. Rhythmic suckling responses to perioral stimulation can clearly be elicited in decerebrate rat pups (Thexton and Griffiths 1979; Kornblith and Hall 1979); however, descriptions vary in the adult decerebrate rat. One study described both licking (rhythmic) responses to water or milk applied to the snout area or between the lips, and nibbling responses to solid objects making perioral contact (Woods 1964). Several other studies, however, specifically deny rhythmic oral responses to mechanical or fluid stimulation of the lip area in the adult decerebrate rat (Lovick 1972; Thexton and Griffiths 1979). Cats are organized somewhat differently; decerebrate kittens do not suckle normally but at weaning will respond to perioral stimulation with ingestive sequences (Bignall and Schramm 1974). The lack of responsiveness to perioral stimulation is not due to a general oromotor deficit since jaw opening occurs during spontaneous grooming and biting in both the decerebrate rat and cat (Berntson and Micco 1976).

Observations of maldeveloped human infants with either an anencephalic or hydrencephalic condition suggest that the human brainstem has a similar sensorimotor capacity. In particular, different gustatory stimuli can evoke a wide range of orofacial responses that include sucking, licking, salivation, swallowing, spitting and vomiting (Steiner 1973). Although the topography for human drinking (sucking) differs from that of rodents (licking/lapping: Chapter 15), the neural control of these responses in different species may have a similar central organization.

The remainder of this chapter will concentrate on each of three components of the ingestive consummatory response for which there is strong experimental evidence for a brainstem substrate. First, brainstem pathways involved in intraoral transport, i.e. the rhythmic production of licking and mastication are described. This is followed by a description of brainstem structures involved with swallowing. The next section describes possible pathways through which gustatory and intraoral tactile stimuli might influence licking and swallowing. The brainstem substrates for ingestive responses include the nucleus of the solitary tract, the trigeminal sensory complex, regions of the medullary and pontine reticular formation, the parabrachial nuclei, and the oromotor nuclei (trigeminal, facial, hypoglossal, ambiguous nuclei).

Nevertheless, experimental studies indicate that specific neural circuits contained within the brainstem are associated with each of these three sensorimotor components that constitute the consummatory aspect of drinking. The final section of this chapter suggests possible pathways through which forebrain structures that comprise the appetitive components of drinking might interface with brainstem systems.

Brainstem Rhythm Generator

Peripheral oral somatosensory receptors play a critical role in both drinking and mastication. Jaw opening, a prerequisite for both these behaviours, is compromised in the periorally deafferented rat and may reflect damage to both oromotor reflex and appetitive (motivational) pathways (Zeigler et al. 1985). Indeed, trigeminal afferent fibres have a relatively direct effect on trigeminal motoneurons. On the basis of latency measurements, a jaw-opening reflex initiated by either perioral or intraoral stimulation minimally requires two synapses (Mason et al. 1985). Primary afferent fibres innervating the oral regions synapse on second-order neurons in the trigeminal sensory complex which in turn project bilaterally to excite jaw opener motoneurons and inhibit jaw closers (Fig. 16.1a). An ipsilateral jaw-closing reflex can be elicited via muscle spindle afferents in jaw-closing muscles through a monosynaptic pathway (reviewed in Zeigler et al. 1985).

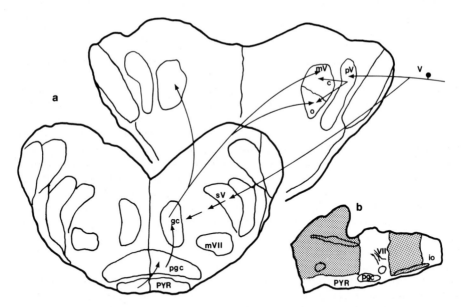

Fig. 16.1. a Schematic representation of pathways producing jaw-opening from either peripheral sensory input via the trigeminal nerve (V) or from central cortical stimulation in the guinea pig. The bilaterality of the peripherally induced jaw-opening reflex via interneurons in pV is not shown. **b** The minimal brainstem substrate necessary (non-shaded region) to elicit rhythmic oral movement from cortical stimulation. Abbreviations: c, jaw closer motoneuron pool in trigeminal motor nucleus; io, inferior olive; gc, nucleus gigantocellularis; mVII, facial nucleus; o, jaw opener motoneuron pool in trigeminal motor nucleus; mV, trigeminal motor nucleus; pgc, nucleus paragigantocellularis; pV, principal trigeminal nucleus; PYR, pyramidal tract; sV, spinal trigeminal nuclei; V, trigeminal nerve. (Based on Nozaki et al. 1986a.)

Rhythmic oromotor movements such as licking or mastication, however, are not simply composed of the alternating activation of these reflexes by peripheral stimuli. Rather, both mastication and licking are centrally programmed (Dellow and Lund 1971). Stimulation of the anterolateral orbital (masticatory) cortex in a paralysed preparation elicits rhythmic activity in the hypoglossal and trigeminal motor nerves. These data indicate that licking and mastication are not contingent on peripheral receptors, either for response initiation or for producing a co-ordinated motor pattern (Sumi 1970; Dellow and Lund 1971).

Reticular Formation Location

A central pattern generator (CPG) for rhythmic oromotor activity, assigned to the hindbrain on the basis of decerebration experiments, has been localized more precisely to the brainstem reticular formation (RF). Studies with guinea pigs transected the brainstem at different levels to determine the minimal substrate necessary for cortically evoked ororhythmic activity (Nozaki et al. 1986a; Chandler and Tal 1986). Both these studies concluded that the medial reticular formation (including nucleus gigantocellularis and nucleus paragigantocellularis) extending from the caudal pons (at the caudal level of the trigeminal motor nucleus) through the medulla to the level of the anterior inferior olive was necessary for rhythmic jaw opening from cortical stimulation. Both studies also indicated that activation of the CPG is primarily from the contralateral cortex but that the output of the CPG drives the trigeminal motor nuclei bilaterally (Fig. 16.1a). In the guinea pig, the medial reticular formation implicated in pattern generation receives direct projections from the orbital (masticatory) cortex (Nozaki et al. 1986b) and cells in the medial pontine and medullary RF in the awake behaving cat respond phasically during both chewing and lapping (Suzuki and Siegel 1985; Hiraba et al. 1988). Short latency activation to electrical stimulation and spike triggered averaging further suggested monosynaptic projections from the medial RF to jaw-closer or tongue protruder motoneurons (Suzuki and Siegel 1985; Hiraba et al. 1988). Previously it was demonstrated in the cat that cells in the medial RF activated by cortical stimulation projected either to the jaw-closing motoneuron pool in the motor trigeminal nucleus (mV) or to the jaw-opening motorneuron pool (Nozaki et al. 1983). The differential projection to jaw-opener and jaw-closer motoneurons from medial RF cells presumably mediates orbital cortex evoked inhibition in jaw-closer motoneurons simultaneous with excitation of jaw-opener motoneurons (Nakamura and Kubo 1978). An additional synapse in nucleus paragigantocellularis (PGC) may be interposed between the descending orbital cortex projection and nucleus gigantocellularis (GC) neurons (Nozaki et al. 1986b). A diagram of the pathways implicated in rhythmic jaw movement in the guinea pig is summarized in Fig. 16.1a.

It is not entirely clear that the neural circuit for rhythmic oral movements, derived from experiments in cat and guinea pig can be considered a general mammalian model. Although there is considerably less experimental work locating a brainstem CPG in rat, work to date indicates a somewhat different anatomical location within the reticular formation. For example, injections of horseradish peroxidase into either of two cortical zones in rat from which rhythmic jaw movements were elicited, labelled brainstem reticular cells including the supratrigeminal, intratrigeminal and juxtatrigeminal regions, areas quite far lateral to GC and PGC (Zhang and Sasamoto 1990). Label in the medial part of the brainstem RF was depicted further caudally at

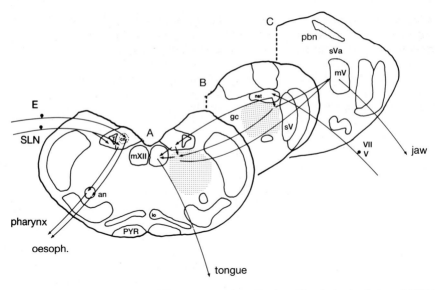

Fig. 16.2. Pathways mediating different stages of swallowing. Electrical stimulation of NST in site receiving soft palate afferent fibres via facial (VII) or trigeminal (V) nerves in section B produces motor activity in lingual and trigeminal muscles via polysynaptic intrasolitary and subjacent reticular formation pathways. At more caudal levels (A) superior laryngeal nerve and oesophageal afferent fibres influence pharyngeal and oesophageal motoneurons in the ambiguous nucleus via different NST subnuclei. Shaded areas depict medullary reticular formation also implicated in a masticatory rhythm generator in rat. Abbreviations: an, ambiguous nucleus; cn, central subdivision of NST; E, afferent fibres innervating oesophagus; gc, nucleus gigantocellularis; i, intermediate subdivision of NST; io, inferior olive; mV, motor trigeminal; mXII, hypoglossal nucleus; nst, nucleus of solitary tract; pbn, parabrachial nuclei; SLN, superior laryngeal nerve; sVa, supratrigeminal area; sV, spinal trigeminal nuclei; T, solitary tract; VII, facial nerve afferent fibres.

the level of the hypoglossal nucleus, corresponding to the shaded area in Fig. 16.2. Neurons responding phasically to cortically evoked rhythmic jaw movement in rat were also distributed laterally in the RF, including the lateral boundary of GC and further laterally in the parvicellular reticular formation (Moriyama 1987). This lateral region of the RF from which phasic oral related activity could be recorded corresponds to a region of premotor neurons that project to all the oromotor nuclei in rat (Travers and Norgren 1983) and is also depicted by the shading in Fig. 16.2.

Licking Versus Chewing

Although considerable evidence points to a brainstem reticular formation substrate necessary for generating rhythmic oral activity, it is not clear to what extent drinking and mastication share this central pattern generator. If those species that lick in order to drink are considered, a common brainstem CPG for both mastication and drinking can be proposed.

Licking and mastication both require the co-ordinated action of lingual and masticatory muscles; however, licking is typically faster than chewing and the degree of vertical jaw opening is smaller during licking (Thexton and Crompton 1989). The degree of jaw opening may be influenced both by the peripheral oral stimulus initiating rhythmic oral activity and by a central pattern generator. For example, during rhythmic oromotor activity elicited by cortical stimulation, the excitability of the jaw opening reflex (JOR) is variable. During the jaw closing phase of rhythmic chewing, the JOR is reduced in amplitude to low threshold afferent stimulation but reflex excitability is unaffected in response to high threshold stimulation (Lund 1984). The implication is that high-threshold (nociceptive) stimuli can interrupt an ingestive sequence through reflex pathways but that low-threshold innocuous stimuli do not interfere with the rhythm produced from the central pattern generator.

Thexton and McGarrick (1987) observed that the excitability of the JOR was maintained during "chewing" movements elicited with peripheral stimulation but that the JOR was reduced during fluid-induced licking. It would appear then, that drinking or mastication may be produced, in part, by gating the degree of jaw-opening through some interaction between the peripheral stimulus and a central pattern generator. In fact, such an interaction is evident in brainstem neurons. Cortically evoked rhythmic movement modulates the excitability of neurons in the trigeminal sensory nucleus. These second-order sensory neurons are part of the afferent limb of the JOR (see above) (Olsson et al. 1986). Thus, one function of the CPG may be to gate oral reflex pathways in addition to driving lingual and masticatory motoneurons.

In addition to the problem of identifying neural mechanisms that produce either a masticatory or licking type response appropriate to the stimulus, there is the additional problem of relating this neural mechanism to swallowing. There is considerable evidence for a CPG for swallowing anatomically distinct from a CPG for mastication. Moreover, swallowing functionally interrupts the CPG for licking (Chapter 15). To the extent that human drinking involves a sucking/swallowing movement that occurs together in one cycle (Chapter 15), it is unclear whether to identify a neural mechanism for human drinking exclusively with a substrate for swallowing or whether to identify it with a substrate for rhythmic licking. Nevertheless, because swallowing is an integral component of drinking in all animals, and has a well-defined substrate in experimental animals, it is covered separately below.

Swallowing

Deglutition, like licking and mastication, can be evoked by appropriate cortical electrical stimulation, is not contingent on peripheral stimulation (centrally programmed), and can be elicited in decerebrate preparations (brainstem substrate) (Miller 1952). In addition both licking and swallowing require co-ordinated trigeminal and lingual movement. Nevertheless, the central pattern generator or "centre" for swallowing located in the caudal NST (Miller 1982; Jean 1984) appears distinct from brainstem circuits for rhythmic oral movement such as licking.

Nucleus of the Solitary Tract

Intraoral sites from which swallowing can most easily be evoked include the posterior tongue, fauces and epiglottis (Miller 1982). Afferent fibres of the superior laryngeal (SLN) and glossopharyngeal (IX) nerves from sites in the posterior oral cavity terminate within the caudal NST (e.g. Norgren 1985; Altschuler et al. 1989). Cells at this level of the NST (see Fig. 16.2 A) respond to stimuli that elicit swallows, including mechanical, gustatory and water stimuli applied to the posterior tongue, soft palate and epiglottis (Storey 1968; Sweazey and Bradley 1988, 1989). In addition, some caudal NST neurons are particularly sensitive to a moving tactile stimulus, as might occur during the accumulation and transport of a bolus (Sweazey and Bradley 1989). Although water is a particularly effective fluid stimulus for laryngeal responses, the location of these receptors suggests a role in protective airway reflexes rather than ingestion (Storey 1968; Sweazey and Bradley 1988).

The caudal NST is also implicated in programming or sequencing the motor action of swallowing (Miller 1982; Jean 1984). In studies of swallowing evoked by trains of electrical stimuli applied to the SLN, cells in the caudal NST respond with greater intensity when a swallow occurs, rather than simply following each electrical pulse. The latency of activity of these cells relative to the onset of swallowing, corresponds to the buccal, pharyngeal and oesophageal stages of swallowing; however, these cells do not require a muscle contraction for their activity (Jean 1984; Kessler and Jean 1985).

Although swallowing appears as a programmed response, the stimulus properties of the bolus may influence the actual motor pattern of swallowing (Jean 1984). In monkey, trigeminal jaw-closer muscles are more active during the swallowing of solids than during swallowing of either saliva or water (McNamara and Moyers 1973). Jean (1984), moreover, has shown that distension of the pharynx excites cells in the NST active during the pharyngeal phase of swallowing and simultaneously inhibits cells driving the oesophageal phase. NST cells active in the oesophageal phase of swallowing are likewise excited by a bolus in the oesophagus, indicating that the sensory properties of the bolus influence peristalsis. Overall, less time is required for fluids to reach the stomach than for solids (Miller 1982).

Anatomical and pharmacological evidence further suggests separate NST mechanisms for co-ordinating the pharyngeal and oesophageal stages of deglutition

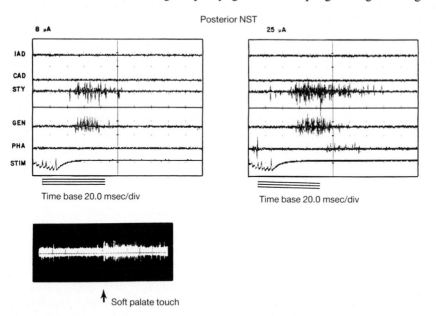

Fig. 16.3. Microstimulation of NST in location responsive to soft palate (*lower left*) evokes long latency response in lingual musculature (*upper left*). At same location and slightly higher current, a pharyngeal contraction is recruited (*right*). STY, styloglossus; GEN, genioglossus; PHA, pharyngeal. (From Travers et al. (1988).)

(Miller 1982). Lesions placed in the caudal NST can abolish the oesophageal component of swallowing but leave the buccopharyngeal stage intact. Several recent studies using micropneumophoresis of different excitatory amino acids indicated that the subnucleus ventralis and intermedialis of the caudal NST control the pharyngeal stage of swallowing and the subnucleus centralis controls the oesophageal component (Bieger 1984; Hashim and Bieger 1989). It is possible that the lingual component of the buccal phase is under separate NST control as well. Electro-myographic recordings following electrical stimulation of the NST demonstrated that the lingual component of swallowing can be elicited independently of pharyngeal activity (Fig. 16.3) (Travers et al. 1988).

Direct projections from the caudal NST to oromotor nuclei potentially control some of the components of deglutition. For example there are direct projections from the caudal NST that receive superior laryngeal nerve (SLN) and IX afferent fibres (e.g. Norgren 1985; Altschuler et al. 1989) to oesophageal motoneurons in the NA (Bieger 1984). Direct projections to mXII from the caudal NST have also been described (Borke et al. 1983; Travers and Norgren 1983). Nevertheless, both anatom-ical and physiological studies indicated a lack of direct projections to trigeminal motor neurons active during swallowing (Travers and Norgren 1983; Amri et al. 1984).

Interneurons: Medullary Reticular Formation

Neurons in the ventrolateral reticular formation adjacent to the nucleus ambiguous (NA) and lateral to the hypoglossal nucleus (mXII) constitute a second set of interneurons active during swallowing (Jean 1984; Kessler and Jean 1985; Car and Amri 1987; Amri and Car 1988; Travers and Jackson 1989). Cells in and adjacent to NA respond later during swallowing than cells in the dorsal group, and cells in the ventral group project to both the trigeminal and hypoglossal motor nuclei (Amri et al. 1984; Amri and Car 1988). Cells lateral to mXII in the RF that are active during the buccopharyngeal stage of swallowing respond later during a swallow than cells in the overlying solitary nucleus (Kessler and Jean 1985) but respond earlier during a swallow than cells in the adjacent mXII. This suggests an interneuron role for cells in this part of the RF between the "sensory" NST and the motor mXII (Car and Amri 1987). Neuroanatomical tracing studies further indicated that this region of the RF projects to all of the oromotor nuclei producing swallowing (Travers and Norgren 1983). Chronic recording in the rat from the RF lateral to mXII reveals cells with both rhythmic (licking) and swallowing related activity in response to fluid stimulation (Travers and Jackson 1989; Fig. 16.4). Thus, cells in the RF lateral to mXII and ventral to the NST are also likely to be involved in co-ordinating swallowing.

Interneurons: Pons

Neurons in the supratrigeminal nucleus represent another pool of interneurons in the brainstem swallowing pathway (Jean et al. 1975; Car and Amri 1982). Cells in the supratrigeminal area active during SLN-evoked swallowing do not respond when muscle activity is paralysed (Car and Amri 1982). Unlike their medullary counterparts, these reticular neurons are not part of the central programming pathway, but rather, represent sensory feedback from active muscles. Cells in the supratrigeminal area receive monosynaptic afferent input from proprioceptors in the jaw-closer muscles (e.g. Luschei 1987).

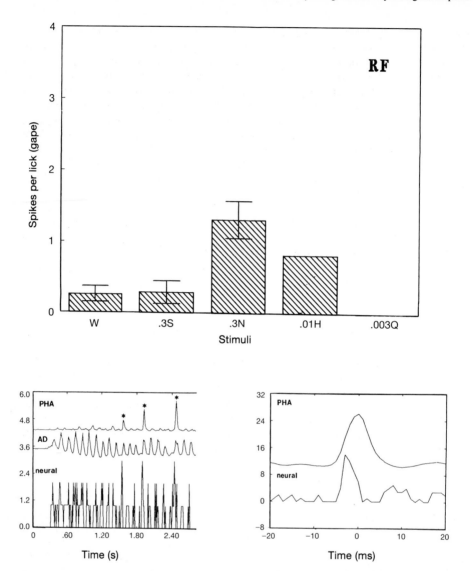

Fig. 16.4. Response of neuron in the reticular formation lateral to mXII to 0.3м-NaCl (*upper*) also shows rhythmic responses during licking and increased activity during swallows (asterisks *lower left*). Averaged unit activity during pharyngeal component of swallowing is also shown (*lower right*). W, water; S, sucrose; N, NaCl; H, hydrochloric acid; Q, QHCl; PHA, integrated pharyngeal EMG; AD, integrated anterior digastric EMG; neural, unit histogram (*left*) or accumulated total (*right*). (From Travers and Jackson (1989).)

Motor Nuclei

Although the evidence from experimental animals suggests separate neural brainstem mechanisms for generating licking and swallowing, both responses use many of the same muscles. Single unit recording from motoneurons provides evidence of how these separate responses are generated. Studies in the hypoglossal motor nucleus, for example, show that swallowing pre-empts or disrupts ongoing rhythmic licking. In both acute and chronic preparations, the period between rhythmic (licking) bursts in hypoglossal neurons increases when a swallow occurs (Sumi 1970; Travers 1989). Only a subset of mXII cells displaying a bursting pattern during licking increased or decreased their activity during a swallow. Virtually all mXII cells, however, discharged in a highly synchronous fashion during a swallow regardless of overall increases or decreases in activity. Cells in mXII classified as lingual protruders were active prior to the peak of the pharyngeal activity, cells classified as lingual retractors were active following the peak of the pharyngeal activity (Travers 1989). Thus, the "interruption" of licking during a swallow is a dynamic change in both the magnitude and timing of mXII neurons.

Several possible sensorimotor pathways involved in different stages of swallowing are depicted in Fig. 16.2. Electrical or pharmacological stimulation of different sites in the NST that receive afferent fibres from the VIIth, IXth and Xth nerves differentially influence lingual (and trigeminal), pharyngeal and oesophageal components of swallowing.

Central Representation of Intraoral Receptors

As discussed earlier, decerebrate preparations demonstrate varying degrees of sensory control over brainstem consumatory responses. The effectiveness of perioral stimulation in decerebrates is particularly variable, and may reflect an appetitive function of the trigeminal system, therefore requiring a forebrain substrate for its full expression. Intraoral stimulation including both gustatory and tactile factors, however, is particularly potent in decerebrates. It ought, then, to be possible to connect intraoral orosensory systems of the brainstem with oromotor systems.

Nucleus of the Solitary Tract

Fibres innervating gustatory receptors of the tongue, palate, pharynx and larynx terminate centrally in the nucleus of the solitary tract (NST). Numerous studies in a variety of species confirm an afferent organization in which anteriorly located peripheral structures have a central representation rostral in the nucleus and increasingly caudal oropharyngeal structures have caudal representations (reviewed in Norgren 1985). Mechanoreceptors from these same intraoral structures terminate in both the sensory trigeminal complex and the NST (e.g. Norgren 1985; Altschuler et al. 1989). The central representation of mechanoreceptors in the NST follows a similar distribution to gustatory receptors. Although there is some convergence between gustatory and somatosensory receptors in the NST, many cells are modality specific and there is evidence of spatial segregation of the two modalities within the nucleus (Travers and Norgren 1988).

Receptive Field Properties

Mechanical, gustatory and thermal stimuli all influence licking behaviour (Chapter 15). Although each of these modalities is represented within the NST, the role of

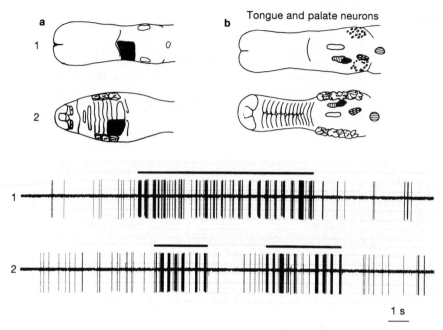

Fig. 16.5. *Top:* Receptive fields (**a**) and associated responses for neuron in rat NST to mechanical stimulation of either the tongue (**a**1) or the appositional hard palate (**a**2) (SP Travers, personal communication). Neurons further caudal in the sheep NST also have receptive fields (**b**) on the tongue and appositional palate. (From Sweazey and Bradley (1989).) *Bottom:* neural responses to stroking the tongue (1) or palate (2) in areas designated in **a**.

specific sensory neurons in guiding ingestive behaviour is unknown (Norgren 1985; Zeigler et al. 1985). Nevertheless, the response properties of some NST neurons correlate well with stimuli that are particularly effective in modifying licking behaviour. For example, licking responses to sucrose or low to midrange concentrations of sodium chloride are very similar (Travers and Norgren 1986). Both these gustatory stimuli produce longer ingestive sequences of licking and swallowing than does a similar volume of water alone. A class of cells in the rat NST, specifically sensitive to stimulation with sucrose applied to the palate, also respond to sodium chloride applied to the anterior tongue (Travers et al. 1986). It has not yet been established, however, that these cells play a particular role in licking behaviour. Other cells in the NST with combined gustatory and thermal sensitivity may also have a specific role in generating licking responses (Ogawa et al. 1988).

Licking is also influenced by the location of an intraoral stimulus. The hard palate, in particular, is a low threshold site for eliciting ororhythmic responses (Van Willigen and Weijs-Boot 1984). Significant proportions of cells in the NST respond to mechanical stimulation of both the tongue and that portion of the hard or soft palate in apposition to the tongue (S. Travers, personal communication; Fig. 16.5a; Sweazey and Bradley 1989; Fig. 16.5b) and mechanoreceptive cells with convergent input from the posterior tongue and soft palate are anatomically positioned in the caudal NST to provide a central substrate for triggering swallowing (Sweazey and Bradley 1989).

Efferent Projections of Anterior NST

The precise pathways through which the anterior, gustatory zone of the NST influences oromotor nuclei are unknown. An ascending sensory pathway from the NST to the parabrachial nuclei (Norgen and Leonard 1973) originates predominantly from cells dorsal within the nucleus (Whitehead and Savoy 1987; Travers 1988), corresponding to the central subdivision of the anterior NST (Whitehead 1988). A second population of cells located more ventrally within the NST have local medullary projections coursing both caudally within the NST and ventrolaterally through the reticular formation in the vicinity of nucleus ambiguous (NA), the facial nucleus and the hypoglossal nucleus (mXII) (Norgren 1978; Travers 1988). These local projections from the gustatory zones of the NST do not overlap extensively with the "minimal" central pattern generator substrate defined by knife-cut experiments in guinea pigs (Fig. 16.1b) but do overlap with brainstem projections originating from cortical sites in which ororhythmic responses were evoked in rat. Thus, the RF lateral to mXII receives descending projections from the rat masticatory area (Zhang and Sasamoto 1990), local projections from the anterior (gustatory) NST (Norgren 1978; Travers 1988), and projects to all the oromotor nuclei (Travers and Norgren 1983). It has recently been demonstrated that cells in the RF lateral to the hypoglossal nucleus have both gustatory responsiveness and motor-related responses associated with both licking and swallowing (Travers and Jackson 1989; Fig. 16.4).

Parabrachial Nuclei

Neurons in the parabrachial nuclei (PBN) have many of the same characteristics as NST cells. For example, cells in the medial zone of the lateral parabrachial nuclei that receive projections from the anterior (gustatory) NST have appositional receptive fields (palate and tongue) and multimodal (gustatory and mechanoreceptive) sensitivities (Hayama et al. 1987). In addition, cells in the PBN receive convergent inputs from both gustatory and hepatic vagal afferents (Hermann and Rogers 1985). PBN cells with convergent inputs from both visceral and oral receptors could mediate brainstem compensatory ingestive responses based on metabolic signals (Hermann and Rogers 1985; Grill 1986). Evidence for satiety-induced responses, for example, can be obtained from the decerebrate preparation. Descending projections from the PBN to medullary structures including the ventrolateral medulla (Saper and Loewy 1980) further indicate how cells in the PBN could form part of the local brainstem circuitry involved in ingestive behaviour.

Sensory Trigeminal Complex

Intraoral mechanoreceptor representation is dorsal and medial in both the principal and spinal trigeminal nuclei (Kruger and Michel 1962; Sessle and Greenwood 1976). Unlike cells found in the NST, cells in the spinal trigeminal complex respond to stimulation of the tooth pulp and periodontal ligament in addition to the tongue, palate, larynx and pharynx (Kruger and Michel 1962; Sessle and Greenwood 1976). In the principal trigeminal sensory nucleus, cells with intraoral receptive fields on the posterior tongue and palate show less convergence from intraoral receptive fields than cells in the NST (Sweazey and Bradley 1987). A detailed comparison of intraoral mechanoreceptor characteristics between the NST and trigeminal sensory complex is lacking; however, trigeminal neurons project to the contralateral thalamus, while NST neurons reach the ipsilateral thalamus via the PBN (Norgren 1985). Local projections from the trigeminal sensory complex reach the medial RF implicated in pattern generation through short axon pathways (Fig. 16.1a) (Scheibel and Scheibel 1958). These trigeminoreticular projections may mediate medial reticular formation responses to intraoral sensory stimulation, both from periodontal and intraoral mucosal receptors (Ermirio et al. 1989). The small intraoral receptive fields for these neurons suggest they might provide sensory cues associated with the location of food in the mouth (Ermirio et al. 1989).

In contrast to cells in the principal trigeminal nucleus, cells in the spinal trigeminal complex (pars caudalis), showed a great deal of convergence, with 50% of the cells responding to stimulation of two or more divisions of the trigeminal nerve (Amano et al. 1986). The predominance of high-threshold mechanoreceptors (nociceptors) in the nucleus caudalis, however, suggests that convergence may be a substrate for referred pain rather than intraoral guidance during feeding (Amano et al. 1986).

In summary, the anterior NST and gustatory responsive PBN provide a brainstem substrate for the identification of gustatory stimuli in the decerebrate preparation. The dual somatosensory representation of intraoral structures in the NST and trigeminal sensory complex requires further investigation as to their relative roles in initiating rhythmic oromotor behaviour. Microstimulation experiments in the NST and trigeminal sensory nuclei do not produce rhythmic licking responses in the acute preparation (Travers et al. 1988). Moreover, chronic recording studies in the NST of the awake behaving monkey (Scott et al. 1986) and the decerebrate rat (e.g. Mark et al. 1988) have not described rhythmic responses indicative of a CPG, unlike the rhythmic responses obtained in the RF (see p. 261). In contrast, more caudal regions of the NST, at sites receiving afferent input from the glossopharyngeal and vagus nerves are more reflexogenic and co-ordinated swallowing responses can be elicited by microstimulating these sites (Miller 1982). Rhythmic responses in the principal trigeminal sensory nucleus probably reflect contact with the substrate or proprioceptive feedback, rather than the activity of a CPG (Lund 1984). In addition to the influence of direct corticobulbar projections and brainstem sensory nuclei on rhythmic oral activity, several sites within the midbrain have been proposed as an interface through which forebrain structures might influence a brainstem CPG.

Midbrain Projections

A number of studies implicate various midbrain structures in drinking behaviour (for review see White 1986). Hypothalamic sites in which angiotensin II (Ang II) injections were effective in eliciting drinking, project to the ventral tegmental area (VTA) and midbrain central grey (CG) (Swanson et al. 1978). Previous studies had demonstrated osmoresponsive neurons in the central grey (Malmo 1976) and drinking in response to infusions of Ang II injected into the central grey (Swanson and Sharpe 1973). In addition, lesions in the ventral tegmental area disrupted drinking elicited by infusion of Ang II into the preoptic area (Kucharczyk and Mogenson 1975). The question arises as to whether these hypothalamic–midbrain projections constitute a descending control pathway to a brainstem CPG.

Microstimulation of either the VTA or CG were not particularly effective midbrain sites for influencing rhythmic oral activity generated by cortical microstimulation (Chandler and Tal 1987). Rhythmic oral activity elicited from stimulating the ventral midbrain (caudal to the VTA) apparently results from current spread to the adjacent pyramidal tract (Tal 1987). Other sites in the midbrain distant from the pyramidal tract were effective in modulating the brainstem CPG but these sites do not overlap with the descending hypothalamic projections to the CG and VTA (Swanson et al. 1978). For example, microstimulation of sites ventral and lateral to the CG disrupted both the amplitude and timing of cortically evoked rhythmic oral activity or inhibited it entirely (Chandler and Tal 1987). In addition, midbrain sites effective in modulating rhythmic oral activity project to those parts of the medullary and pontine reticular formation necessary to the CPG (Chandler and Tal 1987).

Thus there appears to be a mismatch between midbrain sites implicated in drinking behaviour having hypothalamic connections and midbrain sites effective in modulating ororhythmic activity with descending projections to the pons and medulla. Although descending projections from the CG to the medulla have been proposed, for example CG projections to the nucleus raphe magnus, these projections have typically been associated with nociceptive function (e.g. Dostrovsky et al. 1982). More recent studies demonstrating CG projections to the subretrofacial area suggest a CG role in visceral vascular efferent pathways (Carrive et al. 1989).

Recording, lesion and pharmacological studies also provide compelling evidence for a vital role for the substantia nigra (SN) in drinking behaviour (reviewed in White 1986). In general, "The function of the dopaminergic NSB (nigrostriatal-bundle) neurons in this scheme is facilitatory. This means that the function of these neurons is unrelated to the stimuli or responses that produce the observed behaviour" (White 1986, p. 23). A non-specific role for the SN is evident in chronic recording studies that show cell activity, either increasing (rarely) or decreasing (more often) during drinking, that is not phasically related to the ongoing motor activity (e.g. Joseph et al. 1985). The facilitatory effect of dopamine on rhythmic oral activity is also evident in studies showing that apomorphine injected i.v. in anaesthetized guinea pigs induces rhythmic oral activity with similar motor patterns to drinking (Gerstner et al. 1989). The question arises, however, as it did with a hypothalamic–VTA/CG pathway, as to the anatomical substrate between the midbrain pathway and a brainstem CPG for drinking. Similar to the VTA and the CG, the SN was not a particularly low-threshold site for modifying oral rhythmic activity using electrical micro-stimulation (Chandler and Tal 1987). Projections from the SN to the superior colliculus (Beckstead et al. 1979) may provide an indirect pathway to the brainstem RF implicated in the CPG. Other studies have demonstrated direct SN projections to the medullary RF (Schneider et al. 1985; Chronister et al. 1988). It has also been proposed that the substantia nigra might influence oromotor nuclei indirectly via the PBN. Schneider (1986) demonstrated in the cat that the SN projects to the PBN and that the PBN projects both to the sensory trigeminal complex and to areas adjacent to the oral motor nuclei. The existence of at least two independent forebrain pathways with influence over a brainstem CPG was shown by superior colliculus lesions that disrupted apomorphine-induced rhythmic oral movement but left cortically induced ororhythmic activity intact (Chandler and Goldberg 1984).

Conclusions

The intent of this chapter was to focus on local brainstem pathways through which perioral and intraoral sensory receptors control the motor components of ingestion. These sensorimotor pathways include neurons in the NST, PBN, trigeminal nuclei, reticular formation and all the oromotor nuclei. The location and function of interneurons in these pathways is, perhaps, best understood for swallowing. Despite the variety of nomenclatures for divisions of the NST (Whitehead 1988), it is increasingly apparent that specific regions of the NST control specific components of swallowing. It is less clear how subdivisions further rostral in the NST might interact with a central pattern generator for licking and mastication. It is equally unclear through what pathways the spinal trigeminal complex and parabrachial nuclei influence these behaviours.

Hindbrain nuclei including the NST and PBN are targets for osmo- and sodium interoceptors via vagal, spinal and circumventricular pathways that function to control fluid balance (e.g. Norgren 1985). Nevertheless, common manipulations that influence water intake in normal animals have no effect on ingestion in chronically decerebrate preparations (Grill and Miselis 1981). This implies that the critical signals that govern regulatory drinking require forebrain participation. As discussed in Chapters 5 and 7, neurons in the ventral forebrain respond to osmotic challenge and hormones regulating fluid balance. Neural mechanisms for producing ingestion responses and for switching between ingestion and rejection responses, however, exist in the brainstem. For instance, after a gavage feeding, chronic decerebrate rats reject intraoral infusions of sucrose sooner than when they are food-deprived (Grill and Norgren 1978). Thus, the hindbrain can alter the ingestive behaviour elicited by an oral stimulus as a function of internal state. It is at least possible, that in normal animals, the forebrain is required to detect fluid deficits, but that the sensorimotor circuits of the hindbrain are the final common path, both for producing ingestion and rejection behaviour, and for switching from one behaviour to the other, regardless of the stimulus that triggers the change. Hindbrain structures including the NST, PBN and the reticular formation are targets for descending limbic pathways implicated in fluid regulation (Holstege 1987). A major task is to understand how these descending interoceptive pathways interface with the brainstem neural apparatus organizing the consummatory response of drinking.

Acknowledgements: I wish to thank Drs Mark Whitehead and Keith Alley for helpful comments during the preparation of this manuscript. I would especially like to thank Dr Susan Travers for contributing to Figure 16.5. This work was supported by NS 24889.

References

Altschuler SM, Bao X, Bieger D, Hopkins DA, Miselis RR (1989) Viscerotopic representation of the upper alimentary tract in the rat: sensory ganglia and nuclei of the solitary and spinal trigeminal tracts. J Comp Neurol 283:248–268

Amano N, Hu JW, Sessle BJ (1986) Responses of neurons in feline trigeminal subnucleus caudalis (medullary dorsal horn) to cutaneous, intraoral, and muscle afferent stimuli. J Neurophysiol 55: 227–243

Amri M, Car A (1988) Projections from the medullary swallowing center to the hypoglossal motor nucleus: a neuroanatomical and electrophysiological study in sheep. Brain Res 441:119–126

Amri M, Car A, Jean A (1984) Medullary control of the pontine swallowing neurones in sheep. Exp Brain Res 55:105–110

Beckstead RM, Domesick VB, Nauta WJH (1979) Efferent connections of the substantia nigra and ventral tegmental area in the rat. Brain Res 175:191–217

Berntson GG, Micco DJ (1976) Organization of brainstem behavioral systems. Brain Res Bull 1:471–483

Bieger D (1984) Muscarinic activation of rhombencephalic neurones controlling oesophageal peristalsis in the rat. Neuropharmacology 23:1451–1464

Bignall KE, Schramm L (1974) Behavior of chronically decerebrated kittens. Exp Neurol 42:519–531

Borke RC, Nau ME, Ringler RL Jr (1983) Brain stem afferents of hypoglossal neurons in the rat. Brain Res 269:47–55

Car A, Amri M (1982) Etude des neurones deglutiteurs pontiques chez la brebis: I. Activite et localisation. Exp Brain Res 48:345–354

Car A, Amri M (1987) Activity of neurons located in the region of the hypoglossal motor nucleus during swallowing in sheep. Exp Brain Res 69:175–182

Carrive P, Bandler R, Dampney RAL (1989) Viscerotopic control of regional vascular beds by discrete groups of neurons within the midbrain periaqueductal gray. Brain Res 493:385–390

Chandler SH, Goldberg LJ (1984) Differentiation of the neural pathways mediating cortically induced and dopaminergic activation of the central pattern generator (CPG) for rhythmical jaw movements in the anesthetized guinea pig. Brain Res 323:297–301

Chandler SH, Tal M (1986) The effects of brain stem transections on the neuronal networks responsible for rhythmical jaw muscle activity in the guinea pig. J Neurosci 6:1831–1842

Chandler SH, Tal M (1987) Brain-stem perturbations during cortically evoked rhythmical jaw movements: effects of activation of brain-stem loci on jaw muscle cycle characteristics. J Neurosci 7:463–472

Chronister RB, Walding JS, Aldes LD, Marco LA (1988) Interconnections between substantia nigra reticulata and medullary reticular formation. Brain Res Bull 21:313–317

Dellow PG, Lund JP (1971) Evidence for central timing of rhythmical mastication. J Physiol 215:1–13

Dostrovsky JO, Hu JW, Sessle BJ, Sumino R. (1982) Stimulation sites in periaqueductal gray, nucleus raphe magnus and adjacent regions effective in suppressing oral-facial reflexes. Brain Res 252:287–297

Ermirio R, Guggeri P, Blanchi D, Cogo CE, Bergaglio M (1989) Effects of intraoral mechanoreceptor stimulation on reticular formation neurones in the rabbit. Arch Ital Biol 127:1–11

Gerstner GE, Goldberg LJ, De Bruyne K (1989) Angiotensin II-induced rhythmic jaw movements in the ketamine-anesthetized guinea pig. Brain Res 478:233–240

Grill HJ (1986) Caudal brainstem contributions to the integrated neural control of energy homeostasis. In: Ritter RC (ed) Feeding behavior: neural and hormonal controls. Academic Press, New York, pp 103–129

Grill HJ, Miselis RR (1981) Lack of ingestive compensation to osmotic stimuli in chronic decerebrate rats. Am J Physiol 240:81–86

Grill HJ, Norgren R (1978) Chronically decerebrate rats demonstrate satiation but not bait shyness. Science 201:267–269

Grill HJ, Schulkin J, Flynn FW (1986) Sodium homeostasis in chronic decerebrate rats. Behav Neurosci 100:536–543

Hashim MA, Bieger D (1989) Excitatory amino acid receptor-mediated activation of solitarial deglutitive loci. Neuropharmacology 28:913–921

Hayama T, Ito S, Ogawa H (1987) Receptive field properties of the parabrachio-thalamic taste and mechanoceptive neurons in rats. Exp Brain Res 68:458–465

Hermann GE, Rogers RC (1985) Convergence of vagal and gustatory afferent input within the parabrachial nucleus of the rat. J Auton Nerv Syst 13:1–17

Hiraba K, Taira M, Sahara Y, Nakamura Y (1988) Single-unit activity in bulbar reticular formation during food ingestion in chronic cats. J Neurophysiol 60:1333–1349

Holstege G (1987) Some anatomical observations on the projections from the hypothalamus to brainstem and spinal cord: an HRP and autoradiographic tracing study in the cat. J Comp Neurol 260:98–126

Jean A (1984) Control of the central swallowing program by inputs from the peripheral receptors. A review. J Auton Nerv Syst 10:225–233

Jean A, Car A, Roman C (1975) Comparison of activity in pontine versus medullary neurones during swallowing. Exp Brain Res 22:211–220

Joseph JP, Boussaoud D, Biguer B (1985) Activity of neurons in the cat substantia nigra pars reticulata during drinking. Exp Brain Res 60:375–379

Junquera J, Lanzagorta-Sanchez G, Mejia-Perez BE, Racotta R (1984) Motor activity in decerebrate rats: spontaneous and nutrient-induced changes. Am J Physiol 247:R945–R952

Kessler JP, Jean A (1985) Identification of the medullary swallowing regions in the rat. Exp Brain Res 57:256–263

Kornblith CL, Hall WG (1979) Brain transections selectively alter ingestion and behavioral activation in neonatal rats. J Comp Physiol Psychol 93:1109–1117

Kruger L, Michel F (1962) A single neuron analysis of buccal cavity representation in the sensory trigeminal complex of the cat. Arch Oral Biol 7:491–503

Kucharczyk J, Mogenson GJ (1975) Separate lateral hypothalamic pathways for extracellular and intracellular thirst. Am J Physiol 228:295–301

Lovick TA (1972) The behavioural repertoire of precollicular decerebrate rats. J Physiol (Lond) 226: 4P–6P

Lund JP (1984) Sensorimotor integration in the control of mastication. In: Klienberg I, Sessle B (eds) Oro-facial pain and neuromuscular dysfunction: mechanisms and clinical correlates, vol 52. Advances in biosciences. Pergamon Press, Oxford, pp 51–65

Luschei ES (1987) Central projections of the mesencephalic nucleus of the fifth nerve: an autoradiographic study. J Comp Neurol 263:137–145

Malmo RB (1976) Osmosensitive neurons in the rat's dorsal midbrain. Brain Res 105:105–120

Mark GP, Scott TR, Chang F-C T, Grill HJ (1988) Taste responses in the nucleus tractus solitarius of the chronic decerebrate rat. Brain Res 443:137–148

Mason P, Strassman A, Maciewicz R (1985) Is the jaw-opening reflex a valid model of pain. Brain Res Rev 10:137–146

McNamara JA Jr, Moyers RE (1973) Electromyography of the oral phase of deglutition in the rhesus monkey (Macaca mulatta). Arch Oral Biol 18:995–1002

Miller AJ (1982) Deglutition. Physiol Rev 62:129–184

Moriyama Y (1987) Rhythmical jaw movements and lateral pontomedullary reticular neurons in rats. Comp Biochem Physiol 86A:7–14

Nakamura Y, Kubo Y (1978) Masticatory rhythm in intracellular potential of trigeminal motoneurons induced by stimulation of orbital cortex and amygdala in cats. Brain Res 148:504–509

Norgren R (1978) Projections from the nucleus of the solitary tract in the rat. Neuroscience 3:207–218

Norgren R (1985) Taste and the autonomic nervous system. Chem Senses 10:143–161

Norgren R, Leonard CM (1973) Ascending central gustatory pathways. J Comp Neurol 150:217–238

Nozaki S, Enomoto S, Nakamura Y (1983) Identification and input–output properties of bulbar reticular neurons involved in the cerebral cortical control of trigeminal motoneurons in cats. Exp Brain Res 49:363–372

Nozaki S, Iriki A, Nakamura Y (1986a) Localization of central rhythm generator involved in cortically induced rhythmical masticatory jaw-opening movement in the guinea pig. J Neurophysiol 55: 806–825

Nozaki S, Iriki A, Nakamura Y (1986b) Role of corticobulbar projection neurons in cortically induced rhythmical masticatory jaw-opening in the guinea pig. J Neurophysiol 55:826–845

Ogawa H, Hayama T, Yamashita Y (1988) Thermal sensitivity of neurons in a rostral part of the rat solitary tract nucleus. Brain Res 454:321–331

Olsson KA, Sasamoto K, Lund JP (1986) Modulation of transmission in rostral trigeminal sensory nuclei during chewing. J Neurophysiol 55:56–75

Saper CB, Loewy AD (1980) Efferent connections of the parabrachial nucleus in the rat. Brain Res 197:291–317

Scheibel ME, Scheibel AB (1958) Structural substrates for integrative patterns in the brain stem reticular core. In: Jasper HH et al. (eds) Reticular formation of the brain. Little Brown, Boston, pp 31–55

Schneider JS (1986) Interactions between the basal ganglia, the pontine parabrachial region, and the trigeminal system in cats. Neuroscience 19:411–425

Schneider JS, Manettto C, Lidsky TI (1985) Substantia nigra projection to medullary reticular formation: relevance to oculomotor and related motor functions in the cat. Neurosci Lett 62:1–6

Scott TR, Yaxley S, Sienkiewicz ZJ, Rolls ET (1986) Gustatory responses in the nucleus tractus solitarius of the alert cynomologous monkey. J Neurophysiol 55:182–200

Sessle BJ, Greenwood LF (1976) Inputs to trigeminal brain stem neurons from facial, oral, tooth pulp and pharyngolaryngeal tissues: I responses to innocuous and noxious stimuli. Brain Res 117:211–226

Steiner JE (1973) The gustofacial response: observation on normal and anencephalic newborn infants. In: Bosma JF (ed) Oral sensation and perception: development in the fetus and infant. US Department of Health, Education and Welfare, Bethesda, pp 254–278

Storey AT (1968) A functional analysis of sensory units innervating epiglottis and larynx. Exp Neurol 20:366–383

Sumi T (1970) Activity in single hypoglossal fibers during cortically induced swallowing and chewing in rabbits. Pflügers Arch 314:329–346

Suzuki SS, Siegel JM (1985) Reticular formation neurons related to tongue movement in the behaving cat. Exp Neurol 89:689–697

Swanson LW, Sharpe LG (1973) Centrally induced drinking: comparison of angiotensin II- and carbachol-sensitive sites in rats. Am J Physiol 225:566–573

Swanson LW, Kucharczyk J, Mogenson GJ (1978) Autoradiographic evidence for pathways from the medial preoptic area to the midbrain involved in the drinking response to angiotensin II. J Comp

Neurol 178:645–660

Sweazey RD, Bradley RM (1987) Responses of lamb spinal trigeminal nucleus neurons to mechanical, thermal and chemical stimulations of the oral cavity and epiglottis. Soc Neurosci Abstr p 779

Sweazey RD, Bradley RM (1988) Responses of lamb nucleus of the solitary tract neurons to chemical stimulation of the epiglottis. Brain Res 439:195–210

Sweazey RD, Bradley RM (1989) Responses of neurons in the lamb nucleus tractus solitarius to stimulation of the caudal oral cavity and epiglottis with different stimulus modalities. Brain Res 480:133–150

Tal M (1987) Neural basis for initiation of rhythmic digastric activity upon midbrain stimulation in the guinea pig. Brain Res 411:58–64

Thexton AJ, Crompton AW (1989) Effect of sensory input from the tongue on jaw movement in normal feeding opossum. J Exp Zool 250:233–243

Thexton AJ, Griffiths C (1979) Reflex oral activity in decerebrate rats of different age. Brain Res 175:1–9

Thexton AJ, McGarrick J (1987) Effect of experimentally elicited rhythmic oral activity on the linguodigastric reflex in the lightly anesthetized rabbit. Exp Neurol 96:104–117

Travers JB (1988) Efferent projections from the anterior nucleus of the solitary tract of the hamster. Brain Res 457:1–11

Travers JB (1989) Hypoglossal unit activity during licking and swallowing. J Dent Res 68:309

Travers JB, Jackson LM (1989) The effects of gustatory stimulation on neurons in and adjacent to the hypoglossal nucleus. Chem Senses 14:756

Travers JB, Norgren R (1983) Afferent projections to the oral motor nuclei in the rat. J Comp Neurol 220:280–298

Travers JB, Norgren R (1986) Electromyographic analysis of the ingestion and rejection of sapid stimuli in the rat. Behav Neurosci 100:544–555

Travers SP, Norgren R (1988) Oral sensory responses in the nucleus of the solitary tract. Soc Neurosci Abstr p 1185

Travers SP, Pfaffmann C, Norgren R (1986) Convergence of lingual and palatal gustatory neural activity in the nucleus of the solitary tract. Brain Res 365:305–320

Travers JB, Waltzer R, Travers S (1988) Sensory-motor mapping of the solitary nucleus and adjacent structures. Chem Senses 13:741

Van Willigen JD, Weijs-Boot J (1984) Phasic and rhythmic responses of the oral musculature to mechanical stimulation of the rat palate. Arch Oral Biol 29:7–11

White NM (1986) Control of sensorimotor functions by dopaminergic nigrostriatal neurons: influence on eating and drinking. Neurosci Biobehav Rev 10:15–36

Whitehead MC (1988) Neuronal architecture of the nucleus of the solitary tract in the hamster. J Comp Neurol 276:547–572

Whitehead MC, Savoy LD (1987) The solitary nucleus of the hamster. Ann N Y Acad Sci 510:707–709

Woods, JW (1964) Behavior of chronic decerebrate rats. J Neurophysiol 27:635–644

Zeigler HP, Jacquin MF, Miller MG (1985) Trigeminal orosensation and ingestive behavior in the rat. Prog Psychobiol Physiol Psychol 11:63–195

Zhang G, Sasamoto K (1990) Projections of two separate cortical areas for rhythmical jaw movements in the rat. Brain Res Bull 24:221–230

Section V

Behavioural Organization of Drinking

Chapter 17

Learning, Thirst and Drinking

P.C. Holland

Introduction

Animals' ingestion of fluids is known to be affected by many factors, some involving events of the external environment, some involving detection of internal changes, some relating to immediate fluid needs, and some not related to fluid deficits. Many of the mechanisms used to detect, integrate, and act on these kinds of information are modifiable by rather specific experience. This chapter concerns a few of these learned influences on thirst and drinking, mostly in rats.

In discussing the involvement of learning in thirst, no distinction is made between "primary" and "secondary" drinking, nor between "regulatory" and "non-regulatory" drinking. There is no a priori reason to assume that the operation of a regulatory system based on detectable fluid deficits does not involve learning, nor to assume that learning must be involved in the control of drinking that makes no reference to fluid deficit. The evolution of drinking strategies is less concerned with mechanism than outcome: the use of an easily detectable token stimulus that is not causally bound to fluid deficits, but is reasonably correlated with them, may prove to be as adaptive as the use of a less easily detectable stimulus that is causally related to those deficits. These often ill-conceived distinctions should not be allowed to limit the search for the effects of experience on thirst and drinking.

If we are to understand how drinking can be affected by experience, there must be an appreciation of the variety of ways in which behaviour in general can be modified by learning experiences. The next section provides a brief tutorial on selected aspects of learning that may be especially relevant to the study of thirst and drinking (see Rescorla and Holland 1982, for a brief, but broader view of contemporary learning theory). This is followed by a consideration of the functions of internal stimulus aspects of thirst in the learned control of drinking. The final section discusses the learned control of drinking by oral and external cues. Only a limited amount of research is reviewed here: the chapter is intended more as an outline of plausible functions of learning in thirst and drinking, and suggestions for future research.

Overview of Learning

To adapt to relatively short-term variations in their environment, animals must be sensitive to the flow of events around them. In their attempts to specify how animals learn about these events (both internal and external), contemporary learning theorists often distinguish between various kinds of experiences with events and between various consequences of those experiences for the animals' subsequent behaviour (e.g. Rescorla and Holland 1976; Rescorla 1988).

Kinds of Experience

Learning experiences are most commonly classed as being "non-associative" or "associative", as a function of whether an event occurs without restriction, or in some relation to other events, respectively. Simple presentation of an event provides the organism with the chance to acquire what is perhaps the most elementary learning about a stimulus: that it exists and has certain properties. For example, on first encountering a novel, flavoured liquid, water-deprived rats are hesitant to consume that fluid. However, with repeated exposure to that fluid, their "neophobic" reactions will habituate: they will consume more of that fluid than rats that have not been exposed to it.

In addition to this non-associative, "recognition learning", animals must be sensitive to relations among the various individual events in their worlds, especially those providing cause–effect information (e.g. Tolman and Brunswik 1935). Two classes of this associative learning are typically distinguished. First, there is a sensitivity to relations between events and the organism's own behaviour, as commonly studied in operant conditioning. For example, if delivery of the flavoured fluid is made contingent on a rat's pressing a lever, lever pressing will become more frequent. Second, there is a sensitivity to relations between events that are independent of behaviour, as in Pavlovian conditioning. For example, if delivery of the flavoured liquid is preceded by another event, say a tone, rats will show anticipatory licking and general activity increases during the tone.

This chapter stresses the involvement of Pavlovian (event–event) contingencies in thirst-related behaviour. This emphasis reflects both the research interests of the author and the dominance of Pavlovian conditioning in contemporary learning theory (e.g., Dickinson 1980; Mackintosh 1983).

Consequences of Experience: Measurement and "What is Learned?"

Much of the history of learning theory was given over to debates concerning the content of learning. Typically, researchers addressed dichotomies like "is learning S-S or S-R"? But it is important to note that each of the experiences just described may have a variety of effects on the organism, each of which may influence behaviour in different ways. The conclusions about the nature of learning may be crucially influenced by what aspect of the animal's learned behaviour is assessed.

This section introduces four common assessment procedures which illustrate a variety of ways in which learning experiences may affect later behaviour. These techniques are discussed in the context of non-associative learning experiences, and then the conceptually more involved case of the content and measurement of associative learning is discussed in the next section.

The most common assessment of learning as a consequence of simple exposure to a stimulus is the measurement of the response elicitation power of that stimulus. On their initial presentation, most stimuli used in learning experiments elicit measurable, unlearned responses, but simple exposure to a stimulus often results in its reduced ability to elicit those responses (habituation). Second, exposure to a stimulus may alter its emotional significance or value. For example, repeated presentation of loud noises or mild electric shocks not only reduces the ability of those events to elicit responses (habituation) but also diminishes their abilities to punish rats' lever pressing. Third, the associability or attention-grabbing powers of an event may be affected by simple exposure. For instance, repeated exposure to a flavour cue makes it more difficult for rats to subsequently learn a Pavlovian relation between that flavour and illness. Finally, simple exposure to an event may modify its modulation of ongoing behaviour of an organism. For example, the ability of an electric shock to modulate a variety of normal activities of rats (e.g. drinking, exploration and even the action of the immune system) is reduced by prior exposure (e.g., Lysle et al. 1987).

Although many modern learning theorists at least tacitly assume that all of these assessment procedures provide different windows on a common learning process, it is important to recognize that each of these consequences of learning may affect some target behaviours in different ways. Furthermore, there are numerous examples of the dissociation of some of these measures. For example, some manipulations are known to have different effects on the consumption of a frequently presented flavoured liquid and the reduced ability of that flavour to participate later in flavour–illness associations (e.g. Braveman and Jarvis 1978).

Associative Learning

S-S and S-R Associations in Pavlovian Conditioning

Historically, Pavlovian conditioning was described as the substitution of a previously neutral stimulus, the conditioned stimulus (CS), into an existing unconditioned reflex system. Thus, the more prosaic CS comes to control some of the neural/behavioural processes (unconditioned responses or URs) originally engendered only by the biologically significant unconditioned stimulus (US). Early psychologists described this newly acquired power of the CS as a consequence of the formation of associations between internal representatives of the CS and US (that is, convergence between the neural pathways activated by those events).

"S-R" and "S-S" approaches differed as to the locus of that association or convergence. S-R theorists argued that the site of plasticity was nearer the motor output end of the US–UR reflex path, that associative connections were formed between CS–receptor impulses and UR–effector mechanisms. S-S theorists argued that associative connections were formed between CS–receptor impulses and more upstream processing activity in the US–UR path. Thus, the S-R vs. S-S distinction essentially concerned how much of the machinery of the US–UR reflex system is engaged by the CS alone after conditioning. Casually speaking, how much of the US–UR reflex does the animal apprehend in the presence of the CS alone? Does the licking that emerges during a tone that predicts water delivery occur because the tone comes to activate directly some licking pattern oscillator, or because the tone arouses a surrogate perception of water on the tongue, which in turn sets off licking?

Considerable evidence suggests that, under some circumstances, CSs generate conditioned behaviour by activating substantial portions of US-processing pathways, including those with sensory attributes. Under those circumstances, the ability of the CS to elicit a conditioned response (CR) depends on the continued ability of the US to elicit behaviour. For example, consider a water-deprived rat that received pairings of one tone CS (T1) with a water US of one flavour (F1), and another tone

(T2) with water laced with a different flavouring (F2). When (after completion of tone–flavour training, and in the absence of T2) the rat's acceptance of F2 was altered by pairing F2 with a toxin, or by satiation with F2, subsequent tests of responding to the CSs showed maintained CRs to T1, but little responding to T2 (Holland 1990a). Thus, conditioned responding to T2 alone was apparently mediated by a pathway that included processing of the sensory/perceptual attributes of F2.

Not all conditioning involves such S-S association, however. For example, Holland (1990a) and Holland and Rescorla (1975) noted circumstances under which conditioned responding to a CS was maintained despite nearly complete post-training devaluation of its reinforcer. Thus, it is important to recognize that the real issue is not whether learning is S-R or S-S, but when it is S-R and when it is S-S. For example, under some conditions of learning, conditioned behaviour originally based on a fluid deficit may persist long after that deficit is removed but, under other circumstances, it may not.

Response Elicitation in Pavlovian Conditioning

The four assessment procedures discussed above can also be identified in Pavlovian conditioning. However, in this associative paradigm an additional distinction can be made. Not only can CS–US pairings allow the CS to activate processing systems previously activated only by the US, but they can also modify behavioural processes normally engaged by the CS. The most prominent example is that of "alpha conditioning" (e.g., Hull 1934; Holland 1990b), in which CS–US pairings augment the orienting response that was elicited by the CS prior to conditioning. Not only do bell–water pairings endow the bell with the ability to elicit licking responses, they also enhance the original orienting response to the bell.

CS Value in Pavlovian Conditioning

This same distinction between acquisition of new control (transfer) and modification of existing control may also apply to changes in CS value, albeit less clearly. For instance, Garcia et al. (1985) distinguished between the consequences of flavour–shock and flavour–illness learning. Both types of pairings produce learned suppression of consumption of flavoured fluids. But in the former case, the flavoured fluids acquire the aversive properties of the shock USs: the conditioned flavour makes them afraid, suppressing consumption. Conversely, with flavour–illness pairings, "hedonic shifts" – changes in the animal's evaluation of the flavour CS – occur; the rat suppresses consumption because the flavour has come to taste bad. This claim is supported by the results of taste reactivity tests (e.g., Pelchat et al. 1983): rats exhibit orofacial responses consistent with changed evaluations of the flavour CS when it is paired with toxin, but not when it is paired with shock (neither the toxin nor the shock alone generates any of those orofacial responses).

CS Associability in Pavlovian Conditioning

Changes in the CS's associability as a consequence of conditioning play a major role in many recent learning theories. Those theories have in common the notions that the acquisition of CS–US associations is affected by the extent to which the subject attends to the CS, and that prior and contemporaneous learning experiences can influence that attention. For example, Mackintosh (1975) argued that stimuli that are good predictors of USs, relative to other cues, acquire enhanced attention-grabbing powers, and hence are more readily associable with other events, whereas relatively poorer predictors lose their normal attention-grabbing powers. Consequently, the effects of CS–US pairings can critically depend on an organism's prior experience with the various events occurring during learning. For instance, prior conditioning with one cue, A, can block conditioning of another cue, X, if a compound of both cues (AX) is subsequently paired with the same reinforcer (Kamin 1968).

CS Modulation in Pavlovian Conditioning

Considerable emphasis has been placed on the acquired modulatory powers of Pavlovian CSs. Perhaps the simplest example of this modulation is illustrated by the commonly used "conditioned suppression" procedure. In this procedure, the experimenter measures the ability of a CS to modulate ongoing behaviour of the subject. In some cases, that behaviour is fairly arbitrary and is maintained by explicit operant contingencies; for example, rats' lever pressing may be reinforced with water on some schedule of reinforcement. In other cases, the ongoing behaviour is a naturally occurring consummatory response, maintained by potent motivational manipulations, e.g., drinking of water-deprived rats. Learning about, say, tone–shock relations is indexed by the tone's ability to suppress or enhance that lever-pressing or drinking.

A common concern was whether these learned modulatory effects were the result of intrinsic motivational interactions, such that, for example, the conditioned fear response elicited by a tone paired with shock depressed drinking by reducing thirst in some unspecified way (e.g. Konorski 1967; Rescorla and Solomon 1967), or by generating more specific competing responses or expectancies. Decades of research provided some support for each of these consequences.

There has been considerable interest in a more specific kind of modulatory power of Pavlovian CSs (e.g., Holland 1983). Under some circumstances, CSs can come to modulate the effectiveness of associations between other events. For example, consider a discrimination in which a tone CS is reinforced with food delivery when it is preceded by a light, but not when the tone occurs in the absence of the light. Rats readily learn to respond to the tone only when it is preceded by the light. Several kinds of evidence indicate that the solution of this discrimination involves the light's acquisition of the power to modulate the tone's ability to elicit a CR, as if gating an association between tone and food. Interestingly, this ability of the light to set the occasion for responding based on the tone–food association is largely independent of the light's ability to elicit CRs based on its own association with the food. For example, if the light is repeatedly presented in the absence of the tone or food, its ability to elicit its own CRs extinguishes, but its ability to set the occasion for responding to the tone is unaffected (Holland 1989b).

Similarly, in many cases, inhibitory control in conditioning may involve such modulatory functions (e.g., Holland 1985; Rescorla 1985). Suppose in the previous example that the tone was instead reinforced when it was presented alone, but non-reinforced when it was preceded by the light. Rats readily learn to respond to the tone only when it is presented alone. As in the previous case, considerable evidence suggests that the light acts by modulating the action of an association between the tone and the US. Similarly, the light's ability to set the occasion for non-responding to the tone is largely independent of its own associations with the US: the light maintains its powers to suppress responding normally elicited by the tone, even after the light itself is given the ability to elicit a CR, by pairing it with food (Holland 1989c).

The distinction between the eliciting and occasion setting powers of cues is analogous to that between the action of neurotransmitters and neuromodulators: a CS may elicit activity in a pathway directly, or it may alter the threshold for the activation of that pathway by another cue (occasion setting). The CS raises that threshold in the case of negative or inhibitory modulation, but lowers it in the case of positive or excitatory modulation.

Learning and the Stimulus Properties of Thirst

During water deprivation, humans report potent sensations, chiefly localized in the mouth and throat, but also including general weakness, ringing of ears, and so forth, especially as the deprivation becomes more severe (e.g., Wolf 1958). It is not surprising, then, that the simplest point of contact between the studies of learning and of thirst is the consideration of sensations arising from fluid deficits functioning as stimuli like any other stimuli.

Within the Hullian drive-reduction theory of learning prevalent in the middle of this century, the stimulus aspects of drive assumed major importance. Essentially, Hull (e.g. 1933) claimed that organisms adapted to biological needs through S-R learning, reinforced by the reduction of psychological drives closely tied to those needs. Like any other drive, thirst (the psychological state linked to fluid deficit) was assumed to both provoke potent "drive stimuli", e.g. a dry mouth, and elicit general activity and exploratory behaviour. Eventually, some behaviour would be successful at obtaining water. Because the coincidence of the internal thirst-provoked stimuli and successful water-procuring activities was followed by thirst reduction, S-R associations between thirst-provoked cues and those activities would be formed. Consequently, water-procuring behaviours would be repeated when the organism was thirsty in the future. However, because activities that failed to procure water were not followed by thirst reduction, associations between thirst stimuli and those behaviours would extinguish. Similarly, in the absence of thirst, any pre-existing associations between non-thirst (e.g. hunger) cues and water-procuring behaviours would extinguish (because in the absence of thirst, water had no thirst reducing, hence reinforcing, powers). Consequently, organisms would learn to perform water-procuring behaviours when they were thirsty, but not to perform those activities when they were not thirsty.

Thus, simply by allowing thirst cues to elicit behaviour (both unconditionally and conditionally), and to modulate the effectiveness of various events as reinforcers, Hull provided a plausible account for the selective aspects of motivated behaviour without recourse to innate action patterns linking fluid deficits to appetitive behaviour. Furthermore, by postulating that the various properties of drives like thirst could themselves be conditioned to other stimuli, that is, that thirst had reinforcement value, Hull extended the domain of motivated behaviour to situations in which primary drive states or primary reinforcers were absent. This section considers the eliciting, reinforcing and modulating functions of thirst cues, parallel to the discussion of the previous section.

Thirst Cues as Elicitors of Behaviour

Thirst Cues as Elicitors of CRs

Within a theory like Hull's, organisms' abilities selectively to use internal motivational cues as conditioned stimuli assumed paramount importance. In classic studies, Hull (1933) and Leeper (1935) showed that rats could indeed learn differential behaviour solely on the basis of internal motivational conditions. On some days, rats were deprived of food but allowed access to water, and on other days they were deprived of water but not food. On "hungry" days they were given food after right turns in a T-maze, and on "thirsty" days they were given water for turning left in the same maze. The rats gradually came to act appropriately.

There has been relatively little investigation of animals' use of thirst cues as signals for other events or behaviours. This is unfortunate, because these kinds of conditioning experiments might be useful as a means of substituting in animals for the verbal reports of humans about their thirst sensations: conditioning procedures could be used to give animals a way of telling us whether some manipulation affects stimulus aspects of thirst, apart from measures of consumption.

A good example of this strategy in the case of food deprivation and hunger was provided by Davidson (1987). While under one level of food deprivation, rats were placed in a conditioning chamber and given brief, mild electric shocks, but while under another level of deprivation, they were not shocked when placed in the same chamber. After only a few trials in each deprivation state, the rats exhibited freezing responses in the chamber when they were in the first deprivation state, and not when they were in the second. Presumably, the fear or freezing responses were conditioned to the appropriate "hunger" cues.

The effects of putative hunger-affecting manipulations could then be examined by examining the effects of those manipulations on the rats' freezing responses in the experimental chamber. For example, in one experiment (Davidson et al. 1988) one group of rats was shocked when 23 h deprived and not shocked when only 6 h deprived. A second group of rats was shocked when they were 6 h deprived but not when they were 23 h deprived. Presumably, the first group learned to freeze when very hungry, and the second to freeze when they were only a little hungry.

Then, when the rats were all 6 h deprived, Davidson administered insulin, a treatment known to increase food consumption, and thought to increase hunger sensations. Insulin injection increased freezing in the rats in the first group (who were originally shocked under high hunger) but reduced freezing in the rats in the second group (who were shocked under low, but not high, hunger). Thus, insulin injection made the 6 h deprivation state cues more similar to those of 23 h deprivation than to those of 6 h deprivation. Interestingly, administration of cholecystokinin (CCK) at doses known to reduce food consumption had no such differential effects on freezing in the two groups. This outcome suggests that, unlike insulin, CCK alters food consumption without altering internal hunger-level cues used as CSs by the rats, a conclusion consistent with other modes of analysis of the effects of CCK (e.g., Ettinger et al. 1986).

Davidson's procedure could be readily adapted to answer questions about the effects of various manipulations on thirst cues, and their relation to drinking. For example, would rats discriminate between thirst cues induced by normal deprivation, and manipulations specifically inducing cellular dehydration, or hypovolaemia (cf.

Burke et al. 1972)? Would anaesthetizing the mouth (which in many species reduces water consumption, e.g., Wolf 1958) remove thirst cues that control freezing in this situation? Would manipulations known to induce drinking in the absence of fluid deficits (e.g., Falk 1964; Kissileff 1973) also induce thirst cues? There has been substantial investigation of the relation of thirst sensations (especially the dry mouth) and water consumption in both humans and non-humans; the use of such a non-consummatory measure of thirst cues could help remove the circularity of most of our measures of thirst sensation in non-verbal organisms.

Thirst Cues as Elicitors of URs

Early theorists, like Hull, assumed that thirst, hunger, or any other deficit-based motivational state would similarly impel creatures to action/exploration. Studies of "random activity" of rats under water or food deprivation (e.g., Shirley 1929) supported this claim (but see page 288). At the same time, behaviour theorists also recognized specificity of action of deprivational states (e.g., Kendler 1946); those states might preferentially elicit responses appropriate to the current state. For example, pigeons make topographically distinct pecking movements for seizing grain and water. In the absence of either grain or water, hungry pigeons make more food pecks, and thirsty pigeons make more water pecks (e.g., Jenkins and Moore 1973).

Thirst Cues as Reinforcers

Clearly, water deprivation produces biologically significant changes in organisms, changes accompanied by detectable stimulus consequences. Thus, those thirst cues should be capable of functioning as USs in conditioning experiments. That is, neutral cues paired with states of thirst should acquire new behavioural properties.

Mimetic Thirst Conditioning

One of Hull's original postulates concerned the acquisition of secondary drive states. That is, cues consistently associated with a primary drive state, such as thirst, should acquire the properties of those primary states. The simplest test of this hypothesis was to attempt to condition thirst to neutral cues, such as particular contexts. Although there were early reports of failure to obtain conditioned thirst or drinking (e.g., Novin and Miller 1962), several investigators have reported this phenomenon (e.g., Solomon and Swanson 1955, described in Mowrer 1960; Seligman et al. 1970; Weisinger 1975). For example, Solomon and Swanson (1955) deprived rats of water in their home cages for 23 h, then removed them to a watering box with 0.5 h access to water, and finally placed them in a third, neutral box for 0.5 h before returning them to their home cages. Thus, the rats never drank fluids in either the home cage or neutral box, but were thirsty in the former and sated in the latter. Then, drinking was examined in the home cage and neutral boxes, while the rats were water-sated. The subjects drank more in the home cage than in the neutral box. Solomon and Swanson interpreted this outcome as the conditioning of thirst to the home cage.

However, it could be argued that the home cage cues controlled drinking directly by virtue of their forward relation to drinking in the watering box. Any context-dependent drinking then may have more reflected the conditioning of drinking responses than of an antecedent drive-stimulus state. A procedure like Davidson's

(1987) might be used to determine if the rats "feel thirsty" during predictors of thirst states. After stimulus–thirst pairings, that level of thirst could be used as a predictor of the presence or absence of shock. If control over those thirst cues was acquired by the stimulus, then that stimulus might later substitute for thirst as a shock cue.

Compensatory Thirst Conditioning

It is often difficult to predict the CR to be conditioned with USs that produce wide-ranging biological adjustment of the animal. For example, consider a cue that reliably predicts insulin injection. That injection rapidly causes reduced blood sugar levels. However, detection of that hypoglycaemia engages processes which actively re-elevate blood sugar. Which is most likely to be associated with the CS cue – the hypoglycaemic state, the mechanism that caused the initial hypoglycaemic response to the injection, or the hyperglycaemic regulatory response to the hypoglycaemic state (cf. Wikler 1973)? Some evidence suggests, as with simple food conditioning described earlier, that any of these are possible, depending on the circumstances (Flaherty et al. 1987).

At first glance, this may not seem like a problem for the case of CSs associated with water deficits: such a CS should induce drinking regardless of whether it has become associated with the primary thirst cues (which would themselves make the rat drink) or the normal "regulatory" response to thirst cues (drinking). However, induction of thirst in the absence of water has a variety of effects on the animal. For example, suppose the concentrating of urine was directly conditionable (e.g., Bykoff 1957). If a conditioned context elicited this CR when a rat was placed in that context in the absence of water deprivation, the rat would become hypervolaemic, and respond by drinking less than usual. Varied outcomes in drive conditioning experiments might be related to conditioning of different components of the varied responses to fluid deprivation, depending on particular stimulus parameters. (There is considerable evidence that different CRs can be conditioned with complex USs, depending on many physical and temporal properties of the CS; for example, Holland 1984.)

An interesting example is provided by Poulos and Hinson (1984), who administered scopolamine to water-deprived rats in a particular experimental context. This cholinergic antagonist suppressed drinking of freely available water but, after repeated context–scopolamine pairings, placement of the rat in that context after administration of a placebo produced enhanced drinking. This polydipsic response was specific to the experimental context and was abolished by repeated placement of the rat in that context without the drug (extinction). Furthermore, the rats acquired a context-dependent tolerance to the normal adipsic effects of scopolamine, as would be anticipated if the conditioned polydipsic response compensated for the unconditioned adipsic response.

Thirst Cues as Modulators

A common role attributed to thirst is that of modulating the effectiveness of other events. For example, the heart of Hull's performance model was a multiplicative relation between habit (both unconditioned and conditioned behaviours) and drive level: drive determined the gain on the stimulus–response connection. Likewise,

Tinbergen (1951) assumed that deprivation acted by producing internal stimulus and hormone changes that modulated the effectiveness of releaser stimuli in instigating fixed action patterns. Other investigators suggested that deprivation state directed attention to particular classes of stimuli, thus channelling behaviour toward appropriate goal objects (e.g. Kendler 1946).

In fact, many investigators described the apparent unconditioned elicitation of behaviour by thirst cues as a special case of modulation: drive itself may not impel the organism to action, but rather lower the threshold for responding to environmental stimuli. For example, Campbell (1960) found that water-deprived rats were more active than water-sated rats in the presence of cues paired with water delivery, but found no evidence for a difference in the baseline activity rates of those two groups of rats. Similarly, the differential occurrence of food- and water-pecks in hungry or thirsty pigeons is especially obvious in the presence of cues for food or water, or food and water themselves (Jenkins and Moore 1973).

Thirst as an Occasion-Setter

As early behaviour theorists (e.g., Hull 1945) recognized, thirst is unlikely to act in the same fashion as a typical discrete CS used in conditioning experiments. Unlike those CSs, thirst cues (as produced by gradual water deprivation) have gradual onset and are more or less continuously present over long periods of time in which performance of particular responses would not be reinforced. For example, the rat in Leeper's (1935) experiment is also thirsty in its home cage, but is unlikely to begin engaging in right-turn responses there, nor do thirst cue–right-turn associations extinguish, despite the long periods in which thirst cues are not accompanied by thirst reduction from water consumption. Hull's solution of this problem was to invoke the concept of stimulus patterning or "afferent neural interaction": "receptor impulses" from maze cues interact with those from thirst, generating unique sets of "afferent impulses". Thus, the right-turn response is attached to the unique stimulus pattern "choice point + thirst", and the left-turn response is attached to the pattern "choice point + hunger".

Later investigators (e.g. Holland 1983) have suggested that motivational stimuli may act by setting the occasion (see p. 283) for the action of explicit associations between events. That is, rather than forming unique stimulus patterns with explicit cues, thirst cues might serve to modulate the effectiveness of explicit CS–US associations in eliciting behaviour. Experiments with explicit, external cues (e.g., Holland 1989a) have empirically distinguished between patterning and occasion-setting functions, and suggest that the latter function is favoured when the two cues of a compound are dissimilar (as thirst and external cues might be). Furthermore, what we know about the action of explicit occasion-setters is consistent with some of the troublesome aspects of thirst cues' functioning as CSs. For example, the ability of thirst cues to modulate the effectiveness of some other cue would not be anticipated to be affected by the lack of overall correlation between thirst and water delivery, or even an overall negative correlation (Holland 1989c).

Although little explicit research has been conducted with deprivation state cues, those cues have much in common with general context cues, which have been extensively investigated. Most of that research suggests that context cues function as occasion-setters, rather than eliciting CSs, in Pavlovian conditioning experiments (e.g., Bouton and Swartzentruber 1986).

Thirst as a Modulator of Incentive Value

Thirst has been assumed not only to modulate the elicitation of behaviour by learned and unlearned stimuli, but also to modulate the incentive and reinforcement values of those events. For example, humans usually report that water tastes better when they are water-deprived (Rolls and Rolls 1982), as well as wanting it more and being more willing to work for it.

As mentioned above within the Hullian framework, drive modulated the reinforcement value of an event by determining (along with features of those events themselves) the amount of drive reduction possible with that event. Other investigators (e.g. Tolman 1932; Sheffield and Roby 1950; Young 1955; Cabanac 1971) presumed that deprivation state directly affected incentive value of stimulus aspects of the event, rather than via differential drive reduction.

Modern evidence supports the latter view. Deprivation level affects reinforcement value of sham-drinking of water (Mook and Wagner 1988), lapping at an airstream (Oatley and Dickinson 1970), and licking a cold spout (Mendelson and Chillag 1970) in much the same way as it affects reinforcement value of true drinking. At the same time, lever-pressing is readily reinforced by the opportunity to sham-drink, but only with difficulty if the response produces water intragastrically (Epstein 1960) or intravenously (Nicolaïdis and Rowland 1974). Given that the incentive value of particular oral cues like flavour and temperature can also be easily and dramatically affected by learning (as in flavour aversion conditioning, in which a particular flavour is paired with a noxious substance, Garcia et al. 1985), the "final common path" of palatability deserves close attention in the study of drinking.

Regardless of how the value of a fluid is modulated, it is worth noting that changes in that fluid's ability to elicit consumption or reinforce conditioning may often be accompanied by parallel changes in the behavioural properties of other cues that had previously been associated with the fluid. For example, Holland (1990a) presented water-deprived rats with pairings of a tone with wintergreen and a noise with peppermint-flavoured solution. Those auditory cues consequently acquired activity increases and monitoring of the site of liquid delivery. Next, one solution was made unpalatable by pairing it with lithium chloride or encouraging its unlimited consumption. Later examination of conditioned responding elicited by those auditory cues alone showed dramatic losses in activity and goal responding to the cue whose fluid partner had been devalued, but not to the other auditory cue.

Conversely, under other circumstances, learned responses ultimately based on valued fluids may be immune to those modulatory effects. In an unpublished experiment analogous to that of Holland and Rescorla (1975; see p. 281), I found that activity increases conditioned to auditory cues paired directly with water in thirsty rats were reduced after water-satiation, but similar second-order CRs, established by pairing a tone with a light that had been previously paired with water, were untouched by satiation. Indeed, unlike water-deprived rats, the water-satiated rats were more likely to respond in the presence of the second-order cues than in the presence of the first-order cues. Interestingly, the water-satiated rats were also more likely to consume water in the presence of the second-order cues than in the presence of the first-order cues. Thus, learning can also provide an avenue for the maintenance of habitual appetitive and consummatory behaviour, even in the absence of the appropriate "thirst" state.

Learning and the Control of Drinking by Oral and External Cues

The control of drinking by oral cues provides considerable adaptive advantage: compared to the postabsorptive consequences of drinking, response feedback is more immediate, and smaller samples are effective (Epstein 1960). Furthermore, animals are capable of distinguishing considerable variety in those cues (flavour, temperature and texture all play a role), making possible considerable selectivity of behaviour. Similarly, control of drinking and predrinking behaviour by external cues can enhance (or diminish) animals' opportunities to make contact with fluids. Most relevant here is the view of many theorists (e.g., Chapter 4; Teitelbaum 1966) that behaviour controlled by external and oral cues is especially modifiable by learning.

Because thirst-motivated appetitive and consummatory behaviour involves many different stimulus and response elements, and may be ultimately based on a variety of postingestional events as well (e.g., Hall and Blass 1977), the effects of experience can be complex. Control of various response components may shift to new stimulus features, and existing control over behaviour may be modulated. Unfortunately, at present there is relatively little evidence to indicate what portions of drinking sequences are plastic, under what circumstances.

Recognition, Habituation and Satiation

A considerable amount of data suggests that simple exposure to oral cues can substantially affect drinking. For example, as noted previously, exposure to a particular flavour (or even water itself) reduces neophobic responses and slows subsequent conditioning of that fluid (e.g., Ghent 1957; Domjan 1977). Similarly, Hunt and Smith (1967) found that chicks required exposure to water to recognize it as water; initial contact with water involved food-pecks, which were replaced with characteristic water-pecks only after exposure.

Conversely, in the short run, simple exposure can act to reduce consumption of a flavour; that is, there is apparently an oral component to satiation. For example, after drinking one flavour to the point of refusal, rats readily accept a different (but familiar) flavour (e.g. Holland 1988; see Rolls et al. 1981, for analogous data with humans). Although selective satiation phenomena like these can sometimes be attributed to postingestional factors (e.g., Mook et al. 1986), the use of a variety of arbitrarily chosen flavours make those factors unlikely influences in my experiments. Likewise, even sham-drinking is reduced by a recent history of sham-drinking in the same context (von Vort and Smith 1987), and gastric water loads suppress sham-drinking sooner if the gastric load is preceded by sham-drinking (Mook et al. 1988) than if it isn't. Thus, at least some portion of satiation effects can be attributed to habituation of oral cues.

Other oral stimuli besides taste may habituate as well. For example, Epstein (1973) suggested that desalivate rats act as if the ability of "dry mouth" cues to elicit drinking behaviour habituates. Those rats drink less than normal rats after a variety of treatments known both to desiccate the mouth and to induce drinking.

Conditioning

The very sequential nature of drinking behaviour and its postingestive consequences suggests the plausibility of substantial contributions of associative learning to the control of drinking by external and oral cues. For example, it is easy to imagine the conditioning of various late-sequence responses (e.g. swallowing) to stimuli encountered earlier in the ingestive sequence, and early-sequence responses (e.g., licking) to external stimuli (e.g., Debold et al. 1965). Water consumption itself is readily conditionable by operant procedures. If the opportunity to engage in an artificially restricted activity (e.g., running) is made contingent on water consumption, rats will greatly increase that consumption (e.g., Premack 1965); overhydration is prevented by normal renal function.

Moreover, approach responses can be conditioned to reliable external signals of water (like odours or contexts associated with water sources), by virtue of either operant, response-reinforcer, or Pavlovian, stimulus-reinforcer, relations. These approach responses can make it more likely that the animal makes contact with the fluid in the first place (e.g., Holland 1984). And, as described on p. 286, external and oral cues may also acquire drive–stimulus properties, which may themselves serve many functions in controlling ingestive behaviour.

Furthermore, positive or negative postingestive consequences of drinking particular fluids can alter the palatability of those fluids. Indeed, some researchers, loathe to accept the evolution of drinking control mechanisms not based directly on current fluid needs, suggested that all oral control was derived from its learned association with postabsorptive consequences early in development. On the whole, studies of the ontogeny of drinking (see Chapter 3) do not support this extreme claim, but there has been remarkably little systematic study of the developmental origins of stimulus control over drinking.

Nevertheless, it is known that conditioning can endow initially neutral, arbitrarily chosen flavours with a variety of behavioural control properties of fluids that already possess significant oral features. For example, arbitrary grocery-store food flavours paired with salty, sweet or bitter solutions can acquire the consummatory, orofacial responses appropriate to those latter flavours (Holland 1990a). Furthermore, after their pairing with significant flavours, the consumption of arbitrary flavours becomes sensitive to motivational manipulations that normally affect consumption of their significant partners. For instance, Fudim (1978) found that rats exposed to almond+salt and banana+sweet solutions showed a marked preference for almond-alone over banana-alone when a salt deficit was induced. Even external stimuli, like tones, can acquire control over these consummatory responses through Pavlovian conditioning, under certain circumstances (Holland 1990a). Interestingly, these effects seem to involve "S-S" conditioning, that is, rather than simply controlling ingestive responses, the tone or flavour CSs come to activate sensory processing of the (absent, but associated) significant flavoured fluid (Rescorla and Freberg 1978; Holland 1990a).

The ability of arbitrary external or oral cues to acquire other properties of already-significant cues also deserves further study. For example, highly palatable flavours are known to temper the effectiveness of various postingestional cues in terminating drinking (e.g., Rolls et al. 1978; Mook et al. 1988). Would arbitrary flavours or external cues also be capable of acquiring this modulatory capability?

In a related vein, Mook (1988) proposed that oral cues like flavour may selectively activate different postingestive mechanisms that control consumption. For example, the presence of a sweet taste might switch-in regulatory systems based on hunger and feeding, salty tastes, salt-regulation systems, and water tastes, thirst systems. In support of this claim, Mook et al. (1988) found that water drinking by thirsty rats was reduced proportionally by gastric water load alone, but saccharin drinking was not; instead, "oral metering" systems seemed more important.

It would be of interest to examine the pervasiveness of this system-shifting. Are these shifts accompanied by corresponding drive-stimulus changes? A procedure like Davidson's (p. 285) might be used to determine if discriminable drive-stimulus changes were induced by these different flavours. Can initially neutral stimuli gain control over these shifts, if they are paired with sweet, salt, and so forth? The attachment of such system shifts to arbitrary stimuli as a result of conditioning might have profound consequences for the normal and pathological control of human drinking behaviour.

Finally, conditioning that takes place in systems other than thirst and drinking must be recognized as potent sources of control, albeit indirect, over thirst and drinking. For example, much has been made of the interactions between hunger/ eating and thirst/drinking systems (see Chapter 18); it is not surprising that many of these interactions occur among conditioned stimuli and behaviours based on these systems (e.g., Jenkins and Moore 1973; Ramachandran and Pearce 1987). Similarly, pain or stress, and their learned anticipation, may depress (e.g. Rescorla and Solomon 1967) or enhance (e.g. Amsel and Maltzman 1950; Antelman et al. 1976) drinking. Furthermore, the periodic non-reinforcement of food-rewarded operant behaviour, and certain schedules of periodic food delivery, often produce dramatic excess drinking ("schedule-induced polydipsia", Falk 1964; Kissileff 1973).

The Future

We know a little about how the initiation and termination of drinking can be affected by associative and non-associative learning, although systematic study, guided by an appreciation of modern learning theory and data, has been lacking. The application of learning theorists' rather detailed knowledge of how perceptual and informational variables affect the learning of excitatory and inhibitory relations among events, to problems of thirst and drinking, is likely to be fruitful. The realization that learning has a variety of consequences for behaviour besides the elicitation of simple responses may be especially important.

An issue of some concern is that the demonstration that an animal can use some learning strategy in a carefully contrived laboratory experiment does not necessarily mean that it does use that strategy under normal circumstances. For example, the possession of efficient kidneys may minimize the importance of subtle learning effects in maintaining fluid balance in most circumstances. Moreover, it must be realized that there are likely to be important effects of experience that do not fit easily in the paradigms of the learning theorist. The examination of experiential effects early in development is likely to be especially illuminating: control mechanisms that are fairly rigid in the adult may be quite plastic early in life. Further strengthening the case for developmental study is the possibility that some of those experiential effects may be self-terminating: once a solution is found, plasticity is lost.

The investigation of the roles of learning in thirst and drinking has just begun. The involvement of multiple, relatively well-understood, biological control systems, which are readily susceptible to experimental manipulation, makes this an ideal system to study, for understanding the functions of learning in the integration of significant, complex behaviour systems of animals.

References

Amsel A, Maltzman I (1950) The effect upon generalized drive strength of emotionality as inferred from the level of consummatory response. J Exp Psychol 40:563–569

Antelman SM, Rowland NE, Fisher AE (1976) Stimulation bound ingestive behavior: a view from the tail. Physiol Behav 17:743–748

Bouton ME, Swartzentruber D (1986) Analysis of the associative and occasion-setting properties of contexts participating in a Pavlovian discrimination. J Exp Psychol [Anim Behav] 12:333–350

Braveman NS, Jarvis PS (1978) Independence of neophobia and taste aversion learning. Anim Learn Behav 6:406–412

Burke GH, Mook DG, Blass EM (1972) Hyperreactivity to quinine associated with osmotic thirst in the rat. J Comp Physiol Psychol 78:32–39

Bykoff KM (1957) The cerebral cortex and the internal organs. Chemical Publishing Co, New York

Cabanac M (1971) Physiological role of pleasure. Science 173:1103–1107

Campbell BA (1960) Effects of water deprivation on random activity. J Comp Physiol Psychol 53:240–241

Davidson TL (1987) Learning about deprivation intensity stimuli. Behav Neurosci 101:198–208

Davidson TL, Flynn FW, Grill HJ (1988) Comparison of the interoceptive consequences of CCK, LiCl, and satiety in rats. Behav Neurosci 102:134–140

Debold RC, Miller NE, Jensen DO (1965) Effect of strength of drive determined by a new technique for appetitive classical conditioning of rats. J Comp Physiol Psychol 59:102–108

Dickinson A (1980) Contemporary learning theory. Cambridge University Press, Cambridge

Domjan M (1977) Attenuation and enhancement of neophobia for edible substances. In: Barker LF, Best MR, Domjan M (eds) Learning mechanisms in food selection. Baylor University Press, Waco, TX, pp 151–179

Epstein AN (1960) Water intake without the act of drinking. Science 131:497–498

Epstein AN (1973) Epilogue: retrospect and prognosis. In: Epstein AN, Kissileff HR, Stellar E (eds) The neuropsychology of thirst: new findings and advances in concepts. Winston, Washington DC, pp 315–322

Ettinger RH, Thompson S, Staddon JER (1986) Cholecystokinin, lithium chloride and feeding regulation in rats. Physiol Behav 36:801–809

Falk JL (1964) Studies on schedule-induced polydipsia. In: Wayner MJ (ed) Thirst. Pergamon Press, New York, pp 95–116

Flaherty CF, Grigson PS, Brady A (1987) Relative novelty of conditioning context influences directionality of glycemic conditioning. J Exp Psychol [Anim Behav] 13:144–149

Fudim OK (1978) Sensory preconditioning of flavors with a formalin-produced sodium need. J Exp Psychol [Anim Behav] 4:276–285

Garcia J, Lasiter PS, Bermudez-Rattoni F, Deems DA (1985) A general theory of aversion learning. Ann NY Acad Sci 43:8–21

Ghent L (1957) Some effects of deprivation on eating and drinking behavior: J Comp Physiol Psychol 50:172–176

Hall WG, Blass EM (1977) Orogastric determinants of drinking in rats: interaction between absorptive and peripheral controls. J Comp Physiol Psychol 91:365–373

Holland PC (1983) Occasion-setting in Pavlovian feature positive discriminations. In: Commons ML, Herrnstein RL, Wagner AR (eds) Quantitative analyses of behavior: discrimination processes, vol 4. Ballinger, New York, pp 183–206

Holland PC (1984) The origins of Pavlovian conditioned behavior. In: Bower G (ed) The psychology of learning and motivation, vol 18. Prentice-Hall, Englewood Cliffs, NJ, pp 129–173

Holland PC (1985) The nature of conditioned inhibition in serial and simultaneous feature negative discriminations. In: Miller RR, Spear NE (eds) Information processing in animals: conditioned inhibition. Erlbaum, Hillsdale, NJ, pp 267–297

Holland PC (1988) Excitation and inhibition in unblocking. J Exp Psychol [Anim Behav] 14:261–279

Holland PC (1989a) Acquisition and transfer of conditional discrimination performance. J Exp Psychol [Anim Behav] 15:154–165

Holland PC (1989b) Feature extinction enhances transfer of occasion setting. Anim Learn Behav 17:269–279

Holland PC (1989c) Transfer of negative occasion setting and conditioned inhibition across conditioned and unconditioned stimuli. J Exp Psychol [Anim Behav] 15:311–328

Holland PC (1990a) Event representation in Pavlovian conditioning: image and action. Cognition 37: 105–131

Holland PC (1990b) Forms of memory in Pavlovian conditioning. In: McGaugh JL, Weinberger NM, Lynch G (eds) Brain organization and memory: cells, systems, and circuits. Oxford University Press, New York

Holland PC, Rescorla RA (1975) The effect of two ways of devaluing the unconditioned stimulus after first- and second-order appetitive conditioning. J Exp Psychol [Anim Behav] 1:355–363

Hull CL (1933) Differential habituation to internal stimuli in the albino rat. J Comp Psychol 16:255–273

Hull CL (1934) Learning: II. The factor of the conditioned reflex. In: Murchison C (ed) A handbook of general experimental psychology. Clark University Press, Worcester, MA, pp 382–455

Hull CL (1945) Discrimination of stimulus configurations and the hypothesis of afferent neural interaction. Psychol Rev 52:133–142

Hunt GL, Smith WJ (1967) Pecking and initial drinking responses in young domestic fowl. J Comp Physiol Psychol 64:230–236

Jenkins HM, Moore BR (1973) The form of the auto-shaped response with food or water reinforcers. J Exper Anal Behav 20:163–181

Kamin LJ (1968) Attention-like processes in classical conditioning. In: Jones MR (ed) Miami symposium on the prediction of behavior: aversive stimulation. University of Miami Press, Coral Gables, FL, pp 9–32

Kendler HH (1946) The influence of simultaneous hunger and thirst upon the learning of two opposed spatial responses in the white rat. J Exp Psychol 36:212–220

Kissileff HR (1973) Nonhomeostatic controls of drinking. In: Epstein AN, Kissileff HR, Stellar E (eds) The neuropsychology of thirst: new findings and advances in concepts. Winston, Washington, DC, pp 163–198

Konorski J (1967) Integrative activity of the brain. University of Chicago Press, Chicago

Leeper R (1935) The role of motivation in learning: a study of the phenomenon of differential motivational control of the utilization of habits. J Gen Psychol 46:3–40

Lysle DT, Lyte M, Fowler H, Rabin BS (1987) Shock-induced modulation of lymphocyte reactivity: suppression, habituation, and recovery. Life Sci 41:1805–1814

Mackintosh NJ (1975) A theory of attention: variations in the associability of stimuli with reinforcement. Psychol Rev 82:276–298

Mackintosh NJ (1983) Conditioning and associative learning. Oxford University Press, Oxford

Mendelson J, Chillag D (1970) Tongue cooling: a new reward for thirsty rodents. Science 170: 1418–1421

Mook DG (1988) On the organization of satiety. Appetite 11:27–39

Mook DG, Wagner S (1988) Sham drinking of glucose solution in rats: some effects of hydration. Appetite 10:71–87

Mook DG, Dreifuss S, Keats PH (1986) Satiety for glucose in the rat: the specificity is postingestive. Physiol Behav 36:897–901

Mook DG, Wagner S, Schwartz LA (1988) Glucose sham drinking in the rat: satiety for the sweet taste without the sweet taste. Appetite 10:89–102

Mowrer OH (1960) Learning theory and behavior. Wiley, New York

Nicolaïdis S, Rowland N (1974) Long-term self-intravenous 'drinking' in the rat. J Comp Physiol Psychol 87:1–15

Novin D, Miller NE (1962) Failure to condition thirst induced by feeding dry food to hungry rats. J Comp Physiol Psychol 55:373–374

Oatley K, Dickinson A (1970) Air drinking and the measurement of thirst. Anim Behav 18:259–265

Pelchat M, Grill HJ, Rozin P, Jacobs J (1983) Quality of acquired responses to tastes by Rattus norvegicus depends on type of associated discomfort. J Comp Physiol Psychol 97:140–153

Poulos CX, Hinson RE (1984) A homeostatic model of Pavlovian conditioning: tolerance to scopolamine-induced adipsia. J Exp Psychol [Anim Behav] 10:75–89

Premack D (1965) Reinforcement theory. In: Levine D (ed) Nebraska symposium on motivation: 1965. University of Nebraska Press, Lincoln, NE

Ramachandran R, Pearce JM (1987) Pavlovian analysis of interactions between hunger and thirst. J Exp Psychol [Anim Behav] 13:182–192

Rescorla RA (1985) Conditioned inhibition and facilitation. In: Miller RR, Spear NE (eds) Information processing in animals: conditioned inhibition. Erlbaum, Hillsdale, NJ, pp 299–326

Rescorla RA (1988) Behavioral studies of Pavlovian conditioning. Ann Rev Neurosci 11:329–352

Rescorla RA, Freberg L (1978) The extinction of within-compound flavor associations. Learn Motiv 9:411–427

Rescorla RA, Holland PC (1976) Some behavioral approaches to the study of learning. In: Bennett E, Rozensweig MR (eds) Neural mechanisms of learning and memory. MIT Press, Cambridge, MA, pp 165–192

Rescorla RA, Holland PC (1982) Behavioral studies of associative learning in animals. Ann Rev Psychol 33:265–308

Rescorla RA, Solomon RL (1967) Two process learning theory: relationships between classical conditioning and instrumental learning. Psychol Rev 74:151–182

Rolls BJ, Rolls ET (1982) Thirst. Cambridge University Press, Cambridge

Rolls BJ, Wood RJ, Stevens RM (1978) Effects of palatability on body fluid homeostasis. Physiol Behav 20:15–19

Rolls BJ, Rolls ET, Rowe EA (1981) Sensory specific satiety in man. Physiol Behav 27:137–142

Seligman, MEP, Ives CE, Ames H, Mineka S (1970) Conditioned drinking and its failure to extinguish: avoidance, preparedness, or functional autonomy? J Comp Physiol Psychol 71:411–419

Sheffield FD, Roby TB (1950) Reward value of a non-nutritive sweet taste. J Comp Physiol Psychol 43:471–481

Shirley M (1929) Spontaneous activity. Psychol Bull 26:341–365

Teitelbaum P (1966) The use of operant methods in the assessment and control of motivational states. In: Honig WK (ed) Operant behavior: areas of research and application. Appleton-Century-Crofts, New York

Tinbergen N (1951) The study of instinct. Clarendon Press, Oxford

Tolman EC (1932) Purposive behavior in animals and men. Appleton-Century, New York

Tolman EC, Brunswik E (1935) The organism and the causal texture of the environment. Psychol Rev 42:43–77

van Vort W, Smith GP (1987) Sham feeding experience produces a conditioned increase of meal size. Appetite 11:54–61

Weisinger RS (1975) Conditioned and pseudoconditioned thirst and sodium appetite. In: Peters G, Fitzsimons JT, Peters-Haefeli L (ed) Control mechanisms of drinking. Springer-Verlag, Berlin, pp 148–154

Wikler A (1973) Conditioning of successive adaptive responses to the initial effects of drugs. Cond Reflex 8:193–210

Wolf AV (1958) Thirst: physiology of the urge to drink and problems of water lack. Thomas, Springfield IL

Young PT (1955) The role of hedonic processes in motivation. In: Nebraska symposium on motivation. University of Nebraska Press, Lincoln NE

Chapter 18

Effects of Eating on Drinking

F.S. Kraly

Introduction

The relation between eating and drinking for adult mammals is defined by three characteristics. First, eating and drinking tend to occur together: they are episodic but temporally contiguous behaviours, entrained to environmental cues and occurring mainly during the waking hours. Second, eating and drinking appear to be reciprocal: sipping interrupts eating. Eating can have dehydrational consequences that elicit drinking, and drinking can disinhibit eating by preparing the palate or by removing dehydration. Third, eating and drinking serve homeostasis: ingestion will preclude and/or repair deficits in fuels, fluids or electrolytes.

This close relation between eating and drinking invites hypotheses concerning whether one behaviour can activate physiological mechanisms for eliciting the other. The evidence that drinking behaviour can elicit eating comes mainly from experiments that examine the effects of dehydration or rehydration on eating and food intake in rats. Such experiments generally find that dehydration or dipsogenic stimuli inhibit food intake (Gutman and Krausz 1969; Hsiao 1970; Rolls 1975), and that rehydration shortens the latency to initiate eating (Kakolewski and Deaux 1970; Deaux and Kakolewski 1971) and increases food intake (Hsiao and Trankina 1969; Hsiao and Smutz 1976).

Despite such findings, there is no strong argument for the pre-eminence of hydrational control of spontaneous eating in mammals (Le Magnen 1985). In contrast, much can be said about the role of eating behaviour in the control of spontaneous drinking.

The Importance of Food-Related Drinking

Food-related drinking dominates the spontaneous drinking behaviour of humans (Phillips et al. 1984; de Castro 1988; Engell 1988) and other mammals (Kraly 1984a), accounting for greater than 70% of daily fluid intake. Despite this notable contribution to the water economy, relatively little is known about the physiological variables that intervene between eating and drinking behaviour. The textbooks instead provide us with a physiology of thirst that explains much about drinking behaviour activated to repair deficits in cellular or extracellular fluid compartments. Little work has been done, however, to explore the extent to which such controls to repair homeostasis contribute to spontaneous fluid intake.

Investigation into this aspect can be difficult because experimental strategies designed to study spontaneous drinking behaviour can be cumbersome; they must accommodate the study of behaviour that is marked by episodic short bursts (Kissileff 1969a), and that is shaped by learning (Fitzsimons and Le Magnen 1969), the palatability of the fluid (Rolls et al. 1978) and the cost of a drink (Marwine and Collier 1979). Because it is easier to have experimental control over drinking elicited by water deprivation or injection of agents that are dipsogenic, the study of physiological and psychological processes governing spontaneous, ordinary drinking is a task relatively untouched. Getting to this task is possible through the study of drinking elicited by eating due to its prominence among the behaviours of mammals and its amenability to experimental strategies used successfully in the study of postprandial satiety for food (Kraly 1984a). Moreover, the problem of food-related drinking encourages the fitting of new findings into a conceptualization (Fig. 18.1) for the neuroendocrine control of drinking elicited by the stimulus properties of food and the dehydrational consequences of eating, modified by learning that stimulus properties of food may signal postprandial dehydration.

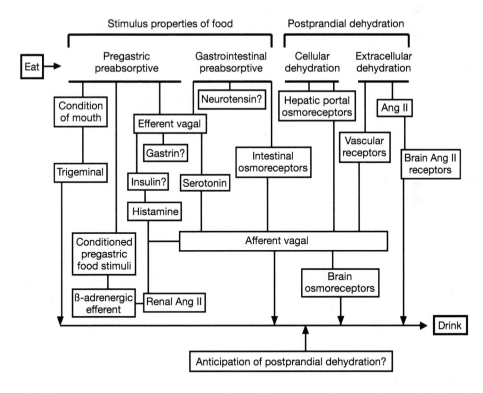

Fig. 18.1. A speculative summary of putative signals and neural substrates which may be activated sequentially to interact in control of the initiation and maintenance of food-related drinking. Abbreviation: Ang II, angiotensin II.

Stimulus Properties of Food

Food can provide visual and olfactory stimulation prior to ingestion. The properties of ingested food include mechanical, chemical, temperature and osmotic, each of which can be detected by specialized receptors distributed along the mucosa from the mouth through the intestine. What can be said about the potential for the stimulus properties of food to elicit drinking?

Pregastric Properties

Food taken into the mouth and swallowed in rats eating with an open gastric fistula, so-called pregastric food-contingent stimulation, is sufficient to elicit drinking (Kraly 1983b) in the absence of detectable systemic dehydration (Kraly 1990). Which pregastric stimulus properties of food are important for drinking?

Oral Factors

The discomforts caused by extreme spiciness, heat or the ability of a food to create a dry mouth are well-known to provoke fluid intake for the purposes of relieving the pain or facilitating chewing. These phenomena, though familiar to us all, have not received the attention of studies on the physiology of drinking. Despite no systematic study of the mechanisms by which oral stimulus properties of food can elicit drinking, there are relevant findings. First, trigeminal deafferentation produces profound decreases in water intake in rats (Jacquin and Zeigler 1983). Although decreases in ingestion in such lesioned rats are not specific to fluid intake, the findings suggest that the trigeminal mediation of oral cues is important for the ability of food in the mouth to elicit drinking. Second, pharmacological pharyngolaryngeal deafferentation inhibits water intake without apparent effect on eating in rats (Miyaoka et al. 1987a), and laryngectomy in humans results in decreased reported incidence of drinking water with meals (Miyaoka et al. 1987b). Third, the surgical (Epstein et al. 1964; Kissileff 1969b) or pharmacological (Chapman and Epstein 1970) prevention of salivation in otherwise neurologically intact rats produces a pattern of so-called prandial drinking which appears to facilitate the chewing and swallowing of dry food. This phenomenon reveals that the nervous system is ready for the task of wetting a dry mouth. Fourth, pregastric stimulation by saline (or water) and systemic signals for plasma osmolality converge in the hypothalamus providing a potential neural substrate for a rapid drinking response to a salty food (Nicolaïdis 1969). Finally, humans report a preponderance of oral cues for thirst (i.e., various kinds of oral discomfort) when drinking around mealtime (Phillips et al. 1984). Apart from these findings, little else is known about how stimulus properties of food in the mouth elicit drinking.

Neuroendocrine Factors

Pregastric food-contingent stimulation has peripheral neuroendocrine consequences including the release of pancreatic insulin and gastric histamine and gastrin, each of which are reported to be dipsogenic.

Insulin. Although pregastric stimulation in the rat sham-feeding with open gastric fistula is sufficient for the vagally mediated (Berthoud and Jeanrenaud 1982; Berthoud et al. 1983; Ionescu et al. 1983) release of pancreatic insulin, there is only little indirect evidence for a role for endogenous insulin in food-related drinking. Exogenous insulin is dipsogenic in humans (Vijande et al. 1989) and rats (Novin 1964; Booth and Pitt 1968). Insulin-induced drinking in rats can be inhibited by systemic antagonism of receptors for histamine (Kraly et al. 1983), consistent with the finding that exogenous insulin can release gastric mucosal histamine (Ekelund et al. 1982). Together these findings permit the hypothesis that the eating-elicited release of pancreatic insulin activates, sensitizes or potentiates a histaminergic (see below) control of food-related drinking.

Gastrin. The eating-elicited release of gastrin, which interacts with histamine in the control of secretion of gastric acid (Grossman 1981), may be important for food-related drinking. Exogenous gastrin is dipsogenic in pigs (Houpt et al. 1986), but its apparent failure to elicit drinking in rats (Anderson and Houpt 1989) tempers enthusiasm for a hypothesis concerning a direct role for gastrin in food-related drinking.

Histamine. Eating provokes the vagally mediated (Beaven et al. 1968) release of histamine from endocrine-like non-mast cells (Soll et al. 1981; Soll and Berglindh 1987) in the gastric mucosa of rats. Histamine, also found in human gastric mucosa (Hakanson and Sundler 1987), elicits drinking when injected subcutaneously in rats. The dipsogenicity of systemic histamine depends primarily on peripheral H_1 and H_2 histamine receptors (Kraly 1983a; Kraly and Arias 1990) in rats but only H_2 receptors in pigs (Houpt et al. 1986).

Pregastric food-contingent stimulation, isolated in the sham-feeding rat (a) activates abdominal vagal efferents (Berthoud and Jeanrenaud 1982) which should release gastric mucosal histamine (Beaven et al. 1968), and (b) elicits drinking (Kraly 1984b). Such drinking is abolished by combined antagonism of peripheral H_1 and H_2 receptors (Kraly 1984b), demonstrating that the dipsogenicity of all pregastric stimulus properties of liquid food depend upon endogenous histamine.

Gastric and Intestinal Properties

Gastrointestinal food-contingent stimulation (by infusion of liquid food into the stomach) is sufficient to elicit drinking in rats (Kraly 1990). Which gastrointestinal stimulus properties of food are important for drinking?

Neurotensin

Eating appears to release neurotensin from stomach and intestine in humans (Mashford et al. 1978) and rats (Carraway and Leeman 1976; George et al. 1987). Because exogenous neurotensin is dipsogenic (Stanley et al. 1983) and releases histamine from mast cells in rats (Carraway et al. 1982), a role can be hypothesized for gastrointestinal neurotensin activating or interacting with a histaminergic control of food-related drinking. This hypothesis has not been evaluated.

Serotonin

Preabsorptive distension of the intestinal lumen elicits the vagally mediated release of serotonin from gastrointestinal mucosal endocrine cells (Rubin and Schwartz 1983; Gronstad et al. 1987) and drinking in rats (Davis et al. 1975; Bernstein and Vitiello 1978). Because exogenous serotonin elicits drinking in rats (Lehr and Goldman 1973; Kikta et al. 1981), a hypothesis for serotoninergic control of food-related drinking is reasonable. Antagonism of peripheral receptors for serotonin inhibits food-related drinking under some experimental conditions, but not others (Kraly 1990). More work is necessary to evaluate a serotoninergic hypothesis.

Gastrointestinal Osmoreceptors

Food-contingent stimulation of osmoreceptors in the intestine (Mei 1983) might contribute to drinking as these receptors inform the brain through vagal afferents (Mei and Garnier 1986) about the osmotic properties of ingesta. This hypothesis has not been tested directly, but it is consistent with findings that selective gastric vagotomy (Jerome and Smith 1982b) or combined transection of gastric and coeliac vagal branches (Kraly 1978) inhibits drinking elicited by cellular dehydration and drinking elicited by eating (Kraly et al. 1975, 1978; Kraly and Miller 1982).

Summary

This consideration of the stimulus properties of ingested food as potential stimuli for drinking behaviour encourages hypotheses for oropharyngeal and gastrointestinal mechanoreceptors, chemoreceptors, thermoreceptors and osmoreceptors together with peripheral neuroendocrine signals including insulin, gastrin, histamine, neurotensin and serotonin as factors in control of food-related drinking. Among these testable hypotheses, the one for a vagally mediated peripheral histaminergic control of drinking has received the most attention (Kraly 1985, 1990).

Pharmacological antagonism of peripheral histamine receptors in rats (using doses of antagonists with demonstrated pharmacological and behavioural selectivity) inhibits food-related drinking (without inhibiting food intake) by approximately 25%–60% depending on the experimental paradigm (Kraly and June 1982; Kraly 1983b; Kraly and Specht 1984), whereas antagonism of histamine receptors in brain fails to produce comparable effects (Kraly and Arias 1990). The finding that as much as 60% of drinking elicited by eating can be attributed to a histaminergic control (Kraly and Specht 1984) emphasizes the potential importance of preabsorptive food-contingent stimulation in the control of spontaneous drinking behaviour in rats.

Dehydrational Consequences of Eating

Relative dehydration can result from the movement of fluid into the gastrointestinal tract within minutes after food is swallowed (Lepkovsky et al. 1957; Oatley and Toates 1969). Does local or systemic dehydration contribute to the control of food-related drinking?

Cellular Dehydration

Plasma osmolality can increase within 10–60 minutes after eating dry food in rats (Almli and Gardina 1974; Deaux et al. 1970), dogs (Rolls et al. 1980) and ponies (Houpt et al. 1986). The site(s) of the osmoreceptors relevant for food-related drinking has not been specified, but it is possible that they are in brain (see Epstein 1983) or in the portal circulation (Haberich 1968; Adachi et al. 1976; Stoppini and Baertschi 1984) where the osmoreceptors activate hepatic vagal afferents (Kobashi and Adachi 1986, 1988).

Extracellular Dehydration and Angiotensin II

Hypovolaemia can follow eating of dry food in rats (Almli and Gardina 1974; Nose et al. 1986), ponies (Houpt et al. 1986) and horses (Clarke et al. 1988). Moreover, renal sympathetic nerve activity increases within seconds of the onset of eating in cats (Matsukawa and Ninomiya 1987), and plasma renin activity increases within 1 hour after a meal of dry food in rats (Mann et al. 1980; Rowland et al. 1987), sheep (Blair-West and Brook 1969) and humans (Brandenberger et al. 1985, 1987).

These findings suggest that the renal renin–angiotensin system plays a role in food-related drinking. Recent work using captopril to block the synthesis of angiotensin II (Ang II) in brain and/or periphery reveals a role for peripheral angiotensin: blockade of synthesis of Ang II inhibits food-related drinking by approximately 35% in the rat (Kraly and Corneilson 1990).

Summary

Postprandial dehydration can occur and contribute to food-related drinking. The relative importance of dehydration for drinking elicited by eating is questioned, however, by two kinds of findings. First, intragastric or intravenous infusions of water, in amounts sufficient to preclude systemic dehydration, fail to abolish drinking elicited by eating in rats (Fitzsimons 1957; Kissileff 1969b, 1973; Rowland and Nicolaïdis 1976). Second, humans (Phillips et al. 1984) and young pigs (Houpt and Anderson 1990) consume the majority of their daily fluids around mealtime without detectable dehydration.

Interaction of Stimulus Properties of Food and Postprandial Dehydration

Food-related drinking is inhibited by antagonism of histamine receptors or by blockade of synthesis of Ang II (see above), but histaminergic antagonism and inhibition of Ang II synthesis are not additive in their effects on food-related drinking in rats (Kraly and Corneilson 1990). This finding, together with the fact that blockade of synthesis of Ang II can attenuate or abolish drinking elicited by exogenous histamine (Evered and Robinson 1984; Kraly and Corneilson 1990), permits speculation that Ang II to some degree mediates a histaminergic mechanism for food-related drinking. For example, histamine could activate an Ang II mechanism through the putative endocrine and/or neurocrine role of histamine (Banks et al. 1978) in kidney by

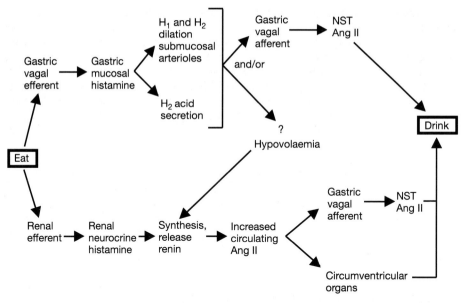

Fig. 18.2. The working hypothesis for interaction of histamine and angiotensin II (Ang II) in the control of food-related drinking. Eating can activate gastric vagal efferents to elicit neurocrine cholinergic and paracrine histaminergic stimuli for acid secretion; histamine's H_1 and H_2 receptor-mediated effects on dilation of submucosal arterioles are supportive and permissive for the H_2-mediated secretion of gastric acid. These paracrine effects of histamine may activate gastric vagal afferents projecting to nucleus of solitary tract. Moreover, movement of fluid into the gastrointestinal tract during eating and/or postprandial redistribution of blood flow to favour the mucosal layer of the gut at the expense of the systemic circulation may contribute to the onset of hypovolaemia resulting in activation of the renal renin–angiotensin system to elicit drinking through circulating Ang II. Eating can also activate renal nerve and then renal neurocrine histamine resulting in activation of the renin–angiotensin system to elicit drinking through circulating Ang II. Ang II could elicit drinking through gastric vagal afferents and nucleus of solitary tract in addition to circumventricular sites in brain (e.g., subfornical organ). (For further explanation see Kraly (1990).) Abbreviations: Ang II, angiotensin II; NST, nucleus of solitary tract.

increasing the synthesis and release of renin (Gerber and Nies 1983; Radke et al. 1985). Alternatively, histamine (by virtue of its role in the secretion of gastric acid) could provoke hypovolaemia secondary to the postprandial redistribution of blood flow favouring the mucosal layer of the gut (Bond et al. 1979) at the expense of the systemic circulation. Either the action of histamine in the kidney or hypovolaemia would result in increased genesis of Ang II. This hypothesis describing a histamine–Ang II interaction in the control of food-related drinking is presented in greater detail in Fig. 18.2.

This working hypothesis, quite testable in its various claims, links a histaminergic control of food-related drinking that appears to be activated in advance (Kraly 1990) of postprandial disturbances to body fluid homeostasis to an Ang II mechanism that is traditionally (Epstein 1983) presumed to be part of the renal response to deficit in extracellular fluid. Such an interactive function may represent a neuroendocrine link between a physiological control (histamine) activated to preclude fluid deficit and a physiological control (Ang II) activated to repair deficit. This link may permit the preabsorptive pregastric stimulus properties of food to signal through histamine a potential fluid deficit. For example, the greater metabolic need for water for rats maintained on a high protein diet is reflected in increased water intake (Fitzsimons and Le Magnen 1969) and plasma renin activity (Paller and Hostetter 1986). Water intake is increased on the first day of eating the protein-rich food (following maintenance on a carbohydrate-rich food), but the drinking of this extra water is not temporally associated with eating until the second or third day on the protein-rich diet (Fitzsimons and Le Magnen 1969). This delay in associating drinking (to meet the increased metabolic demand) with eating a high protein diet has been attributed (Fitzsimons and Le Magnen 1969) to the time needed by the rat to learn to relate pregastric stimulus properties of food to the challenge to fluid homeostasis that follows the digestion of that food. Thus, the spontaneous pattern of eating and drinking may to some degree be attributable to learning that certain oropharyngeal food cues predict future fluid deficit. Given that such learning might require a histaminergic signal activated by pregastric food-contingent stimulation and a renal renin–angiotensin mechanism activated by fluid deficit, this working hypothesis (Fig. 18.2) may provide a starting point for experiments aimed at understanding how the brain and physiology mobilize behaviour in anticipation of a physiological deficit.

Learning Processes in Food-Related Drinking

Learning that one can avoid the dehydrational consequences of ingesting, for example, a familiar salty food, could provoke the ingestion of fluid before eating or early in a meal, soon enough to preclude all or part of the anticipated postprandial dehydrational consequences of eating. What is the evidence for a role for learning in food-related drinking?

Learning about Stimulus Properties of Food

The stimulus properties of pregastric food-contingent stimulation during sham-feeding are sufficient to elicit the secretion of gastric acid (Weingarten and Powley 1980) and drinking behaviour in advance of systemic dehydrational consequences of eating in rats (Kraly 1990). Moreover, such secretory and drinking responses can be conditioned. This conditioned cephalic phase of acid secretion (Weingarten and Powley 1981) depends on the activation of peripheral histamine receptors for its acquisition and expression (Kraly 1990). In contrast, the acquisition and expression of conditioned water intake (i.e., drinking in anticipation of pregastric stimulus properties of food) does not depend on activation of histamine receptors (Kraly 1990). Another point of contrast between conditioned secretion and water intake is the finding that blockade of vagal efferents, using the peripheral anticholinergic atropine methyl nitrate, prevents expression of a conditioned secretory response without effect on conditioned water intake (Kraly 1990).

Taken together these results suggest the vagal cholinergic and histaminergic control of the acquisition and maintenance of the conditioned cephalic phase of acid secretion, apparently separate from the physiological control of conditioned water intake. The conditioned water intake appears to depend at least in part on Ang II, because captopril-induced blockade of synthesis of Ang II inhibits the expression of conditioned water intake without effect on conditioned secretion (Kraly 1990). A role for Ang II in drinking in anticipation of pregastric stimulus properties of food is consistent with the finding that the anticipation of sham-feeding in dogs can elicit ß-adrenergic sympathetic activity (Skinner and Randall 1985; Kostarczyk 1986) which should be sufficient to activate the renal renin–angiotensin system. These findings suggest a new type of role for Ang II in drinking – drinking in anticipation of eating, prior to a systemic fluid deficit.

In summary, the pregastric stimulus properties of food can serve as unconditional stimuli for eliciting gastric acid secretory and drinking responses in the absence of systemic dehydration in rats. These pregastric unconditional food stimuli can, when paired with neutral stimuli, support the formation of conditioned secretory and drinking responses. The extent to which such conditioned drinking responses contribute to spontaneous food-related drinking remains to be studied.

Learning about Dehydrational Consequences of Eating

Spontaneous drinking may in part reflect learning to relate pregastric stimulus properties of food to postprandial dehydrational consequences of eating (see above). The claim (Fitzsimons and Le Magnen 1969) that rats can learn which oropharyngeal food cues signal forthcoming dehydration has not been evaluated. Failure to evaluate experimentally this reasonable and long-held supposition is a problem. Recent attempts at examining whether systemic injections of histamine, Ang II or hypertonic saline (each intended to mimic a different putative signal for food-related drinking) could serve as unconditional stimuli in paradigms designed to demonstrate conditioned water intake have produced interesting findings. For example, subcutaneous histamine (or intraperitoneal hypertonic saline) can serve as an unconditional stimulus for conditioned initiation of drinking (i.e., latency to drink) in response to a salient conditioned food stimulus without evidence for conditioned water intake (i.e., 30-min intake subsequent to presentation of the conditioned stimulus; see Kraly 1990). Thus, it is possible to demonstrate conditioned appetitive behaviour without conditioned consummatory behaviour in rats drinking around mealtime.

Although these findings are preliminary and carry some of the burden common to negative findings – that of being never totally convincing – the failure to demonstrate conditioned water intake agrees with a small but persistent literature testifying to the difficulty of ever conditioning drinking. In general, when conditioned water intake is observed, or rats may learn to drink to anticipate fluid loss or need, the phenomenon is rather unimpressive in its magnitude if it is there at all (Novin and Miller 1962; Weissman 1972; D'Amato 1974; Zamble et al. 1980; see also Weingarten 1985; Turkkan 1989). Moreover, when conditioned water intake can be observed postprandially, subsequent to the ingestion of the presumed conditioned stimulus (food), rats do not learn to drink prior to a meal in anticipation of its dehydrational consequences (Lucas et al. 1989).

In summary, as reasonable as it may seem that humans and rats attend to the stimulus properties of food in a way that supports learning to drink to avoid dehydrational consequences of eating, there is no convincing evidence that mammals do this nor is it clear how learning is important for spontaneous food-related drinking.

Neural Substrates

The stimulus properties of ingested food activate digestive processes as food moves along from the mouth into the lumen of the gastrointestinal tract. The preabsorptive food-contingent stimuli of pregastric, gastric and intestinal origin are joined by postabsorptive consequences of eating to co-ordinate digestion of food. This process can be initiated by the anticipation of eating, for which the brain must have a role. Thus, the brain, autonomic nervous system and gastrointestinal neuroendocrine physiology, as they are activated to control digestive processes, may simultaneously control food-related drinking.

Parasympathetic

The central role of vagal innervation of the abdominal viscera for integrating the physiological mechanisms that process ingested food is well known (Makhlouf 1974). Vagal efferents mediate the ability of ingested food to activate many of the putative signals for drinking (see above); this predicts the finding that blockade of vagal efferents using atropine methyl nitrate inhibits drinking elicited by eating in rats (Kraly 1990). In addition, vagal afferents play a prominent role in drinking (Fig. 18.1), a claim supported by three types of findings. First, various types of selective (and combined selective) transections of abdominal vagus abolish or inhibit food-related drinking in rats (Kraly et al. 1975, 1978, 1986; Smith and Jerome 1983). Second, abdominal vagotomy can inhibit drinking elicited by either pregastric (Kraly 1984a) or gastrointestinal (Kraly 1990) food-contingent stimulation in rats. Third, abdominal vagotomy inhibits the ability of many of the putative signals for food-related drinking to elicit drinking when these signals are injected systemically, including histamine (Kraly and June 1982; Kraly and Miller 1982), serotonin (Simansky et al. 1982), cellular dehydration in rats (Kraly et al. 1975; Kraly 1978; Jerome and Smith 1982b) and humans (Kral 1983) and Ang II (Jerome and Smith 1982a; Simansky and Smith 1983). These lesion-induced effects on drinking are generally attributed to the loss of vagal afferents because peripheral cholinergic blockade of vagal efferents using atropine methyl nitrate generally fails to mimic the effects of vagotomy (e.g., see Kraly and June 1982; Kraly and Miller 1982).

In summary, the abdominal vagus appears to be important for the abilities of stimulus properties of food and postprandial dehydrational consequences of eating to elicit drinking. How afferent vagal signals become functionally integrated with somatosensory trigeminal and gustatory signals (Hamilton and Norgren 1984; Norgren et al. 1989) for food-related drinking remains unknown.

Sympathetic

The relative importance of sympathetic efferents and afferents for the control of food-related drinking is completely unknown. The recent finding (Kraly and Corneilson 1990) that Ang II participates in drinking elicited by eating raises the possibility that sympathetic efferent ß-adrenergic activation of the renal renin–angiotensin system or perhaps of local mesenteric arterial Ang II (Nakamaru et al. 1986; see also Levens 1985) makes a contribution to spontaneous drinking behaviour. Moreover, speculation on a role for sympathetic afferents follows the evidence of splanchnic mediation of signals from osmoreceptors in intestinal mesentery or portal vein (Kwon and Baertschi 1989).

Brain

How does the brain detect and organize signals for food-related drinking? Afferent vagal function for various abdominal sensory fields is represented in medulla (Appia et al. 1986; Cechetto 1987), in particular the nucleus of the solitary tract and area postrema (Kalia and Mesulam 1980; Contreras et al. 1982; Leslie et al. 1982; Gwyn et al. 1985; Miceli and Malsbury 1985: Norgren and Smith 1988). Integration of visceral, trigeminal and gustatory stimuli occurs (Norgren et al. 1989), but its relevance for behaviour remains unknown (Norgren 1983). Further integration of stimuli of peripheral origin by hypothalamic, limbic, thalamic and cortical mechanisms (Norgren and Grill 1982; Sawchenko 1983; Kannan and Yamashita 1985) is likely to be important for the spontaneous pattern of drinking. Despite this pregnant neuroanatomy, virtually nothing is known about how the brain controls drinking elicited by eating. Those lesions in the brain that appear to disrupt drinking elicited by eating (e.g., zona incerta; Walsh and Grossman 1973; Evered and Mogenson 1976) seem to do so due to sensorimotor impairment making ingestion difficult (Evered and Mogenson 1977).

In summary, neuroanatomical and neurophysiological facts describing relations between brain and signals from peripheral sites provide an arena for future experimental study of how the brain organizes drinking as the brain simultaneously co-ordinates postprandial neuroendocrine processes.

Formulation

The stimulus properties of food, in advance of postprandial dehydrational consequences of eating, appear to be capable of signalling the initiation and maintenance of drinking behaviour. Numerous hypotheses concerning the food-contingent stimulus control of drinking can be posited (Fig. 18.1). Each of these hypotheses can be tested. Of the various neuroendocrine factors which appear to play a role in food-related drinking, gastric mucosal histamine has received the most attention in the laboratory. The fact that a histaminergic mechanism in the rat may account for as much as 60% of food-related drinking, which itself accounts for the vast majority of spontaneous fluid intake, emphasizes the potential importance of stimulus properties of food for eliciting drinking behaviour.

The signals for drinking generated by the stimulus properties of food are likely to interact with dehydrational consequences of eating to elicit drinking. Consider, for example, the putative role for Ang II in food-related drinking. Ang II, traditionally considered to signal drinking in response to deficit in extracellular fluid, may interact with a histaminergic mechanism activated in advance of fluid deficit (Fig. 18.2). This working hypothesis provides a point of experimental attack for the study of the interaction of a physiological process to preclude deficit and a physiological process to repair deficit. The study of this problem may help the understanding of how the brain and peripheral neuroendocrine mechanisms organize behaviour intended to avoid physiological troubles.

Such drinking to avoid dehydration could require learning about the predictive value of stimulus properties (i.e., tastes, textures, etc.) of those foods which have postprandial dehydrational consequences. Although there is preliminary evidence that stimulus properties of food can serve as conditioned stimuli, there is no demonstration that learning plays a prominent role in spontaneous drinking behaviour.

Should learning about dehydrational consequences of eating not be necessary for drinking in advance of fluid deficit, then what mechanisms account for the fact that animals and humans drink in the absence (or in advance) of dehydration? Mechanisms for circadian timing of behavioural and physiological processes could ensure homeostasis in advance of it being challenged (Moore-Ede, 1986). The circadian pattern of ingestion for rats suggests two independent, coupled oscillators for eating and drinking, entrained to environmental cues (Spiteri 1982; Mori et al. 1983; Strubbe et al. 1986). The extent to which these oscillators reflect diurnal variation in the functional activity of gastrointestinal neuroendocrine processes (e.g., Jorde and Burhol 1985) or in the potency of food-related dipsogenic stimuli (e.g., Oishi et al. 1987) is unknown. It is possible, however, that these circadian factors, together with the ability to learn that it is important to drink when the price of fluid is right, due to the economics of variable costs and shifting availability of fluids (Marwine and Collier 1979; Collier 1986), may be relatively more important for the control of spontaneous drinking than is learning about dehydration.

The quality of available fluids undoubtedly also contributes in an important way to spontaneous ingestion. Palatability, through gustatory afferents known to interact with signals from trigeminal nerve and abdominal vagal afferents (e.g., Hamilton and Norgren 1984), can provoke intake of enough fluid to ensure fluid homeostasis and even to challenge homeostasis with a surfeit of water (Rolls et al. 1978; Toates 1979). How the various factors which define "quality" of fluids (e.g., sweetness, temperature) contribute to fluid intake at mealtime remains unknown.

In summary, fluid intake at mealtimes may be governed by factors as diverse as the palatability of the fluid source, the relative cost or availability of fluid, the stimulus properties of food, ingested food's capacity to dehydrate, and circadian patterns of ingestion. These factors (palatability, availability, cost, time) are held constant in most experiments designed to study how the stimulus properties and dehydrational potential of food control drinking. Consequently, nearly nothing has been done to assess the relative contribution of these other factors to food-related drinking. Moreover, little or nothing is known about the termination of drinking elicited by eating. This important issue merits study, because it may not be safe to assume that the signals for termination of drinking in response to dehydration (Epstein 1983; Smith 1986) are signals that terminate drinking around mealtime.

In conclusion, there are numerous testable hypotheses concerning the stimulus properties of food and dehydrational consequences of eating for the control of drinking behaviour. Experimentally examining some of these hypotheses could provide a good start toward integrating what is known about eating-induced physiological signals with mechanisms for circadian rhythms, learning about the costs of fluids and the unpleasant price of dehydration, and the attraction of drinking fluids because they taste good. Such an integration holds the promise of a more satisfying understanding of the physiology of spontaneous drinking behaviour.

References

Adachi A, Niijima A, Jacobs HL (1976) An hepatic osmoreceptor mechanism in the rat: electrophysiological and behavioral studies. Am J Physiol 231:1043–1049

Almli RC, Gardina J (1974) Ad libitum drinking of rats and vascular osmolality changes. Physiol Behav 12:231–238

Anderson CR, Houpt TR (1989) Pentagastrin fails to stimulate water drinking in rats. Appetite 12: 195–196

Appia F, Ewart WR, Pittam BS, Wingate DL (1986) Convergence of sensory information from abdominal viscera in the rat brain stem. Am J Physiol 251:G169–G175

Banks RO, Fondacaro JD, Schwaiger MM, Jacobson ED (1978) Renal histamine H_1 and H_2 receptors: characterization and functional significance. Am J Physiol 235:F570–F575

Beaven MA, Horakova Z, Severs WB, Brodie BB (1968) Selective labeling of histamine in rat gastric mucosa: application to measurement of turnover rate. J Pharmacol Exp Ther 161:320–328

Bernstein IL, Vitiello MV (1978) The small intestine and the control of meal patterns in the rat. Physiol Behav 20:417–422

Berthoud H-R, Jeanrenaud B (1982) Sham feeding-induced cephalic phase insulin release in the rat. Am J Physiol 242:E280–E285

Berthoud H-R, Niijima A, Sauter J-F, Jeanrenaud B (1983) Evidence for the role of the gastric, coeliac and hepatic branches in vagally stimulated insulin secretion in the rat. J Auton Nerv Syst 7:97–110

Blair-West JR, Brook AH (1969) Circulatory changes and renin secretion in sheep in response to feeding. J Physiol (Lond) 204:15–30

Bond JH, Prentiss RA, Levitt MD (1979) The effects of feeding on blood flow to the stomach, small bowel, and colon of the conscious dog. J Lab Clin Med 93:594–599

Booth DA, Pitt ME (1968) The role of glucose in insulin-induced feeding and drinking. Physiol Behav 3:447–453

Brandenberger G, Follenius M, Muzet A, Ehrhart J, Schieber JP (1985) Ultradian oscillations in plasma renin activity: their relationships to meals and sleep stages. J Clin Endocrinol Metab 61:280–284

Brandenberger G, Simon C, Follenius M (1987) Night–day differences in the ultradian rhythymicity of plasma renin activity. Life Sci 40:2325–2330

Carraway R, Leeman SE (1976) Characterization of radioimmunoassayable neurotensin in the rat. J Biol Chem 251:7045–7052

Carraway R, Cochrane DE, Lansman JB, Leeman SE, Paterson BM, Welch HJ (1982) Neurotensin stimulates exocytotic histamine secretion from rat mast cells and elevates plasma histamine levels. J Physiol (Lond) 323:403–414

Cechetto DF (1987) Central representation of visceral function. Fed Proc 46:17–23

Chapman HW, Epstein AN (1970) Prandial drinking induced by atropine. Physiol Behav 5:549–554

Clarke LL, Ganjam VK, Fichtenbaum B, Hatfield D, Garner HE (1988) Effect of feeding on renin–angiotensin–aldosterone system of the horse. Am J Physiol 254:R524–R530

Collier G (1986) The dialogue on strategy between the economist and the resident physiologist. Appetite 7:88–89

Contreras RJ, Beckstead RM, Norgren R (1982) The central projections of the trigeminal, facial, glossopharyngeal and vagus nerves: an autoradiographic study in the rat. J Auton Nerv Syst 6: 303–322

D'Amato MR (1974) Derived motives. Ann Rev Psychol 25:83–106

Davis JD, Collins BJ, Levine MW (1975) Peripheral control of drinking: gastrointestinal filling as a negative feedback signal, a theoretical and experimental analysis. J Comp Physiol Psychol 89: 985–1002

Deaux E, Kakolewski JW (1971) Character of osmotic changes resulting in the initiation of eating. J Comp Physiol Psychol 74:248–253

Deaux E, Sato E, Kakolewski JW (1970) Emergence of systemic cues evoking food-associated drinking. Physiol Behav 5:1177–1179

de Castro JM (1988) A microregulatory analysis of spontaneous fluid intake by humans: evidence that the amount of liquid ingested and its timing is mainly governed by feeding. Physiol Behav 43: 705–714

Ekelund M, Hakanson R, Hedenbro J et al. (1982) Effects of insulin on serum gastrin concentrations, effect of acid secretion and histamine mobilization in the rat. Acta Physiol Scand 114:17–29

Engell D (1988) Interdependency of food and water intake in humans. Appetite 10:133–141

Epstein AN (1983) The neuropsychology of drinking behavior. In: Satinoff E, Teitelbaum P (eds) Handbook of behavioral neurobiology, Plenum Press, New York, pp 367–423

Epstein AN, Spector D, Samman A, Goldblum C (1964) Exaggerated prandial drinking in the rat without salivary glands. Nature 201:1342–1343

Evered MD, Mogenson GJ (1976) Regulatory and secondary water intake in rats with lesions of the zona incerta. Am J Physiol 230:1049–1057

Evered MD, Mogenson GJ (1977) Impairment of fluid ingestion in rats with lesions of the zona incerta. Am J Physiol 233:R53–R58

Evered MD, Robinson MM (1984) Increased or decreased thirst caused by inhibition of angiotensin-converting enzyme in the rat. J Physiol (Lond) 348:573–588

Fitzsimons JT (1957) Normal drinking in rats. J Physiol (Lond) 138:39P

Fitzsimons JT, Le Magnen J (1969) Eating as a regulatory control of drinking in the rat. J Comp Physiol Psychol 67:273–283

George JK, Albers HE, Carraway RE, Ferris CF (1987) Neurotensin levels in the hepatic-portal circulation are inversely related to the circadian feeding cycle in rats. Endocrinology 121:7–13

Gerber JG, Nies AS (1983) The role of histamine receptors in the release of renin. Br J Pharmacol 79: 57–61

Gronstad KO, Zinner MJ, Nilsson O, Dahlstrom A, Jaffe BM, Ahlman H (1987) Vagal release of serotonin into gut lumen and portal circulation via separate control mechanisms. J Surg Res 43:205–210

Grossman MI (1981) Regulation of gastric acid secretion. In: Johnson LR (ed) Physiology of the gastrointestinal tract. Raven Press. New York, pp 659–669

Gutman Y, Krausz M (1969) Regulation of food and water intake in rats as related to plasma osmolarity and volume. Physiol Behav 4:311–313

Gwyn DG, Leslie RA, Hopkins DA (1985) Observations on the afferent and efferent organization of the vagus nerve and the innervation of the stomach in the squirrel monkey. J Comp Neurol 239:163–175

Haberich FJ (1968) Osmoreception in the portal circulation. Fed Proc 27:1137–1141

Hakanson R, Sundler F (1987) Localisation of gastric histamine: immunocytochemical observations. Med Biol 65:1–7

Hamilton RB, Norgren R (1984) Central projections of gustatory nerves in the rat. J Comp Neurol 222:560–577

Houpt TR, Anderson CR (1990) Spontaneous drinking: is it stimulated by hypertonicity or hypovolemia? Am J Physiol 258:R143–R148

Houpt TR, Weixler LC, Troy DW (1986) Water drinking induced by gastric secretagogues in pigs. Am J Physiol 251:R157–R164

Hsiao S (1970) Reciprocal and additive effects of hyperoncotic and hypertonic treatments on feeding and drinking in rats. Psychonom Sci 19:303–304

Hsiao S, Smutz ER (1976) Thirst-reducing and hunger-inducing effects of water and saline by stomach-tubing vs. drinking in rats. Physiol Psychol 4:111–113

Hsiao S, Trankina F (1969) Thirst–hunger interaction: I. Effects of body-fluid restoration on food and water intake in water-deprived rats. J Comp Physiol Psychol 69:448–453

Ionescu E, Rohner-Jeanrenaud F, Berthoud H-R, Jeanrenaud B (1983) Increases in plasma insulin levels in response to electrical stimulation of the dorsal motor nucleus of the vagus nerve. Endocrinology 112:904–910

Jacquin MF, Zeigler HP (1983) Trigeminal orosensation and ingestive behavior in the rat. Behav Neurosci 97:62–97

Jerome C, Smith GP (1982a) Gastric or coeliac vagotomy decreases drinking after peripheral angiotensin II. Physiol Behav 29:533–536

Jerome C, Smith GP (1982b) Gastric vagotomy inhibits drinking after hypertonic saline. Physiol Behav 28:371–374

Jorde R, Burhol PG (1985) Diurnal profiles of gastrointestinal regulatory peptides. Scand J Gastroenterol 20:1–4

Kakolewski JW, Deaux E (1970) Initiation of eating as a function of ingestion of hypoosmotic solutions. Am J Physiol 218:590–595

Kakolewski JW, Deaux E, Christensen J, Case B (1971) Diurnal patterns in water and food intake and body weight changes in rats with hypothalamic lesions. Am J Physiol 221:711–718

Kalia M, Mesulam M (1980) Brain stem projections of sensory and motor components of the vagus complex in the cat: II. Laryngeal, tracheobronchial, pulmonary, cardiac, and gastrointestinal branches. J Comp Neurol 193:467–508

Kannan H, Yamashita H (1985) Connections of neurons in the region of the nucleus tractus solitarius with the hypothalamic paraventricular nucleus: their possible involvement in the neural control of the cardiovascular system in rats. Brain Res 329:205–212

Kikta DC, Threatte RM, Barney CC, Fregly MJ, Greenleaf JE (1981) Peripheral conversion of L-5-hydroxytryptophan to serotonin induces drinking in rats. Pharmacol Biochem Behav 14:889–893

Kissileff HR (1969a) Food-associated drinking in the rat. J Comp Physiol Psychol 67:284–300

Kissileff HR (1969b) Oropharyngeal control of prandial drinking. J Comp Physiol Psychol 67:309–319

Kissileff HR (1973) Non-homeostatic controls of drinking. In: Epstein AN, Kissileff HR, Stellar E (eds) Neuropsychology of thirst: new findings and advances in concepts. Winston, Washington DC, pp 163–198

Kobashi M, Adachi A (1986) Projection of nucleus tractus solitarius units influenced by hepatoportal afferent signal to parabrachial nucleus. J Auton Nerv Syst 16:153–158

Kobashi M, Adachi A (1988) A direct hepatic osmoreceptive afferent projection from nucleus tractus solitarius to dorsal hypothalamus. Brain Res Bull 20:487–492

Kostarczyk E (1986) Autonomic correlates of alimentary conditioned and unconditioned reactions in the dog. J Auton Nerv Syst 17:279–288

Kral JG (1983) Behavioral effects of vagotomy in humans. J Auton Nerv Syst 9:273–281

Kraly FS (1978) Abdominal vagotomy inhibits osmotically induced drinking in the rat. J Comp Physiol Psychol 92:999–1013

Kraly FS (1983a) A probe for a histaminergic component of drinking in the rat. Physiol Behav 31: 229–232

Kraly FS (1983b) Histamine plays a part in induction of drinking by food intake. Nature 301:65–66

Kraly FS (1984a) Physiology of drinking elicited by eating. Psychol Rev 91:478–490

Kraly FS (1984b) Preabsorptive pregastric vagally mediated histaminergic component of drinking elicited by eating in the rat. Behav Neurosci 98:349–355

Kraly FS (1985) Histamine: a role in normal drinking. Appetite 6:153–158

Kraly FS (1990) Drinking elicited by eating. In: Epstein AN, Morrison A (Eds) Progress in psychobiology and physiological psychology, vol 14. Academic Press, New York, pp 67–133

Kraly FS, Arias RL (1990) Histamine in brain may have no role for histaminergic control of food-related drinking in the rat. Physiol Behav 47:5–9

Kraly FS, Corneilson R (1990) Angiotensin II mediates drinking elicited by eating in the rat. Am J Physiol 258:R436–R442

Kraly FS, June KR (1982) A vagally mediated histaminergic component of food-related drinking in the rat. J Comp Physiol Psychol 96:89–104

Kraly FS, Miller LA (1982) Histamine-elicited drinking is dependent upon gastric vagal afferents and peripheral angiotensin II in the rat. Physiol Behav 28:841–846

Kraly FS, Specht SM (1984) Histamine plays a major role for drinking elicited by spontaneous eating in rats. Physiol Behav 33:611–614

Kraly FS, Gibbs J, Smith GP (1975) Disordered drinking after abdominal vagotomy in rats. Nature 258:226–228

Kraly FS, Smith GP, Carty WJ (1978) Abdominal vagotomy disrupts food-related drinking in the rat. J Comp Physiol Psychol 92:196–203

Kraly FS, Miller LA, Hecht ES (1983) Histaminergic mechanism for drinking elicited by insulin in the rat. Physiol Behav 31:233–236

Kraly FS, Jerome C, Smith GP (1986) Specific postoperative syndromes after total and selective vagotomies in the rat. Appetite 7:1–17

Kwon SC, Baertschi AJ (1989) Mesenteric nerves mediate plasma vasopressin (AVP) response to splanchnic osmoreceptor stimulation in conscious rats. Soc Neurosci Abst 15:661

Lehr D, Goldman W (1973) Continued pharmacologic analysis of consummatory behavior in the albino rat. Eur J Pharmacol 23:197–210

Le Magnen J (1985) Hunger. Cambridge University Press, Cambridge

Lepkovsky S, Lyman R, Fleming D, Nagumo M, Dimick M (1957) Gastrointestinal regulation of water and its effect on food intake and rate of digestion. Am J Physiol 188:327–331

Leslie RA, Gwyn DG, Hopkins DA (1982) The central distribution of the cervical vagus nerve and gastric afferent and efferent projections in the rat. Brain Res Bull 8:37–43

Levens NR (1985) Control of intestinal absorption by the renin–angiotensin system. Am J Physiol 249: G3–G15

Lucas GA, Timberlake W, Gawley DJ (1989) Learning and meal-associated drinking: meal-related deficits produce adjustments in postprandial drinking. Physiol Behav 46:361–367

Makhlouf GM (1974) The neuroendocrine design of the gut. Gastroenterology 67:159–184

Mann JFE, Johnson AK, Ganten D (1980) Plasma angiotensin II: dipsogenic levels and angiotensin-generating capacity of renin. Am J Physiol 238:R372–R377

Marwine A, Collier G (1979) The rat at the waterhole. J Comp Physiol Psychol 93:391–402

Mashford ML, Nilsson G, Rokaeus A, Rosell S (1978) The effect of food ingestion on circulating neurotensin-like immunoreactivity (NTLI) in the human. Acta Physiol Scand 104:244–246

Matsukawa K, Ninomiya I (1987) Changes in renal sympathetic nerve activity, heart rate, and arterial blood pressure associated with eating in cats. J Physiol (Lond) 390:229–242

Mei N (1983) Recent studies on intestinal vagal afferent innervation. Functional implications. J Auton Nerv Syst 9:199–206

Mei N, Garnier L (1986) Osmosensitive vagal receptors in the small intestine of the cat. J Auton Nerv Syst 16:159–170

Miceli MO, Malsbury CW (1985) Brainstem origins and projections of the cervical and abdominal vagus in the golden hamster: a horseradish peroxidase study. J Comp Neurol 237:65–76

Miyaoka Y, Sakaguchi T, Yamazaki M, Shingai T (1987a) Changes in water intake following pharyngolaryngeal deafferentation in the rat. Physiol Behav 40:369–371

Miyaoka Y, Sawada M, Sakaguchi T, Hasegawa A, Shingai T (1987b) Differences in drinking behavior between normal and laryngectomized man. Percept Mot Skills 64:1088–1090

Moore-Ede MC (1986) Physiology of the circadian timing system: predictive versus reactive homeostasis. Am J Physiol 250:R735–R752

Mori T, Nagai K, Nakagawa H (1983) Dependence of memory of meal time upon circadian biological clock in rats. Physiol Behav 30:259–265

Nakamaru M, Jackson EK, Inagami T (1986) ß-Adrenoceptor-mediated release of angiotensin II from mesenteric arteries. Am J Physiol 250:H144–Hl48

Nicolaïdis S (1969) Early systemic responses to orogastric stimulation in the regulation of food and water balance. Functional and electrophysiological data. Ann NY Acad Sci 157:1176–1203

Norgren R (1983) Afferent interactions of cranial nerves involved in ingestion. J Auton Nerv Syst 9: 67–77

Norgren R, Grill H (1982) Brain-stem control of ingestive behavior. In: Pfaff DW (ed) The physiological mechanisms of motivation. Springer-Verlag, New York, pp 99–131

Norgren R, Smith GP (1988) Central distribution of subdiaphragmatic vagal branches in the rat. J Comp Neurol 273:207–223

Norgren R, Nishijo H, Travers SP (1989) Taste responses from the entire gustatory apparatus. Ann NY Acad Sci 575:246–263

Nose H, Morita M, Yawata T, Morimoto T (1986) Continuous determination of blood volume on conscious rats during water and food intake. Jpn J Physiol 36:215–218

Novin D (1964) The effects of insulin on water intake in the rat. In: Wayner MJ (ed) Thirst in the regulation of body water. Pergamon Press, Oxford, pp 177–184

Novin D, Miller NE (1962) Failure to condition thirst induced by feeding dry food to hungry rats. J Comp Physiol Psychol 55:373–374

Oatley K, Toates FM (1969) The passage of food through the gut of rats and its uptake of fluid. Psychonom Sci 16:225–226

Oishi R, Itoh Y, Nishibori M, Saeki K (1987) Feeding-related circadian variation in *tele*-methylhistamine levels of mouse and rat brains. J Neurochem 49:541–547

Paller MS, Hostetter TH (1986) Dietary protein increases plasma renin and reduces pressor reactivity to angiotensin II. Am J Physiol 251:F34–F39

Phillips PA, Rolls BJ, Ledingham JG, Morton JJ (1984) Body fluid changes, thirst and drinking in man during free access to water. Physiol Behav 33:357–363

Radke KJ, Selkurt EE, Willis LR (1985) The role of histamine H_1 and H_2 receptors in the canine kidney. In: Berlyne GM (ed) Renal physiology. Karger, Basel, pp 100–111

Rolls BJ (1975) Interaction of hunger and thirst in rats with lesions of the preoptic area. Physiol Behav 14:537–543

Rolls BJ, Wood RJ, Stevens RM (1978) Palatability and body fluid homeostasis. Physiol Behav 20: 15–19

Rolls BJ, Wood RJ, Rolls ET (1980) Thirst: the initiation, maintenance, and termination of drinking. In: Sprague JM, Epstein AN (eds) Progress in psychobiology and physiological psychology, vol 9. Academic Press, New York, pp 263–321

Rowland N, Nicolaïdis S (1976) Metering of fluid intake and determinants of ad libitum drinking in rats. Am J Physiol 231:1–8

Rowland NE, Caputo FA, Fregly MJ (1987) Body fluid shifts and elevation in plasma renin activity accompany meal-associated drinking in rats. Fed Proc 46:1434

Rubin W, Schwartz B (1983) Identification of the serotonin-synthesizing endocrine cells in the rat stomach by electron microscopic radioautography and amine fluorescence. Gastroenterology 84:34–50

Sawchenko PE (1983) Central connections of the sensory and motor nuclei of the vagus nerve. J Auton Nerv Syst 9:13–26

Simansky KJ, Smith GP (1983) Acute abdominal vagotomy reduces drinking to peripheral but not central angiotensin II. Peptides 4:159–163

Simansky KJ, Jerome C, Santucci A, Smith GP (1982) Chronic hypodipsia to intraperitoneal and subcutaneous hypertonic saline after vagotomy. Physiol Behav 28:367–370

Skinner TL, Randall DC (1985) Behaviorally conditioned changes in atrio-ventricular transmission in awake dogs. J Auton Nerv Syst 12:23–34

Smith GP (1986) Peripheral mechanisms for the maintenance and termination of drinking in the rat. In: de Caro G, Epstein AN (eds) The physiology of thirst and sodium appetite. Plenum Press, New York, pp 265–277

Smith GP, Jerome C (1983) Effects of total and selective abdominal vagotomies on water intake in rats. J Auton Nerv Syst 9:259–271

Soll AH, Berglindh T (1987) Physiology of isolated gastric glands and parietal cells: receptors and effectors regulating function. In: Johnson LR (ed) Physiology of the gastrointestinal tract, 2nd edn. Raven Press, New York, pp 883–909

Soll AH, Lewin KJ, Beaven MA (1981) Isolation of histamine-containing cells from rat gastric mucosa: biochemical and morphologic differences from mast cells. Gastroenterology 80:717–727

Spiteri NJ (1982) Circadian patterning of feeding, drinking and activity during diurnal food access in rats. Physiol Behav 28:139–147

Stanley BG, Hoebel BG, Leibowitz SF (1983) Neurotensin: effects of hypothalamic and intravenous injections on eating and drinking in rats. Peptides 4:493–500

Stoppini L, Baertschi AJ (1984) Activation of portal-hepatic osmoreceptors in rats: role of calcium, acetylcholine and cyclic AMP. J Auton Nerv Syst 11:297–308

Strubbe JH, Spiteri NJ, Alingh Prins AJ (1986) Effect of skeleton photoperiod and food availability on the circadian pattern of feeding and drinking in rats. Physiol Behav 36:647–651

Toates FM (1979) Homeostasis and drinking. Behav Brain Sci 2:95–139

Turkkan JS (1989) Classical conditioning: the new hegemony. Behav Brain Sci 12:121–179

Vijande M, Marin B, Brime J et al. (1989) Water drinking induced by insulin in humans. Appetite 12:243

Walsh LL, Grossman SP (1973) Zona incerta lesions: disruption of regulatory water intake. Physiol Behav 11:885–887

Weingarten HP (1985) Stimulus control of eating: implications for a two-factor theory of hunger. Appetite 6:387–401

Weingarten HP, Powley TR (1980) A new technique for the analysis of phasic gastric acid responses in the unanesthetized rat. Lab Anim Sci 30:673–680

Weingarten HP, Powley TR (1981) Pavlovian conditioning of the cephalic phase of gastric acid secretion in the rat. Physiol Behav 27:217–221

Weissman A (1972) Elicitation by a discriminative stimulus of water-reinforced behavior and drinking in water-satiated rats. Psychonom Sci 28:155–156

Zamble H, Baxter DJ, Baxter L (1980) Influences of conditioned incentive stimuli on water intake. Can J Psychol 34:82–85

Commentary

Thrasher: The hypothesis that eating behaviour is a strong determinant of spontaneous drinking is a reasonable hypothesis in many species. However, the first definition of a characteristic of the relation between eating and drinking in adult mammals: "First, eating and drinking tend to occur together . . . " (page 296) seems to eliminate most mammalian species. For example, grazing species in all continents tend to eat or chew more or less continuously but usually visit waterholes at daily or longer intervals. Also carnivores in the wild engorge themselves on a kill and then sleep for 12–18 hours before seeking water. Finally, wild rodents in temperate climates usually do not rest next to a water source. They gather food for storage and consumption in their nest but have to expose themselves to predators to find water. The tendency to eat and drink together seems limited to the laboratory rats, domestic pets, some classes of farm animals and humans.

Therefore, it seems important to note that presumed mechanisms which mediate the effects of feeding behaviour on the control of spontaneous drinking, reported in this chapter, may or may not apply to species other than rats in a laboratory experiment.

Chapter 19

Inhibitory Controls of Drinking: Satiation of Thirst

J.G. Verbalis

Introduction

Most studies of thirst have focused on the initiation of water intake and the neural mechanisms responsible for this vital behaviour. Less attention has been paid to the stimuli and mechanisms that terminate a bout of drinking and limit fluid ingestion. In contrast, studies of food intake have long emphasized the importance of satiety mechanisms to caloric homeostasis. In considering the factors underlying this historical difference in emphasis on stimulatory versus inhibitory systems between these two ingestive behaviours, several unique aspects of water homeostasis should be recalled.

First, water ingestion is not physiologically required except to replace accrued deficits, and so it is appropriate that this behaviour be primarily under stimulatory control in response to manifestations of body water deficiency. Second, because most mammals do not possess large reserves of excess water, in contrast to the often substantial caloric reserves stored in adipose tissue, it is important that water deficits be replaced quickly and completely when animals have access to water. Premature activation of inhibitory mechanisms terminating drinking might adversely interfere with successful correction of conditions threatening survival. Third, water homeostasis represents a balance between intake and excretion of water, and the latter is exquisitely controlled via vasopressin-mediated water excretion by the kidneys. Given the substantial diluting ability of the kidneys, any excesses of ingested water can be easily and rapidly excreted in normal animals. Consequently, finely tuned behavioural mechanisms to limit water intake to only the amounts required to replace deficits are not essential under most physiological conditions.

As a result of such considerations, regulation of water intake has been viewed by many as a simple single-loop feedback system in which thirst is absent until a specific stimulus occurs, and water intake then continues until the ingested fluid corrects the deficit thereby eliminating the stimulus to continued drinking. In this model "satiation" of thirst would be predicted to occur only after rehydration returns plasma osmolality and volume to normal ranges. Although this interpretation adequately accounts for many experimental and clinical observations, it represents an oversimplified view of a complex behaviour that is known to be regulated more finely than this.

This chapter will review data regarding inhibitory mechanisms that limit fluid intake in mammals, including assessments of the relative importance of such mechanisms to both physiological and pathophysiological situations. However, just as inhibition of feeding is not necessarily synonymous with satiety for food, inhibition of drinking does not always reflect satiety for water either. Later sections will therefore attempt to integrate our present knowledge of specific inhibitory mechanisms into a co-ordinated scheme of thirst satiation. Wherever possible emphasis will be placed on pertinent results obtained since the publication of several excellent texts that have addressed this issue (Fitzsimons 1979; Rolls and Rolls 1982).

Parenteral Factors Inhibiting Fluid Ingestion

Because correcting the perturbations in body fluid balance that initially stimulate fluid intake is of obvious importance to achieving water homeostasis, postabsorptive factors that act to inhibit drinking will be considered first, i.e. postingestional factors beyond the gastrointestinal tract.

Osmolality

Maintenance of extracellular and intracellular osmolality within relatively narrow limits is the most important function of water homeostasis. Hyperosmolar animals continue drinking, at different rates depending upon the species (Adolph 1967), until their plasma osmolality again reaches normal ranges. According to the single-loop model, this behaviour can be explained by cessation of drinking once osmoreceptor neurons no longer provide excitatory input to brain areas controlling fluid ingestion (see Chapter 5). Thus, in this model fluid ingestion would end only after water deficits have been corrected.

Upon closer scrutiny this simple interpretation fails to account for several important observations. Perhaps most striking is the fact that in most species drinking ceases even before complete absorption of the ingested water and consequently prior to correction of plasma osmolality (Rolls et al. 1980a) Thus, preabsorptive factors must also be of importance for this behaviour. A second potential problem is that inhibition of drinking behaviour is considered a passive phenomenon in this model, and simply represents the absence of excitatory osmotic stimulation. This seems an unlikely regulatory process for defending plasma osmolality against decreases as well as increases from normal ranges. Although hypo-osmolality can be avoided by renal water excretion resulting from inhibition of vasopressin secretion (Robertson 1977), several lines of evidence indicate that osmotic dilution generates a direct inhibitory signal to fluid intake.

The first stems from studies of induced hypo-osmolality in rats. The "safety-valve" feature of renal water excretion by the kidneys can be overridden by continuously infusing arginine vasopressin (AVP), or the synthetic vasopressin agonist 1-desamino-8-D-arginine vasopressin (dDAVP), thereby blocking water excretion even in the presence of dilute plasma osmolality. Rats infused with AVP or dDAVP and given free access to water decrease their water intakes to levels equivalent to their daily insensible water losses and thereby avoid becoming significantly hypo-osmolar. These results are consistent with clinical observations that most patients treated with dDAVP do not become significantly hypo-osmolar despite the induced antidiuresis (see Chapter 30). However, hypo-osmolality can be produced in antidiuretic rats if fluid intakes are increased either by offering more palatable dextrose

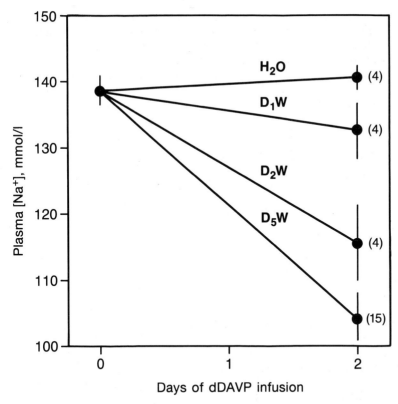

Fig. 19.1. Plasma sodium concentrations of rats infused with dDAVP subcutaneously for two days. Animals were given either water or solutions of dextrose water (1%, 2% or 5% w/v, as indicated) to drink, but were not fed. The figures in parentheses indicate the number of animals studied in each group. (From Verbalis (1984).)

solutions to drink (Verbalis 1984; Fig. 19.1), or by using a liquid formula so the ingested fluid becomes the sole source of calories (Verbalis and Drutarosky 1988). Under such conditions the animals continue drinking and achieve severe degrees of hypo-osmolality (plasma sodium concentrations of 100–110 mmol/l). Similar results have been reported for shorter periods using saccharin-flavoured drinking solutions (Rolls et al. 1978). Inhibition of fluid intake by hypo-osmolality is therefore only partial, since it can be overcome by increasing the palatability of the ingested fluids.

A second example of osmotic inhibition of thirst occurs during volume-regulatory fluid intake in response to hypovolaemia induced by subcutaneous injection of polyethylene glycol (PEG) (Stricker 1966). When offered isotonic saline to drink, PEG-treated rats continue drinking until the hypovolaemia is repaired (Stricker and Jalowiec 1970). However, when offered water, or when water is administered intragastrically, net fluid intakes are much lower despite the continued presence of severe degrees of hypovolaemia (Stricker 1969; Fig. 19.2). This blunting of fluid intake coincides with the production of plasma hypo-osmolality as a result of renal retention of the ingested water. Note again, however, that this represents a partial inhibition, since some fluid ingestion continues although at markedly reduced rates in comparison to normonatraemic rats.

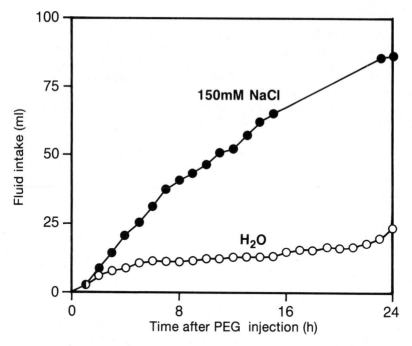

Fig. 19.2. Fluid intakes of rats for 24 hours following subcutaneous injection of 30% polyethylene glycol solution. The solid circles show the intakes when isotonic saline was given as the only drinking solution and the open circles the intakes when water was given as the only drinking solution. (From Stricker and Jalowiec (1970).)

Extracellular Fluid Volume

In view of the symmetrical nature of hyperosmotic stimulation and hypo-osmotic inhibition of thirst, similar results might be expected with regard to changes in extracellular fluid volume. Hypovolaemia and hypotension are major stimuli to fluid intake in many species (Fitzsimons 1961; Stricker 1966; Ramsay and Thrasher 1986a), and therefore it would be appropriate to expect that hypervolaemia and hypertension might directly inhibit fluid intake. Several lines of experimental evidence are in support of this.

The best demonstration of such an inhibition is found in studies using chronically implanted balloons in the left atrium of dogs (Moore-Gillon and Fitzsimons 1982). When inflated, the animals drank significantly less water in response to acute stimuli such as hypertonic sodium chloride (water intakes were decreased approximately 60% compared to the same animals with the balloon deflated), and longer (24 hour) periods of inflation significantly decreased spontaneous water ingestion by 30%–50% without affecting food intakes. These data therefore demonstrate a substantial global inhibition of fluid intake as a result of mimicking hypervolaemic conditions. In combination with earlier data that vagosympathectomy enhanced drinking to some stimuli (Sobocinska 1969), it has been proposed that afferent tone from low pressure baroreceptors may exert tonic inhibition over fluid intake (Moore-Gillon and Fitzsimons 1982). Similar studies in rats have shown that distension of a balloon in the right atrium abolished PEG-induced thirst, but unlike in dogs this manoeuvre did not blunt osmotically stimulated drinking (Kaufman 1984). Analogous results were found in sheep after denervation of the atrial baroreceptors (Zimmerman et al. 1981). These studies are therefore more consistent with the interpretation that volume expansion acts primarily to eliminate a hypovolaemic stimulation of thirst rather than to produce a global inhibition of thirst. Whether these represent species differences with respect to baroreceptor inhibition of thirst, or are due to methodological differences between these studies, remains unanswered. Although intravenous infusions of fluids to animals reduces their spontaneous fluid intakes (Nicolaïdis and Rowland 1975), many studies utilizing intravenous fluid administration to expand blood volume of euvolaemic animals have failed to note significant effects on osmotically stimulated drinking in monkeys (Rolls et al. 1980a), dogs (Ramsay et al. 1977) and rats (Corbit 1965). However, in one study expansion of the intravascular volume by 10%–20% with dextran infusions had the effect of elevating the osmotic threshold for drinking in dogs (Kozlowski and Szczepanska-Sadowska 1975), suggesting that larger pathophysiological degrees of volume expansion are capable of inhibiting the onset, and possibly the volume, of fluid ingested in response to osmotic stimuli.

Inhibition of thirst via activation of high pressure arterial baroreceptors has similarly been proposed. Studies of systemic infusion of angiotensin have demonstrated a greater dipsogenic effect when the angiotensin-induced increases in arterial blood pressure were blocked by simultaneous administration of various vasodilators (Robinson and Evered 1987), suggesting that the hypertension inhibited the dipsogenic effects of exogenously infused angiotensin (Evered et al. 1988). Similar results have been obtained using a suprarenal intra-aortic balloon which stimulated endogenous angiotensin release as a result of producing decreased blood pressure distal to the obstruction and variable degrees of hypertension proximal to the obstruction (Fitzsimons and Moore-Gillon 1981). Although such data support a direct inhibitory effect of arterial hypertension on fluid intake, both of these manoeuvres also activate low pressure baroreceptors through increases in right and left atrial pressures. On the other hand, analogous studies using other pressor agents such as phenylephrine did not demonstrate any inhibition of drinking in dehydrated dogs (Ramsay and Thrasher 1986b). Nonetheless, both arterial and cardiac baroreceptors appear to have inhibitory effects on drinking under some circumstances, just as both mediate fluid intake during hypotension (Quillen et al. 1988).

Preabsorptive Factors Influencing Inhibition of Fluid Ingestion

As mentioned previously, the largest weakness of the single-loop model is the long-standing observation that in many species dehydration-induced drinking ceases before sufficient time has passed for absorption of the ingested water, as soon as 5 minutes in dogs (Adolph 1939). Such observations can be explained only by an action of preabsorptive factors to inhibit fluid ingestion.

Oropharyngeal Factors

Multiple studies have now clearly demonstrated that oropharyngeal factors can inhibit water intake, although with marked species differences (see reviews, Fitzsimons

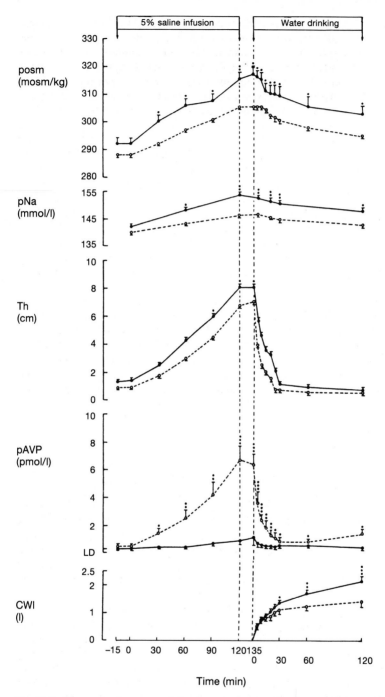

Fig. 19.3. Changes in plasma osmolality (posm), plasma sodium (pNa), thirst (Th), plasma vasopressin (pAVP), and cumulative water intake (CWI) in 13 patients with diabetes insipidus (closed circles) compared to five normal subjects (open circles). All subjects were infused with hypertonic saline and then allowed to drink water ad libitum. (From Thompson et al. (1987).)

1979; Rolls and Rolls 1982; Ramsay and Thrasher 1986b). Man is included among this group, since studies of dehydrated humans (Rolls et al. 1980a) as well as humans infused with hypertonic saline (Thompson et al. 1987) have demonstrated decreases in thirst within 2.5–5 minutes after beginning drinking, whereas significant decreases in plasma osmolality from absorption of the ingested water did not occur until 12.5–25 minutes (Fig. 19.3). Although the importance of oropharyngeal factors appears to be established, the mechanism(s) whereby these effects occur are less certain.

Evidence to date suggests that both oropharyngeal receptors as well as neural reflexes activated by the act of swallowing participate in this inhibition. Simply gargling with water transiently inhibits thirst in dehydrated humans (Seckl et al. 1986; Rolls et al. 1980a). This effect is well known clinically and is the basis for allowing patients who are unable to drink the use of oral irrigants or ice chips in their mouths. On the other hand, the presence of dilute fluids in the oral cavity is not necessary for inhibition of thirst. This has been clearly demonstrated in dogs that drank equivalent volumes of isotonic saline or water following dehydration (Thrasher et al. 1981), as well as in dehydrated humans who reported a decrease in thirst after drinking hypertonic saline (Seckl et al. 1986). These results strongly suggest that the act of swallowing provides an inhibitory stimulus when fluids are ingested. However, swallowing food does not appear to reproduce this result (Ramsay and Thrasher 1986a, b). Consequently, afferent neural traffic from the oropharynx to the brain appears able to differentiate between food and water during ingestion. More specific details of this circuitry remain to be evaluated, but it is worth noting than an inhibition of thirst that is proportional to the volume of ingested fluid, rather than simply to the presence of dilute fluids in the mouth, would better account for the "metering" effect whereby many animals rapidly drink amounts closely approximating their fluid deficit before significant absorption of the ingested fluid (Adolph 1967).

Gastrointestinal Factors

Although the rapidity of decreases in thirst and fluid ingestion after the onset of drinking argues in favour of a prominent role for oropharyngeal factors, other data indicate that gastrointestinal sensing of the ingested fluid is also important. The basis for this is the observation that most animals with open oesophageal or gastric fistulae markedly overdrink (Blass and Hall 1976; Maddison et al. 1980), and this "sham-drinking" continues unabated despite continuous oropharyngeal stimulation from the ingested water. However, conflicting data exist with regard to the importance of gastrointestinal factors, and in large part this appears to be due to substantial species differences in the relative strengths of the various preabsorptive inhibitory stimuli.

The most convincing evidence for gastric distension as an inhibitory stimulus to drinking comes from studies in monkeys where dehydration-induced drinking was allowed to terminate normally, but the ingested fluid was drained via a gastric cannula (Maddison et al. 1980). Such animals began to drink again within minutes, suggesting that the presence of water in the stomach was providing an inhibitory signal to further fluid ingestion. This appears to be purely a mechanical effect, since similar inhibitions of intake have been produced by intragastric distension using a wide variety of substances such as air or hypertonic saline (Adolph 1950), and even balloon inflation (Towbin 1949). Therefore, gastric distension is less specific for thirst than oropharyngeal stimulation, which appears to be capable of discriminating ingestion of fluids from food. Furthermore, some species are less sensitive to gastrointestinal factors. Early studies of dogs with oesophageal fistulae also showed excess sham-drinking, but they then paused for 20–60 minutes before resuming ingestion (Bellows 1939). In more recent studies, dehydrated dogs with open gastric fistulae did not overdrink, but rather ingested equivalent volumes of water in one hour as when the fistulae were closed (Ramsay and Thrasher 1986a,b). Finally, intragastric preloads of water are less effective in inhibiting subsequent drinking in dogs than in other species (Adolph 1950). Consequently, dogs appear to be influenced more by oropharyngeal cues, whereas monkeys are more sensitive to gastric factors.

Data have also been presented in support of intestinal and hepatoportal factors inhibiting drinking, either directly or indirectly via effects on gastric emptying (see review, Rolls and Rolls 1982), but the evidence in support of such effects is more limited. Because such mechanisms would to a large degree depend upon activation of intestinal or hepatoportal osmoreceptors by dilute fluids, they would not easily explain the inhibition of thirst after ingestion of isotonic fluid in dogs or hypertonic fluid in man.

Cognitive Factors

In addition to physiological events related to the ingestion of fluid, a variety of psychological factors clearly exert a large influence on drinking behaviour. For convenience these are grouped together here as "cognitive" factors, and include taste novelty, oral habituation, primary polydipsia, learned behaviours and sensory-specific satiety. Most of these influences act to stimulate rather than inhibit fluid ingestion, with the exception of learned behaviours and sensory-specific satiety.

Learned behaviours can influence drinking as a result of either positive or negative past reinforcement. Arguments supporting a role for past positive experiential learning in mediating the termination of drinking have already been advanced (see Chapters 4 and 17). Although it is quite likely that positive rewards generated by fluid consumption in response to past dehydrational challenges do influence subsequent drinking behaviour, and particularly anticipatory drinking, nonetheless the consistency of drinking in response to experimental challenges of magnitudes never experienced previously by laboratory animals is striking. Furthermore, animals sham-drink large volumes of fluid the first time that gastric cannulae are opened (Maddison et al. 1980). Such observations do not deny an important influence of past learning on limiting fluid intake, but do argue that this must be accompanied by appropriate physiological responses as a consequence of the behaviour to achieve a sustained termination of drinking. Negative learned behaviours include the development of learned taste aversions to novel tastes (Garcia and Ervin 1968). Although this phenomenon can limit drinking of flavoured solutions previously associated with symptoms of nausea, the ubiquitous presence of water throughout animals' lives precludes development of significant taste aversions to this fluid. Sensory-specific satiety has been studied mostly with regard to food intake, but it can also limit ingestion of flavoured fluids just as it does food (Rolls et al. 1980a). However, it is unlikely that water ingestion in response to physiological needs is limited by such mechanisms.

Most of the other cognitive factors have the net effect of stimulating rather than inhibiting fluid ingestion. Nonetheless, it is appropriate to consider such influences in a discussion of drinking because they provide a measure of the relative strengths of various physiological inhibitory stimuli to fluid ingestion. It is noteworthy that all the pre- and postabsorptive inhibitory factors discussed can be overcome by cognitive stimuli that drive ingestive behaviour. Not only will rats increase intakes when fluids are made more palatable (Ernits and Corbit 1973), but rats made antidiuretic with dDAVP will continue to ingest such fluids to the point of extreme hypo-osmolality and the degree of hypo-osmolality achieved is proportional to the palatability of the fluid (Verbalis 1984; Fig. 19.1). Analogous results have been obtained with schedule-induced polydipsia in rats treated with AVP (Stricker and Adair 1966). In these examples drinking continued despite the production of both osmotic dilution and volume expansion, and despite real drinking behaviour sufficient to activate both oropharyngeal and gastrointestinal inhibitory factors. Obviously drinking will not continue indefinitely in the absence of renal excretion until some factor causes inhibition of further intake, but before this happens it is possible to achieve plasma dilutions of 20%–30%. By comparison, it is difficult to imagine a non-physiological stimulus that would cause rats to ingest hyperosmolar solutions in amounts causing equivalent increases of plasma osmolality.

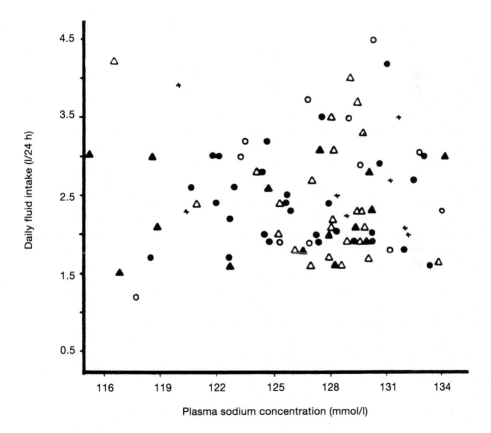

Fig. 19.4. Daily fluid intakes of 91 hospitalized patients with hyponatraemia of varying degrees and aetiologies. Each point represents a single patient: open circles, SIAD; open triangles, cardiac failure; closed circles, volume contraction; closed triangles, cirrhosis; pluses, undiagnosed. (From Gross et al. (1987).)

The logical conclusion to be drawn from these observations is that although direct inhibitory physiological stimuli to thirst and fluid ingestion clearly exist, they are relatively weak in comparison to excitatory stimuli and can be overridden by a variety of non-homeostatic influences. From a teleological point of view this is not surprising. With dehydration the kidneys under the influence of maximal AVP stimulation can conserve fluid to limit further losses, but only drinking can actually restore body osmolality and volume to normal; therefore the mechanisms that stimulate fluid intake must be sensitive and robust. However, when overhydration occurs, normal kidneys in the absence of AVP can excrete large amounts of dilute urine; although drinking is also appropriately inhibited at this time, given the ability of the kidneys to defend against overhydration this behavioural response is not nearly as critical to survival. Thus, it is overly simplistic, but not unreasonable, to view drinking as the primary defence against hyperosmolality and volume depletion, but renal water and sodium excretion as the primary defences against hypo-osmolality and volume expansion. Within this physiological framework, the evolutionary development of strong inhibitory controls of fluid intake was not particularly beneficial and therefore likely did not occur.

Although clinical disorders are discussed by subsequent authors (see Chapters 26, 29 and 30), it is relevant here to consider the occurrence of hyponatraemia in humans. This represents the most common electrolyte disorder in modern medical practice, and a large number of such cases are caused by inappropriate AVP secretion (Zerbe et al. 1980; Verbalis 1990). Such patients obviously do not need water but they continue to drink nonetheless. Some hyponatraemic patients manifest polydipsia (Whitaker et al. 1979; Goldman et al. 1988), but such cases generally represent exceptions since most hyponatraemic patients maintain fluid intakes commensurate with the normal population. Analysis of daily fluid intakes of 91 hyponatraemic patients showed an average fluid intake of 2.4 ± 0.2 litres in 24 hours (Gross et al. 1987; Fig. 19.4), which does not differ appreciably from earlier measured intakes of medical students or hospitalized cardiac patients (mean fluid intakes of 2.4 and 2.8 1/24 h, respectively; Holmes 1964), or more recent studies of adult subjects (mean fluid intake of 2.1 1/24 h; de Castro 1988). This, therefore, probably illustrates a clinical parallel to what has become apparent from animal studies: osmotic inhibition of thirst is a relatively weak phenomenon and easily overcome by a variety of non-homeostatic stimuli causing drinking. In man, similar to animals, these include prandial drinking, oral habituation to various beverages, pleasurable sensations from palatable fluids, social interactions promoting fluid ingestion, and mouth dryness as a result of local factors, all of which combined account for the major part of spontaneous human fluid ingestion (see Chapters 18 and 21). By themselves such stimuli are benign and simply lead to more frequent production of dilute urine to excrete the increased fluids ingested. However, in the presence of pathological conditions that impair renal water excretion they are a major factor contributing to the substantial incidence of hyponatraemia in medical practice, and consequently are of clinical importance as a cause of morbidity and mortality in man.

Integration: Transient Versus Sustained Inhibition of Drinking

The above summary allows some tentative conclusions to be drawn about how the various inhibitory stimuli of drinking are integrated, and about their significance for various types of drinking behaviour. Several investigators have proposed the terminology of "temporary" or "rapid" satiety to characterize the preabsorptive factors that limit fluid intake, and "permanent" satiety to describe the cessation of drinking eventually resulting from correction of the underlying stimulus stimulating fluid intake (Fitzsimons 1979; Ramsay and Thrasher 1986a,b). This classification has considerable merit and emphasizes the essential differences between these two types of inhibitory signals: preabsorptive factors act rapidly to terminate intake before ingested fluid has been completely absorbed, but also dissipate fairly rapidly leading to resumed drinking if not soon replaced by postabsorptive inhibitory factors resulting from absorption of the ingested fluid and correction of the initial deficits causing thirst. However, to avoid the semantic issue of whether a temporary relief of thirst really represents "satiety", not to mention the potentially greater problem of whether "permanent" can be used appropriately to describe a state in which additional fluid might be ingested for other reasons, the terminology of transient versus sustained inhibition of drinking will be used here. Because the various inhibitory signals described earlier are generated in temporal sequence during normal drinking, their overlapping nature acts to produce a continuous inhibitory signal to limit additional ingestion (Rolls and Rolls 1982). This co-ordinated progression of inhibitory signals is shown schematically in Fig. 19.5, and except for the addition of oropharyngeal metering is not substantively different from the well-known overlapping phases of gastrointestinal and postabsorptive satiety that control food intake. What probably accounts for the significant behavioural differences among species is not entirely distinct neural pathways that inhibit drinking, but rather the relative intensities, degrees of overlap and timing of these various phases. It has been suggested that similar differences in the mechanisms for sensing the volume of ingested fluid may account for the well-known differences in the rates of drinking among various species; animals with sensitive oropharyngeal sensing mechanisms (e.g., dogs) have the ability to ingest fluids rapidly, whereas those more dependent on gastrointestinal and postabsorptive mechanisms (e.g., rats) by necessity must drink more slowly (see Chapter 13).

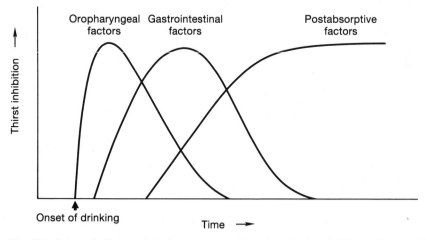

Fig. 19.5. Schematic diagram depicting the onset and duration of various inhibitory signals to continued fluid ingestion following initiation of drinking in response to body fluid deficits. Although each signal by itself is capable of terminating ingestion (depending upon the species), it is the overlapping nature of these sequentially activated mechanisms that produces and sustains the inhibition of further water ingestion.

However, regardless of the speed or intensity of individual or combined preabsorptive factors, failure to normalize body fluid homeostasis prevents the achievement of sustained inhibition and allows resumption of drinking behaviour. Conversely, correction of fluid deficits intravenously prior to drinking would be expected to create a state of sustained inhibition prematurely, thereby eliminating subsequent ingestion. This has been proven experimentally in dogs using combined intracarotid infusions of hypotonic fluid to normalize the osmolality of blood perfusing central osmoreceptors along with intravenous infusions of isotonic saline to abolish hypovolaemia (Ramsay et al. 1977).

The benefits of such a system are obvious: fluid deficits can be corrected quickly without a long reiterative process of drinking, awaiting absorption, and then assessing adequacy of the ingestion; it also avoids unnecessary overdrinking as a result of continuing to ingest fluids until the delayed absorption eventually results in normalization of body fluid deficits, and thereby optimizes economical use of limited environmental resources. Although preventing body "overhydration" has also been mentioned as a reason for the existence of transient inhibition, this seems of lesser importance since any excess ingestion can be handled fairly quickly by normal kidneys. The real evolutionary advantage of such a system is more likely to be simply that drinking can be terminated quickly, allowing the smallest possible period of vulnerability to predation during the time spent drinking (Fitzsimons 1979; Ramsay and Thrasher 1986a,b).

Parallel Effects of Inhibitory Stimuli on Secretion of Hormones Involved in Body Fluid Homeostasis

In response to body fluid deficits not only is thirst stimulated, but obviously cardiovascular and renal mechanisms are also called into play to promote fluid and solute conservation. Many of these responses are mediated humorally via endocrine secretions, most prominently vasopressin which promotes renal water conservation, and aldosterone which promotes renal sodium conservation. As sustained inhibition of drinking occurs, it would be appropriate that stimulated secretion of these hormones diminish to allow excretion of any excess ingested fluid, and it is of interest to consider the time course of these systemic hormonal changes in relation to drinking. Several other hormones promote natriuresis and/or diuresis, particularly atrial natriuretic peptide and in some species oxytocin, and consideration of the relation of the secretion of these hormones to thirst inhibition is also useful for a more complete picture of the co-ordination of endocrine and behavioural responses.

Vasopressin

The hormone of greatest relevance for water homeostasis is vasopressin (AVP). Secretion of this peptide by the pituitary is stimulated by hyperosmolality and hypovolaemia, and consequently AVP secretion correlates temporally with the onset of thirst. AVP has a short half-life (10–20 minutes in man; Lauson 1974), so plasma levels are capable of responding quickly to changes in stimulated secretion. Although return of plasma AVP concentrations to unstimulated levels has long been known to accompany the restoration of body fluid deficits, recent studies have unequivocally demonstrated that preabsorptive inhibitory signals are similarly accompanied by a rapid decrease in plasma AVP levels (Thrasher et al. 1981), and this effect has subsequently been demonstrated in several species including man (Geelen et al. 1984; Thompson et al. 1987 see Fig. 19.3). The rapidity of the decline of the AVP levels is consistent with a complete suppression of pituitary secretion and parallels the cessation of fluid intake. Occurrence of this phenomenon in dogs drinking isotonic saline or sham-drinking with open gastric fistulae (Thrasher et al. 1981), and in humans drinking hypertonic saline (Seckl et al. 1986), suggests causation by reflexes related to the act of swallowing rather than the effect of dilute fluids on oropharyngeal or gastrointestinal receptors. However, as with thirst inhibition, when the initial inhibition is not appropriately followed by postabsorptive correction of the fluid deficits, then AVP secretion is stimulated again. Thus, AVP secretion and thirst are closely related not only during stimulation but also during inhibition of thirst. AVP secretion is clearly not required for normal thirst, as demonstrated by the polydipsia of animals and patients with diabetes insipidus, but both processes must be regulated by very similar afferent inputs. To what degree these neural pathways intersect and overlap remains to be fully answered. In this regard it is noteworthy that some pre-absorptive stimuli that decrease thirst apparently do not cause suppression of AVP secretion, specifically wetting the mouth with water in the absence of swallowing (Seckl et al. 1986). Although cooling the mouth has been shown transiently to suppress stimulated AVP levels in dehydrated humans, this is a weaker effect inasmuch as it failed to suppress hypertonic saline-stimulated AVP secretion (Salata et al. 1987). However, colder fluids appear to quench thirst better in both animals (Kapatos and Gold 1972) and man, so in this instance changes in AVP secretion again parallel thirst. Additional work is needed in this area, but a consistent functional linkage between thirst and pituitary AVP secretion has been established unequivocally.

Renin–Angiotensin–Aldosterone

Similar to AVP, the renin–angiotensin–aldosterone system is potently activated during hypovolaemia, and endogenous angiotensin II appears to some degree to potentiate baroreceptor inputs stimulating thirst (Mann et al. 1987; see Chapters 6 and 9). Consequently, activation of this system also correlates well with thirst stimulation during hypovolaemia. However, studies of plasma renin activity (PRA) following rehydration in several species have demonstrated an absence of any suppression (Thrasher et al. 1981; Geelen et al. 1984; Seckl et al. 1986), stimulated activity (Blair-West et al. 1979), or a delayed suppression coinciding with absorption of isotonic saline (Thrasher et al. 1981). Several factors likely account for these diverse findings, but for the purpose of this discussion the most important relates to the control of secretion of renin. Renin is secreted by the juxtaglomerular apparatus in the kidney primarily in response to decreases in renal perfusion and tubular sodium delivery, or increases in renal sympathetic nerve activity (Davis and Freeman 1976). Suppression of PRA therefore requires restoration of normal, or even supranormal, renal perfusion in conjunction with non-stimulated sympathetic activity. This combination of events is unlikely to occur until substantial restoration of body fluid deficits has been achieved, rather than in response to preabsorptive stimuli. Thus, unlike AVP which is synthesized and secreted from hypothalamic neurons regulated by multiple stimulatory and inhibitory afferent neural inputs, the control mechanisms for hypovolaemia-induced renin secretion are more dependent on slower changes in renal haemodynamics. Furthermore, because aldosterone is the main regulator of renal sodium conservation, even a rapid suppression of PRA would not produce an equivalent effect on this hormone due to the combined plasma half-lives of renin, angiotensin and aldosterone. Hence, whatever the contribution of the peripheral renin–angiotensin system to hypovolaemic thirst, suppression of this system does not appear to play any role in the inhibition of thirst.

Atrial Natriuretic Peptide

Atrial natriuretic peptide (ANP) is a 28 amino acid peptide secreted by the heart in response to atrial distension (Needleman and Greenwald 1986). Although the physiological importance of ANP as an endogenous diuretic and natriuretic hormone remains controversial, it seems likely that it does contribute significantly to such effects in response to acute volume expansions (Schwab et al. 1986; Hirth et al. 1986). Plasma ANP levels are generally decreased during dehydration, but to date this peptide has not been carefully studied during the rehydration phase. Because the kidney is relatively insensitive to the natriuretic and diuretic effects of ANP except during states of volume expansion (Metzler and Ramsay 1989), and since animals and man generally do not overhydrate themselves during rehydration (Phillips et al. 1984; Rolls et al. 1980b), it seems unlikely that systemic secretion of this hormone plays any physiological role during rehydration. However, given reports that central administration of ANP is inhibitory to drinking (Nakamura et al. 1985) and sodium appetite (Antunes-Rodrigues et al. 1986), elevated plasma levels of this peptide might in some way act synergistically with baroreceptor suppression to inhibit drinking during pathological states of volume expansion.

Oxytocin

Oxytocin (OT) is secreted together with AVP in response to hyperosmolality and hypovolaemia in some species, particularly the rat (Stricker and Verbalis 1986). Secretion of this peptide is inversely correlated with salt appetite in rats, and it has been proposed that central OT secretion may cause inhibition of sodium intake (see Chapter 11). Complementary to this potential action, peripherally OT produces natriuresis via direct renal effects (Balment et al. 1980; Verbalis et al. 1991). And similar to AVP, osmotic dilution as well as the act of drinking suppress OT secretion in rats (Stricker and Verbalis 1986). Interestingly, recent studies have shown that intracerebroventricular administration of antisera to OT, as well as to AVP and angiotensin II, reduced water intake in dehydrated rats (Franci et al. 1989); a peptide possessing effects of enhancing water intake while simultaneously inhibiting solute (sodium) intake and promoting urinary sodium excretion would represent an ideal candidate in the body's defence against hyperosmolality via mechanisms such as dehydration natriuresis (see Chapter 2).

Relation of Central to Peripheral Secretion

Although peripheral hormone secretion complements drinking behaviour to achieve water homeostasis, it is also noteworthy that all of the above hormonal systems are represented in the brain and each has been implicated in the central control of fluid ingestion. The data supporting potential central actions of these and other brain peptides have already been extensively summarized (see Chapter 7), but it merits re-emphasis that in each case the postulated central action is complementary to the hormone's known peripheral actions (systemic AVP is antidiuretic and central AVP has been implicated as a dipsogen; peripheral angiotensin promotes water and sodium retention and central angiotensin administration stimulates thirst and sodium appetite; systemic ANP enhances diuresis and natriuresis and central administration of ANP inhibits drinking and sodium appetite; systemic OT is natriuretic and pituitary OT secretion is inversely correlated with sodium appetite). Therefore it is reasonable to anticipate that future studies of the role of central secretion of these and other peptides will be instrumental in understanding how the brain and periphery interact not only to stimulate but also to inhibit fluid intake.

Mechanisms of Drinking Inhibition During Thirst Satiation

In considering the potential involvement of inhibitory controls of drinking in the physiological regulation of fluid intake, it is useful to differentiate "regulatory" (also called "need-induced" or "primary") drinking in response to disturbances in body fluid homeostasis, from "non-regulatory" (also called "need-free" or "secondary") drinking (Nicolaïdis and Rowland 1975; Fitzsimons 1979; Rolls and Rolls 1982). Because the various physiological mechanisms that produce both transient and sustained drinking inhibition during regulatory fluid ingestion appear to exert relatively little if any inhibitory effects on non-regulatory fluid consumption, it seems unlikely that such mechanisms directly cause the generation of strong inhibitory signals to fluid ingestion. Rather, it is more likely that they act indirectly to diminish or eliminate the hyperosmolar and/or hypovolaemic excitatory stimuli to drink created as a result of body fluid deficits. This analysis leads to the conclusion that the postabsorptive inhibitory stimuli to fluid ingestion generated by hypo-osmolality, and to some degree volume expansion as well, probably do not play a significant role in normal satiety of regulatory thirst. Instead, it seems more likely that these inhibitory mechanisms are in place primarily to defend against larger more pathological degrees of overhydration, although even this probably constitutes a relatively weak inhibition since overhydration can be efficiently dealt with by renal excretion of the excess fluid.

Thus, despite several weaknesses of the single-loop model of fluid ingestion, the concept that thirst satiation occurs primarily as a result of abolition of excitatory stimuli to fluid ingestion generated by body fluid deficits still appears to me to account best for most experimental and clinical observations to date. If this is the case, then it follows that any stimulus which inhibits osmoreceptor or baroreceptor activity in response to fluid deficits should produce a state of "satiety" for fluids. Because satiety is a subjective sensation, its presence can only be inferred from animal studies. However, it is of interest that studies of thirst following rehydration in dehydrated humans have consistently shown marked decreases in thirst ratings within minutes of drinking, and far before absorption of the ingested fluid (Rolls and Rolls 1982; Thompson et al. 1987; Fig. 19.3). Thus, preabsorptive inhibitory factors are not only capable of terminating drinking, but also of creating subjective feelings of thirst satiety in humans. These observations support the concept that thirst satiation occurs when excitatory inputs to the brain generated by physiological fluid deficits are subsequently inhibited, regardless of the mechanisms by which this is accomplished.

Summary

The harmonious co-ordination of systemic endocrine secretion with both stimulation and inhibition of regulatory drinking accomplishes the integrated physiological and behavioural response crucial to restoration and maintenance of normal body fluid homeostasis. Under these conditions, thirst satiety appears to occur primarily through inhibition, either transiently via preabsorptive factors or in a more sustained manner via postabsorptive factors, of the excitatory input to the brain generated by stimuli arising from body fluid deficits. Parallel changes in peripheral secretion of hormones regulating body fluid homeostasis can be viewed as being "protective" against overhydration, but this effect most likely simply reflects the analogous controls of such endocrine secretion and thirst. Although non-regulatory drinking is not inhibited by the pre- and postabsorptive factors causing thirst satiety during regulatory drinking, this behaviour can be readily accommodated by appropriate physiological endocrine and renal responses that allow excess ingested fluid to be excreted in the normal animal. However, when these responses are impaired, the combination of continued non-regulatory drinking with relatively weak inhibitory mechanisms to limit fluid ingestion can, and does, lead to pathological derangements of body fluid homeostasis in both animals and man.

References

Adolph EF (1939) Measurements of water drinking in dogs. Am J Physiol 125:75–86

Adolph EF (1950) Thirst and its inhibition in the stomach. Am J Physiol 161:374–386

Adolph EF (1967) Regulation of water intake in relation to body water content. In: Code CF (ed) Alimentary canal. American Physiological Society, Washington DC, pp 163–171 (Handbook of physiology, vol 1, sect 6)

Antunes-Rodrigues J, McCann SM, Samson WK (1986) Central administration of atrial natriuretic factor inhibits saline preference in the rat. Endocrinology 118:1726–1728

Balment RJ, Brimble MJ, Forsling ML (1980) Release of oxytocin induced by salt loading and its influence on renal excretion in the male rat. J Physiol (Lond) 308:439–449

Bellows RT (1939) Time factors in water drinking in dogs. Am J Physiol 125:87–97

Blair-West JR, Brook AH, Gibson A et al. (1979) Renin, antidiuretic hormone and the kidney in water restriction and rehydration. J Physiol (Lond) 294:181–194

Blass EM, Hall WG (1976) Drinking termination: interactions among hydrational, orogastric, and behavioural controls in rats. Psychol Rev 83:356–374

Corbit JD (1965) Effect of intravenous sodium chloride on drinking in the rat. J Comp Physiol Psychol 60:397–406

Davis JO, Freeman RH (1976) Mechanisms regulating renin release. Physiol Rev 56:1–54

de Castro J (1988) A microregulatory analysis of spontaneous fluid intake by humans: evidence that the amount of liquid ingested and its timing is mainly governed by feeding. Physiol Behav 3:705–714

Ernits T, Corbit JD (1973) Taste as a dipsogenic stimulus. J Comp Physiol Psychol 83:27–31

Evered MD, Robinson MM, Rose PA (1988) Effect of arterial pressure on drinking and urinary responses to angiotensin II. Am J Physiol 254:R69–R74

Fitzsimons JT (1961) Drinking by rats depleted of body fluids without increases in osmotic pressure. J Physiol (Lond) 159:297–309

Fitzsimons JT (1979) The physiology of thirst and sodium appetite. Cambridge University Press, Cambridge

Fitzsimons JT, Moore-Gillon MJ (1981) Renin-dependence of drinking induced by partial aortic obstruction in the dog. J Physiol (Lond) 320:423–433

Franci CR, Kozlowski GP, McCann SM (1989) Water intake in rats subjected to hypothalamic immunoneutralization of angiotensin II, atrial natriuretic peptide, vasopressin, or oxytocin. Proc Natl Acad Sci USA 86:2952–2956

Garcia J, Ervin FR (1968) Gustatory-visceral and telereceptor-cutaneous conditioning: adaptation in external and internal milieus. Commun Behav Biol 1:389–415

Geelen GL, Keil LC, Kravik SE et al. (1984) Inhibition of plasma vasopressin after drinking in dehydrated humans. Am J Physiol 247:R968–R971

Goldman MB, Luchins DJ, Robertson GL (1988) Mechanisms of altered water metabolism in psychotic patients with polydipsia and hyponatremia. N Engl J Med 318:397–403

Gross PA, Pehrisch H, Rascher W et al. (1987) Pathogenesis of clinical hyponatremia: observations of vasopressin and fluid intake in 100 hyponatremic medical patients. Eur J Clin Invest 17:123–129

Hirth C, Stasch JP, John A et al. (1986) The renal response to acute hypervolemia is caused by atrial natriuretic peptides. J Cardiovasc Pharmacol 8:268–275

Holmes JH (1964) Thirst and fluid intake problems in clinical medicine. In: Wayner MJ (ed) Thirst. Pergamon Press, Oxford, pp 57–75

Kapatos G, Gold RM (1972) Tongue cooling during drinking: a regulator of water intake in rats. Science 176:685–686

Kaufman S (1984) Role of right atrial receptors in the control of drinking in the rat. J Physiol (Lond) 349:389–396

Kozlowski S, Szczepanska-Sadowska E (1975) Mechanisms of hypovolaemic thirst and interactions between hypovolaemia, hyperosmolality and the antidiuretic system. In: Peters G, Fitzsimons JT, Peters-Haefeli L (eds) Control mechanisms of drinking. Springer-Verlag, New York, pp 25–35

Lauson HD (1974) Metabolism of the neurohypophysial hormones. In: Knobil E, Sawyer WH (eds) The pituitary gland and its neuroendocrine control. American Physiological Society, Washington DC, pp 287–393 (Handbook of physiology, vol 4, sect 7)

Maddison S, Wood RJ, Rolls ET et al. (1980) Drinking in the rhesus monkey: peripheral factors. J Comp Physiol Psychol 94:365–374

Mann JFE, Johnson AK, Ganten D et al. (1987) Thirst and the renin angiotensin system. Kidney Int 32:S27–S34

Metzler CH, Ramsay DJ (1989) Atrial peptide potentiates renal responses to volume expansion in conscious dogs. Am J Physiol 256:R284–R289

Moore-Gillon MJ, Fitzsimons JT (1982) Pulmonary vein-atrial junction stretch receptors and the inhibition of drinking. Am J Physiol 242:R452–R457

Nakamura M, Katsuura G, Nakao K et al. (1985) Antidipsogenic action of alpha-human atrial natriuretic polypeptide administered intracerebroventricularly in rats. Neurosci Lett 58:1–6

Needleman P, Greenwald JE (1986) Atriopeptin: a cardiac hormone intimately involved in fluid, electrolyte, and blood pressure homeostasis. N Engl J Med 314:828–834

Nicolaïdis S, Rowland N (1975) Systemic versus oral and versus oral and gastrointestinal metering of fluid intake. In: Peters G, Fitzsimons JT, Peters-Haefeli L (eds) Control mechanisms of drinking. Springer-Verlag, Berlin, pp 14–21

Phillips PA, Rolls BJ, Ledingham JGG, Morton JJ (1984) Body fluid changes, thirst and drinking in man during free access to water. Physiol Behav 33:357–363

Quillen EW Jr, Reid IA, Keil LC (1988) Cardiac and arterial baroreceptor influences on plasma vasopressin and drinking: In: Cowley AW Jr, Liard J-F, Ausiello DA (eds) Vasopressin: cellular and integrative functions. Raven Press, New York, pp 405–411

Ramsay DJ, Thrasher TN (1986a) Hyperosmotic and hypovolemic thirst. In: de Caro G, Epstein AN, Massi M (eds) The physiology of thirst and sodium appetite. Plenum Press, New York, pp 83–96

Ramsay DJ, Thrasher TN (1986b) Satiety and the effects of water intake on vasopressin secretion. In: de Caro G, Epstein AN, Massi M (eds) The physiology of thirst and sodium appetite. Plenum Press, New York, pp 301–307

Ramsay DJ, Rolls BJ, Wood RJ (1977) Thirst following water deprivation in dogs. Am J Physiol 232: R93–R100

Robertson GL (1977) Vasopressin function in health and disease. Recent Prog Horm Res 33:333–385

Robinson MM, Evered MD (1987) Pressor action of intravenous angiotensin II reduces drinking response in rats. Am J Physiol 252:R754–R759

Rolls BJ, Rolls ET (1982) Thirst. Cambridge University Press, Cambridge

Rolls BJ, Wood RJ, Stevens RM (1978) Effects of palatability on body fluid homeostasis. Physiol Behav 20:15–19

Rolls BJ, Wood RJ, Rolls ET (1980a) Thirst: the initiation, maintenance, and termination of drinking. In: Sprague JM, Epstein AN (eds) Progress in psychobiology and physiological psychology. Academic Press, New York, pp 263–321

Rolls BJ, Wood RJ, Rolls ET (1980b) Thirst following water deprivation in humans. Am J Physiol 239:R476–R482

Salata RA, Verbalis JG, Robinson AG (1987) Cold water stimulation of oropharyngeal receptors in man inhibits release of vasopressin. J Clin Endocrinol Metab 65:561–567

Schwab TR, Edwards BS, Heublein DM et al. (1986) Role of atrial natriuretic peptide in volume-expansion natriuresis. Am J Physiol 251:R310–R313

Seckl JR, Williams TDM, Lightman SL (1986) Oral hypertonic saline causes transient fall of vasopressin in humans. Am J Physiol 251:R214–R217

Sobocinska J (1969) Effect of cervical vagosympathectomy on osmotic reactivity of the thirst mechanism in dogs. Bull Acad Pol Sci 17:265–270

Stricker EM (1966) Extracellular fluid volume and thirst. Am J Physiol 211:232–238

Stricker EM (1969) Osmoregulation and volume regulation in rats: inhibition of hypovolemic thirst by water. Am J Physiol 217:98–105

Stricker EM, Adair ER (1966) Body fluid balance, taste, and postprandial factors in schedule-induced polydipsia. J Comp Physiol Psychol 62:449–454

Stricker EM, Jalowiec JE (1970) Restoration of intravascular fluid volume following acute hypovolemia in rats. Am J Physiol 218:191–196

Stricker EM, Verbalis JG (1986) Interaction of osmotic and volume stimuli in regulation of neurohypophyseal secretion in rats. Am J Physiol 250:R267–R275

Thompson CJ, Burd JM, Baylis PH (1987) Acute suppression of plasma vasopressin and thirst after drinking in hypernatremic humans. Am J Physiol 252:R1138–R1142

Thrasher TN, Nistal-Herrera JF, Keil LC et al. (1981) Satiety and inhibition of vasopressin secretion after drinking in dehydrated dogs. Am J Physiol 240:E394–E401

Towbin EJ (1949) Gastric distention as a factor in the satiation of thirst in esophagostomized dogs. Am J Physiol 159:533–541

Verbalis JG (1984) An experimental model of the syndrome of inappropriate antidiuretic hormone secretion in the rat. Am J Physiol 247:E540–E553

Verbalis JG (1990) Inappropriate antidiuresis and other hypo-osmolar states. In: Becker KL (ed) Principles and practice of endocrinology and metabolism. J. B. Lippincott, Philadelphia, pp 237–247

Verbalis JG, Drutarosky MD (1988) Adaptation to chronic hypoosmolality in rats. Kidney Int 34: 351–360

Verbalis, JG, Mangione M, Stricker EM (1991) Oxytocin produces natriuresis in rats at physiological plasma concentrations. Endocrinology (in press)

Whitaker MD, McArthur RG, Corenblum B et al. (1979) Idiopathic, sustained, inappropriate secretion of ADH with associated hypertension and thirst. Am J Med 67:511–515

Zerbe RL, Stropes L, Robertson GL (1980) Vasopressin function in the syndrome of inappropriate antidiuresis. Ann Rev Med 31:315–327

Zimmerman MB, Blaine EH, Stricker EM (1981) Water intake in hypovolemic sheep: effects of crushing the left atrial appendage. Science 211:489–491

Commentary

Epstein: You make the welcome point that the area is neglected relative to research on mechanisms of thirst arousal.

Tachykinins are potent and selective suppressors of thirst, and may have a role in natural satiety. de Caro et al. (1988) discuss neurokinin A and B which are the most interesting of the three mammalian tachykinins.

This chapter does not consider the possibility that the brain itself is the source of satiety rather than a mere responder to inputs from the periphery. The brain contains a set of endogenous agents that have powerful effects on thirst (arousal and suppression). These include angiotensin, ANF (and BNF), and the tachykinins. Peripheral events may arouse or suppress the activity of these agents (and others) resulting in thirst satiety by a variety of mechanisms that are endogenous to the brain. This is, in principle, what Chapters 11 and 19 are suggesting in the oxytocinergic hypothesis for salt intake arousal; why not generalize it? After all, it is cerebral angiotensin activated by peripheral sodium loss that is the likely synergist with aldosterone in the arousal of salt intake in the rat (Sakai and Epstein 1990; Sakai et al. 1990).

References

de Caro G, Perfumi M, Massi M (1988) Tachykinins and body fluid regulation. In: Epstein AN, Morrison AR (eds) Progress in psychobiology and physiological psychology. Plenum Press, New York, vol 13, pp 31–61

Sakai RR, Epstein AN (1990) The dependence of adrenalectomy-induced sodium appetite on the action of angiotensin II in the brain of the rat. Behav Neurosci 104:167–176

Sakai RR, Chow SY, Epstein AN (1990) Peripheral angiotensin II is not the cause of sodium appetite in the rat. Appetite (in press)

Verbalis: This chapter focuses primarily on the mechanisms by which signals generated in the periphery interact with the brain to control the inhibition of drinking behaviour. Obviously, multiple neuropeptides and neurotransmitters are intimately involved with the central circuitry that regulates thirst and drinking behaviour, but this is the subject of other chapters in this volume (see Chapters 7,9,10,13 and 15). My comments, therefore, concentrated on those hormones whose systemic secretion is known to influence body fluid homeostasis, and should not be construed as ignoring others that may be important central regulators of both stimulation and inhibition of drinking. Tachykinins and endogenous opioids are two such potential classes, and undoubtedly still others will be implicated by future studies. Because so much more needs to be learned about how the brain processes and integrates peripheral signals to regulate drinking behaviour, we must not limit ourselves to fixed ideas about brain function but we should keep an open mind about the pathways and neuro-transmitters that work together to accomplish this feat.

With regard to the concept of the brain as the "source" of satiety, I would object to the notion that one can define any single area as being the sole controller of a homeostatic behaviour. The brain does not just respond to inputs from the periphery; it processes and integrates these with other information to arrive at co-ordinated behaviours of benefit to the whole animal. However, in the absence of signals from peripheral physiological conditions, the brain of even the most intelligent animals would have a difficult time regulating body fluids. I would, therefore, prefer to continue to view thirst satiety as a manifestation of central integration of peripheral inputs in which both the brain and the periphery are essential to achieve homeostasis.

Nicolaïdis: Evidence exists for a possible mechanism of water satiation. Specific water and salt (hypertonic saline) afferents converge on anterior hypothalamic neurons that also sense hypotonicity or hypertonicity of the plasma. This convergence of both sensory and systemic information onto the same neurons provides a possible mechanism for the disappearance of thirst when water is consumed before it is absorbed across the intestinal wall (Nicolaïdis 1969).

Reference

Nicolaïdis S (1969) Early systemic responses to orogastric stimulation in the regulation of food and water balance. Functional and electrophysiological data. Ann NY Acad Sci 157:1176–1203

Verbalis: Thank you for bringing up this point. The fact that both thirst and AVP secretion are transiently inhibited by preabsorptive factors as well as in a more sustained manner by absorption of ingested fluid argues in favour of a common locus of action for both types of stimuli. The anterior hypothalamus is a logical place for such an inhibitory action, given the importance of this region for integration and expression of osmotically stimulated drinking. Your studies demonstrating convergence of oropharyngeal and osmoreceptive afferents on the same interneurons clearly provides a potential neuronal mechanism that could participate in this response.

Nicolaïdis: Although most data favour the idea that water satiety is due to the neutralization of thirst, there are observations favouring the idea that some additional active quenching mechanism may exist. The extremely unpleasant state resulting from imposed overconsumption of water (for medical or other reasons) pleads in favour of this view.

Verbalis: I have no real problem with the possibility that the act of drinking may co-activate central reward (or in the case of excessive ingestion, aversive) pathways to explain the differences in the perceived pleasantness of ingested water relative to the volume ingested. My main point is that inhibition of the excitatory inputs producing thirst and drinking is for the most part adequate to account for the observed inhibition of drinking in most species, but certainly not that this is the only factor involved. However, with regard to the unpleasant effects produced by water loading in humans, although subjects ingesting a formal water load do complain of the water tasting "bad", they also frequently become nauseated as a result of the gastric distension resulting from the rapid ingestion of 1200–2000 ml of water over 15–30 minutes. Furthermore, the subjects' reports of difficulties with drinking more water usually begin far in advance of much absorption of the ingested water. Finally, it is quite possible to fluid load bypassing the stomach, and we tend to do this frequently with intravenous infusions in hospitalized patients. Such patients generally do not complain of any similar aversive feelings toward water or other fluids, and the tendency of some such patients to become hypo-osmolar further supports a lack of strong inhibitory controls of drinking. Consequently, I find it difficult to ascertain whether these unpleasant states produced by water-loading reflect effects of overhydration or less specific gastrointestinal effects. Nonetheless, I want to emphasize that I have supplied multiple examples for direct hypo-osmolar and hypervolaemic inhibition of thirst. I certainly do not by any means deny they exist, but simply argue that they represent relatively weak inhibitions. Furthermore, human studies (Rolls et al. 1980; Phillips et al. 1984) demonstrating that people generally do not overdrink, as reflected by excretion of a dilute urine, also indicate that overhydration does not play a significant role in the normal satiety of thirst.

References

Phillips PA, Rolls BJ, Ledingham JGG, Morton JJ (1984) Body fluid changes, thirst and drinking in man during free access to water. Physiol Behav 33:357–363
Rolls BJ, Wood RJ, Rolls ET et al. (1980) Thirst following water deprivation in humans. Am J Physiol 239:R476–R482

Ramsay: Many of the arguments presented in this chapter show that thirst inhibition is not always synonymous with satiety. However, the mechanisms involved in terminating deficit-induced drinking are frequently different from those which initiate it. For example, if the increase in plasma osmolality which occurs during intravenous infusion of hypertonic sodium chloride is prevented from affecting cerebral osmoreceptors, dogs will not drink. This is inhibition of thirst, not satiety. Although permanent satiety must entail elimination of thirst stimuli, in the short-term thirst inhibition and satiety cannot be equated.

Verbalis: I quite agree that one cannot always equate inhibition of drinking with thirst satiety. However, at the risk of possibly arguing a semantic point, I disagree that bathing the osmoreceptors of dehydrated animals with hypotonic fluid, as you have done, is not equivalent to inducing a state of satiety artificially. I interpret satiety as the sense of having ingested sufficient fluid not to desire any more, rather than the state of actually having ingested sufficient water to achieve body fluid homeostasis. If osmoreceptor-mediated functions such as drinking and AVP secretion are inhibited by the act of drinking (even drinking hypertonic saline), and human subjects report accompanying marked decreases in thirst (which they clearly do prior to absorption of the ingested fluid), then why is this not satiety, albeit a transient form of satiety in the absence of subsequent correction of the underlying fluid deficits? For this reason I would argue that it is difficult to assess satiety in animal studies, and we should discuss such results simply in terms of inhibition of drinking. Only in human studies can satiety be accurately assessed, and in this case it will depend on the subjective state of the individual regardless of his or her actual physiological condition.

Ramsay: One can distinguish experimentally between those factors which initiate deficit-induced drinking and those which stop it. In dehydrated dogs, selective removal of hyperosmolality or hypovolaemia before being offered access to water reduce subsequent intake. This allows identification of factors which start drinking. Dogs are rapid drinkers, as are camels and sheep, and drink their deficits accurately, before absorption of ingested water has altered blood composition. Dehydrated dogs with open gastric fistulae drink equivalent volumes to intact dehydrated dogs, again showing the unimportance of blood composition in bringing this rapid drinking process to an end, and the importance of oropharyngeal inputs. Thus the inputs which stop drinking – or cause satiety – are different from those which start it. Unless these oropharyngeal inputs are followed by restoration of blood composition, the stimuli which initiated drinking in these dehydrated animals will again cause drinking. Indeed, a dehydrated dog with an oesophageal fistula will drink continuously. The difference between thirst and satiety is not just a matter of semantics but of concept and mechanism.

Booth: This discussion seems to me to be ignoring the logical link between the concept of satiety and the concepts of thirst, hunger and appetite. Satiety is not any one mechanism in the brain or periphery or a sensation only in human consciousness, any more than thirst is. Satiety is simply the lack of the tendency to ingest, induced by effects of ingestion, the lack of disposition that is as easy or difficult to measure in other animals as it is in people. Which effects of ingestion inhibit the disposition and where each effect is signalled are the scientific questions about satiety, not how a physiological manipulation of ingestion affects some hypothetical entity or global measures such as thirst ratings or water intake. Reduction of intake by manipulation prior to the drinking does not identify initiating factors; it confounds facilitatory influences with factors that modulate the operation of other sorts of terminating factors, perhaps at a central integration stage. The differences between the facilitatory and inhibitory influences at the start, at the end, and after the end of the drink can only be determined by direct behavioural and physiological measurements at each time of interest.

Verbalis: This exchange makes it apparent to me that the concept of "satiety" is interpreted differently by various investigators. Although I am in substantial agreement with most of the points raised by Drs Ramsay and Booth, I still feel that it is valid to interpret any situation or manipulation that decreases perceived thirst as being satiating, and that any accurate differentiation between simple inhibition of drinking and the production of satiety can only be made in human subjects who are able to tell us when an inhibition of drinking is actually accompanied by a decrease in thirst, which is what we humans generally associate with the sensation of satiety.

Chapter 20

Social Influences on Fluid Intake: Laboratory Experiments with Rats, Field Observations of Primates

B.G. Galef, Jr

Introduction

To undertake a review of research in the area of social influences on mammalian fluid intake is a daunting prospect. Little work has been done on social influences on drinking (at least of the non-alcoholic sort) in mammals and the results of much of what has been done are negative. Consequently, a straightforward review of social influences on mammalian fluid intake is not yet possible. I propose, instead, to proceed indirectly, first by reviewing processes known to support social influences on mammalian feeding behaviour (an area where there are many positive findings), then by considering whether the social processes that influence food selection or food intake might also influence fluid selection or fluid intake.

In the Laboratory

Effects of the Presence of Adult Rats at an Ingestion Site

Adult rats can influence weanling young to eat one food in preference to others by simply eating at one location rather than at another. The presence of adults at a potential feeding site attracts weanlings from a distance to that site and markedly increases the probability that weanlings will begin to eat whatever food is located there (Galef and Clark 1971; Galef 1981). In both laboratory (Galef and Clark 1971) and field situations (Steiniger 1950), maturing wild rat pups will, for weeks, eat only those foods to which the adults of their colony have introduced them.

Such local enhancement (Thorpe 1963) of a foraging site also has the potential to induce young rats to drink from one water source rather than from another. However, the only experiments undertaken to examine directly effects of drinking site utilization by adult rats on pups' choice of drinking site (Galef 1978) revealed little if any adult influence on rat pups' drinking site selection. In general, rat pups seeking food appeared more strongly motivated to affiliate with conspecifics than did rat pups seeking water (Galef 1978).

Effects of Residual Olfactory Cues at an Ingestion Site

Adult rats mark both foods that they are eating and the area around those foods with residual olfactory cues that make marked foods and marked feeding sites more attractive to juveniles seeking food than are unmarked foods or unmarked feeding sites (Galef and Heiber 1976; Galef and Beck 1985). There is no apparent reason why rats should not mark drinking sites in the same way that they mark feeding sites. However, that possibility has not yet been examined.

Flavour Cues in the Milk of Rat Dams

The results of several studies (Galef and Henderson 1972; Galef and Sherry 1973) are consistent with the hypothesis that (a) the milk of a lactating rat contains cues reflecting the flavour of her diet and (b) such flavour cues in mother's milk can influence pups' selection of foods at weaning. Although it has not been investigated, there is reason to expect that learned preferences for orally ingested, flavoured fluids, like preferences for intraperitoneally injected, flavoured fluids (Le Magnen and Tallon 1968, Martin and Alberts 1979) could be readily transmitted from mothers to their young via flavour cues incorporated in mother's milk. However, mother's milk may not provide an efficient medium for communication between dams and their young about relatively flavourless fluids, such as water. The effects of flavour cues in mother's milk on fluid selection by weanling pups will be limited to induction of enhanced preferences for substances the flavours of which pups can detect in mothers' milk.

Whether such enhanced preferences for flavoured substances would be useful to weanlings attempting to maintain fluid balance in natural settings depends on circumstance. If the main source of fluids for a population was widely separated water holes, flavour cues in mother's milk would probably not be of much use to juveniles in maintaining fluid balance. If on the other hand a population were garnering most of its water from flavoured fruit, then enhanced preference for the flavour of fruit, contingent upon experience of its flavour in mother's milk, might be a factor in maintenance of fluid balance by pups.

Olfactory Cues on the Breath of Rats

Galef and Wigmore (1983) and Posadas-Andrews and Roper (1983) discovered independently that after a naive rat (an observer) interacted for a few minutes with a recently fed conspecific (a demonstrator) the observer exhibited an enhanced preference for the diet its demonstrator had eaten (see also Strupp and Levitsky 1984; Galef 1989). Both olfactory cues escaping from the digestive tract of demonstrators and the smell of bits of food clinging to the fur and vibrissae of recently fed demonstrators permit observers to identify foods that their respective demonstrators have eaten (Galef and Stein 1985). However, simple exposure to the smell or taste of a food does not suffice to enhance an observer's preference for that food (Galef et al. 1985; Galef 1989; Heyes and Durlach 1990). Observers' preferences for foods are enhanced only by smelling a food in contiguity with rat-produced odours (Galef and Stein 1985) emitted by demonstrators. These demonstrator-produced odours are probably volatile sulphur compounds, like carbon disulphide (Galef et al. 1988), a chemical constituent of rat breath that, when added to a food, increases the preferences of both rats and mice for that food (Bean et al. 1988; Mason et al. 1989). Social influences acting via rat breath can (a) facilitate identification by the naive of foods containing needed nutrients (Beck and Galef 1989), (b) facilitate identification and avoidance of toxins (Galef 1986a, b), and (c) act as aids in deciding where to search for foods (Galef and Wigmore 1983; Galef et al. 1987).

Interaction of observer rats with demonstrators that have drunk a flavoured liquid enhances an observer's preference for the flavoured liquid, just as exposure to a demonstrator that ate a flavoured food enhances an observer's preference for the flavoured food (Galef et al. 1985, Galef and Stein 1985). Thus, as with flavour cues in the milk, cues on rat breath might be important in directing fluid selection by populations of rats that obtain fluids from fruits or other plant materials, but probably not in populations that subsist on free water.

Socially Induced Aversions

A naive rat (a subject) that ingests an unfamiliar food and then interacts with an ill rat (a "poison partner") develops an aversion to the unfamiliar food it ate prior to interacting with its poison partner (Coombes et al. 1980; Lavin et al. 1980). It is, however, not obvious how this "poison-partner effect" might enhance avoidance of either toxic foods or toxic, flavoured fluids in natural circumstances. If an unfamiliar food eaten by an individual before it interacted with a poison partner were safe, then subsequent avoidance of that food would be counterproductive. If, on the other hand, an unfamiliar food ingested before interaction with a poison partner were toxic, then information received from the ill individual would be redundant; aversion learning would occur even in the absence of a poison partner (Galef et al. 1983).

Gemberling (1984) examined the situation in which the rat ingesting an unfamiliar, flavoured solution was a lactating female and the poison partners with which the dam then interacted were infants. It is possible for a lactating dam to eat something that has no ill effects on her, but causes distress to her suckling young and it would clearly be adaptive for dams to avoid eating foods that distressed their offspring. Gemberling found that lactating dams that had ingested a novel saccharin solution shortly before they interacted with young that had been injected with an illness-inducing agent subsequently showed an aversion to saccharin.

Competitive Enhancement of Intake

Harlow (1932) reported that rats would both eat more food and show greater weight gain when fed for 1 h/day in pairs than when fed for 1 h/day in isolation. He suggested that such social facilitation of food intake depended on interacting subjects engaging in unrestrained, active competition for food (see Clayton 1978, for a review). Whether socially enhanced water intake would also be observed in rats competing for fluids is not known.

Summary

Results of laboratory studies indicate (a) that the feeding behaviour and food choices of rats can be modified by social interactions among rats and (b) that a number of different behavioural processes can play a role in such social modification of feeding behaviour. Although there are not a great deal of relevant data, social influences on feeding-site selection seem more robust than social influences on drinking-site selection. The fact that all foods have smells and flavours associated with them, whereas free water, possibly the most important source of fluids for many mammals, probably does not have a readily detected volatile component may result in social influence being less important in directing fluid intake than in directing food intake in natural circumstances.

In the Field

Unobtrusive observations in natural situations rarely permit analysis of behavioural mechanisms underlying food or fluid choice. However, field studies do provide evidence of possible social influences on ingestive behaviour that need to be analysed under controlled conditions (Galef 1984). The literature contains many observations of differences in the food preferences of different social groups of conspecific animals living in natural circumstances. However, we have only the crudest understanding of how these locale-specific behaviours develop, though there is reason to suspect that some may be the result of social learning (Galef 1990).

Examples of Locale-Specific Differences in Food Selection by Primates

Nishida (1987, see also Goodall 1986) has catalogued differences in food selection by chimpanzees at Mahale and Gombe, two field sites a few tens of miles apart along the east shore of Lake Tanganyika in Tanzania. For example, unripe seeds of *Saba florida* are neglected by Mahale chimpanzees, whereas those at Gombe eat them often (Nishida et al. 1983). Gombe chimpanzees eat *Dorylus* ants, rarely eat *Crematogaster* ants, and totally avoid *Camponotus* ants. Mahale chimpanzees eat the last two frequently and reject the first completely.

Although such differences between local populations in diet selection are well established, there are many possible explanations for the development of such locale-specific feeding habits. Differences in the relative availability of foods (Galef 1990; Nishida 1987), differences in the details of the characteristics of foods themselves in different areas (Nishida 1987), differences in alternatives to particular items (Gaulin and Kurland 1976), as well as differences between populations in socially transmitted food preferences could all produce locale-specific differences in food choice.

Field Observations of Apparent Social Learning About Foods by Primates

Occasionally, observations of the behaviour of free-living animals are sufficient to suggest that social learning may truly be involved in the development of a unique pattern of ingestive behaviour. One of the more convincing cases is Whitehead's (1986) observations of infant howling monkeys (*Allonata palliata*) feeding for the first time on seasonally available leaves and fruits in the forests of Costa Rica. When feeding on leaves, infant howling monkeys (a) looked at a parent before eating, (b) fed only when a parent fed, (c) ate only what a parent ate, and (d) were subject to parental intervention if they chose to eat an "incorrect" leaf.

Although it seems likely that such interactions between adults and young would result in social transmission of food selection patterns between generations, there is no evidence that the consequences of mother–young interaction on the food choices of young last for more than a few minutes (Galef 1990). It is known both that there are differences among troops of primates in the foods they eat and that social interactions occur between mothers and young that could produce differences in later food selection by the young; however it is not known if the two observations are related.

Hauser's (1988) observations of vervet monkeys at Amboseli, Kenya provide some of the most convincing evidence of social transmission of a feeding pattern available in the field literature. During a period of drought, the vervets were forced to subsist on *Acacia*, a plant they normally avoided. In September 1983, an adult female (BA) was observed to dip dry *Acacia* pods into the well of an *Acacia* tree containing a viscous exudate. Eating of exudate and the use of pods to secure exudate had not been observed during six preceding years of observations. Eight days after BA began pod-dipping, BA's two juvenile sons exhibited the behaviour; a day later two other troop members joined in, etc. The behaviour of dipping pods in *Acacia* exudate disappeared in 1984 when the drought broke and both free surface water and a diversity of plant material were again available to the troop. It is not known whether troop members had been eating or drinking *Acacia* exudate previously.

In Conclusion

Finding safe, adequate foods and sufficient liquids to sustain life are challenges facing all mammals. Although individuals can learn independently where and what to eat or drink, each young mammal interacts for an extended period with an adult caretaker whose survival and successful reproduction indicate that the caretaker has had access to adequate food and drink. There is every reason to expect the naive to exploit the knowledgeable as sources of information about the nature and location of needed nutrients. The evidence suggests that they do.

Both field observations of social interactions surrounding ingestive behaviour and experimental analyses of social influences on choice of ingesta indicate that study of conspecific influences on eating and drinking can provide insight into the development of adaptive ingestive repertoires by mammals (Galef and Beck 1990). Although to date most such observations and experiments directly involve feeding behaviours, they may also be relevant to understanding the ways in which animals come to locate free-standing sources of water and to select items to ingest that provide water for the maintenance of fluid balance.

Acknowledgement. The author gratefully acknowledges financial support from the Natural Sciences and Engineering Research Council and McMaster University Research Board during the preparation of this review.

References

Bean NJ, Galef BG Jr, Mason JR (1988) At biologically significant concentrations, carbon disulfide both attracts mice and increases their consumption of bait. J Wildl Manage 52:502–507
Beck M, Galef BG Jr (1989) Social influences on the selection of a protein-sufficient diet by Norway rats. J Comp Psychol 103:132–139
Clayton DA (1978) Socially facilitated behavior. Q Rev Biol 53:373–392

Coombes S, Revinsky SH, Lett BT (1980) Long-delay taste-aversion learning in an unpoisoned rat: exposure to a poisoned rat as the unconditioned stimulus. Learn Motiv 11:256–266

Galef BG Jr (1978) Differences in affiliative behavior of weanling rats selecting eating and drinking sites. J Comp Physiol Psychol 92:431–438

Galef BG Jr (1981) The development of olfactory control of feeding site selection in rat pups. J Comp Physiol Psychol 95:615–622

Galef BG Jr (1984) Reciprocal heuristics: a discussion of the relationship of the study of learned behavior in laboratory and field. Learn Motiv 15:479–493

Galef BG Jr (1986a) Social identification of toxic diets by Norway rats (*R. norvegicus*). J Comp Psychol 100:331–334

Galef BG Jr (1986b) Social interaction modifies learned aversions, sodium appetite and both palatability and handling-time induced dietary preference in rats. J Comp Psychol 100:432–439

Galef BG Jr (1988) Communication of information concerning distant diets in a social, central-place foraging species: *Rattus norvegicus*. In: Zentall TR, Galef BG Jr (eds) Social learning: psychological and biological perspectives. Lawrence Earlbaum, Hillsdale NJ

Galef BG Jr (1989) Enduring social enhancement of rats' preferences for the palatable and the piquant. Appetite 13:81–92

Galef BG Jr (1990) Tradition in animals: field observations and laboratory analyses. In: Bekoff M, Jamieson D (eds) Interpretation and explanation in the study of behavior: a comparative approach. Westview, Boulder

Galef BG Jr, Beck M (1985) Aversive and attractive marking of toxic and safe foods by Norway rats. Behav Neural Biol 43:298–310

Galef BG Jr, Beck M (1990) Diet selection and poison avoidance by mammals individually and in social groups. In: Stricker EM (ed) Handbook of neurobiology, vol 10. Plenum Press, New York

Galef BG Jr, Clark MM (1971) Social factors in the poison avoidance and feeding behaviors of wild and domesticated rat pups. J Comp Physiol Psychol 75:341–357

Galef BG Jr, Heiber L (1976) The role of residual olfactory cues in the determination of feeding site selection and exploration patterns of domestic rats. J Comp Physiol Psychol 90:727–739

Galef BG Jr, Henderson PW (1972) Mother's milk: a determinant of the feeding preferences of weaning rat pups. J Comp Physiol Psychol 78:220–225

Galef BG Jr, Sherry DF (1973) Mother's milk: a medium for the transmission of cues reflecting the flavor of mother's diet. J Comp Physiol Psychol 83:374–378

Galef BG Jr, Stein M (1985) Demonstrator influence on observer diet preference: analysis of critical social interactions and olfactory signals. Anim Learn Behav 13:31–38

Galef BG Jr, Wigmore SW (1983) Transfer of information concerning distant foods in rats: a laboratory investigation of the 'information centre' hypothesis. Anim Behav 31:748–758

Galef BG Jr, Wigmore SW, Kennett DJ (1983) A failure to find socially mediated taste-aversion learning in Norway rats (*R. norvegicus*). J Comp Psychol 97:358–363

Galef BG Jr, Kennett DJ, Stein M (1985) Demonstrator influence on observer diet preference: effects of simple exposure and the presence of a demonstrator. Anim Learn Behav 13:25–30

Galef BG Jr, Mischinger A, Malenfant SA (1987) Hungry rats' following of conspecifics to food depends on the diets eaten by potential leaders. Anim Behav 35:1234–1239

Galef BG Jr, Mason JR, Preti G, Bean NJ (1988) Carbon disulfide: a semiochemical mediating socially-induced diet choice in rats. Physiol Behav 42:119–124

Gaulin SJC, Kurland JA (1976) Primate predation and bioenergetics. Science 191:314–315

Gemberling GA (1984) Ingestion of a novel flavor before exposure to pups injected with lithium chloride produces a taste aversion in mother rats (*Rattus norvegicus*). J Comp Psychol 98:285–301

Goodall J (1986) The chimpanzees of Gombe. Harvard University Press, Cambridge, MA

Harlow HF (1932) Social facilitation of feeding in the albino rat. J Genet Psychol 41:211–221

Hauser MD (1988) Invention and social transmission: new data from wild vervet monkeys. In: Byrne RW, Whiten A (eds) Machiavellian intelligence. Clarendon, Oxford, pp 327–343

Heyes CM, Durlach PJ (1990) "Social blockade" of taste aversion learning in Norway rats (*R. norvegicus*): is it a social phenomenon? J Comp Psychol 104:82–87

Lavin MJ, Freise B, Coombes S (1980) Transferred flavor aversions in adult rats. Behav Neural Biol 28:15–33

Le Magnen J, Tallon S (1968) Préférence alimentaire du jeune rat induite par l'allaitment maternel. C R Soc Biol 162:387–390

Martin LT, Alberts JR (1979) Taste aversions to mother's milk: the age-related role of nursing in acquisition and expression of a learned association. J Comp Physiol Psychol 93:430–445

Mason JR, Bean NJ, Galef BG Jr (1989) Attractiveness of carbon disulfide to wild Norway rats. In: Crabb AL, Marsh RE (eds) Proceedings of the thirteenth vertebrate pest conference. University of California Press, Davis, pp 95–97

Nishida T (1987) Local traditions and cultural transmission. In: Smuts BB, Cheney DL, Wraugham RN, Struhsaker TT (eds) Primate societies. University of Chicago, Chicago, pp 462–474

Nishida T, Wraugham RW, Goodall J, Uehara S (1983) Local differences in plant-feeding habits of chimpanzees between the Mahale Mountains and Gombe National Park, Tanzania. J Hum Evol 12: 467–480

Posadas-Andrews A, Roper TJ (1983) Social transmission of food preferences in adult rats. Anim Behav 31:265–271

Steiniger F (1950) Beitrage zur Soziologie und sonstigen Biologie der Wanderratte. Z Tierpsychol 7: 356–379

Strupp BJ, Levitsky DA (1984) Social transmission of food preferences in adult hooded rats (*Rattus norvegicus*). J Comp Psychol 98:257–266

Thorpe WH (1963) Learning and instinct in animals, 2nd edn. Methuen, London

Whitehead JM (1986) Development of feeding selectivity in mantled howling monkeys, *Alouatta palliata*. In: Else JG, Lee PC (eds) Primate ontogeny, cognition and social behaviour. Cambridge University Press, Cambridge, pp 105–107 (Proceedings of the 10th congress of the International Primatological Society, vol 3)

Commentary

de Castro: Social factors profoundly influence fluid intake in humans but do so secondarily as a result of a marked facilitation of food intake. The number of other people present strongly correlates with the amount of fluid ingested and also with the amount of food ingested. When both the number of people present and the amount of food ingested are used as predictors of fluid intake in a multiple linear regression, then the effect of the number of people vanishes while the effect of solids ingested remains. This implies that the social effect of fluid intake occurs secondarily to the influence of food intake. (see chapter 21.)

Section VI
Determinants of Human Fluid Intake

Chapter 21

Bout Pattern Analysis of Ad Libitum Fluid Intake

J.M. de Castro

Introduction

Experimentation vs. Observation

The traditional approach to studying any particular behaviour is to isolate the target behaviour, eliminating or holding constant as many alternative behaviours as possible. The experimenter then manipulates the organism or the environment to ascertain the influence of an independent variable on the target behaviour. This research approach has led to the conclusion that body fluid homeostasis, in particular the defence of the intracellular (Gilman 1937; Fitzsimons 1961a) and extracellular (Fitzsimons 1961b; Stricker 1968) fluid compartments, is the primary determinant of fluid intake. This traditional approach relies on the experimenter's good judgement as to which factors are irrelevant and thus should be eliminated or controlled and which are important and need to be manipulated and studied. The outcome, at least to some extent, may be preordained by the biases and preconceptions of the experimenter as expressed in the choice of variables for control or manipulation. Indeed, the most potent and influential factor determining the amount eaten in a meal, the number of other people present (de Castro and de Castro 1989), was missed because in attempting to design an experimentally analysable situation social factors were removed and controlled. The contention expressed in the present chapter is that a similar situation exists with the study of fluid intake regulation; the most important factors in determining the behaviour are removed and controlled in the experiments.

Models and Actual Intakes

The first indication that there might be something wrong with body fluid homeostasis as the primary explanation for fluid intake came as a result of an attempt to use the Toates and Oatley (1970) model of thirst and body fluid homeostasis to predict actual fluid intake by rats (de Castro, unpublished). The actual fluid intakes of the rat, recorded with drinkometers and monitored by computer were entered into the Toates and Oatley (1970) equations to predict at any point in time the levels of intracellular and/or extracellular fluid depletion/repletion. These calculated levels were then used in univariate and multivariate regression analyses to predict when and how much the rat would drink on the next occasion. As it turned out none of the variables calculated from the model were capable of significantly predicting subsequent fluid intake. At the time, I concluded that the model must be inaccurate, and abandoned this line of inquiry. I now realize that the model was not the problem but rather the underlying assumption that body fluid homeostasis provides the primary signal for ad libitum drinking.

Non-Homeostatic Drinking

It has long been recognized that fluid intake occurs without any clearly defined deficit in either the intracellular of extracellular fluid compartments (Kraly 1984). Such non-homeostatic drinking (Rowland 1977; Bolles 1979; Toates 1979) is quite common and occurs in response to palatability factors (Rolls et al. 1978), to scheduling factors (Falk 1966), or even to anticipation of future deficits (Fitzsimons and Le Magnen 1969). Fluid intake can also be markedly altered by ecological conditions (Rowland 1977; Collier 1986). It is clear, then, that fluid intake can be precipitated by either homeostatic or non-homeostatic mechanisms. It is not clear, however, which influences are primary and which operate only rarely or in specialized circumstances. Phillips et al. (1984) observed water intake along with the contents of the blood and the urine in healthy men during their working hours and could find no changes in body fluid variables associated with the spontaneous ingestion of water. Additionally, humans mostly produce less than optimally concentrated urine which implies that their drinking is greatly in excess of homeostatic needs (Rolls and Rolls 1982; see Chapter 23).

Bout Pattern Analysis

In order to investigate whether ad libitum fluid intake is controlled by homeostatic or non-homeostatic mechanisms the pattern of fluid intake spontaneously expressed by normal subjects, under baseline conditions was analysed (de Castro 1988). During the course of the day an organism emits a stream of behaviour consisting of periods of time during which single behaviours are initiated, expressed and then terminated, termed a bout, followed by the initiation of a new behaviour which is expressed and later terminated and so on. In order to perform a bout pattern analysis, every occurrence of a behaviour, its magnitude and time of occurrence, has to be recorded over a long continuous period of time. The investigator does not attempt to alter or control the situation, but merely arranges to measure the level of as many variables as possible. Variables to control and study need not be selected as the behaviours emitted will define the variables for study.

The primary dependent measures, bout size (ml ingested for fluid intake) and the interbout interval (min from the end of the last bout until the beginning of the next bout), are then calculated. An attempt is then made to relate the magnitude of these two factors to preceding conditions derived from the prior behavioural stream. The measured preceding conditions are then correlated with the bout size or the interbout interval. To the extent that a factor consistently and significantly correlates with the dependent variable then the factor must be either a part of or associated with the mechanisms responsible for the control of the behaviour.

Causal Analysis of Bout Patterns

Although the research is observational and causation cannot be conclusively demonstrated, this kind of analysis can be used to test for potential causal influences on regulatory behaviours. In order to demonstrate that a variable is the cause of the level of another, three conditions must be met; the putative cause must precede the effect in time, the two factors must be correlated, and possible third factor responsibility for the covariation must be eliminated. Clearly the first two conditions are met in the bout pattern analysis. The results of such an analysis can then be appropriately used to test causal inferences. If a putative causal factor does not precede or is not correlated with its purported effect then a causal interpretation can be rejected. In addition, bout pattern analysis may be employed to eliminate third factors as potential spurious covariates. If a third factor is not correlated with both the putative causal factor and the effect then it cannot be producing the covariation. Even when the third factor is intercorrelated with the putative cause and effect factors, it can be eliminated as an explanation by entering the significant predictors into a multiple regression analysis predicting the dependent variable. In the multiple regression the influence of a factor on the dependent variable is ascertained only after the influence of all the other factors have been mathematically removed. Thus if a factor is associated with the dependent variable only secondarily via an association with a measured third factor then, when the third factor's influence is removed in the multiple regression, the first factor's prediction of the dependent variable should vanish. If, on the other hand, the factor has a primary influence on the dependent variable then its influence will remain whatever other factor is partialled out.

As an example, the amount of food ingested in a meal is positively correlated with the duration of the prior interval and negatively with the amount of food energy estimated to be present in the stomach at the start of the meal. When both factors are entered into a multiple regression analysis predicting meal size, only the estimated premeal stomach content maintains a significant prediction (de Castro and Kreitzman 1985). This indicates that the prior interval duration is associated with the meal size secondarily because it provides the time for the stomach to empty. Hence, bout pattern analysis can be used as a method to reject factors from consideration as causes and can further be employed to eliminate third factors as explanations. It can thus be used to make a convincing, if not compelling, case for a causal connection between two observed variables.

The Diary Self-Report Technique

The prerequisite for bout analysis is continuous measurement of the amounts and timing of fluid intake and other surrounding influences. Humans can easily provide this information. They can record, in a diary, on an ongoing basis, their behaviour, their feelings, and the environmental context. This simple approach has not often been employed because of the widespread suspicion that it is inaccurate and unreliable. However, this bad reputation of the self-report method stems historically from investigations using a 24-hour recall procedure wherein the subject is asked to record everything they ate or drank over the prior 24-hour period. This technique has been found to produce fairly inaccurate results (Krantzler et al. 1982). The diary self-report method, on the other hand, requires the subjects to carry the diary with them and record their intake at the time it occurs. This technique has been found to have very good reliability and validity (Adelson 1960; Gersovitz et al. 1978; Krantzler et al. 1982; St Jeor et al. 1983).

In our studies, we add a couple of nuances to promote accuracy further. First, the subjects are recruited with the understanding that they will receive a detailed nutritional composition analysis of their reported diets as a reward for participation. They are informed that the accuracy of this report is exactly linked to the accuracy of their recordings. Subjects are given a small pocket-sized diary and are instructed to record for a 7-day period, in as detailed a manner as possible, every item that they either drink or eat, the time they eat it, the amount they eat, and how the food was prepared. They are also instructed to record their thirst and hunger at the beginning of each bout on seven point, full-thirsty and full-hungry scales. As a further check on accuracy, each subject is asked to identify two individuals with whom they will be eating during the recording period. At the conclusion of the recording these individuals are contacted and asked to verify the diary entries. These procedures result in highly detailed and accurate diary records and after hundreds of verifications we have yet to discover a disconfirmation of a reported entry.

Patterns of Fluid Intake in Humans

Separating Food from Fluid Intake

One of the main obstacles to studying spontaneous fluid intake in the natural environment is separating food from fluid intake since many foods are liquids (e.g., soups), many liquids contain significant amounts of food energy (e.g., milk shake), and even solid foods contain varying amounts of water. In performing a bout pattern analysis of human fluid intake, no attempt was made to draw absolute distinctions but rather three different ways of classifying fluid intakes were employed (de Castro 1988). In the first analysis, food and fluid intakes were not differentiated, intake was classified according to its water content and data analyses were performed on simple total water contents of the bouts. The second analysis characterized the water intake in terms of its surplus over that required for digestion of the food (excess fluid = total fluid -1.1 x total solid; see Toates (1978) for discussion). In the third analysis, arbitrary distinctions were made between food and fluid intakes and only the data for those events classified as "drinks" were analysed.

Predictors of Fluid Intake

As predictors of the amount of fluid ingested in the bout, a number of prior conditions were employed including the thirst and hunger self-ratings, the duration of the prior interbout interval, the amounts estimated to be present in the stomach prior to the meal of total fluid, excess fluid, and food energy (see de Castro 1988 for description of the calculations), and the amount of solid ingested in the bout. As predictors of the duration of the subsequent interbout interval, the three different measures of the fluid content of the bout were employed together with the food energy content of the bout and the estimated stomach contents of total fluid, excess fluid and food energy.

Human Fluid Intake Results

The results of the analyses performed on the self-reported intakes of 36 adult humans (de Castro 1988) indicated that the regulation of fluid intake under ad libitum conditions is secondary to the regulation of food intake. Indeed, the majority of fluid was ingested when there was very little perceived need, in the sense that thirst was on the average rated to be only slight prior to a draught, even when fluid was ingested without food. Regardless of which measure of fluid intake was used, the amount of food energy ingested in the bout produced the strongest correlations, accounting for over 50% of the variance in the amount of fluid ingested. The other variables only produced small correlations with the amount of fluid ingested, and when the amount of food ingested was paired with the other correlated factors in multiple regressions predicting amount of fluid ingested, the influences of the other factors vanished. Food intake became the sole significant predictor. The amount of fluid ingested, then, would appear to be primarily determined by food intake.

Regression analysis predicting the duration of the interval following the bout suggested, similarly, that the amount of food ingested and not fluid was the strongest determining factor. Neither the amount of water ingested nor the amount of liquid in the stomach at the end of the bout was significantly related to the period of time until fluid was ingested again. Only the amount of food ingested along with the water and the estimated content of food in the stomach at the end of the bout were significant predictors of the duration of the postbout interval. Hence, not only the amount but also the timing of fluid ingestion would appear to be determined by eating in human beings.

Patterns of Fluid Intake in Rats

Monitoring of Bout Patterns

These results were quite striking but the modern environment of humans is admittedly unlike that at any other time or of any other species. Therefore, it is important to extend these analyses with a different species under completely different conditions. To accomplish this the feeding and drinking behaviour of hooded rats was continuously monitored in individual enclosed chambers which were equipped with pellet-sensing eatometers and lick-sensitive drinkometers (de Castro 1989). A microcomputer system was programmed to recognize and record each meal or draught, control the 12/12-hour light/dark cycle, and continuously monitor core temperature via a telemetry system (de Castro and Brower 1977). The amounts of food and fluid in the stomach were estimated with a computer model (Booth 1978; Toates 1978). Since solid and liquid intakes were clearly separated the analyses were greatly simplified.

Rat Fluid Intake Results

With humans in their natural environments, intakes appear to be controlled primarily by adjusting the amounts ingested in the bouts (de Castro and Kreitzman 1985). Rats, in the laboratory environment, on the other hand, tend to regulate by adjusting the duration of the interval following the bout (Le Magnen and Tallon 1968; de

Castro 1981). It is not surprising then that the amount of fluid ingested was only weakly correlated with either the predraught core temperature or the stomach contents of solid and water. On the other hand, the amount drunk was positively correlated with the amount of solid ingested. The multiple regression analyses, predicting the amount of water ingested indicated that the estimated prebout stomach contents of solid inhibits drinking whereas anomalously the stomach contents of liquid facilitates drinking. This pattern of results is identical to that observed for predicting the amount eaten in a meal, wherein the estimated prebout stomach contents of solid was found to inhibit eating whereas the stomach contents of liquid facilitated eating. This suggests that the amount of liquid ingested is primarily responsive to factors associated with caloric intake.

Since the rat's primary mode of regulation is through the adjustment of the interval between bouts, it was not surprising that much clearer and stronger relationships were apparent between the prior conditions and the duration of the postdraught interval. Multiple regression analyses predicting the postdraught interval demonstrated that the amount of food in the stomach at the end of the draught and not the amount of fluid is the primary predictor of the interval duration. This suggests that drinking bouts for rats, as for humans, are primarily associated with food intake, the amount and timing of fluid intake determined not by factors associated with fluid balance but rather with factors associated with energy balance.

Bout Nutrient Compositions and Fluid Intake

If the amount of fluid ingested is determined by food intake then an analysis of the differential abilities of various components of the meal to influence the amount of fluid ingested might suggest the nature of the underlying mechanism. If body fluid defence was the primary mechanism involved then the amount of fluid ingested with the meal should be related to the dehydrating properties of the ingested nutrients. Since the rats ate a uniform diet, the composition of the meals did not vary and thus rat bout patterns could not be analysed for composition effects on fluid intake. On the other hand, the varied diet of humans affords an opportunity to perform just such an analysis.

Multiple linear regressions were performed using the intake of humans of carbohydrate, fat, protein and sodium in the meal as independent variables predicting either the total fluid in the bout, the excess fluid in the bout, or the amount of fluid ingested in the form of "drinks". The beta coefficients from these regressions are presented in Fig. 21.1. Beta coefficients are standardized weighting factors which reflect the importance of the factor in predicting the amount of fluid intake with the other predictors held mathematically constant. The results for total fluid and excess fluid intake have to be interpreted cautiously since fat in food is generally accompanied by very little water. This is probably responsible for the facts that the beta coefficients for fat are significant and negative when predicting either total water or excess water but positive when predicting "drink" fluid only. Clearly the strongest association with fluid intake, regardless of whether total, excess, or "drink" fluid is the dependent variable, is the amount of carbohydrate ingested in the bout. Protein would appear to have the next strongest association with fluid intake followed by sodium and lastly by fat.

With the exception of fat, the differential effectiveness of the nutrients on fluid intake would appear to be inversely related to their osmotic properties. Sodium, the

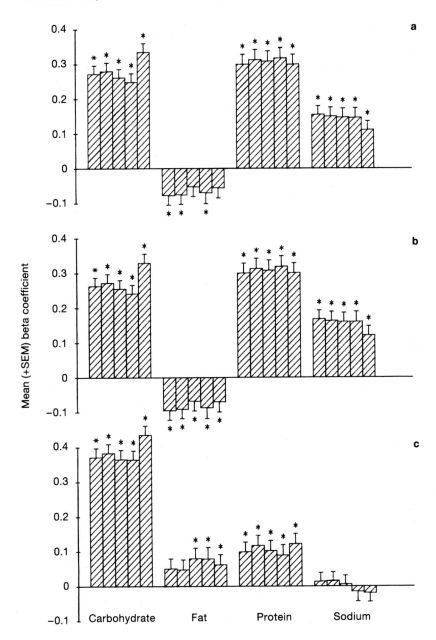

Fig. 21.1. Mean (+ SEM) ß coefficients (standardized partial regression) from the multiple linear regression predictions of the amount of fluid ingested in human meals. **a** represents the ß coefficients for the prediction of the total amount of fluid in food and drink ingested in the meal, on the basis of the amount of carbohydrate, fat, protein and sodium ingested in the bout. **b** and **c** represent the ß coefficients from the equivalent multiple regression predicting fluid ingested in excess of that required for digestion (**b**) and the amount of fluid ingested in the form of "drinks" (**c**). The first bar of each set of five represents the meal definition of minimum 15 minute IMI and 50 kcal size; the second, 45 min/50 kcal; the third, 45 min/ 100 kcal; the fourth, 45 min/200 kcal; and the fifth, 90 min/50 kcal. Asterisks (*) indicate a value that is significantly (P <0.05) different from zero as assessed by a t test.

molecule with the greatest osmotic effect, has the least influence on fluid intake and if fluid ingested in "drink" form only is considered sodium intake appears to have no influence at all. Protein has a greater dehydrating effect than carbohydrate (Fitzsimons and Le Magnen 1969) yet carbohydrate has a comparable impact on total and excess fluid intake and a decidedly larger effect on fluid ingested in "drink" form. These results suggest that it is not the osmotic effects of ingested nutrients that are responsible for fluid intake in a bout. Hence, once again, it would appear that body fluid balance is not responsible for regulating the ad libitum ingestion of liquids in humans.

Conclusions

It is clear from these analyses that, for both humans in their natural environment and rats in isolated laboratory environments, fluid balance homeostasis is not directly related to ad libitum fluid intake regulation. The timing and amounts of fluids ingested are primarily related to food intake. This would explain why there are no changes in the solute concentrations of the body fluids observed in association with water intake in humans (Phillips et al. 1984). This is not to imply that body fluid homeostasis may not be indirectly involved. Intake of fluids might be viewed as anticipatory; occurring in order to prevent a deficit from developing (Fitzsimons and Le Magnen 1969). Regardless, something other than the levels of fluids in the intracellular and extracellular compartments at the time of a bout would appear to be responsible for regulating the amount and timing of fluid intake.

Due to the efficiency of the kidney, organisms have the luxury of only having to regulate fluid intake such that minimum requirements are met. Within limits, no regulation of the maximum amount of fluid ingested is required. Whatever excess is ingested can easily be removed by the kidney. No comparable condition exists with food intake. To some extent brown adipose tissue can burn off excess calories but its capability is limited. Therefore, food intake must be regulated both to ensure that minimum requirements are met and to ensure that they are not significantly exceeded. Hence, food intake must be more tightly regulated than fluid intake. It is thus not surprising that the loosely regulated fluid intake follows along with the tightly regulated food intake.

This is not to imply that caloric balance is any more tightly regulated than fluid balance but rather that energy balance is regulated by adjustments to intake while fluid balance is primarily achieved by regulation of excretion. This suggests that both rats and humans principally regulate the amount and timing of food intake. Fluid is then ingested together with the food in excess of the requirements for digestion or body fluid maintenance. This insures that minimum requirements are met. Excesses are then eliminated by the kidney, which actually performs body fluid regulation.

This interpretation does not necessarily apply to other ecological conditions. The data for the present study were obtained from primarily middle-class Americans living in a modern urban environment where food and fluids are readily and abundantly available and from laboratory rats under ad libitum conditions. In situations where fluids are not available ad libitum or where obtaining water might be difficult or dangerous, tighter regulation of fluid intake independent of food intake may well occur (Rowland 1977; Collier 1986). The system has evolved to adapt to a variety of circumstances and the results of the present analyses only give a glimpse of how it operates under conditions of abundance.

References

Adleson SF (1960) Some problems in collecting dietary data from individuals. J Am Dietet Assoc 36: 453–461

Bolles RC (1979) Toy rats and real rats: nonhomeostatic plasticity in drinking. Behav Brain Sci 2:103

Booth DA (1978) Prediction of feeding behavior from energy flows in the rat. In: Booth DA (ed) Hunger models. Academic Press, London, pp 227–278

Collier G (1986) The dialog between the house economist and the resident physiologist. Nutr Behav 3:926

de Castro JM (1981) The stomach energy content governs meal patterning in the rat. Physiol Behav 26:795–798

de Castro JM (1988) A microregulatory analysis of spontaneous fluid intake by humans: evidence that the amount of liquid ingested and its timing is mainly governed by feeding. Physiol Behav 43: 705–714

de Castro JM (1989) The interaction of fluid and food intake in the spontaneous feeding and drinking patterns of rats. Physiol Behav 45:861–870

de Castro JM, Brower E (1977) Simple, reliable, and inexpensive telemetry system for continuous monitoring of small animal core temperature. Physiol Behav 19:331–333

de Castro JM, de Castro ES (1989) The presence of other people is associated with enlarged meal sizes and disruption of postprandial regulation in the spontaneous eating patterns of humans. Am J Clin Nutr 50:237–247

de Castro JM, Kreitzman SN (1985) A microregulatory analysis of spontaneous human feeding patterns. Physiol Behav 35:329–335

Falk JL (1966) Production of polydipsia in normal rats by an intermittent food schedule. Science 133: 195–196

Fitzsimons JT (1961a) Drinking by nephrectomized rats injected with various substances. J Physiol (Lond) 155:563–579

Fitzsimons JT (1961b) Drinking of rats depleted of body fluid without increase in osmotic pressure. J Physiol (Lond) 159:297–309

Fitzsimons JT, Le Magnen J (1969) Eating as a regulatory control of drinking. J Comp Physiol Psychol 67:273–283

Gersovitz M, Madden JP, Smicikalas-Wright H (1978) validity of the 24-hour dietary recall and seven-day record for group comparisons. J Am Dietet Assoc 73:48–55

Gilman A (1937) The relation between blood osmotic pressure, fluid distribution and voluntary water intake. Am J Physiol 120:323–328

Krantzler NJ, Mullen BJ, Schultz HG, Grivetti LE, Holden CA, Meiselman HL (1982) The validity of telephoned diet recalls and records for assessment of individual food intake. Am J Clin Nutr 36: 1234–1242

Kraly FS (1984) Physiology of drinking elicited by eating. Psychol Rev 91:478–490

Le Magnen J, Tallon S (1968) L'effet du jêune préalable sur les caractéristiques temporelles de la prise d'aliments chez le rat. J Physiol (Paris) 60:143–154

Phillips PA, Rolls BJ, Ledingham JGG, Morton JJ (1984) Body fluid changes, thirst and drinking in man during free access to water. Physiol Behav 33:357–363

Rolls BJ, Rolls ET (1982) Thirst. Cambridge University Press, Cambridge

Rolls BJ, Wood RJ, Stevens RM (1978) Effects of palatability on body fluid homeostasis. Physiol Behav 20:15–19

Rowland, N (1977) Regulatory drinking: do physiological substrates have an ecological niche? Biobehav Rev 1:261–272

St Jeor ST, Guthrie HA, Jones MB (1983) Variability of nutrient intake in a 28 day period. J Am Dietet Assoc 83:155–162

Stricker EM (1968) Some physiological and motivational properties of the hypovolemic stimulus for thirst. Physiol Behav 3:379–385

Toates FM (1978) A physiological control theory of the hunger–thirst interaction. In: Booth DA (ed) Hunger models. Academic Press, New York, pp 347–374

Toates FM (1979) Homeostasis and drinking. Behav Brain Sci 2:95–139

Toates FM, Oatley K (1970) Computer simulation of thirst and water balance. Med Biol Engng 8:71–87

Chapter 22

Individual and Cultural Factors in the Consumption of Beverages

H. Tuorila

Introduction

Research on food acceptance is based on the notion that nutrients need to be in a palatable form in order to be ingested (Solms and Hall 1981). Plain nutritive substances, e.g. sucrose or triglycerides, are seldom acceptable per se. The same does not fully apply with fluids. Water is drunk as such, provided that it is safe and acceptable in sensory terms (Zoeteman 1978). Yet, most water is drunk as beverages rather than as plain water. In addition to water, these products contain major nutrients (e.g. sucrose in soft drinks, fat in milk), physiological stimulants (e.g. caffeine in coffee and tea; ethanol in beer and liquors) and food additives (e.g. artificial flavours and colours in soft drinks). As a matter of fact, many beverages (e.g. milk) provide more nutrients and less water than some solid foods (e.g. fruit and vegetables). Thus, the division of products into "beverages" and "foods" appears to be based on the method of intake – whether it can be better described as "drinking" or "eating" (see also Chapters 3 and 21). This chapter first examines consumption trends in various beverages. Next, the sensory qualities of beverages are considered, and the choices made by consumers between traditional types of products and new alternatives are discussed. Finally, the context of consumption is discussed. The review focuses primarily on non-alcoholic beverages, particularly on soft drinks, milk, and coffee, which account for major market shares in many Western industrialized countries.

Trends in the Beverage Market

Cultural Variations

National consumption statistics are considered as rough indicators of "cultural" variations, as will be illustrated by a few examples.

According to several sources (Woodroof and Phillips 1981; Jacobson 1986; Brewery Statistics 1989), the annual consumption of beverages in many countries is around or above 500 l per capita. This corresponds to an average daily consumption of 1.4 l. Consumption figures are, however, variable, and some European consumption statistics (Euromonitor 1988) imply somewhat lower figures.

Table 22.1. Market shares (litre per capita) of beverages in various countries: the USA consumption from 1985 (Bunch 1987; Putnam 1987), European figures from 1986 (Euromonitor 1988). The volumes of coffee, tea and liquors have been estimated from dry substance with dilution ratios given

Beverage	USA	GB	FRG	Italy	Finland
Soft drinks	173	91	78	47	31
Milk	102	126	55	68	104
Beer	89	108	147	2	63
Mineral water	?	2	62	50	8
Fruit juice	28	16	27	4	29
Wine [a]	16	11	23	80	5
Liquors [b]	–	(1.7)	(6.3)	(3.7)	(11.6)
Liquors, litres (1:3)	–	5	19	11	35
Coffee (kg)	–	(1.0)	(7.4)	(4.4)	(12.0)
Coffee, litres (1:13)	98	13	96	57	156
Tea (kg)	–	(2.5)	(0.3)	(0.1)	(0.2)
Tea, litres (1:16)	25	40	5	2	3

[a] In the US data includes liquors.
[b] At 100% alcohol equivalent.

Table 22.1 lists the market shares of eight rough categories of beverages in the USA and in four European countries representing the south (Italy), middle (Great Britain, Federal Republic of Germany) and north (Finland) of Europe. Consumption of soft drinks appears to be the most characteristic feature of the US beverage culture. Beer, accompanied by milk (in Great Britain) and coffee (in Germany) are typical of middle Europe, and coffee and milk are the most popular beverages in the north. In the south (Italy), wine and milk are the most common beverages. In both Germany and Italy (and also in Belgium and France) mineral waters are also popular. Of the countries listed, Great Britain is the only one where tea accounts for a notable share of the beverage market.

While the international statistics (Euromonitor 1988) have the undeniable advantage that they give comparable figures, their interpretation is problematic. Estimating the volumes of coffee and tea drunk per capita from kilograms of the product is rough since the strength of these beverages varies with the type of product and with cultural and individual preferences. Even more important is the fact that, in some countries, instant coffee and, in others, ground coffee lead the market (Anonymous 1988a). The amount of instant coffee needed to make a cup of coffee is considerably less than that needed for ground coffee. Furthermore, the Euromonitor (1988) data are not fully consistent with other available statistics. For example, the UNESDA (1988) statistics report lower consumption of soft drinks in Great Britain (59 vs. 91 l per capita), and Food Balance Sheet (Agricultural Economics Research Institute 1989) report considerably higher consumption of milk in Finland (176 vs. 104 l per capita). The problem with the latter two sources of statistics is that they are fragmentary (from the international viewpoint) and the fact that each typically focuses on a particular type of beverage at a time. A further difficulty is that some products, particularly milk, are not necessarily consumed as beverages but are often also used for cooking (Tuorila 1987).

Nevertheless, Table 22.1 shows the large international variation, thus concurring with the point made by Rozin and Vollmecke (1986) that culture is the major determinant of food preferences. The impact of culture is even more striking when the findings are compared with consumption habits in non-Western societies. For

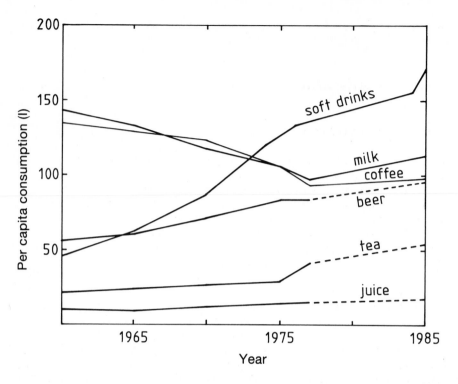

Fig. 22.1. Consumption of various beverages in the USA in 1960–1985. Figures from Shanken (1979), except for milk, soft drinks and coffee in 1984–85 which are from Bunch (1987). Dashed line, predicted values (Shanken 1979).

example, in China, soup is the most popular "beverage" at meals, followed by tea and water (Newman and Ludman 1984).

Consumption over Time

Figure 22.1 shows the historical development in the consumption of beverages in the USA. For beer, tea and juice, the trends from 1977 onwards are predicted estimates (Shanken 1979), but they agree well with actual data from 1982 (Jacobson 1986) and 1985 (Putnam 1987). The most striking feature is the rapid increase in the consumption of soft drinks, which seem to substitute for coffee and milk. The total consumption of beverages has grown by 15% in 15 years, reaching 531 l per capita in 1985 (Bunch 1987). The trend in the Netherlands, for example, has been identical (455 vs. 532 l from 1970 to 1982, Alexander 1989). The increased consumption of various beverages may partly reflect the increasing supply and variety of commercial beverages and partly problems with tapwater quality, which creates pressure towards the consumption of other fluids. In a German survey on beverage preferences, a vast majority of respondents ($n = 789$) held a negative attitude towards drinking tapwater, whereas mineral water was regarded very positively by the same population (Diehl et al. 1985). In an extensive Dutch study (Zoeteman 1978), 320 ml of tapwater was drunk daily by people who thought it tasted good, but only 150 ml by those who thought the taste was bad.

Sensory Quality of Beverages

Composition of Beverages

Each food and beverage has certain sensory attributes which serve as a cue for product identification and as a source of sensory stimulation. Some of the attributes, based on favourable innate dispositions, are more readily accepted than others. Rozin (1982) hypothesized about the existence of pathways, some simple, some complex, for the development of liking for a product. According to his interpretation, exposure to sweetness simply and easily leads to liking, whereas a complex set of influences, including repeated exposures, peer pressure and the detection of the pharmacological effect, are involved in the development of liking for coffee. In an analysis by Booth and Blair (1988) (from which social factors were excluded) perceptions related to the composition of coffee (coffee, sugar, whitener), perceived properties of the beverage (hot, sweet, strong) and cognitions related to after-effects (fattening, bad for teeth) provide a complex network predicting a momentary choice.

The sensory attributes of foods and beverages are commonly divided into three categories: appearance, texture, and flavour (or aroma and taste). An American survey collected ratings of the relative importance of appearance, texture and flavour in a variety of foods and beverages (Schutz and Wahl 1981). Most beverages, including coffee, cola drinks, lemon juice, beer, red table wine, apricot nectar and tomato juice, scored high in flavour, but lower in texture and appearance. The result with texture is understandable since the textural properties of beverages, even though they differ under close examination (Szczesniak 1979), vary only slightly when compared with many solid foods.

In a study in which colour (colourless or coloured), flavour (added artificial flavours) and sweetness (normal or lowered) of soft drinks were systematically varied, taste turned out to be the most critical attribute of the overall liking (Tuorila-Ollikainen et al. 1984). Normal sweetness was best liked in samples in which added flavour was also present (Tuorila-Ollikainen and Mahlamäki-Kultanen 1985). Some other studies have shown that colour is important in the anticipation of flavour. In the absence of the appropriate colour, aroma is difficult to identify (DuBose et al. 1980). Thus, colour seems to be of importance for cognitive orientation with respect to beverages.

Visual cues are also important for the recognition of milks with varying fat contents (Pangborn and Dunkley 1964; Tuorila 1986). If visual inputs are removed, the discrimination between varying fat levels is based on textural properties (Mela 1988). Milk as such does not provide as much sensory variety as many other beverages, but there is a large number of cultured and flavoured liquid products based on milk. One of the latest innovations is carbonated milk, aimed at taking advantage of this favourably perceived attribute of soft drinks (Yau et al. 1989).

Sensory quality of coffee is highly variable due to many factors, including the origin of coffee beans, roasting, and preparation. Although experts observe subtle differences between coffee beverages (Pangborn 1982), consumers characterize coffee beverages only with a few attributes, e.g. aroma, strength and bitterness (Griffiths et al. 1986). Cultural differences have been observed in the appreciation of and finickiness towards the flavour of coffee (Rozin and Cines 1982).

Serving Temperature

Temperature influences on beverages are described in detail by Engell and Hirsch (Chapter 24). Therefore only two relevant reports on beverages, dealing with cultural conventions, will be briefly summarized. Cardello and Maller (1982) found that lemonade, milk and coffee were favoured at the temperatures at which they are normally consumed. Thus, the hedonic ratings of lemonade and milk decreased markedly when the temperature was raised to room temperature or above. Coffee was liked best hot (ca. 50°C) but it was also liked cold, a finding obviously due to the American habit of drinking coffee both hot and cold. Zellner et al. (1988) likewise found that each beverage has a characteristic optimum serving temperature. Subjects were more negative about unusual beverage/temperature combinations when verbally described than when tasted as samples in a sensory test. Their preferences for certain temperatures could be modified by creating expectations as to which temperature they felt to be appropriate. Thus, cultural factors seem to guide preferences for serving temperatures.

Consumer Choices Within Beverage Categories

Traditional and "New" Alternatives

Beverages add substantial amounts of energy to an average diet; according to a Swedish estimate the figure is as high as 15%–20% (Wheeler 1987). Dairy fat in milk and coffee cream, sugars in soft drinks and juice, and ethanol in beer and other alcoholic beverages are examples. Coffee and tea are major sources of caffeine (Shirlow 1983; Scott et al. 1989). With the present consumer concern about harmful substances in foods, beverages from which the supposedly harmful components have been removed are increasingly competing with traditional beverages for the market.

Soft Drinks

In the mid-1980s, artificially sweetened "diet" sodas accounted for about one-quarter of the US soda market. Their proportion is expected to rise to 50% by 1995 (Duxbury 1988; Houghton 1988). The increasing use of aspartame as a sweetener in these beverages is obviously a major reason for the popularity, as the taste of aspartame resembles that of sucrose (Larson-Powers and Pangborn 1978). The total consumption of soft drinks is much lower in Europe than in the USA, but the trend toward low-calorie beverages is apparent there, too (Hendley and Seymour 1988). In 1988, low-calorie drinks accounted for 15% of the British soft drinks market (Anonymous 1989). With such extensive changes in consumption figures, any data on consumers of regular vs. diet sodas are rapidly outdated. Nevertheless, a few figures will hopefully serve to highlight current trends.

According to the US national data from 1977/78, the most typical consumers of regular sodas were teenaged girls and young men, whereas low-calorie sodas were typically drunk by adults (Morgan et al. 1985). Three-quarters of teenagers consumed soft drinks daily, and the average consumption among all subjects was 200–300 ml (Guenther 1986). The variation in the amount drunk could not be explained by demographic factors. In the mid-1980s, teenagers, too, showed an increasing tendency to drink "diet" and caffeine-free sodas (Andres 1987). According to recent statistics, white, educated and/or females are more likely to use low-calorie cola beverages than black, less educated and/or males (National Soft Drink Association 1989).

Affective responses (liking for a taste; Lewis et al. 1989; Tuorila et al. 1990) and cognitive factors (commitment to use or not to use low-sugar products; Lewis et al. 1989) are involved in the selection of regular vs. diet sodas; however, no research has been conducted on how liking or commitment develops. Consumers rated functional properties (thirst quenching, compatability with other items on the menu) highest in their "own" type of soda (Tuorila et al. 1990). Recent experimental data suggest true differences in thirst quenching so that beverages containing sucrose quench thirst less than beverages containing aspartame (Rolls et al. 1990).

Milk

In the USA, the proportion of whole milk in relation to total milk consumption decreased from 84% in 1965 to 49% in 1985 (Bunch 1987). Whole milk was replaced primarily by low-fat milk (Young et al. 1986; Bunch 1987) whereas the consumption of non-fat milk remained virtually unchanged (about 5% of the total milk consumption) (Young et al. 1986). The consumption of low-fat milk has also increased in most European countries (EC 1986), but the strength of this trend varies from one country to another (Wilbey 1988). For example, in France most of the milk consumed is of the low-fat type, whereas in Great Britain, low-fat milk came onto the market only a few years ago and accounted for less than 10% of total consumption in the mid-80s (Wilbey 1988). In Finland, most of the milk drunk is low-fat, and furthermore, the consumption of non-fat milk has grown rapidly so that 15% of adults were drinking it in 1988 (Niemensivu et al. 1988).

In sensory tests of liking, people prefer the type of milk to which they are accustomed (Tuorila 1987). This applies to all fat levels, including non-fat milk. Unfortunately, there is no research on the habituation process, or on the development of liking.

Women have shown more interest in the consumption of low-fat milks than men (Shepherd 1988; Niemensivu et al. 1988). Apart from the pressure exerted by health education, the shift towards low-fat alternatives can be explained by availability and the process of urbanization. For instance, teenagers living on farms drink strikingly more whole and regular-fat milks than do their urbanized peers (Prättälä 1988). In an American study, liking and commitment to eat/not to eat low-fat foods were the most important predictors of consumption of low-fat or whole milk (Lewis et al. 1989). In another study, beliefs related to nutrient contents, functionality and weight concern were also found to be significant predictors of consumption (Tuorila and Pangborn 1988). In a recent British survey, beliefs related to nutritional factors and to sensory quality were both important predictors for the consumption of low-fat milk (Shepherd 1988).

Coffee

In recent decades, coffee drinking has declined in the USA, but the tendency in many European countries, including Sweden (Wheeler 1987), the Netherlands (Matze et al. 1988) and Great Britain (Anonymous 1988a), has been quite the reverse. With ground coffee, the alternatives are traditional brewed and percolated varieties; with instant coffee the alternatives are caffeinated and decaffeinated. The Scandinavian countries typically favour ground coffee, whereas in Great Britain, 90% of coffee is of the instant variety (Anonymous 1988a). In Australia, too, the majority of subjects in a large consumer survey drank instant coffee (Shirlow 1983). Decaffeinated coffee accounts for 8% of the market in Great Britain (Anonymous 1988a) and 17.5% in the USA (Anonymous 1988b).

Published studies offer meagre, if any, information on consumer choices of a particular type of coffee and reasons leading to the choice. In an American survey, subjects preferring caffeinated coffee rated most of its attributes, including taste, positively, whereas those preferring non-caffeinated beverages rated caffeinated products negatively (Page and Goldberg 1986). The consumption of caffeine (from coffee and other beverages) by young women was associated with their attitude that it is hard to limit coffee consumption (Guiry and Bisogni 1986). In a most interesting experiment, Griffiths et al. (1986) found that, after withdrawal symptoms such as perceived sleepiness and headache, caused by a lack of caffeinated coffee, subjects gradually developed a taste for decaffeinated coffee after having been exposed to it as the only option for two days. The subjects even reported it as having a stimulant effect. Cines and Rozin (1982) found that subjects drinking decaffeinated coffee reported it made them feel more alert in the morning; however, the feeling of alertness was stronger among drinkers of caffeinated coffee. Kozlowski (1976) observed a tendency to compensate for the low caffeine content of coffee by increasing the amount drunk ad libitum. Thus, immediate deprivation of caffeine seems to be followed by an adjustment in a short time interval, if no other choices are available.

Context of Use

Consumption at Meals

The consumption of beverages varies in the course of the day and primarily takes place during meals (Chapter 21). American teenagers drank milk at breakfast, lunch and supper, and also as a snack. When chosen, coffee or tea were drunk at breakfast or supper, respectively. Juice was typically drunk at breakfast, and soft drinks were preferred at lunch and dinner and as snacks, but not at breakfast (Stults et al. 1982; Guenther 1986). American adults typically favoured soft drinks as snacks, except for young women who typically drank their soft drinks at lunch (Morgan et al. 1985).

In an American study on college students that monitored only the beverages drunk at dinner, cola beverages and milk were preferred; hot tea and coffee were rarely selected (Khan 1980).

In an adult Swedish population, coffee was typically drunk on any occasion, including as a snack. Milk was preferred at meals, but not as a snack. Juice was part of breakfast and beer part of dinner; carbonated beverages were drunk as snacks (Qvist 1987).

Other Context Factors

Perceptions concerning the appropriate beverages for a particular situation vary. Bruhn and Schutz (1986) found that non-dairy beverages (soft drinks, coffee, tea, beer) were considered appropriate for social situations, but milk was not. In a study by Rozin and Cines (1982), American Italians considered the social factors related to coffee consumption as being the most important reason for drinking coffee, whereas wake-up effects and the menu context were most important reasons for American Jews.

In a survey on German consumers, the most typical occasion to drink plain water was in the context of taking medicines (Elmadfa and Huehn 1985).

Certain non-ingestive behaviours accompany the drinking of beverages. A typical example is the correlation found between the consumption of coffee and cigarette smoking (Istvan and Matarazzo 1984). Whether there is a true interaction between these behaviours or only a correlation due to a more general behavioural tendency ("life style") is not fully clear. A study by Emurian et al. (1982) suggests that smoking is most probable immediately after drinking coffee.

Summary and Conclusions

1. Cultural variation in the types of beverage consumed is considerable. There seems to be an increasing tendency to drink beverages rather than plain water.
2. Flavour is considered to be the most important sensory attribute of beverages, but it interacts with colour and appearance characteristics, which are important cues for the identification of a product. Also, each beverage has a characteristic optimum temperature at which it is expected to be served.
3. Market segmentation is proceeding rapidly in the field of beverages. The trend is toward lower energy contents and fewer substances perceived as harmful to health. Thus artificially sweetened soft drinks, low-fat or non-fat milk and decaffeinated coffee are competing for the market with "traditional" products. When selecting between alternative products, people often give liking as the most important reason for their choice, but there are hardly any experimental data concerning the development of liking.
4. The contexts in which various beverages are drunk vary by meal, social situation and culture.
5. The above data are from scientific journals but also, to a great extent, from less formal marketing studies and from statistical sources, which do not always concur. Furthermore, the validity of consumer studies is often confined to a particular country or culture. The present information must therefore be considered as a rough and practically oriented outline of trends. For a thorough analysis, each culture or subgroup should be studied as its own system.

References

Agricultural Economic Research Institute (1989) Balance sheet for food commodities. Year 1988. Helsinki
Alexander A (1989) How to promote milk. Food Processing no.1:37–38
Andres C (1987) Third annual national study. Teenage food survey. Food Processing no. 6:21–25

Anonymous (1988a) Mixed fortunes in hot beverages. The Grocer April 16:41–44
Anonymous (1988b) Soluble and decaf coffee fight for greater consumption share. World Coffee and Tea 28 (11):8–9
Anonymous (1989) Focus on soft drinks. The Grocer May 13:99–158
Booth DA, Blair AJ (1988) Objective factors in the appeal of a brand during use by the individual customer. In: Thomson DMH (ed) Food acceptability. Elsevier, London pp 329–346
Brewery Statistics (1989) Food Industry Federation. Helsinki (in Finnish)
Bruhn CM, Schutz HG (1986) Consumer perceptions of dairy and related-use foods. Food Technol 40 (1):79–85
Bunch K (1987) Highlights of 1985 food consumption data. Natl Food Rev 36:1–5
Cardello AV, Maller O (1982) Acceptability of water, selected beverages and foods as a function of serving temperature. J Food Sci 47:1549–1552
Cines BM, Rozin P (1982) Some aspects of the liking for hot coffee and coffee flavor. Appetite 3:23–34
Diehl JM, Elmadfa I, Walter B (1985) Structure and distribution of beverage preferences in adults. Akt Ernähr 10:34–41 (English summary)
DuBose CN, Cardello AV, Maller O (1980) Effect of colorants and flavorants on identification, perceived flavor intensity, and hedonic quality of fruit-flavored beverages and cake. J Food Sci 45:1393–1399, 1415
Duxbury DD (1988) Aspartame sweetener propels diet beverage market. Food Processing no. 5:54–55
EC – Economic Commission for Europe (1986) Volume V. The milk and dairy products market. United Nations, New York (Agricultural Review for Europe No. 28, 1984 and 1985)
Elmadfa I, Huehn W (1985) Consumption and preference of drinking water. In: Diehl JM, Leitzmann C (eds) Measurement and determinants of food habits and food preferences. Wageningen, pp 129–135 (Euronut Report 7)
Emurian HH, Nellis MJ, Brady JV, Ray RL (1982) Event time-series relationship between cigarette smoking and coffee drinking. Addict Behav 7:441–444
Euromonitor (1988) European marketing data and statistics 1988. London
Griffiths RR, Bigelow GE, Liebson IA (1986) Human coffee drinking: reinforcing and physical dependence producing effects of caffeine. J Pharm Exp Ther 239:416–425
Guenther PM (1986) Beverages in the diets of American teenagers. J Am Dietet Assoc 86:493–499
Guiry VC, Bisogni CA (1986) Caffeine knowledge, attitudes, and practices of young women. J Nutr Educ 18:16–22
Hendley BG, Seymour V (1988) Markets for low-calorie foods. In: Birch GG, Lindley MG (eds) Low-calorie products. Elsevier, London, New York, pp 1–9
Houghton HW (1988) Low-calorie soft drinks. In: Birch GG, Lindley MG (eds) Low-calorie products. Elsevier, London, New York, pp 11–21
Istvan J, Matarazzo JD (1984) Tobacco, alcohol, and caffeine use: a review of their interrelationships. Psychol Bull 95:301–326
Jacobson RE (1986) Review of current and future consumption trends of milk and dairy products. J Dairy Sci 69:1447–1453
Khan MA (1980) Observations on beverage preferences of college students at specific times. J Am Dietet Assoc 77:56–57
Kozlowski LT (1976) Effects of caffeine on coffee drinking. Nature 264:354–355
Larson-Powers N, Pangborn RM (1978) Descriptive analysis of the sensory properties of beverages and gelatins containing sucrose or synthetic sweeteners. J Food Sci 43:47–51
Lewis CJ, Sims LS, Shannon B (1989) Examination of specific nutrition/health behaviors using a social cognitive model. J Am Dietet Assoc 89:194–202
Matze M, Jacobs NJM, Cremers SBL, Katan MB (1988) Coffee and health I. Consumption, caffeine, wakefulness and decaffeinated coffee. Voeding 49:7–11 (English summary)
Mela DJ (1988) Sensory assessment of fat content in fluid dairy products. Appetite 10:37–44
Morgan KJ, Stults VJ, Stampley GL (1985) Soft drink consumption patterns of the U.S. population. J Am Dietet Assoc 85:352–354
National Soft Drink Association (1989) Statistics on the consumption of soft drinks 1988. Washington, DC
Newman JM, Ludman EK (1984) Chinese elderly: food habits and beliefs. J Nutr Elderly 4 (2):3–13
Niemensivu H, Berg MA, Piha T, Puska P (1988) Health behaviour among Finnish adult population Spring 1988. Publications of the National Public Health Institute B 4, Helsinki
Page R, Goldberg R (1986) Practices and attitudes toward caffeinated and non-caffeinated beverages. Health Educ 17(5):17–21
Pangborn RM (1982) Influence of water composition, extraction procedures, and holding time and temperature on quality of coffee beverage. Lebensm Wiss Technol 15:161–168
Pangborn RM, Dunkley WL (1964) Difference-preference evaluation of milk by trained judges. J Dairy

Sci 47:1414–1416

Prättälä R (1988) Socio-demographic differences in fat and sugar consumption patterns among Finnish adolescents. Ecol Food Nutr 22:53–64

Putnam J (1987) Food consumption. Natl Food Rev 37:2–9

Qvist I (1987) Drinks as seen from the viewpoint of the producer. I. Consumption patterns and sugar and alternative sweeteners. Näringsforskning 31:140–143 (English summary)

Rolls BJ, Kim S, Fedoroff IC (1990) Effect of drinks sweetened with sucrose or aspartame on hunger, thirst and food intake in man. Physiol Behav 48:19–26

Rozin P (1982) Human food selection: the interaction of biology, culture and individual experience. In: Baker LM (ed) The psychobiology of human food selection. Ellis Horwood, Chichester, pp 225–254

Rozin P, Cines BM (1982) Ethnic differences in coffee use and attitudes to coffee. Ecol Food Nutr 12:79–88

Rozin P, Vollmecke TA (1986) Food likes and dislikes. Ann Rev Nutr 6:433–456

Schutz HG, Wahl OL (1981) Consumer perception of the relative importance of appearance, flavor and texture to food acceptance. In: Solms J, Hall RL (eds) Criteria of food acceptance. How man chooses what he eats. Forster Verlag, Zurich, pp 97–116

Scott NR, Chakraborty J, Marks V (1989) Caffeine consumption in the United Kingdom: a retrospective study. Food Sci Nutr 42F:183–191

Shanken MR (1979) The changing world of the soft drink industry. Impact Newsletter 3 and 4. Data reviewed in: Woodroof JG, Phillips GF (eds) Beverages: carbonated and noncarbonated. Avi, Westport

Shepherd R (1988) Belief structure in relation to low-fat milk consumption. J Human Nutr Diet 1: 421–428

Shirlow MJ (1983) Patterns of caffeine consumption. Human Nutr Appl Nutr 37A:307–313

Solms J, Hall RL (eds) (1981) Criteria of food acceptance. How man chooses what he eats. Forster Verlag, Zurich

Stults VJ, Morgan KJ, Zabik ME (1982) Children's and teenagers' beverage consumption patterns. School Food Service Res Rev 6 (1):20–25

Szczesniak AS (1979) Classification of mouthfeel characteristics of beverages. In: Sherman P (ed) Food texture and rheology. Academic Press, New York, pp 1–20

Tuorila H (1986) Sensory profiles of milks with varying fat contents. Lebensm Wiss Technol 19: 344–345

Tuorila H (1987) Selection of milks with varying fat contents and related overall liking, attitudes, norms and intentions. Appetite 8:1–14

Tuorila H, Pangborn RM (1988) Prediction of reported consumption of selected fat-containing foods. Appetite 11:81–95

Tuorila H, Pangborn RM, Schutz HG (1990) Choosing a beverage: comparison of preferences and beliefs related to the reported consumption of regular vs. diet sodas. Appetite 14:1–8

Tuorila-Ollikainen H, Mahlamäki-Kultanen S (1985) The relationship of attitudes and experiences of Finnish youths to their hedonic responses to sweetness in soft drinks. Appetite 6:115–124

Tuorila-Ollikainen H, Mahlamäki-Kultanen S, Kurkela R (1984) Relative importance of color, fruity flavor and sweetness in the overall liking of soft drinks. J Food Sci 49:1598–1600, 1603

UNESDA (1988) Boissons rafraichissantes sans alcool en 1988. Statistics of the European Soft Drinks Association, Brussels

Wheeler B (1987) Consumption of drinks in Sweden. Näringsforskning 31:123–126 (English summary)

Wilbey RA (1988) Technical problems in the development of low-calorie dairy products. In: Birch GG, Lindley MG (eds) Low-calorie products. Elsevier, London, New York, pp 31–42

Woodroof JG, Phillips GF (eds) (1981) Beverages: carbonated and noncarbonated. Avi, Westport

Yau NIN, McDaniel MR, Bodyfelt FW (1989) Sensory evaluation of sweetened flavored carbonated milk beverages. J Dairy Sci 72:367–377

Young CW, Miller JK, Freeman AE (1986) Production, consumption and pricing of milk and its components. J Dairy Sci 69:272–281

Zellner DA, Stewart WF, Rozin P, Brown JM (1988) Effect of temperature and expectations on liking for beverages. Physiol Behav 44:61–68

Zoeteman BCJ (1978) Sensory assessment and chemical composition of drinking water. A study based on the situation in the Netherlands. Dissertation, University of Utrecht

Commentary

Saltmarsh: Is there evidence that when people drink without being under obvious thirst stress they choose their beverage on the basis of anticipated pleasantness?

Tuorila: I do believe – and it has also been discussed in this volume – that much of our drinking takes place under conditions in which we do not even realize that we are thirsty. In these cases habitual drinking (for instance in the context of eating, Chapter 21) and drinking based on the pleasantness of a particular beverage, or the act of drinking, become important. In my own studies on soft drinks (Tuorila et al. 1990) and milk (Tuorila 1987) liking has turned out to be one of the best correlates of consumption. Thus, expected pleasantness should be one of the most powerful predictors of consumption.

References

Tuorila H (1987) Selection of milks with varying fat contents and related overall liking, attitudes, norms and intentions. Appetite 8:1–14
Tuorila H, Pangborn RM, Schutz HG (1990) Choosing a beverage: comparison of preferences and beliefs related to the reported consumption of regular vs. diet sodas. Appetite 14:1–8

Chapter 23

Alcohol- and Caffeine-Beverage Consumption: Causes Other Than Water Deficit

R.M. Gilbert

The Meaning of Thirst

Thirst is an ancient word of unchanged meaning. The *thyrste* and *thurste* of ninth-century Old English are recognizable; the contemporary *Durst* of German, *dorst* of Dutch, and *tørst* of Scandinavian languages suggest an even older common root. Thirst is without synonyms in its literal sense. In popular use it refers to the desire to drink. Metaphorically, thirst is one of many words for craving. It was used in this way in the ninth century. (A thirst for knowledge and a hunger for knowledge are not quite synonymous; the former denotes an eager desire, the latter a compelling desire.)

Ordinary speech allows for three kinds of use of "thirst" in relation to liquids. A thirsty woman can be one who reports a dry mouth and other consequences of having too little water in her body. She might also be someone who drinks a lot. A man who is said to have a great thirst could be one with a large appetite for beer.

Science has no special term for thirst; a term might emerge as debate sharpens about its meaning. In science, thirst has come to be associated with the notion of bodily water deficit: the deficit causes a motivational state, thirst, that causes behaviour, drinking, that can remedy the deficit. The simplicity of this conceptualization is marred by three kinds of data about humans and some other species. One kind, discussed in Chapter 24, is that bodily water deficit is an insufficient condition for the restoration of water balance even when palatable liquid is available. Another kind of data, discussed in Chapters 18 and 21, indicates that the timing and amount of drinking is normally determined mostly by food intake. The third kind of data that complicates a simplistic notion of thirst as a regulator of water balance and drinking concerns the evident roles of many other determinants of drinking – the subject of this chapter.

If thirst, meaning a motivational state associated with water deficit, is not a reliable cause of drinking, and drinking is not a reliable indicator of thirst, what is the most useful meaning of thirst? One suggestion is that of Booth (Chapter 4) that thirst is the disposition to drink whether determined by water deficit or any other cause. A thirsty person is one for whom, in the language of behaviourists, palatable liquid is a reinforcer.

An alternative is to confine the term thirst to occasions when water deficit causes a liquid to be a reinforcer, as Robertson does in Chapter 30. But because of the powerful entrainment of drinking to eating, a more useful limitation of the term may be to occasions of water or energy deficit. Such use allows easy distinction between drinking attributable to homeostatic disturbances and drinking that results from other causes.

The point of this preamble is to define the scope of this chapter, which considers causes of drinking other than water deficit and other than food ingestion.

The Importance of Other Causes of Drinking

These causes are of at least three kinds:

1. Reinforcers in the beverage, including sweeteners and drugs.
2. Reinforcers associated with the beverage, including sexual opportunity and esteem.
3. Expectations of reinforcement – expectations that ultimately depend on occasional confirmation but mostly appear to act independently.

A fourth category of cause is availability, which may denote simple proximity or may refer to the effort required to secure the fluid. It is of special interest as a determinant of drinking during the expression of other reinforced behaviour, including conversation, smoking and eating.

One or more causes of these types may determine a person's drinking completely; no recourse to notions of fluid deprivation, energy or water imbalance, or thirst may be required. For example, an alcoholic may regularly consume enough liquid to meet physiological requirements through his search for alcohol. Non-metaphorical thirst may be rarely important in the maintenance of his water balance.

Perhaps more commonly, water deficit may determine at least part of what is drunk each day, directly or indirectly; other causes determine the remainder. But in all cases where thirst determines less drinking than is required to replace lost water, the other causes assume significance. They deserve understanding as vital components of the body's daily economy.

Also, knowledge of the roles of the other causes of drinking allows for greater precision in the delineation of thirst. Little is known about these other causes, and even less about their respective contributions to normal and abnormal human drinking. Most of the remainder of this chapter comprises discussion of them with some speculation about their impact. The focus is on beverages containing alcohol or caffeine, both because they comprise much of the fluid consumed in industrialized societies and because they provide for a full range of other causes of drinking.

The Importance of Drug-Containing Beverages in Human Fluid Consumption

The remarkable attempt of Ershow and Cantor (1989) to determine usual patterns of total water and tapwater consumption in the United States was marred by gross under-reporting of the use of beverages containing alcohol. The extent of the under-reporting is evident if the daily consumption figures for beer, wine, and liquor reported in that study are compared with estimates of per capita consumption of alcohol in the United States based on sales.

Ershow and Cantor's data (from the 1977–78 Nationwide Food Consumption Survey) suggest that the per capita consumption of alcohol in alcoholic beverages by residents of the United States aged 20–64 years was about 1.6 litres a year, an average that hides wide variability. This amount was the result of consumption each day of beer containing 58 g of water, wine containing 7 g of water, and spirits, liquor or mixed drinks containing 5 g of water. The average of 1.6 litres a year was estimated by assuming beer to be 5% alcohol by volume, wine to be 12%, and the beverages in the other category to be 30%.

Estimates of consumption of the alcohol for the same period based on sales of alcoholic beverages are in the order of 8.3 litres of alcohol per person per year (e.g., Smart 1989). If it can be assumed that average consumption by the 44% of persons not aged 20–64 years is no more than half that of persons aged 20–64, the per capita consumption of the latter group can be estimated to be at least 10.6 litres a year, or 6.6 times the amount suggested by Ershow and Cantor's (1989) analysis. Under-reporting of alcohol use is common in surveys (e.g., Lemmens et al. 1988), but it is usually less than what is apparent here.

Ershow and Cantor's data on consumption of coffee and tea are more consistent with sales data. Average tea use was reported as involving consumption of 152 g of water a day (i.e., 152 ml); estimates based on sales and other data point to the average person drinking just under a cup of weak-to-medium strength tea a day – about 35 mg of caffeine in 150 ml of water. Average coffee use was reported as involving consumption of 395 g of water a day; sales and other data suggest consumption is an average of about two cups a day with a third of the total being decaffeinated – ingestion of some 125 mg of caffeine in 380 ml of water (Gilbert 1984).

The amounts reported by Ershow and Cantor for soft drink consumption also seem consistent with sales figures. A reasonable estimate of the proportion of consumed carbonated soft drinks containing caffeine is 70% (Gilbert 1984).

Table 23.1 sets out the conclusions about total water consumption and its excess over physiological requirement that can be drawn from the data of Ershow and Cantor, after adjusting for under-reporting of alcohol-beverage consumption and

Table 23.1. Estimates of average total daily water intake of US residents aged 20–64 years through drinking, 1977–78, and of amounts through drinking required to maintain water balance by sedentary US adults

	Volume (ml)	% of total
Estimate of total water intake through drinking:		
As plain water[a]	670	32
As beverages without alcohol or caffeine[b]	370	18
As beverages containing alcohol[b]	470	23
As beverages containing caffeine[b]	550	27
Total water intake through drinking	2060	100
Estimate of amount needed through drinking:		
Water requirement[c]	1500	100
Less: water combined with food[b]	500	33
Less: water from metabolism[c]	300	20
Amount needed through drinking	700	47
Excess of intake over amount needed	1360	

[a] Estimate of Ershow and Cantor (1989).
[b] Estimate of Ershow and Cantor modified as explained in the text.
[c] Estimate of Robertson (Chapter 30).

after distributing soft drinks according to the above-noted estimate, and from other data. The essential conclusions are these. The average sedentary US adult in a temperate climate drinks about three times as much as is necessary for physiological reasons. Half of what is drunk contains a drug: alcohol or caffeine.

Understanding the causes of consumption of large amounts of alcohol and caffeine-containing beverages should thus be considered central to our understanding of the regulation of human drinking. But, in the words of Engell and Hirsch (Chapter 24, p. 382), the study of human fluid intake is "sometimes murky and always difficult". I shall start by first considering the control of consumption of alcohol and caffeine solutions in the rat.

Ways to Cause Rats to Drink Large Amounts of Alcohol or Caffeine Solution

Breed Them for it

When food and water are freely available, most rats shun a 10% alcohol solution. By mating alcohol-drinking rats, an alcohol-preferring line can be developed (Li et al. 1981). After one such exercise in selective breeding, rats of the resulting alcohol-preferring line each drank some 25 ml of the 10% solution a day, equivalent for a human on a weight-for-weight basis to six or seven 750-ml bottles of wine.

Most rats also shun caffeine solutions. Selective breeding for heavy caffeine use has not been tried. Individual differences observed during studies of oral caffeine self-administration (Atkinson and Enslen 1976) suggest that development of caffeine-preferring lines might be possible.

Rear in an Enriched Environment

Rats reared in an enriched environment appear to drink more of an alcohol solution than rats reared in standard cages, with or without a male or female partner (Rockman et al. 1988). The effect of this procedure on caffeine consumption has not been investigated.

Restrict the Food Supply

Rats drink less water when they are fed less (Gilbert and Sherman 1970), but consumption of alcohol solution increases (Stiglick and Woodworth 1984). The increased drinking may be partly for the solution's energy value, which is about 7 kcal (29.4 kJ) for each 10 ml of a 10% by volume alcohol solution (Guthrie et al. 1990). Rats receiving food energy providing only 70 kcal (294 kJ) a day rather than the 100 kcal (420 kJ) a day they normally ingest can thus substantially augment their energy intake by drinking alcohol solution. The effect of food restriction may appear only when the rats are otherwise well-nourished (Gilbert 1979).

Food restriction increases consumption of caffeine solutions (Heppner et al. 1986). Because intake of quinine solutions did not increase under the same conditions of food restriction, the authors concluded that the increased caffeine consumption was due to a specific interaction between reduced body weight and the drug rather than to increased preference for bitter tastes. Drinking of solutions of other non-nutritious drugs is also enhanced by food deprivation.

Space the Feeding

Delivery of single food pellets at regular intervals to food-restricted rats engenders consumption of remarkable quantities of water (Falk 1972) or of alcohol solution (Gilbert 1976b; Falk and Tang 1988), whichever is available. This schedule-induced polydipsia may be understood as occurring when the spacing of the pellets is such as to cause a generalized motivational state that expresses itself in drinking if drinking is possible (Segal 1972). The procedure has not been used to cause excessive drinking of caffeine solutions. Schedule induction is not specific to spaced food delivery (see below).

Sweeten the Fluid

Rats drink more water when it is sweetened by sucrose or saccharin. Food restriction enhances the effect (Gilbert 1972). Similar enhancement occurs when alcohol solutions are sweetened (Gilbert 1974). Rats not preferring alcohol solution to water will switch to alcohol solution when the solution is sweetened (Gilbert 1976c). Sweetening does not appear to have been used to increase consumption by rats of caffeine solutions.

Restrict Food, Space Feeding, and Sweeten Fluid – All Together

Rats drink extraordinary amounts of an alcohol solution when their food is restricted, when the restricted ration is given in the form of single pellets at regular intervals, and when the solution is sweetened. In one study, these procedures combined produced a rate of oral intake of pure alcohol of approximately 3 g/kg of body weight per hour (Gilbert 1978) – equivalent to consumption by an average adult man of about three bottles of wine in an hour.

Incorporate Alcohol into a Liquid Diet

The above techniques concern consumption of alcohol solutions that are combined with nothing more than a sweetener. Another much-used technique is to administer alcohol to rats and other species as part of a nutritionally balanced liquid diet. Daily doses of alcohol in the order of 15 g/kg of body weight can be achieved in this way (Lieber and DeCarli 1989). Such doses are higher than can be achieved by other methods involving ingestion. They provide the alcohol equivalent of more than a dozen bottles of wine a day for a human adult. Unlike the above procedures, incorporation of alcohol into a liquid diet has little relevance to the delineation of thirst: it is remote from normal regulation of fluid intake. The method has not been used with caffeine solutions.

Induce Physical Dependence

Alcohol-preferring rats will self-administer alcohol solution directly into their stomachs, which may be evidence of consumption for pharmacological effects (Waller et al. 1984). But seeking the pharmacological consequences of alcohol administration does not imply dependence on the drug. Physical dependence refers to the appearance of a withdrawal syndrome when chronic administration of a drug is discontinued. In the rat, the alcohol withdrawal syndrome includes heightened excitability and seizure-proneness. Regular consumption of an amount of alcohol sufficient to cause physical dependence causes rats to choose alcohol solutions over weak glucose solutions and thus, under these circumstances, an increase in alcohol consumption (Tang and Falk 1977). The effect is weak. In a similar study, physical dependence did not cause increased intake of an alcohol solution when it was the only fluid available (Tang et al. 1982). Le Magnen and Marfaing-Jallat (1984) showed that chronic treatment with intragastrically intubated alcohol can cause a substantial elevation in alcohol consumption, but their rats may not have been physically dependent.

A few reports have described physical dependence of rats on caffeine (Griffiths and Woodson 1988), but none has shown that rats drink more caffeine solution on account of the dependence.

Provide Superior Company

Subordinate male rats in a colony drink more of an available alcohol solution than dominant males (Blanchard et al. 1987; Ellison 1987). Females, included in colonies in the former study, drank even more. In both reports, the heightened consumption was said to be associated with relief of social stress. The consumption of caffeine, which has been found to reduce social activity in rats (File et al. 1988), has not been examined for its susceptibility to social influences or to stress.

Administer Other Drugs

The choice by rats of a 10% alcohol solution over water is markedly enhanced by the addition of caffeine to both fluids, but only when the rats are malnourished (Gilbert 1979). Addition of isohedonic quinine is ineffective; the enhancement of alcohol consumption in malnourished rats is not simply a matter of taste. One report has suggested that intake of caffeine solution by rats can be increased by chronic exposure to nicotine (Schulte-Daxboek and Opitz 1981).

Conclusion

The foregoing overview suggests that successful enhancement of consumption by rats of alcohol solutions and, to a lesser extent, caffeine solutions has been achieved using a variety of genetic, dietary, and environmental manipulations that for the most part bear some similarity to those found in human experience. Many of these techniques have worked with other species. Thus, the regulation of consumption of fluids containing alcohol or caffeine is potentially much more than a matter of fluid or energy balance or thirst, even in rats, and more than a matter of pharmacology.

Causes of Human Consumption of Drug-Containing Beverages

Almost every one of the techniques mentioned above for augmenting consumption of solutions of alcohol and caffeine in rats has a counterpart in a potential cause of human consumption of alcohol- and caffeine-containing beverages; but data are few. This section provides an overview of some of these data. The emphasis is on factors that increase or are believed to increase consumption rather than on factors that reduce or terminate consumption.

Alteration of Mood or Behaviour: the Role of Expectancy

The central finding on the pharmacological control of alcohol use in humans is in a misleadingly titled report by Marlatt et al. (1973). In what has come to be known as the balanced placebo design (Ross and Pihl 1989), subjects were examined during a fake taste test in one or another of four conditions defined by two variables: they were given vodka and tonic or tonic only, and they were told they were receiving vodka and tonic or tonic only. Of interest during the experiment, which was conducted with 32 alcoholics and again with 32 social drinkers, was not the subjects' choices in the taste test but how much they drank of the fluids they were comparing.

The main results are reproduced in Table 23.2, which shows that the alcoholic content of the beverages did not affect consumption during the taste test. Information about alcohol content (correct or false) affected consumption of liquid profoundly: the alcoholics drank 117% more when they were led to expect that the beverages being compared all contained alcohol: the social drinkers drank 90% more. Alcoholics drank more fluid than social drinkers under all conditions, an average of 52% more.

The finding of Marlatt et al. (1973) was essentially repeated by Berg et al. (1981). In a major review, Hull and Bond (1986) concluded that the predominant role of expectancy in alcohol consumption was well established.

A satisfactory account of how expectancy is sustained should be able to reconcile details about consumption with the host of reasons drinkers give for drinking. In most studies that address this issue, the researchers rely on subjects' self-reports on their drinking, thereby potentially compounding the analysis with massive under-reporting of the kind noted above. An exception is the study of Lex et al. (1989) in which 26 women social drinkers resided for 21 days in a research unit where the researchers recorded their alcohol use and their reasons for it. Heavy drinkers (more than five drinks per study day, i.e., more than a bottle of wine or about 1700 ml of beer) gave stronger reasons than moderate drinkers or light drinkers (fewer than 2.5 drinks per study day). Subjects drank mostly to relax and because they liked the taste. Another exception is the work of de Wit et al. (1989), who asked male subjects

Table 23.2. Average amounts of fluid drunk per subject during 15-min taste tests (in ml)

Beverage condition	Alcoholics		Social drinkers	
	Told tonic	Told alcohol	Told tonic	Told alcohol
Given tonic	324	706	275	432
Given alcohol	303	654	176	427

Data from Marlatt et al. (1973).

to choose between an alcoholic beverage and a placebo and found that those who chose the most alcohol reported experiencing stimulant-like effects from the alcohol; other subjects reported primarily sedative-like effects.

Of the potentially relevant pharmacological effects of alcohol, tension reduction has received the most interest as a possible cause of consumption – tension in this sense being synonymous with anxiety. Tests of the role of tension reduction have mostly comprised asking drinkers why they consume alcohol and relating measures of anxiety to consumption or to reported consumption (e.g., Kalodner et al. 1989). A recent review (Young et al. 1990) concluded, "The interaction of pharmacology, expectancy, gender role, and situation suggest that tension reduction is of major importance in understanding the drinking of many, but not all, individuals".

Other pharmacological effects of alcohol that might have a role in its use, at least in males, are increases in assertiveness, aggression, and general arousal, including sexual arousal (Mooney et al. 1987).

Caffeine is an intrinsically easier drug to work with than alcohol. Elucidating the role of pharmacology in the everyday use of caffeine should be a relatively simple proposition: it has no energy value, it can be more readily disguised, and there is little concern about its use. Nevertheless, the role of caffeine in beverage consumption is as much a mystery as that of alcohol. Work on the matter is sparse.

Griffiths and Woodson (1988) discussed the results of a series of laboratory experiments designed to determine whether coffee drinking is maintained by the pharmacological consequences of caffeine administration. This work indicated that the caffeine in coffee can be a reinforcer, but only for some individuals and with reliability only when drinkers are physically dependent on the drug. A subsequent study showed caffeine to be a reliable reinforcer for chronic heavy users of coffee, whether or not they were currently dependent (Griffiths et al. 1989).

An important observation in the study by Marlatt et al. (1973) was that alcoholics drank more whatever the condition. The generalized polydipsia of alcoholism has been noted in other research (e.g., Williams 1977). Perhaps alcoholics are primarily people who drink more, for whatever reason, and only secondarily people who drink more alcohol.

Dependence and Craving

Dependence on alcohol is a result of chronic use of large amounts of the drug: usually an average of five or more drinks a day. It is a medically recognized condition for which a patient must meet three of nine diagnostic criteria, only three of which refer to continued heavy use or the need to sustain it (American Psychiatric Association 1987). Thus dependence and craving (compulsion to drink alcohol) are recognized clinically as separable phenomena. There is no doubt that craving – sometimes associated with notion of loss of control by an individual over his drinking – is part of the cause of the maintenance of high rates of fluid ingestion in some heavy users of alcohol (Marlatt et al. 1988), but craving did not cause them to become heavy drinkers in the first place.

Role of Consumption of Other Drugs: Lifestyle Factors

Heavy uses of beverages containing alcohol or caffeine are often found to be correlated with each other and with heavy use of other drugs. For example, Carmody et al.

(1985) found that use of cigarettes, alcohol-containing beverages, and coffee were all correlated with each other, sometimes weakly, in "healthy, community-living, middle-class Americans", both males and females. Moreover, ex-smokers used more alcohol and caffeine than non-smokers. Zeiner et al. (1985) surveyed 258 alcoholics and found that all but three were abusing another drug – mostly nicotine or caffeine or both, but also "stimulants, tranquilizers, depressants, narcotics or toluene".

Such correlations have led to the suggestion there may be "learned voluntary behaviours that place an individual at a higher risk of developing one or more of the major chronic diseases" and "one or more underlying global or generalizable biobehavioral factors with which to account partially for the observed individual difference in the performances of at least some health-related behaviors", all known as lifestyle risk factors (Istvan and Matarazzo 1984). These concepts have been elaborated into the notion of "an underlying trait such as risk-taking or impulsivity", which has been used to explain the correlations (Soeken and Bausell 1989). Unless independent evidence for such factors is found, their use in explanations of consumption of drug-containing beverages is tautologous.

A partial explanation of the concomitant use of some pairs of drugs could be that one is used to compensate for the effects of the other. Such compensatory effects have been found and proposed as a cause of the increased rate of smoking after alcohol use (Michel and Bättig 1989). Yet there is also an increased rate of smoking after coffee drinking (Emurian et al. 1982), even though nicotine and caffeine tend to have similar rather than antagonistic effects. Casual observation suggests that caffeine-containing beverages are often consumed in order to counteract the effects of alcohol and thus make safe driving possible. The available evidence does not support this belief (Fudin and Nicastro 1988).

Genetic Causes

Twin, family and adoption studies show that the risk of developing alcoholism has a genetic component (Li and Lockmuller 1989). Genetic differences have been found in tolerance to alcohol, and in ability to metabolize the drug without causing adverse side effects (Schuckit 1985; Crabb et al. 1989). However, male social drinkers with an alcoholic close relative have generally been found not to drink more than or respond differently to alcohol from control subjects without such a close relative (de Wit and McCracken 1990). One genetically determined feature that is apparent at low levels of alcohol consumption is the flushing response (Nagoshi et al. 1988). The few twin studies of caffeine use have produced results suggestive of genetic influence (Gurling et al. 1985). The relevance of genetic models of heavy use of alcohol and other drugs remains controversial (Peele 1986).

Palatability

A few studies have indicated that alcohol- and caffeine-containing beverages are consumed for their taste (e.g., Lex et al. 1989), or that taste might be involved in the establishment and maintenance of alcohol-drinking habits (Sherman et al. 1984). Research using rats suggests that sweetening fluids can powerfully increase consumption of them (see above), but there have been few published studies (reviewed by Rolls 1990) directly relevant to the effects of palatability on human consumption of alcohol- and caffeine-containing beverages under everyday conditions.

Three important centres of research on the role of palatability in beverage consumption by humans are represented in this volume. Work by Engell and Hirsch (Chapter 24) suggests that soldiers exercising in a hot climate drink much more water when it is pleasantly flavoured, and that fluid intake at meals can be further enhanced by providing more than one flavoured beverage. Booth (Chapter 4) describes his studies of the sensory influences on drinking. Tuorila (Chapter 22) reviews demonstrations that flavour is the most important determinant of the acceptability of beverages.

Achieving desirable tastes of alcohol- and caffeine-containing beverages in everyday life is a massive enterprise of commerce and connoisseurship. The overall research effort does not seem to complement this enterprise.

Schedule-Induced Drinking

Two studies have compared drinking by human subjects when monetary rewards in a video slot-machine game were delivered non-contingently at intervals of 130 and 190 s. In the first experiment (Doyle and Samson 1985), much more water was consumed when the interval was longer – approximately 170 vs. 280 ml during the 30-minute session. When beer was available, subjects drank more fluid – 380 vs. 540 ml – but the difference was not significant. Before the second comparison of drinking during the two reward regimens, subjects each drank 360 ml of beer (Doyle and Samson 1988). During the regimens, the average amounts of beer drunk were 156 and 284 ml respectively – a significant difference. When the 190-s but not the 130-s interval between monetary rewards prevailed, the pattern of the subjects' drinking was characteristic of the schedule-induced drinking observed in rats (see above), with the maximum rate of drinking occurring at the beginning of each interval.

The authors thus extended to consumption of alcohol-containing beverages previous findings that fluid consumption in humans can be substantially enhanced by manipulating the interval between reinforcement of other behaviour. Schedule induction has been offered as a basis for human excessive behaviour (Gilbert 1976a; Falk 1981), but the occurrence of schedule-induced drinking in everyday life remains to be established, including its role in consumption of caffeine-containing beverages.

Availability

An elegant study of the strong role of physical proximity of fluid on drinking is mentioned by Engell and Hirsch (Chapter 24). Much more water was drunk during a meal in a laboratory setting when water was on the table (an average of 444 ml per subject) than when it was in sight about 6.5 m from the table (197 ml) or out of sight some 13 m away (187 ml). Reid (1981) made an observation that also illustrates the strong influence of availability on fluid consumption, in this case mostly alcoholic beverages. During a flight with stages of 28, 18 and 35 minutes, drinks could be ordered during the first two stages but the drink cart was wheeled down the aisle only during the third stage. The numbers of passengers ordering drinks during the three stages were 4%, 1% and 43% of the respective totals.

Hammer and Vaglum (1989) surveyed 3997 Norwegian women in order to assess the importance of availability and stress as causes of their alcohol consumption. They concluded that availability (reflected in population density and husbands' alcohol use) was a significant factor, but stress (reflected in employment, work stress, nervousness) was not a factor. In a study of risk for osteoporosis, Parent and Krondl (1989) asked healthy Canadian women aged 50–64 years about how their use of caffeine beverages had changed during the past 30 years and about the reason for the change. Of the 56% of respondents reporting an increase in use, 66% gave increased availability as the main reason for the increase. (Of those reporting a decrease in consumption, 48% gave concerns about health as the main reason.)

Rabow et al. (1982) surveyed 580 residents of California to examine alcohol consumption in relation to three aspects of availability: perceived social obligation to serve alcohol, purchase price, and effort required to secure alcoholic beverages. According to the authors, these variables explained most of the variance, with social obligation being the most salient. Analyses of purchase price have shown that alcohol consumption increases with decrease in relative price, i.e., price in relation to disposable income (Godfrey 1989). This relationship has had elegant experimental confirmation (Babor et al. 1978).

Social and Societal Causes

The amounts people drink in bars seems to depend on the size of the group they are with; the larger the group, the more each member drinks (McCarty 1985). A similar finding has been made by de Castro (1990) who found the same relationship during his analysis of diaries kept of eating and drinking and their circumstances. For most people, the amount of alcohol consumed was strongly and positively correlated with the number of people present while drinking, whether they were light or heavy drinkers. A similar finding has been made for the consumption of other fluids (de Castro and de Castro 1989), but this drinking, unlike alcoholic drinking, was found to be secondary to the social facilitation of eating.

In addition to the generalized social facilitation of alcohol consumption just discussed, there is evidence of specific person-to-person influence on amount drunk. Chipperfield and Vogel-Sprott (1988) found that amounts drunk were influenced by what a drinking companion drank and, moreover, that subjects with problem-drinking relatives were more suggestible. These researchers observed young adult males drinking in the company of a confederate who drank 700 ml or 100 ml during a taste test. The male subjects were matched for their own reported alcohol use. Those drawn from families with problem-drinking members drank an average of 324 and 171 ml of beverage, respectively. Those drawn from families without problem-drinking members drank an average of 259 and 238 ml.

The accumulation of social influences such as the modelling demonstrated by Chipperfield and Vogel-Sprott results in significant cultural and national differences in beverage use. These can result in markedly different amounts of particular beverages being consumed with possible profound impacts on total liquid consumption. Some of these differences are discussed by Tuorila (Chapter 22).

Tuorila referred both to the decline in coffee use in the United States and to soft drinks being "the most characteristic feature of the US beverage culture". The extraordinary changes in beverage consumption habits that underlie her observations are worth a further comment. Annual per capita coffee consumption (green bean disappearance) in the United States fell from 7.2 kg in 1960 to 4.2 kg in 1982, a decline by 36%. Liquid drunk as coffee probably fell from an annual per capita total of about 144 to 92 l. (This estimate uses a conversion factor of 20 l of beverage per kilogram of bean, which is more appropriate for the United States than Tuorila's

13 l/kg.) Meanwhile, annual per capita soft drink consumption increased from 45 to 149 l, a rise of 231%. Over the period in question, the average United States resident's annual liquid consumption thus increased by some 52 l on account of the shift from coffee to soft drinks. Overall caffeine consumption fell by some 18% for the same reason (Gilbert 1984).

The changes affected younger members of the population only. During the 1960s and 1970s, the United States changed from being a country in which coffee was a beverage drunk by adults of all age groups to one in which its main consumers were the elderly and middle-aged. Younger people drank soft drinks. These trends have continued since 1982, although at a much slower rate. In Canada, where per capita coffee consumption increased slightly between 1960 and 1982, and where per capita soft drink consumption increased relatively slowly during the same period (Gilbert 1984), the switch from coffee to soft drinks may be about to happen with as much rapidity as it did in the United States in the 1960s.

Little is known about the causes of these massive changes in consumption habits. In the United States, health concerns were not the main factor: coffee use fell most rapidly in the 1960s and reports on the adverse effects of caffeine did not begin to appear in the mass media until the early 1970s. One explanation is that the changes in patterns of caffeine-beverage consumption came about in the United States, and are accelerating in Canada, as a result of advertising (discussed below).

Consideration of societal factors in the use of drug-containing beverages should not ignore the way in which the consumption of any widely available product is distributed in the population. The distribution curve of percentage of users against level of consumption usually reaches an early peak and then subsides into a long tail. This kind of distribution for all fluids is suggested in the data of Ershow and Cantor (1989). It was first noted for alcohol by Ledermann (1956). Its application is controversial (Ravn 1987) but the plausibility of the basic notion remains unchallenged: whatever the circumstances of society, a few people drink a lot, many people drink a little, and the peak level of consumption is close to or at the lowest level. The question of what societal factors contribute to this kind of distribution of consumption has not been addressed.

Advertising and Other Marketing Methods

Of all the potential causes of increased consumption of drug-containing beverages, the role of advertising and other commercial promotion attracts the greatest controversy. A recent comprehensive review of empirical studies of advertising's effect on overall consumption concluded, "in general, the evidence indicates little impact of alcohol advertising on alcohol sales or drinking ... some results are suggestive" (Smart 1988). The review did not address recruitment to drinking by adolescent males, in which advertising may play a part (Aitken et al. 1988). Neither television advertising of alcohol products nor use of them during programmes appears to cause immediate consumption of alcohol, at least in a controlled laboratory setting (Sobell et al. 1986). The controversies over the role of advertising in the use of tobacco products are equally fierce and inconclusive (Raftery 1989). By contrast, little concern is expressed about advertising for coffee and tea and for soft drinks that contain caffeine.

According to Strauss (1989), "Coca-Cola and Pepsi-Cola have got their guns trained on the coffee market [in Canada], intent on knocking it off as the leading drink in the country ... In a bid to replace coffee in the morning, Pepsi A.M., an extra-caffeine version of the cola, is being tested ... Coca-Cola will step up radio commercials next summer for Coke in the morning." Comparison of the United States and Canadian experiences regarding the role of advertising in changing beverage habits could provide valuable information about the degree to which this societal factor can be a cause of total fluid consumption.

The same article by Strauss revealed another marketing method that may be having a profound effect on total caffeine-beverage and fluid consumption in North America. She wrote, "Another strategy is to get us to buy bigger bottles and cans. ... The companies can charge more, yet consumers often buy the same number of containers. 'We know that once a larger package is in the refrigerator, it will be consumed just as quickly as a smaller package,' says (a Coca-Cola official) ... Consumers are trading up from the 280-ml can to the more expensive 500-ml bottle".

Conclusion

A reasonable conclusion from the available empirical evidence is that pharmacological factors play only a minor direct role in causing and maintaining consumption of drug-containing beverages. Many other potential factors have been identified, including expectancy of effects, palatability, schedule induction, availability, and social and societal causes. Pharmacological effects may be involved in maintaining the effectiveness of some of these other factors, as may thirst, but there is no evidence of possible mechanisms.

The huge amount of excessive drinking engaged in by sedentary adults in the United States possibly happens above all on account of the way in which rewarding events occur in everyday life (schedule induction) and the ready availability of extremely palatable fluids.

Concluding Remarks

This necessarily superficial overview has focused on the augmentation of consumption of alcohol- and caffeine-containing fluids. It has not paid enough attention to important aspects of the regulation of the use of these beverages, including factors that serve to reduce their use, such as laws and health scares, factors that stop ingestion once it has started, and factors such as personality traits that might contribute to individual differences in consumption. Also mostly ignored were what may be powerful interactions between the entrainment of drinking to eating and the factors other than thirst. Engell and Hirsch's (Chapter 24) data on water proximity give us a clue as to the possible importance of such interactions. Another omission has been reference to the pharmacologically diuretic effects of both alcohol and caffeine. In the case of alcohol, such an effect can cause elimination of a volume of urine greater than that of the consumed beverage (Gill et al. 1982). Yet another omission concerns factors other than water deficit that cause a dry mouth, such as certain foods and drugs, and thereby augment drinking.

Conspicuous in the overall effort to understand human drinking is a relative lack of work on consumption of beverages containing caffeine. Such work should be prominent because caffeine-containing beverages account for more than a quarter of adult fluid intake in the United States (Table 23.1) and this proportion is increasing.

The only reason for not focusing on this category of beverage would be that caffeine is considered irrelevant to consumption of beverages containing caffeine, a position that is given some weight in the above review.

Although the research related to everyday drinking points to availability and other causes as being more important than pharmacological factors and thirst, the potential importance of the caffeine and alcohol in beverages should not, however, be ignored. More beer is sold than near-beer, and more regular coffee than decaffeinated. These differences in sales are conceivably just a matter of taste, but they are more likely rooted in an indirect way in the pharmacological properties of the products. Likewise, although restoration of water balance may hardly feature in the direct control of everyday drinking by sedentary adults in the United States, its indirect role, perhaps during the greater physiological fragility of childhood or in some other way, should not be discounted.

Reviews of this kind customarily end with a plea for more research. This review is no exception. Drinking is much of what we do each day. Too little is known about it. More research on drinking would help provide a better assessment of the meaning of thirst and its role in fluid consumption.

Acknowledgement. Many thanks to Dr Harold Kalant for a careful reading of a draft of the chapter and some helpful comments.

References

Aitken PP, Leathar DS, Scott AC (1988) Ten- to sixteen-year-olds' perception of advertisements for alcoholic drinks. Alcohol Alcohol 23:491–500

American Psychiatric Association (1987) Diagnostic and statistical manual of mental disorders (DSM-III-R) 3rd edn revised. American Psychiatric Association, Washington DC

Atkinson J, Enslen M (1976) Self-administration of caffeine by the rat. Arzneimittelforschung 26: 2059–2061

Babor TF, Mendelson JH, Greenberg I, Kuehnle J (1978) Experimental analysis of the "happy hour": effects of purchase price on alcohol consumption. Psychopharmacology 58:35–91

Berg C, LaBerg JC, Skutle A, Ohman A (1981) Instructed versus pharmacological effects of alcohol in alcoholics and social drinkers. Behav Res Ther 19:55–66

Blanchard RJ, Hori K, Tom P, Blanchard DC (1987) Social structure and ethanol consumption in the laboratory rat. Pharmacol Biochem Behav 28:437–442

Carmody TP, Brischetto CS, Matarazzo JD, O'Donnell RP (1985) Co-occurrent use of cigarettes, alcohol, and coffee. Health Psychol 4:323–335

Carroll ME, Meiseh RA (1981) Determinants of increased drug self-administration due to food deprivation. Psychopharmacology 74:197–200

Chipperfield B, Vogel-Sprott (1988) Family history of problem drinking among young male social drinkers: modeling effects on alcohol consumption. J Abnorm Psychol 97:423–428

Crabb DW, Edenberg HJ, Bosron WF, Li T-K (1989) Genotypes for aldehyde dehydrogenase deficiency and alcohol sensitivity. The inactive ALDH2² allele is dominant. J Clin Invest 83:314–316

de Castro JM (1990) Social, circadian, nutritional, and subjective correlates of the spontaneous pattern of moderate alcohol intake of normal humans. Pharmacol Biochem Behav 35:923–931

de Castro JM, de Castro ES (1989) The presence of other people is associated with enlarged meal sizes and disruption of postprandial regulation in the spontaneous eating patterns of humans. Am J Clin Nutr 50:237–247

de Wit H, McCracken SG (1990) Ethanol self-administration in males with and without an alcoholic first-degree relative. Alcohol Clin Exp Res 14:63–70

de Wit H, Pierri J, Johanson CE (1989) Assessing individual differences in ethanol preference using a cumulative dosing procedure. Psychopharmacology 98:113–119

Doyle TF, Samson HH (1985) Schedule-induced drinking in humans: a potential factor in excessive alcohol use. Drug Alcohol Depend 16:117–132

Doyle TF, Samson HH (1988) Adjunctive alcohol drinking in humans. Physiol Behav 44:775–779

Ellison G (1987) Stress and alcohol intake: the socio-pharmacological approach. Physiol Behav 40: 387–392

Emurian HH, Nellis MJ, Brady JV, Ray RL (1982) Event time-series relationship between cigarette smoking and coffee drinking. Addict Behav 7:441–444

Ershow AG, Cantor KP (1989) Total water and tapwater intake in the United States: population-based estimates of quantities and sources. Federation of American Societies for Experimental Biology, Bethesda, Maryland

Falk JL (1972) The nature and determinants of adjunctive behaviour. In: Gilbert RM, Keehn JD (eds) Schedule effects: drugs, drinking, and aggression. University of Toronto Press, Toronto, pp 148–173

Falk JL (1981) The environmental generation of excessive behavior. In: Mulé SJ (ed) Behavior in excess: an examination of the volitional disorders. Free Press, New York, pp 313–337

Falk JL, Tang M (1988) What schedule-induced polydipsia can tell us about alcoholism. Alcohol Clin Exp Res 12:577–585

File SE, Baldwin HA, Johnston AL, Wilks LJ (1988) Pharmacol Biochem Behav 30:809–815

Fudin R, Nicastro R (1988) Can caffeine antagonize alcohol-induced performance decrements in humans? Percept Mot Skills 67:375–391

Gilbert RM (1972) Persistence of palatability-induced polydipsia. Psychon Sci 29:55–58

Gilbert RM (1974) Effects of food deprivation and fluid sweetening on alcohol consumption by rats. J Stud Alcohol 35:42–47

Gilbert RM (1976a) Drug abuse as excessive behaviour. Can Psychol Rev 17:231–240

Gilbert RM (1976b) Schedule-induced self-administration of drugs. In: Blackman DE, Sanger DJ (eds) Contemporary research in behavioral pharmacology. Plenum Press, New York, London, pp 289–323

Gilbert RM (1976c) Shifts in the water and alcohol solution intake by rats under conditions of schedule induction. J Stud Alcohol 37:940–949

Gilbert RM (1978) Schedule induction, sweetness, and ethanol consumption. Pharmacol Biochem Behav 8:739–741

Gilbert RM (1979) Augmentation of alcohol consumption by caffeine in malnourished rats. J Stud Alcohol 40:19–27

Gilbert RM (1984) Caffeine consumption. In: Spiller GA (ed) The methylxanthine beverages and foods: chemistry, consumption, and health effects. Alan R. Liss, New York, pp 185–213

Gilbert RM, Sherman IP (1970) Palatability-induced polydipsia: saccharin, sucrose, and water intake in rats, with and without food deprivation. Psychol Rep 27:319–325

Gill GV, Baylis PH, Flear CT, Skillen AW, Diggle PH (1982) Acute biochemical responses to moderate beer drinking. Br Med J [Clin Res] 285:1770–1773

Godfrey C (1989) Factors influencing the consumption of alcohol and tobacco: the use and abuse of economic models. Br J Addict 84:1123–1138

Griffiths RR, Woodson PP (1988) Caffeine physical dependence: a review of human and laboratory animal studies. Psychopharmacology 94:437–451

Griffiths RR, Bigelow GE, Liebson IA (1989) Reinforcing effects of caffeine in coffee and capsules. J Exp Anal Behav 52:127–140

Gurling HMD, Grant S, Dangl J (1985) The genetic and cultural transmission of alcohol use, alcoholism, cigarette smoking and coffee drinking: a review and an example using a log linear cultural transmission model. Br J Addict 80:269–279

Guthrie GD, Myers KJ, Gesser EJ, White GW, Koehl JR (1990) Alcohol as a nutrient: interactions between ethanol and carbohydrate. Alcohol Clin Exp Res 14:17–22

Hammer T, Vaglum P (1989) The increase in alcohol consumption among women: a phenomenon related to accessibility or stress? A general population study. Br J Addict 84:767–775

Heppner CC, Kemble ED, Cox WM (1986) Effects of food deprivation on caffeine consumption in male and female rats. Pharmacol Biochem Behav 24:1555–1559

Hull JC, Bond CF Jr (1986) Social and behavioral consequences of alcohol consumption and expectancy: a meta-analysis. Psychol Bull 99:347–360

Istvan J, Matarazzo JD (1984) Tobacco, alcohol, and caffeine use: a review of their interrelationships. Psychol Bull 95:301–326

Kalodner CR, DeLucia JL, Ursprung AW (1989) An examination of the tension reduction hypothesis: the relationship between anxiety and alcohol in college students. Addict Behav 14:649–654

Le Magnen J, Marfaing-Jallat P (1984) Further study of induced behavioral dependence on ethanol in rats. Alcohol 1:269–273

Ledermann SC (1956) Alcool, alcoolisme, alcoolisation. Institute Nationale d'Etudes Démographiques, Paris. (Cahier no. 29)

Lemmens P, Knibbe RA, Tan F (1988) Weekly recall and diary estimates of alcohol consumption in a general population survey. J Stud Alcohol 49:131–135

Lex BW, Mello NK, Mendelson JH, Babor TF (1989) Reasons for alcohol use by female heavy, moderate, and occasional social drinkers. Alcohol 6:281–287

Li T-K, Lockmuller JC (1989) Why are some people more susceptible to alcoholism? Alcohol Health Res World 13:310–315

Li T-K, Lumeng L, McBride WJ, Waller MB (1981) Indiana selection studies on alcohol-related behaviors. In: McClearn GE, Deitrich RA, Erwin VG (eds) Development of animal models as pharmacogenetic tools. US Government Printing Office, Washington, DC, pp 171–191. (NIAAA Research Monograph no. 6)

Lieber CS, DeCarli LM (1989) Liquid diet technique of ethanol administration: 1989 update. Alcohol Alcohol 24:197–211

Marlatt GA, Demming B, Reid JR (1973) Loss of control drinking in alcoholics: an experimental analogue. J Abnorm Psychol 81:233–241

Marlatt GA, Baer JS, Donovan DM, Kivlahan OR (1988) Addictive behaviors: etiology and treatment. Ann Rev Psychol 39:223–252

McCarty D (1985) Environmental factors in substance abuse: the micro-setting. In: Galizio M, Maisto SA (eds) Determinants of substance abuse: biological, psychological, and environmental factors. Plenum Press, New York, pp 247–282

Michel C, Bättig K (1989) Separate and combined psychophysiological effects of cigarette smoking and alcohol consumption. Psychopharmacology 97:65–73

Mooney DK, Fromme K, Kivlahan DR, Marlatt GA (1987) Correlates of alcohol consumption: sex, age, and expectancies relate differentially to quantity and frequency. Addict Behav 12:235–240

Nagoshi CT, Dixon LK, Johnson RC, Yuen SHL (1988) Familial transmission of alcohol consumption and the flushing response to alcohol in three Oriental groups. J Stud Alcohol 49:261–267

Parent M-E, Krondl M (1989) Effect of current and past caffeine intake on the bone mass of postmenopausal women. Conference paper, Federation of American Societies for Experimental Biology

Peele S (1986) The implications and limitations of genetic models of alcoholism and other addictions. J Stud Alcohol 47:63–73

Rabow J, Schwartz C, Stevens S, Watts RK (1982) Social psychological dimensions of alcohol availability: the relationship of perceived social obligations, price considerations, and energy expended to the frequency, amount, and type of alcoholic beverage consumed. Int J Addict 17:1259–1271

Raftery J (1989) Editorial: advertising and smoking – a smouldering debate? Br J Addict 84:1241–1246

Ravn I (1987) The control-of-consumption approach to alcohol abuse prevention. II. A review of empirical studies. Int J Addict 22:957–979

Reid JR (1981) Study of drinking in natural settings. In: Marlatt GA, Nathan PE (eds) Behavioral approaches to alcoholism. Rutgers Center of Alcohol Studies, New Brunswick, NJ, pp 58–74

Rockman GE, Hall AM, Markert LE, Glavin GB (1988) Influence of rearing conditions on voluntary ethanol intake and response to stress in rats. Behav Neural Biol 49:184–191

Rolls BJ (1990) Palatability and fluid intake. In: Fluid replacement and heat stress. National Academy Press, Washington, DC, pp XIII-1 – XIII-11 (National Academy of Sciences Publication IOM-89-005)

Ross DF, Pihl RO (1989) Modification of the balanced placebo design for use at high blood-alcohol levels. Addict Behav 14:91–97

Schuckit MA (1985) Ethanol-induced changes in body sway in men at high alcoholism risk. Arch Gen Psychiatry 42:375–379

Schulte-Daxboek G, Opitz K (1981) Increased caffeine consumption following chronic nicotine treatment in rats. IRCS Med Sci 9:1062

Segal EF (1972) Induction and the provenance of operants. In: Gilbert RM, Millenson JR (eds) Reinforcement: behavioral analyses. Academic Press, New York, London, pp 1–34

Sherman J, Rusiniak KN, Garcia J (1984) Alcohol-ingestive habits. The role of flavor and affect. Recent Dev Alcohol 2:59–79

Smart RG (1988) Does alcohol advertising affect overall consumption? A review of empirical studies. J Stud Alcohol 49:314–323

Smart RG (1989) Is the postwar drinking binge ending? Cross-national trends in per capita alcohol consumption. Br J Addict 84:743–748

Sobell LC, Sobell MB, Riley DM et al. (1986) Effect of television programming and advertising on alcohol consumption in normal drinkers. J Stud Alcohol 47:333–340

Soeken KL, Bausell RB (1989) Alcohol use and its relationship to other addictive and preventive behaviors. Addict Behav 14:459–464

Stiglick A, Woodworth I (1984) Increase in ethanol consumption in rats due to caloric deficit. Alcohol

1:413–415

Strauss M (1989) Canadian pop units lag parents' efforts. Globe and Mail, Toronto, 16 Nov, p 88

Tang M, Falk JL (1977) Ethanol dependence as a determinant of fluid preference. Pharmacol Biochem Behav 7:471–474

Tang M, Brown C, Falk JL (1982) Complete reversal of chronic ethanol polydipsia by schedule withdrawal. Pharmacol Biochem Behav 16:155–158

Waller MB, McBride WJ, Gatto GJ, Lumeng L, Li T-K (1984) Intragastric self-infusion of ethanol by ethanol-preferring and -nonpreferring lines of rats. Science 255:78–80

Williams RJ (1977) Effects of deprivation and pre-loading on the experimental consumption of tea by alcoholics and social drinkers. Br J Addict 72:31–35

Young RM, Oei TPS, Knight RG (1990) The tension reduction hypothesis revisited: an alcohol expectancy perspective. Br J Addict 85:31–40

Zeiner AR, Stanitis T, Spurgeon M, Nichols N (1985) Treatment of alcoholism and concomitant drugs of abuse. Alcohol 2:555–559

Commentary

Booth: In considering the effects of advertising on soft drinks consumption, we should not neglect technological changes without which ads would likely be useless. The beverage can and the can dispenser have made sodas that much more convenient to drink and "available" than the hot drinks, even given coffee brewers and vending machines. Cooling and carbonation technologies deliver higher palatability too (Chapters 22 and 24).

Chapter 24

Environmental and Sensory Modulation of Fluid Intake in Humans

D. Engell and E. Hirsch

Introduction

In response to a question about individual differences in replenishing fluid losses, Adolph (1964) was quick to point to the critical role of experiential and environmental factors in controlling human fluid intake. He stated (Adolph 1964):

man is the most difficult species to work with because all men are conditioned by past experience in water drinking and by the advertising that has been shot at them over the years. If we force a group of people to hike in the desert for a few hours, until they are ready to drink, we discover all sorts of drinking patterns and preferences. Some refuse water because it is warm, others because it is not flavored with grape juice but something else. Therefore we prefer to use another species where individuals can be raised on a relatively uniform regimen, and in this way obtain less biased subjects.

Despite the tremendous advances that have been made in elucidating the physiological, anatomical and hormonal bases of water balance in the intervening years, too many of us have heeded Adolph's advice and too few researchers have ventured into the sometimes murky and always difficult area of studying the factors that control human fluid intake, particularly outside the laboratory.

This chapter will describe the effects of fluid availability, climate and the sensory properties of the fluid (temperature, taste and variety) on intake and water balance in humans. A complete theory of thirst must account for these factors as well as the physiological mechanisms that underlie drinking behaviour and the regulation of fluid balance.

Our research on human drinking behaviour stems from two sources. First, the military operates in demanding environments (e.g. hot, cold) under stressful working conditions (e.g. wearing protective clothing or being encapsulated in a tank) where fluid intake is likely to be inadequate to meet fluid losses. Thus, one concern of our research programme is to uncover those factors which encourage or hinder fluid intake. Second, our research programme on human feeding behaviour has an applied component where rations are evaluated under field conditions. Our standard methodology for evaluating rations or feeding systems calls for measuring body weight, food acceptability, food intake, water intake and urine concentration during military field training exercises (e.g. Hirsch et al. 1985). In these ration tests, water intake is monitored as a possible explanation for inadequate energy intake due to thirst-induced anorexia (Engell 1988). A satisfactory ration may not be consumed in sufficient quantity to meet energy needs because water intake is too low. Rather than modifying the ration, a more satisfactory solution would be to find ways to encourage higher levels of fluid intake under field conditions. In our efforts to understand how environmental factors affect water consumption these factors are frequently tested under controlled laboratory conditions and then an attempt is made to transport successful findings to the field for validation.

Water Availability

Unlimited Access to Water

The urban-dwelling Westerner generally has abundant access to food and fluid. In a study of spontaneous fluid intake in free-living humans de Castro (1988) found that the majority of fluid was consumed in the absence of perceived need. Similarly, in a laboratory study of human drinking behaviour Phillips et al. (1984) found low levels of subjective thirst as well as the absence of changes in haematocrit, plasma osmolality, sodium, potassium, protein and angiotensin II concentrations preceding fluid ingestion.

More than 20 years before the publication of these data, Holmes (1964) made the point that habitual fluid intake patterns and not hypohydration are the major determinants of daily fluid intake in humans. Comparing oedematous cardiac patients to medical students and medical technology students he found that water represented only 25%–50% of the daily fluid intake in these three groups. He further suggested that it is the desire for coffee or milk rather than thirst which dictates the pattern and amount of fluids consumed (see also Chapter 22). Recent data on meal-associated drinking on humans, which found values ranging from 68% to 78% for the amount of daily fluid consumed at meals (de Castro 1988; Engell 1988), only serve to buttress this point.

Limitations on Water Availability

In contrast to the permissive conditions described in the previous section, water and other fluids are not always readily available to humans. For example, soldiers in the field face much more restrictive conditions of water availability. Depending on their geographical location and their mission, water may be warm, frozen, difficult to obtain, or require treatment with a disinfectant such as iodine before it is safe to consume.

Field Studies

In general, military ration studies find that water intake in the field tends to be low relative to need and urine specific gravities in excess of 1.030 are frequently observed. Although this level of urine concentration is not necessarily indicative of hypohydration it does suggest that water conservation mechanisms are coming into play (Francesconi et al. 1987). These mildly concentrated urines also contrast sharply with the copious quantities of dilute urine that are observed in laboratory studies of human fluid balance.

In the heat, Adolph and his colleagues (1947) have documented in detail that fluid intake fails to match water losses even under conditions where there is ample cool water (Adolph and Wills 1947; Rothstein et al. 1947). This type of voluntary dehydration has also been observed in free-living residents of the Negev desert in Israel (Kristal-Boneh et al. 1988) who are characterized by low, concentrated urine outputs, high haematocrit ratios and a high incidence of kidney disease.

Cold weather ration studies also reveal that fluid intake is rarely sufficient to prevent mild hypohydration. Typically relatively high urine specific gravities are observed when troops are fed either dehydrated packaged rations or more conventional packaged rations that contain water (Wyant and Caron 1983; Engell et al. 1987c; Roberts et al. 1987; Edwards et al. 1989).

In addition to any reduction in fluid intake associated with restricted access to potable water in the cold during field studies, the effect of cold temperature per se on human drinking leads to hypohydration. Negative water balance is observed on exposure to cold under laboratory conditions in humans (Conley and Nickerson 1945; Lennquist et al. 1974; see also Chapter 28).

Several authors have attributed the low level of drinking in the cold to diminished "thirst sensations" (Rogers et al. 1964; Wyant and Caron 1983). To test this notion directly, thirst sensation ratings were compared in the same individuals in a temperate environment (ambient temperature, 21°C; relative humidity, 10%) and in a simulated arctic environment (ambient temperature, −18°C; wind speed, 2.5 mph; relative humidity, 10%) (Engell, unpublished technical report 1985). A thirst scale which had been previously used to assess the intensity of various sensations associated with graded levels of hypohydration (Engell et al. 1987b) was used. Cold served to reduce the magnitude of several sensations associated with thirst in a temperate environment. Sensations such as having a dry mouth or throat or having a bad or chalk-like taste in the mouth were all reduced in the cold. These observations lend support to the suggestion that reduced "sensations of thirst" may contribute to under-drinking in the cold.

Inadequate fluid intake is also frequently observed in ration studies conducted in temperate environments (Hirsch et al. 1985; Francesconi et al. 1987; Popper et al. 1987). For example, in one field study conducted in Hawaii where daytime temperatures range from 21° to 27°C and nighttime temperatures from 2° to 7°C, three groups of infantry troops were fed on different versions of a packaged ration called the Meal-Ready-to-Eat. Fluid intake from all sources (water, food, water added to food) ranged from 3731 ml to 4474 ml in the three groups. Despite these relatively high levels of fluid intake, measures of urine specific gravity showed a high incidence of values that exceeded 1.030 which ranged from an average of 19% in the group with the highest level of fluid consumption to 41% in the group with the lowest level of intake.

This brief overview of fluid balance in military field studies under different climatic conditions reveals that water intake is usually low relative to need. Both the heat and the cold pose special problems in terms of the high rates of water loss (diuresis and insensible water loss in dry, cold climates, and high sweat rates in the heat). The fact that water intake is also low in temperate environments suggests that other factors such as the conditions of water availability, competing activities, or the sensory attributes of the water may influence intake in all environments. The very nature of a military exercise, with little time available for eating and drinking or preparing potable water by melting snow or adding a disinfectant may limit voluntary fluid intake.

Laboratory Studies

Chew (1965) has documented in detail how the patterns of drinking and levels of fluid intake in many animal species vary as a function of the habitat and the availability and sources of water. Under laboratory conditions adult rats (Marwine and Collier 1979) and growing guinea pigs (Hirsch and Collier 1974) show dramatic changes in the pattern of drinking when access to water is contingent on completing a fixed ratio schedule of reinforcement. The adult rats were able to maintain stable levels of food and water intake by adopting patterns of large infrequent bouts of drinking, whereas the guinea pigs showed a small decline in intake at high response costs despite alterations in their pattern of intake. These data clearly show that the pattern of consumption and the level of intake are sensitive to laboratory manipulations that vary the availability of water.

In a laboratory study with humans asked to participate in a luncheon taste test, access to water was systematically manipulated (Engell et al. 1990). Water was available either: (a) on the dining table, (b) about 6m from the table, across the room or, (c) about 13m from the table and out of sight across a hall. Subjects were tested individually and participated in only one condition. All subjects received a 70 ml drinking cup and large portions of palatable food. The subjects drank 444.0 ± 259.6 ml (mean $\pm$ SD) in the first condition and 197.1 ± 100.2 ml and 186.7 ± 115 ml in the latter conditions, respectively. Intake was significantly higher when water was on the table relative to the other two conditions. Food intake was similar in the three conditions. These data demonstrate the powerful role of this relatively minor environmental manipulation on human water intake.

The ease with which fluids are obtained and prepared for consumption has also been shown to affect intake and patterns of drinking in several studies where effort was not explicitly manipulated (Szlyk et al. 1989; Kramer et al. 1990; Lester et al. 1990). A series of laboratory studies compared different versions of a drinking system for troops clothed in protective garments and exercising in moderate temperatures (Szlyk et al. 1989). This combination of protective clothing and exercise produces high sweat rates. One drinking system required decontamination, connection and disconnection with every drink, two-handed operation, and also inhaling against a positive pressure. The second drinking system drew water through tubing directly from the canteen, allowed single-handed operation of a hand-pump and required only one-time connection to the face mask. The hourly rate of fluid consumption during the 50-minute exercise periods, but not during 10-minute rest periods was higher in the group with the hand pump.

In a second study employing a very similar protocol and a larger number of subjects, Szlyk et al. (1990) found that the rate of drinking during the exercise (0.35 l/h vs 0.24 l/h) as well as the rate of drinking over the combined 50-minute exercise and 10-minute rest period (0.42 l/h vs 0.36 l/h) was slightly but not significantly higher in the group with the drinking system that was easier to operate. This group also initiated drinking more frequently during exercise.

A field study designed to evaluate the effectiveness of different devices for heating food in the cold revealed that the different heating devices were also associated with significantly different levels of water intake (Lester et al. 1990). Two of the devices, the canteen cup stand and the ration heater pad, were used by the individual soldier whereas the third, the Optimus Hiker Stove, was shared by a group of five men. Both the amount of water drunk and the total amount of fluid consumed were significantly higher in the group with the canteen cup stand. Although there are several plausible interpretations for this observation, the one currently preferred is that this heating device was readily available and easy to use relative to the others. Water was difficult to heat quickly with the ration heater pad and the detergent used in the heating pad left a scummy film on the outside of the plastic bag. The Optimus stove was equally easy to use as the canteen cup stand but it was not readily available due to the fact that it was shared among several people.

These laboratory and field studies strongly suggest that the availability of water and/or the ease with which it can be obtained play a central role in human drinking behaviour.

Water Temperature

In addition to water availability the sensory characteristics of a fluid may impede the soldiers' water intake in the field. In some instances soldiers need to disinfect the water with iodine tablets which impart an unpleasant taste, and in other instances the temperature of the water may be unsatisfactory. A recent study (Rolls et al. 1990) has shown that fluid temperature not only affects acceptance, but can also affect ratings of thirst.

Studies on the effects of water temperature on intake in animals have shown that there is a linear increase in intake with fluid temperature in the range 6°C – 37°C with intake decreasing above 37°C (e.g. Deaux 1973; Carlisle 1977; Gold and Laforge 1977). The effects of temperature on water intake in humans do not mirror the effects shown in animals (e.g., Adolph and Wills 1947; Rothstein et al. 1947; Sohar et al. 1962). Boulze et al. (1983) found that subjects who were allowed to rehydrate for 10–15 s following exercise or a steam bath consumed more water at 16°C than water samples at three other temperatures ranging from 0° to 50°C and expressed a preference for water at both 5°C and 16°C when the samples were only tasted. Results from another study (Sandick et al. 1984), however, demonstrated preferential drinking of the coldest water available (5°C) when subjects could choose water samples at 5°, 16°, 22° and 38°C following exercise. The subjects in the former study may have felt discomfort from drinking cold water rapidly, whereas subjects in the latter study were permitted to drink leisurely. Alternatively, the discrepancy may be due to cultural differences in temperature expectations between Europeans and Americans. The subjects in the former study were presumably French, whereas those in the 1984 study were from the USA.

The role of beverage temperature and cultural expectations for beverages has been explored in two studies (Cardello and Maller 1982; Zellner et al. 1988). Cardello and Maller (1982) studied the acceptability of water, four fruit-flavoured beverages, and tea at five serving temperatures (3°, 11°, 23°, 36° and 49°C). They also examined the acceptability of three beverages (lemonade, milk and coffee) and ten foods including entrees, vegetables and desserts at five serving temperatures (5°, 13°, 21°, 38° and 57°C). The results showed that the acceptability of both beverages and foods was a function of the temperature at which the item is normally served. The authors suggested that their data reflected culturally based preferences.

The role of expectations in preferences for beverages was directly addressed by Zellner et al. (1988). In a series of three experiments they found that acceptance ratings for unfamiliar juices of different temperatures could be influenced by manipulating individuals' expectation of the appropriateness of temperature. In the first experiment they confirmed the results of Cardello and Maller (1982) and found that their subjects had strong serving temperature preferences for all the beverages tested. In the second experiment, subjects' taste test ratings of beverages at appropriate and inappropriate temperatures were found to be higher than their ratings given to the beverages on a questionnaire (which included a serving temperature description). The authors suggested that the discrepant ratings could be due to learned expectations about appropriateness. In the third experiment, subjects were served novel fruit juices at cold ("appropriate") and room ("inappropriate") temperatures. When subjects were told that the juices were usually served warm, the ratings for the novel juices at room temperature were higher than when no information was given about the beverages.

Flavour and Beverage Variety

The effects of flavour on drinking in animals is striking (Pfaffmann 1961; Epstein 1967; Mook and Blass 1968; Mook and Kenny 1977). Bitter fluids can cause animals to become hypohydrated (Teitelbaum and Epstein 1962; Mook 1963) and sweetened fluids can stimulate drinking when there is no hydrational deficit (Young 1955; Mook 1963; Ernits and Corbit 1973). Sweetened fluids can also produce marked overhydration and hyponatraemia in experimental situations of contrived antidiuresis (see Chapter 19). Although the effects of flavour on fluid consumption has been studied less in humans than in animals, such an effect has been demonstrated. Sohar et al. (1962) showed that flavouring water could prevent voluntary dehydration in men during a desert march. Similarly, we found that flavouring enhanced voluntary intake of uncooled water during a field training exercise in a hot environment (unpublished report, Engell). During this field test, soldiers consumed almost 1.5 l more per day when water was cherry flavoured and sweetened than when the water was not flavoured and sweetened. This study demonstrated that flavouring alone (without cooling) can enhance water intake in humans in a hot environment.

The interactive effects of flavour and temperature on drinking have also been studied in humans (Hubbard et al. 1984). In a climatic chamber study, soldiers were given disinfected (iodine tablets), flavoured disinfected, or untreated water to drink during two simulated 14.5 km desert walks. On one trial the water was 40°C (ambient temperature); on the other, it was 15°C. Flavouring and cooling effects on ad libitum intake were additive and increased intake by over 100%. The average consumption (mean ± SE) over the 6-hour test period ranged from 1389 ml ± 296 ml in the group that had warm (ambient temperature) disinfected water to 3052 ml ± 128 ml in the group drinking cooled, flavoured, disinfected water. Although sweat rates were comparable in all groups, 6-hour weight loss varied significantly among the groups, ranging from 1.03 ± 0.18% with cooled, flavoured disinfected water to 3.21 ± 0.38% in the group drinking warm, disinfected water.

The effects of flavour and temperature on beverage acceptability have also been shown in a taste-testing situation (unpublished report, Engell). Subjects in this test rated the acceptability of flavoured (chocolate, strawberry) milk at four different temperatures (5°, 21°, 38°, 55°C). Both flavour and temperature were found to affect acceptability. A significant interaction was also evidenced: chocolate milk was most acceptable at the high and low temperatures, whereas plain milk was less acceptable than chocolate at all temperatures and decreased in acceptability as a function of temperature; strawberry milk was as acceptable as chocolate at the coldest temperature, but then its acceptability declined with temperature.

Flavour variety can also affect beverage consumption in humans. In a taste test situation, subjects consumed more of a beverage when three flavours were offered in succession than when only one flavour or water were offered (Rolls et al. 1980). The effect of flavour variety on beverage consumption has also been demonstrated where beverages were presented simultaneously with a luncheon meal (Engell et al. 1987a). When four fruit-flavoured beverages were available, subjects consumed 17% more beverage than when only one flavour, each subject's favourite, was available. Beverage variety also enhanced food consumption by 18% during the test meal. This increased food intake is interesting in light of the close association between food and water intake (see Chapters 18 and 21). Beverage variety may also enhance fluid consumption in the elderly (see Chapter 26).

Summary

When potable water is abundant, easily accessible and palatable, individuals drink enough to maintain euhydration. Voluntary fluid intake, however, is not sufficient to prevent dehydration when environmental conditions cause the cost or effort to obtain water to be high or the sensory characteristics of the fluid to be unpleasant. This chapter has described military field tests and laboratory experiments in which fluid intake did not meet physiological needs in healthy individuals. Although these studies illustrate how sensory and environmental factors can impede maintenance of water balance in humans for up to several days, they do not address the relative potency of physiological, sensory and environmental factors over long periods of time or at high levels of dehydration. A comprehensive theory of thirst and fluid balance will have to include how these factors modulate drinking and will also have to meet the challenges and experimental constraints of studying fluid balance in humans.

References

Adolph EF (1964) Regulation of body water content through water ingestion. In: Wayner MJ (ed) Thirst: first international symposium on thirst in the regulation of body water. Macmillan, New York, pp 5–17

Adolph EF, Wills JH (1947) Thirst. In: Adolph EF and Associates. Physiology of man in the desert. Interscience Publishers Inc, New York, pp 241–253

Boulze D, Montastruc P, Cabanac M (1983) Water intake, pleasure and water temperature in humans. Physiol Behav 30:97–102

Cardello AV, Maller O (1982) Acceptability of water, selected beverages and foods as a function of serving temperature. J Food Sci 47:1549–1552

Carlisle HJ (1977) Temperature effects on thirst: cutaneous or oral receptors? Physiol Psychol 5:247–249

Chew RM, (1965) Water metabolism of mammals. In: Mayer RV, Van Gelder RG (eds) Physiological mammalogy. Academic Press, New York, pp 43–178

Conley CL, Nickerson JL (1945) Effects of temperature change on the water balance of man. Am J Physiol 143:373–384

Deaux E (1973) Thirst satiation and the temperature of ingested water. Science 181:1166–1167

de Castro JM (1988) A microregulatory analysis of spontaneous fluid intake by humans: evidence that the amount of liquid ingested and its timing is mainly governed by feeding. Physiol Behav 43:705–714

Edwards JSA, Roberts DE, Morgan TE, Lester LS (1989) An evaluation of the nutritional intake and acceptability of the Meal, Ready-to-Eat consumed with or without supplemental pack in a cold environment. Technical Report, T18-89, US Army Institute of Environmental Medicine, Natick, Massachusetts

Engell D (1988) Interdependency of food and water intake in humans. Appetite 10:133–141

Engell D, Edenberg J, Abrams I (1987a) Effects of variety on beverage intake during a meal. Presented at Neurosciences Satellite Conference, Appetite, Thirst and Related Disorders, November, San Antonio

Engell DB, Maller O, Sawka MN, Francesconi RN, Drolet L, Young AJ (1987b) Thirst and fluid intake following graded hypohydration levels in humans. Physiol Behav 40:229–236

Engell D, Roberts DE, Askew EW, Rose MS, Buchbinder J, Sharp MA (1987c) Evaluation of the Ration, Cold Weather during a 10-day cold weather field training exercise. Technical Report Natick TR-87/030, US Army Natick Research, Development and Engineering Center, Natick, Massachusetts

Engell D, Mutter S, Kramer FM (1990) Impact of effort required to obtain water on human consumption during a meal. Eastern Psychological Association Annual Meeting. 29 March – 1 April, Philadelphia

Epstein A (1967) Oropharyngeal factors in eating and drinking. In: Code CF (ed) Handbook of physiology. American Physiological Society Press, Washington, DC, pp 197–218

Ernits T, Corbit JD (1973) Taste as a dipsogenic stimulus. J Comp Physiol Psychol 83:27–31

Francesconi RP, Hubbard RW, Szlyk PC et al. (1987) Urinary and hematologic indexes of hypohydration. J Appl Physiol 62:1271–1276

Gold RM, Laforge RG (1977) Temperature of ingested fluids: preference and satiation effects. In: Weijnen JWM, Mendelson J (eds) Drinking behavior, oral stimulation, reinforcement and preference. Plenum Press, New York, pp 247–274

Hirsch E, Collier G (1974) Effort as a determinant of intake and patterns of drinking in the guinea pig. Physiol Behav 12:647–655

Hirsch E, Meiselman HL, Popper RD et al. (1985) The effects of prolonged feeding Meal, Ready-to-Eat (MRE) operational rations. Technical Report Natick TR-85/035, US Army Natick Research, Development and Engineering Center, Natick, Massachusetts

Holmes JH (1964) Thirst and fluid intake problems in clinical medicine. MJ Wayner (ed) Thirst: first international symposium on thirst in the regulation of body water. Macmillan, New York, 57–78

Hubbard RW, Sandick BL, Matthew WT et al. (1984) Voluntary dehydration and alliesthesia for water. J Appl Physiol 57:865–875

Kramer FM, Mutter S, Engell D (1990) Situational factors and ration consumption: from the dining hall to the field. US Army Natick Science Symposium, June 5–7

Kristal-Boneh E, Glusman JG, Chaemowitz C, Cassuto Y (1988) Improved thermoregulation caused by forced water intake in human desert dwellers. Eur J Appl Physiol 57:220–224

Lennquist S, Granberg PA, Wedin B (1974) Fluid balance and physical work capacity in humans exposed to cold. Arch Environ Health 29:241–249

Lester LS, Kramer FM, Edinberg J, Mutter S, Engell DB (1990) Evaluation of the canteen cup stand and ration heater pad: effects on acceptability and consumption of the Meal, Ready-to-Eat in a cold weather environment. Technical Report Natick TR-90/008L, US Army Natick Research, Development and Engineering Center, Natick, Massachusetts

Marwine A, Collier G (1979) The rat at the waterhole. J Comp Physiol Psychol 93:391–402

Mook DG (1963) Oral and postingestional determinants of the intake of various solutions in rats with esophageal fistulas. J Comp Physiol Psychol 56:645–659

Mook DG, Blass EM (1968) Quinine aversion thresholds and "finickiness" in hyperphagic rats. J Comp Physiol Psychol 65:202–207

Mook DG, Kenny NJ (1977) Taste modulation of fluid intake. In: Weijnen JWM, Mendelson J (eds) Drinking behavior: oral stimulation, reinforcement and preference. Plenum Press, New York, pp 275–313

Pfaffmann C (1961) The sensory and motivating properties of the sense of taste. In: MR Jones (ed) Nebraska symposium on motivation. University of Nebraska Press, Lincoln, pp 71– 108

Phillips PA, Rolls BJ, Ledingham JCG, Morton JJ (1984) Body fluid changes, thirst, and drinking in man during free access to water. Physiol Behav 33:357–363

Popper R, Hirsch E, Lesher L et al. (1987) Field evaluation of improved MRE, MRE VII, and MRE IV. Technical Report Natick TR-87/027, US Army Natick Research, Development and Engineering Center, Natick, Massachusetts

Roberts DE, Askew EW, Rose MS et al. (1987) Nutritional and hydration status of special forces soldiers consuming the Ration, cold weather or the Meal, Ready-to-Eat ration during a ten day, cold weather training exercise. Technical Report T8-87, US Army Institute of Environmental Medicine, Natick, Massachusetts

Rogers TA, Setlife JA, Klapping JC (1964) Energy lost, fluid and electrolyte balance in subarctic survival situations. J Appl Physiol 19:1–8

Rolls BJ, Wood RJ, Rolls ET (1980) Thirst: the initiation, maintenance, and termination of drinking. In: Sprague JM, Epstein AN (eds) Progress in psychobiology and physiological psychology. Academic Press, New York, pp 263–321

Rolls BJ, Federof IC, Guthrie JF, Laster LJ (1990) Effects of temperature and mode of presentation of juice on hunger, thirst, and food intake in humans. Appetite 15:199–208

Rothstein A, Adolph EF, Wills RH (1947) Voluntary dehydration. In: Adolph and Associates (eds) Physiology of man in the desert. Interscience Publishers Inc, New York, pp 254–270

Sandick BL, Engell DB, Maller O (1984) Perception of drinking water temperature and effects for humans after exercise. Physiol Behav 32:851–855

Sohar E, Kaly J, Adar R (1962) The prevention of voluntary dehydration. UNESCO/India Symposium on environmental and Physiological Psychology. Lucknow, pp 7–12

Szlyk PC, Francesconi RP, Sils IV, Foutch R, Hubbard RN (1989) Effects of chemical protective clothing and masks, and two drinking water delivery systems on voluntary dehydration. Technical Report T14-89, US Army Research Institute of Environmental Medicine, Natick, Massachusetts

Szlyk PC, Sils IV, Tharion WJ et al. (1990) Effects of a modified through-mask drinking system (MDS) on fluid intake during exercise in chemical protective gear. Technical Report T1/90, US Army Institute of Environmental Medicine, Natick, Massachusetts

Teitelbaum P, Epstein AN (1962) The lateral hypothalamic syndrome: recovery of eating and drinking after LH lesions. Psychol Rev 69:74–90

Wyant KN, Caron PC (1983) Water discipline and an arctic ration prototype. Milit Med 148:435–438

Young PT (1955) The role of hedonic preferences in motivation. In: Jones MR (ed) Nebraska Symposium on Motivation. University of Nebraska Press, Lincoln, pp 71–108

Zellner DA, Stewart WF, Rozin P, Brown JM (1988) Effect of temperature and expectation on liking for beverages. Physiol Behav 44:61–68

Chapter 25

Physiological Determinants of Fluid Intake in Humans

B.J. Rolls

Introduction

Thirst, which can be intense and compelling, is associated with a dry, tacky, unpleasant-tasting mouth. Some of the oldest theories concerning the origin of thirst localize the urge to drink in the mouth and throat. Changes in the dryness of the mouth can affect thirst (Adolph, 1947), but water intake does not always depend on local hydration in the mouth and throat. This is clearly demonstrated in sham-drinking animals in which the oropharynx is continually bathed with water, but the fluid drains out of the severed oesophagus and is not absorbed. Such animals drink continuously unless fluid is administered via the lower half of the oesophagus (Maddison et al. 1980). Such water would reverse fluid depletions.

Thus, thirst is of a more general origin than local dehydration of the oral cavity. In the nineteenth century, it became clear that dehydration of the tissues and/or the blood must be an important factor in thirst. For example, in 1832, Latta treated the circulatory collapse which resulted from severe diarrhoea and vomiting in cholera with a mixture of saline and bicarbonate of soda. This mixture gave immediate relief to the patient's symptoms including the excessive thirst.

This chapter reviews what controlled experiments have revealed about the critical body fluid changes associated with thirst and consumption of water in humans. The unique advantage of studying thirst in humans is that the measurement of physiological variables can be combined with simultaneous subjective reports of how an individual feels. Thus, we can gain understanding of both the physiological and psychological parameters associated with the urge to drink.

Cellular Stimuli for Drinking

A number of studies have shown that increases in the effective osmotic pressure of the plasma stimulate drinking in humans. In 1918, Leschke observed that patients being treated for tuberculosis with intravenous hypertonic saline complained of intense thirst. Arden (1934) found that intense thirst occurred after drinking hypertonic sodium chloride but not potassium chloride. In 1947, Holmes and Gregersen infused hypertonic sodium chloride intravenously to treat obstructive arterial disease. They observed that thirst was associated with increases of plasma chloride and with a reduction of salivary flow. They attributed the thirst to the dry mouth and changes in the saliva, which became thick and sticky, rather than to the plasma hypertonicity. Wolf (1950) compared the dipsogenic effects of intravenous infusions of hypertonic saline in male medical students and calculated that thirst was associated with a 1.23% osmotic cellular dehydration.

Wolf did not measure plasma variables, and he simply noted when the subjects first said they were thirsty, but did not confirm that this reported thirst was associated with fluid intake. This is important since subjective reports do not always correlate well with the related behaviour (Hetherington and Rolls 1987). Subsequent studies by other investigators did measure plasma variables, but again relied on the subjects to volunteer that they were thirsty without the availability of water. These studies showed that the plasma osmolality at which subjects first complained of thirst (296–299 mosmol/kg water) was higher than that for the release of vasopressin (280–287 mosmol/kg water) (Baylis and Robertson 1980; Robertson 1983; Zerbe and Robertson 1983). Robertson (1984) suggested that it would be adaptive to have the thirst mechanism activated at a higher level than renal water conservation mechanisms since it would free an individual from the distraction of feeling thirsty and having to seek water. However, a more recent study in which hypertonic saline was infused in ten healthy men found that when thirst was assessed at regular intervals with line rating scales asking specific questions about thirst, the osmotic thirst threshold (281 mosmol/kg) was similar to that for vasopressin release (285 mosmol/kg) (Thompson et al. 1986). Recently Robertson (1989) has maintained that the osmotic threshold for thirst is higher than that for vasopressin. Part of the difference between studies may be that lower thirst thresholds were found when subjects were asked about thirst at regular intervals (Thompson et al. 1986) than when subjects volunteered that they were thirsty (Zerbe and Robertson 1983). The issue is further complicated by the fact that the thresholds can vary from person to person (Robertson 1984) and can be affected by changes in blood volume or pressure (Robertson 1983, 1984), the menstrual cycle (Thompson et al. 1988; Vokes et al. 1988) and pregnancy (Barron 1987). The important practical points to note in relation to this issue are that when fluids are freely available, humans often produce a hypertonic urine in the absence of thirst. However, during dehydration thirst is initiated before urine is maximally concentrated and renal mechanisms are saturated.

In 1985, we reported the first double-blind controlled crossover study of hypertonic saline-induced thirst and vasopressin release in healthy men that combined detailed descriptions of thirst with body fluid analyses and measurements of water intake (Phillips et al. 1985a). Seven healthy young men were infused with 4.8 ml/kg of either hypertonic (0.45 M) or isotonic (0.15 M) saline over 20–25 minutes. Plasma variables and ratings of thirst were measured at intervals throughout the infusions and during the 60 minutes after the infusions when water consumption was allowed.

Since a unique feature of studying thirst in humans is the measurement of subjective changes associated with body fluid alterations, it is worth briefly describing the methodology. In studies of dehydration-induced thirst Rolls et al. (1980) conducted pilot studies in which dehydrated subjects described their thirst. Based on this information five questions were formulated and subjects were asked to make a mark on a 100-mm line (a visual analogue scale) to indicate their responses to the questions at that moment in time. The questions (and the extremes of the possible answers) were: how thirsty do you feel now? (very thirsty–not at all thirsty); how pleasant would it be to drink some water now? (very unpleasant–very pleasant); how dry does your mouth feel now? (not at all dry–very dry); how would you describe the taste in your mouth? (normal–very unpleasant); how full does your stomach feel now? (not at all full–very full). Changes in ratings from the individual subject's baseline were then calculated for each sample time. The changes in these ratings following dehydration were found to correlate well with water intake, but it should not be assumed that this will always be the case. It is important to assess not only rated thirst, but also to measure fluid intake. The use of such rating procedures has been discussed in detail by Hetherington and Rolls (1987). Additional examples of questions that can be asked in relation to thirst can be found in Engell et al. (1987).

We found that the infusion of hypertonic saline significantly increased plasma sodium concentration, plasma osmolality, plasma vasopressin concentration, and ratings of thirst (Phillips et al. 1985a). The changes in ratings first became significant when plasma osmolality had increased by approximately 7 mosmol/kg water and plasma sodium concentration had increased by 4.2 mequiv/l. Dryness of the mouth and an unpleasant taste were consistently and statistically associated with hypertonic saline-induced thirst and it was relief of these that the subjects found satiating. Throughout the 60-minute drinking period after the infusions, water intake was always significantly greater after the hypertonic saline than after the isotonic saline. The amounts consumed were sufficient to return plasma sodium and osmolality and thirst ratings to preinfusion levels.

The cellular mechanisms mediating the osmotic stimulation of thirst and vasopressin secretion in man are unclear. It seems likely that the receptors respond to cellular dehydration rather than to changes in sodium concentration. This conclusion is based on the finding that intravenous hypertonic mannitol, but not urea or glucose, is as effective as hypertonic sodium chloride in stimulating vasopressin secretion and complaints of thirst in humans (Zerbe and Robertson 1983). Since the osmotically active mannitol was effective despite a fall in plasma sodium concentration, it seems that osmoreceptor cellular dehydration, and not sodium receptor response, is the mechanism mediating human osmotic thirst and vasopressin secretion. Even so, more definitive experiments to ensure that the nature and intensity of the thirst were similar with the different osmotically active solutions remain to be performed.

The exact location of the osmoreceptors in humans is unclear; however, the evidence from patients with abnormalities of osmotic thirst and vasopressin secretion suggests that the osmoreceptors are in the anterior hypothalamus and project to the neurohypophysis and/or cortex via the nucleus medianus (Robertson 1983, 1989).

In summary, the osmotic pressure and sodium content of the plasma are normally maintained within a narrow range. This is achieved by the release of vasopressin, which helps to conserve water, and through the ingestion of water to replenish supplies. Thus, osmoreceptors are part of a homeostatic feedback loop which acts to maintain fluid balance. Controlled infusions of hypertonic saline indicate that thirst, drinking, and vasopressin secretion are associated with increases in plasma osmolality and sodium that are well within the physiological range.

Extracellular Thirst Stimuli

The verbal descriptions of thirst and the ratings of thirst associated with hypertonic saline infusion were similar to those occurring after 24-hour water deprivation but were less intense and the amounts of water consumed during rehydration were less (Phillips et al. 1985a). The changes in sodium concentration were similar in the two conditions and so the difference may have been due to decreased extracellular fluid volume associated with water deprivation. However, the evidence that hypovolaemia is a physiological thirst stimulus in humans is much less clear than that for cellular dehydration.

The most direct way to cause hypovolaemia is to remove blood. Although it is commonly supposed that haemorrhage is a thirst stimulus and Cannon (1945) speaks of the "universal call for water" of the wounded man, blood loss has not been found experimentally to be a potent thirst stimulus. Holmes and Montgomery (1951) found no consistent reduction in salivary flow or increase in thirst in 50 blood donors who gave 500 ml or 5% to 10% of their blood volume. Also, Phillips et al. (Chapter 26) did not find an increase in rated thirst or in fluid intake following removal of 500 or 1000 ml of blood in healthy young men. The study of haemorrhage as a thirst stimulus is complicated by the fact that it may be accompanied by shock or dizziness. This would mean that subjects would be lying down and this would shift blood away from the periphery into the central circulation and mask the overall reduction in circulating blood volume.

In a controlled experiment, extracellular fluid volume was depleted by combining consumption of a low salt diet and sweating in radiant heat (McCance 1936). The 28%–38% reduction of the extracellular fluid volume was associated with aberrant flavour and taste and this was interpreted by one subject as thirst. The "thirst" was alleviated by the ingestion of salty food. Unbearable thirst, which was only satisfied by water, came later. Thus, there is no clear evidence that hypovolaemia alone is a physiological thirst stimulus. The depletions required are large and even then do not unequivocally stimulate thirst and drinking.

It was suggested earlier that extracellular fluid volume depletion might contribute to deprivation-induced thirst. The evidence for this is, however, not clear-cut. In one study of 24-hour fluid deprivation, it was found that there was a significant decrease in plasma volume as indicated by plasma protein, but not by haematocrit (Rolls et al. 1980). In a later study (Phillips et al. 1984a), neither plasma protein nor haematocrit indicated significant changes in plasma volume in either young or elderly subjects. Furthermore, plasma angiotensin levels were not affected by dehydration.

Angiotensin II is thought to be one of the components of hypovolaemic thirst. Reductions in plasma volume are sensed by the juxtaglomerular apparatus in the kidneys, which then release the enzyme renin. Renin acts on substrate in the plasma to form angiotensin I, which is converted into angiotensin II. Angiotensin II has been found to be a potent and specific stimulus to drinking in a wide variety of species (see Rolls and Rolls 1982). Because of case reports of excessive thirst in association with high plasma renin levels in men with severe renal disease and hypertension and the alleviation of thirst by nephrectomy (Brown et al. 1969; Rogers and Kurtzman 1973), angiotensin has been thought to stimulate thirst in man.

Phillips et al. (1985b) tested the hypothesis that angiotensin is dipsogenic in man by infusing angiotensin (2–16 ng/min/kg) intravenously in ten healthy young men. Thirst, water intake, a dry mouth and vasopressin secretion were increased in four of the ten subjects. These effects were seen at plasma angiotensin levels well above (7–22 times greater) those measured under physiological conditions associated with thirst and vasopressin secretion such as water deprivation. Similarly, Morton et al. (1976) have found that plasma angiotensin II levels under other physiological conditions associated with thirst, such as haemorrhage and sodium depletion, are not elevated to the degree seen in this experiment.

Whether a further increase in plasma angiotensin II levels or longer angiotensin II infusions would have stimulated thirst and water intake in all subjects is not known. Although equivalent infusions have been found to stimulate drinking effectively in other species, it is possible that humans are relatively resistant to the dipsogenic effects of intravenous angiotensin. It has been found in rats that intravenous angiotensin II is a more potent dipsogen when arterial pressure is at or below normal (Evered et al. 1988). It is possible that the mild increases in blood pressure seen in our subjects during the angiotensin infusions reduced the dipsogenic effect. However, against this idea is that the blood pressure changes were small, an elevation of vasopressin was seen, and there were no differences in the blood pressure responses to angiotensin II in those individuals that drank and those that did not.

Further studies are needed to clarify why only certain individuals respond to intravenous angiotensin II infusions and to determine whether potentiation of angiotensin-induced thirst and vasopressin secretion by other stimuli (e.g. hypovolaemia and hypertonicity) might occur in man. This is important in view of the fluid deficits often associated with pathological conditions in which angiotensin II is elevated.

Although angiotensin may be dipsogenic in some pathological conditions, there is little evidence that it plays a physiological role in thirst in man. In view of the non-significant changes in plasma angiotensin II levels associated with thirst when water is freely available, mild haemorrhage, or 24-hour water deprivation, and the two- to threefold rise associated with sodium depletion (see Phillips et al. 1985b), it seems unlikely that circulating angiotensin II would be involved in physiological thirst under such conditions when increases up to 22-fold fail to be dipsogenic. However, it is possible that measurements of plasma angiotensin II do not accurately reflect local changes around angiotensin II receptors. Also, it is possible that the form of angiotensin infused (angiotensin II-amide, Hypertensin) is not as biologically active for a given plasma level as human angiotensin II.

Nevertheless, it could be argued that it would be maladaptive not only for angiotensin but also for any hypovolaemic stimulus to have a role in normal thirst in humans since simple postural changes can lead to significant changes in plasma volume of 6%–17% (see Greenleaf 1982). This may mean that extracellular volume depletion becomes an important thirst stimulus only in severe dehydration or in some pathological conditions. However, it is hard to draw a firm conclusion since the types of experiments that have been performed have been limited by what is safe and ethical to do in humans. Further discussion of the role of hypovolaemia in thirst in humans can be found in Chapter 27.

Rehydration

Studies of the effects of rehydration following a thirst stimulus can help to understand the factors involved in the termination of drinking. The effects of rehydration with water after a 24-hour fluid deprivation have been studied by Rolls et al. (1980). The deprivation caused significant intracellular and extracellular depletions, thirst and a dry unpleasant tasting mouth. During rehydration, subjects drank 65% of their total intake within 2.5 minutes. The marked decrease in drinking rate and the alleviation of thirst occurred before plasma dilution had become significant. This attenuation of drinking was subjectively attributed to stomach fullness. Presystemic factors may therefore be important for drinking termination in humans.

Drinking has also been found to cause a rapid decrease in ratings of thirst associated with intravenous infusion of hypertonic saline. Plasma vasopressin and thirst fell shortly after drinking started and by 20 minutes had declined significantly despite little change in plasma osmolality (Thompson et al. 1987). The mechanism for this rapid decrease in thirst is not known, but possibilities include gastric distension, cold-sensitive oropharyngeal receptors, or a neuroendocrine reflex mediated by swallowing (see Thompson and Baylis 1988). The amount of fluid ingested per swallow affects the decrease in thirst before there are plasma changes (Williams et al. 1989). Sipping and swallowing 1 ml/kg promptly reduced ratings of thirst; however, drinking 15 ml/kg resulted in a significantly greater and more prolonged reduction in thirst. This suggests that there is some oropharyngeal metering of volume which affects thirst.

It seems likely that in humans, as in other species, the termination of drinking will be affected by oropharyngeal factors, gastric distension, intestinal (and/or hepatic portal) factors, and rehydration of body fluids. As yet the relative importance of these factors is not well understood. There is scope for much more work in this area. For example, there is a need for studies of the effects of selective rehydration of the cellular and extracellular fluid compartments following water deprivation. Also, a comparison of the effects of water administered via different routes (e.g. oral, intra-gastric, or intravenous) would further our understanding of the termination of drinking.

Ad Libitum Drinking

We have seen that humans, like other animals, respond to experimental body fluid deficits with thirst and fluid intake. I would like to consider the relevance that body fluid deficits have for thirst and drinking which occur when water is freely available. To determine whether thirst and drinking occur in response to or precede body fluid deficits, blood samples and thirst ratings were obtained from five healthy young male volunteers at hourly intervals and when they were thirsty during a normal working day (Phillips et al. 1984b). Although there were significant increases in ratings of thirst, pleasantness of drinking water, mouth dryness, and unpleasantness of the taste in the mouth when subjects were thirsty enough to drink compared with intervening intervals, there were no concomitant changes in body fluid variables (haematocrit, plasma osmolality, sodium, potassium, protein and angiotensin II concentrations). Subjects drank mainly in association with eating, a finding that has been confirmed by de Castro (1988) in a study in which subjects used a self-report diary to record all of their food and fluid intake over a 7-day period. The results indicate that during free access to water, humans become thirsty and drink before body fluid deficits develop. It is possible that subjects have learned to associate subtle oropharyngeal cues or other cues such as the time of day or activity levels with impending fluid deficits and thus drink in anticipation of plasma changes.

Conclusions

Compared with other species, there have been relatively few studies of the controls of fluid intake in humans. Studies in which hypertonic sodium chloride was infused indicate that thirst and water intake are stimulated by changes in plasma sodium and osmolality which are within the range seen after exercise, thermal dehydration, or 24–96 hours of fluid deprivation. Extracellular fluid volume depletion, which is a thirst stimulus in some species, in humans probably stimulates drinking only in severe dehydration or in some pathological conditions associated with high plasma angiotensin II levels. Although both cellular and extracellular fluid depletions can stimulate drinking, under normal conditions of free access to water most drinking is associated with eating and occurs in the absence of significant fluid deficits.

Acknowledgement. This work was supported by DK39177.

References

Adolph EF (1947) Physiology of man in the desert. Interscience, New York

Arden F (1934) Experimental observations upon thirst and on potassium overdosage. Aust J Exp Biol Med Sci 12:121–122

Barron WM (1987) Water metabolism and vasopressin secretion during pregnancy. Baillieres Clin Obstet Gynaecol 1:853–857

Baylis PH, Robertson GL (1980) Plasma vasopressin response to hypertonic saline infusion to assess posterior pituitary function. J R Soc Med 73:255–260

Brown JJ, Curtis JR, Lever AF, Robertson JIS, de Wardener HE, Wing AJ (1969) Plasma renin concentration and the control of blood pressure in patients on maintenance haemodialysis. Nephron 6:329–349

Cannon WB (1945) The way of an investigator. A scientist's experiences in medical research. Norton, New York

de Castro JM (1988) A microregulatory analysis of spontaneous fluid intake by humans: evidence that the amount of liquid ingested and its timing is mainly governed by feeding. Physiol Behav 43: 705–714

Engell DB, Maller O, Sawka MN, Francesconi RN, Drolet L, Young AJ (1987) Thirst and fluid intake following graded hypohydration levels in humans. Physiol Behav 40:229–236

Evered MD, Robinson MM, Rose PA (1988) Effect of arterial pressure on drinking and urinary responses to angiotensin II. Am J Physiol 254:R69–R74

Greenleaf JE (1982) Dehydration-induced drinking in humans. Fed Proc 41:2509–2514

Hetherington M, Rolls BJ (1987) Methods of investigating human eating behavior. In: Toates F, Rowland N (eds) Feeding and drinking. Elsevier Science Publishers, Amsterdam, pp. 77–109

Holmes JH, Gregersen MI (1947) Relation of the salivary flow to the thirst produced in man by IV injection of hypertonic salt solution. Am J Physiol 151:252–257

Holmes JH, Montgomery AV (1951) Observations on relation of hemorrhage to thirst. Am J Physiol 167:796

Latta T (1832) Letter from Dr Latta to the Secretary of the Central Board of Health, London, affording a view of the rationale and results of his practice in the treatment of cholera by aqueous and saline injections. Lancet ii:274–277

Leschke E (1918) Ueber die Durstemfindung. Arch Psychiat Nerv 59:773–781

Maddison S, Wood RJ, Rolls ET, Rolls BJ, Gibbs J (1980) Drinking in the rhesus monkey: peripheral factors. J Comp Physiol Psychol 945:365–374

McCance RA (1936) Experimental sodium chloride deficiency in man. Proc R Soc Lond [Biol] 119: 245–268

Morton JJ, Semple PF, Waite MA, Brown JJ, Lever AF, Robertson JIS (1976) Estimation of angiotensin I and II in the human circulation by radioimmunoassay. In: Antoniades HN (ed) Hormones in human blood: detection and assay. Harvard University Press, Cambridge, MA, pp 607–642

Phillips PA, Rolls BJ, Ledingham JGG et al. (1984a) Reduced thirst following water deprivation in healthy elderly men. N Engl J Med 311:753–759

Phillips PA, Rolls BJ, Ledingham JGG, Morton JJ (1984b) Body fluid changes, thirst and drinking in man during free access to water. Physiol Behav 33:357–363

Phillips PA, Rolls BJ, Ledingham JGG, Forsling ML, Morton JJ (1985a) Osmotic thirst and vasopressin release in man: a double-blind cross-over study. Am J Physiol 248:R645–R650

Phillips PA, Rolls BJ, Ledingham JGG, Morton JJ, Forsling ML (1985b) Angiotensin II-induced thirst and vasopressin release in man. Clin Sci 68:669–674

Robertson GL (1983) Thirst and vasopressin function in normal and disordered states of water balance. J Lab Clin Med 101:351–357

Robertson GL (1984) Abnormalities of thirst regulation. Kidney Int 25:460–469

Robertson GL (1989) Discussion, pp 422–425. In: Perez GO Severe hypernatremia with impaired thirst. Am J Nephrol 9:421–434

Rogers PW, Kurtzman NA (1973) Renal failure, uncontrollable thirst and hyperreninemia; cessation of thirst with bilateral nephrectomy. JAMA 225:1236–1238

Rolls BJ, Rolls ET (1982) Thirst. Cambridge University Press, Cambridge

Rolls BJ, Wood RJ, Rolls ET, Lind H, Lind W, Ledingham JGG (1980) Thirst following water deprivation in humans. Am J Physiol 239:R476–R482

Thompson CJ, Baylis PH (1988) Osmoregulation of thirst. J Endocrinol 117:155–157

Thompson CJ, Bland J, Burd J, Baylis PH (1986) The osmotic thresholds for thirst and vasopressin release are similar in healthy man. Clin Sci 71:651–656

Thompson CJ, Burd JM, Baylis PH (1987) Acute suppression of plasma vasopressin and thirst after drinking in hypernatremic man. Am J Physiol 252:R1138–R1142

Thompson CJ, Burd JM, Baylis PH (1988) Osmoregulation of vasopressin secretion and thirst in cyclical oedema. Clin Endocrinol (Oxf) 28:629–635

Vokes TJ, Weiss NM, Schreiber J, Gaskill MB, Robertson GL (1988) Osmoregulation of thirst and vasopressin during the normal menstrual cycle. Am J Physiol 254:R641–R647

Williams TDM, Seckl JR, Lightman SL (1989) Dependent effect of drinking volume on vasopressin but not atrial peptide in humans. Am J Physiol 257:R762–R764

Wolf AV (1950) Osmometric analysis of thirst in man and dog. Am J Physiol 161:75–86

Zerbe RL, Robertson GL (1983) Osmoregulation of thirst and vasopressin secretion in human subjects: effect of various solutes. Am J Physiol 244:E607–E614

Commentary:

McKinley: Similar to rats, dogs and sheep, the subfornical organ and organum vasculosum of the lamina terminalis (OVLT) of the human brain (which lack a blood–brain barrier) are rich in angiotensin receptors as shown by the in vitro binding of ^{125}I-labelled [Sar^1, Ile^8] angiotensin II to sections of human brain (McKinley et al. 1987). This indicates that blood-borne angiotensin II may influence regions of the human brain which may subserve water intake. Additionally, a high concentration of angiotensin-converting enzyme occurs in the human OVLT, indicating the possibility that angiotensin II may be formed there locally from circulating angiotensin I (Chai et al. 1990). It is possible that this locally formed angiotensin II in the OVLT could stimulate thirst.

References

Chai SY, McKenzie JS, McKinley MJ, Mendelsohn FAO (1990) Angiotensin converting enzyme in the human basal forebrain and midbrain visualized by in vitro autoradiography. J Comp Neurol 291: 179–194

McKinley MJ, Allen AM, Clevers J, Paxinos G, Mendelsohn FAO (1987) Angiotensin receptor binding in human hypothalamus: autoradiographic localization. Brain Res 420:375–379

Verbalis: I fully agree with the arguments regarding the relative insensitivity of humans to volaemic stimuli to thirst. Two other points relative to this issue might also be mentioned here. First, vasopressin secretion is also not very potently stimulated by low levels of volume depletion but only by much higher levels in all animals. Second, what hypovolaemic animals and humans need is not only water but solutions containing sodium to restore plasma volume; consequently, what should be precipitated here in addition to thirst is a sodium appetite. Although clinical experience suggests that with pathologically extreme degrees of volume depletion (e.g., haemorrhage, diarrhoea, diabetic ketoacidosis) thirst is in fact precipitated in humans, nonetheless there is no clinical experience to suggest that such patients have any recognizable sodium appetite in comparison with various animal studies which use hypovolaemia as a reliable stimulus to sodium chloride intake. Probably the best example of this is the occurrence of hyponatraemia in humans following diuretic use. In most such cases the hyponatraemia is a dilutional hyponatraemia resulting from ingestion of only water or other dilute fluids in response to the diuretic-induced salt and water losses, certainly not the pattern which would be expected from a stimulated sodium appetite.

Section VII
Variations in Human Fluid Intake

Chapter 26

Thirst and Fluid Intake in the Elderly

P.A. Phillips, C.I. Johnston and L. Gray

Introduction

In all organisms water homeostasis is essential for life and is achieved by a balance between water intake and output. Although water output is regulated closely by renal mechanisms and circulating hormones such as the antidiuretic hormone, arginine vasopressin (AVP), it is only through water intake, controlled by thirst, that water deficits can be replenished. Renal water conservation can only minimize further losses. Should thirst and water intake be diminished, or access to water denied, dehydration and eventually death result.

Alternately, when overhydration exists, thirst and water intake must be diminished to allow body fluid correction through renal free water excretion and insensible water loss. Should thirst and inappropriate fluid intake persist, dangerous overhydration and hyponatraemia may result. The control mechanisms for this balance between thirst and water intake and renal water excretion, are integral to overall water homeostasis, and, as with a number of homeostatic processes such as temperature regulation, blood pressure and salt balance (Phillips et al. 1990), seem to change with age. Thus, although dehydration or hyponatraemia may be seen in all age groups, they are particularly common in the elderly, and may be responsible for significant morbidity and mortality (Anderson et al. 1985; Snyder et al. 1987).

Reduced Thirst with Age

Dehydration is a common cause of fluid and electrolyte disturbance in the elderly occurring in approximately 1% of community hospital admissions (Snyder et al. 1987). It may occur alone, or in the setting of increased water losses such as with surgery, fever, diabetes mellitus, gastrointestinal disease or vomiting (Snyder et al. 1987). Both diminished renal water-conserving ability and diminished thirst contribute to the increased susceptibility of the aged to dehydration.

Reduced urinary concentrating ability is well documented in the elderly (see Epstein 1979; Ledingham et al. 1987) both in response to dehydration or exogenous AVP administration. The reduced urinary concentrating ability following dehydration (Phillips et al. 1984a) is not due to inability to secrete AVP from the neurohypophysis appropriately, as the plasma AVP response to 24-hour water deprivation (Phillips et al. 1984a) or osmotic stimulation with hypertonic saline (Helderman et al. 1978) is maintained or increased in the elderly compared with young control subjects. These findings and the reduced urinary concentrating ability following exogenous AVP

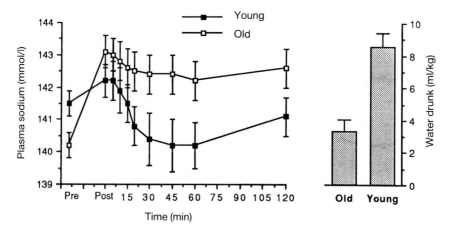

Fig. 26.1. Plasma sodium concentration and total water intake following 24 hours of dehydration in healthy old and young men, showing that only the young men drank sufficient to correct their hypertonicity to predeprivation levels. Water was available during the first 60 minutes of the postdeprivation study period. (Modified from Phillips et al. (1984a).)

administration (Miller and Shock 1953), indicate reduced renal responsiveness to AVP. Whether this is due to reduced glomerular filtration rate secondary to nephron dropout with increased osmotic load and so filtrate flow per remaining nephron, increased renal medullary blood flow and washout of the medullary concentration gradient, or reduced AVP antidiuretic receptor stimulation (as may occur in aged rats (Beck and Yu 1982)) is unknown.

In the face of diminished renal water conservation capacity with age, thirst following dehydration is also reduced in the elderly. After 24 hours of complete water deprivation and consumption of a diet containing less than 70% water by weight, healthy elderly men consumed insufficient water to replenish their body water deficit and return plasma sodium levels to normal (Fig. 26.1) whereas healthy young control subjects drank sufficient water to rehydrate themselves to predeprivation levels (Phillips et al. 1984a). This lack of thirst was despite similar weight loss, and greater increases in plasma sodium and AVP concentrations in the elderly group and was not associated with the usual oropharyngeal symptoms of thirst such as mouth dryness, seen in younger subjects. Similarly, healthy elderly men did not experience increasing thirst over the 7 hours following acute water loading (Crowe et al. 1987) in contrast to significant thirst in the young control subjects. Immediately following the waterload there was suppression of thirst and a fall in plasma sodium and osmolality in both young and old subjects. However, as the excess water was excreted over the following 7 hours (subjects remained fasting throughout) plasma osmolality and sodium concentration increased. As plasma sodium and osmolality rose, thirst ratings increased from their nadir immediately after the load only in the young control subjects (Crowe et al 1987). Furthermore, Miescher and Fortney (1989) demonstrated reduced thirst ratings, slower rehydration and impaired temperature regulation in elderly men undergoing thermal dehydration.

These studies suggest that in the elderly, thirst and water intake responses may be blunted, despite physiological need. Nevertheless, under most circumstances, the water deficit would eventually be replaced by a combination of water intake associated

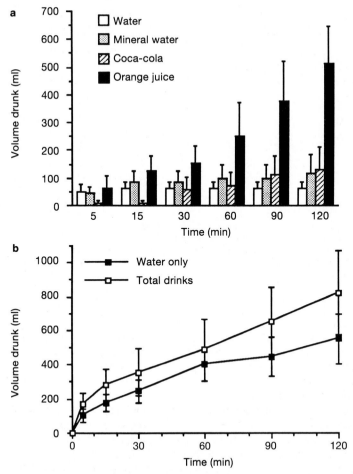

Fig. 26.2. Cumulative intakes of different beverages by healthy elderly men during 2 hours after two periods of 24 hours' dehydration. **a**, intakes of a selection of beverages available ad libitum on one occasion. **b**, total intake of the different beverages consumed after the 24 hour dehydration shown in **a** (□) as well as the intake of room temperature tap water (■) following another 24-hour period of dehydration.

with food and palatable liquids and renal water conservation. However, should excess water loss occur in association with illness, physical incapacity or reduced access to water, then this reduced thirst drive may predispose the elderly to dehydration and hypernatraemia.

Effects of Palatability

In Western societies, most fluid intake is in the form of palatable fluids and not plain water (Chapter 22). Therefore, under most circumstances having access to a selection of palatable fluids may counteract the diminished effect of reduced thirst in the elderly, through fluid intake for taste or other social factors, and not just

Volume drunk (ml)

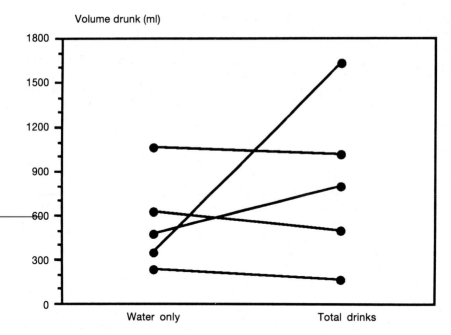

Fig. 26.3. Total fluid intakes in the elderly subjects after two periods of 24 hours' dehydration. On one occasion they drank water only and on the other they drank a variety of palatable fluids (Total drinks). See text for details.

because of thirst. An attempt was made (Phillips, Gray and Johnston, unpublished observations) to examine this by depriving healthy, informed, volunteer elderly men (66–78 years old, $n = 5$) of water on two occasions approximately 3 weeks apart. The men consumed no fluids and ate a diet containing less than 70% water by weight for 24 hours. After the period of dehydration, on one occasion the men had ad libitum access to room temperature tapwater for 2 hours. On the other occasion the men had ad libitum access to all of mineral water (4°C), a carbonated cola drink (Coca-Cola, Australia, 4°C), fresh orange juice (4°C) and room temperature water for 2 hours. The order of offering only room temperature tapwater or a selection of palatable beverages was randomized. On each occasion the men lost equivalent amounts of weight (2% initial body weight), and had identical changes in visual analogue ratings of thirst and mouth dryness. Although the mean intake of the palatable liquids was always greater than the mean intake of water alone, not all subjects drank more of the combined palatable fluids and the difference was not statistically significant. The fluid intakes are shown in Figs. 26.2 and 26.3. There was also no significant difference in the subsequent ratings of thirst and mouth dryness during the periods of rehydration with the different fluid regimens. It should be noted that the differences in mean intake were least early in the rehydration period (e.g. 177 ± 48 ml of water versus 282 ± 95 ml of palatable liquids) when preabsorptive satiety signals were active (see below and Chapter 19), and that the differences in intake tended to increase with time (548 ± 144 ml of water versus 819 ± 247 ml of palatable liquids at the end of the experiment). These results suggest that if fluid access after dehydration is shortlived, then the influence of access to a variety of palatable liquids may not be sufficient to overcome the reduced thirst drive seen in the elderly. However, in the long term, such influences may become more important. This could especially be the case if isotonic palatable beverages are used for rehydration since rehydration of healthy young subjects during exercise in the heat may be more rapid and complete with these fluids than with water alone (Follenius et al. 1989). Since plasma measurements of rehydration were not available in the experiment above, further studies are needed to clarify this.

Possible Mechanisms of the Thirst Deficit

The mechanism of the thirst deficit in the elderly is unclear, but could be at any part of the thirst pathway from the sensors involved in detecting a body water deficit (cerebral osmoreceptors, blood volume/pressure baroreceptors) and the afferent pathways to and within the brain leading to thirst and water intake.

Both thirst and AVP secretion are under similar control mechanisms. Both are mainly stimulated by hypothalamic osmoreceptor neurons (see Chapter 5) which detect changes in body fluid tonicity, and by vascular and cardiac baroreceptors which detect changes in blood pressure and volume (see Chapter 6).

Reduced osmoreceptor sensitivity could account for the changes with age. However, AVP responses to osmotic stimuli are maintained or increased in healthy elderly individuals (Helderman et al. 1978; Phillips et al. 1984a) and it seems unlikely that separate thirst and AVP osmoreceptors would exist and change differently with ageing. Alternatively, pathways from osmoreceptors to the cortex leading to the sensation of thirst could be involved. Miller et al. (1982) have described elderly patients with cerebrovascular disease who were repeatedly hospitalized with dehydration and hypernatraemia due to deficient thirst. All were physically able to obtain water and had no evidence of other hypothalamic–pituitary dysfunction. They suggested that cortical dysfunction secondary to cerebrovascular disease may have led to the patients' profound thirst deficit. In contrast, in the previous studies showing diminished thirst in healthy elderly men (Phillips et al. 1984a; Crowe et al. 1987) there was no clinical evidence of any cerebral or other disorder. A recent finding of altered AVP responses in patients with Alzheimer's disease (Norbiato et al. 1988) showed normal AVP responses to hypotension but reduced AVP osmoreceptor responses to osmotic stimulation compared to age- and sex-matched controls. Since Alzheimer's disease is associated with reduced cognitive function, with reduced communication of a thirst drive, such patients may be at even greater risk of dehydration if AVP responses and renal water conservation are also diminished.

With regard to thirst following hypovolaemia, although high pressure baroreceptor (Gribbin et al. 1971) and low pressure volume receptor (Cleroux et al. 1988) sensitivities decline with age and there is reduced pressure–volume-mediated AVP release in the elderly (Rowe et al. 1982; Bevilacqua et al. 1987) these changes are unlikely to account for the thirst deficit in the elderly. This is because relatively small changes in blood volume ($< 5\%$) occur with water deprivation but large changes are necessary to stimulate thirst even in healthy young subjects (see Chapter 25). For example, during the 5 hours after non-hypotensive single-blind randomized crossover sham haemorrhage ($n = 8$), haemorrhage of 500 ml (approximately 10% of blood volume, $n = 8$), or haemorrhage of 1000 ml (approximately 20% of blood volume, $n = 4$) in fluid-replete fasting recumbent healthy informed volunteer young men 21–23 years old there was no significant difference in thirst ratings. Nor was there any significant difference in room temperature tapwater intake during the last 2 hours of the 5-hour study period when water was made available ad libitum (Phillips, Johnston and Gray, unpublished observations). Although the subjects were recumbent throughout the experiment and were fluid-replete initially, factors which would tend to limit the hypovolaemia sensed centrally, they were observed for several hours. This is important since in animal studies hypovolaemic thirst takes time to become apparent. It, therefore, seems unlikely that the small changes in blood volume occurring after water deprivation would be playing a major role in stimulating thirst, and that blunting of the sensitivity of receptors sensing these deficits in the elderly would contribute to their thirst deficit. Similarly, it has been suggested that reduced activity of the renin–angiotensin system, with reduced hypovolaemia-stimulated plasma angiotensin II levels, could contribute to the hypodipsia of ageing (Yamamoto et al. 1988). However, circulating angiotensin II is not a potent thirst stimulus even in young controls (Phillips et al. 1985b; Chapters 6 and 25) and so this too seems unlikely as the explanation for the reduced thirst in the elderly.

Oropharyngeal sensations such as mouth dryness are commonly associated with thirst (Rolls et al. 1980; Rolls and Rolls 1982; Phillips et al. 1984b, 1985a), and age-related changes in oropharyngeal sensations such as taste are known to occur. In previous studies (Phillips et al. 1984a; Rolls and Phillips 1990) the elderly have shown diminished oropharyngeal symptoms such as mouth dryness following water deprivation when compared with healthy young control subjects. However, the role of oropharyngeal receptors in stimulating thirst is unclear. Oropharyngeal sensations may well just be sensations that come to be associated with body fluid deficits. In contrast, such receptors seem to play a major role in early satiety of thirst and AVP inhibition following drinking before systemic water absorption occurs (Rolls and Rolls 1982; Geelen et al. 1984).

Satiety and Thirst in the Elderly

Not only are the elderly predisposed to dehydration, but they are also predisposed to overhydration and hyponatraemia with significant associated morbidity and mortality (Anderson et al. 1985).

Hyponatraemia in the presence of normal or expanded extracellular fluid volume, occurs because of the inability to excrete the excess free water and because of inappropriate water intake. The reduced ability of the elderly to excrete excess water is well documented (Lindeman et al. 1966; Crowe et al. 1987). This seems due to the decline in glomerular filtration rate and renal function with age and not factors such as inability to suppress plasma AVP (Crowe et al. 1987). This reduced renal water excretory capacity predisposes the elderly to hyponatraemia when inappropriate water input occurs, e.g. with intravenous fluids or in association with food or palatable liquids. Satiety of thirst occurs in the short term in response to oropharyngeal stimuli associated with the act of drinking and gastric distension (Fitzsimons 1979; Rolls and Rolls 1982). In the long term correction of the hypertonicity and/or hypovolaemia leads to removal of the thirst stimulus (Fitzsimons 1979; Rolls and Rolls 1982; Chapter 19). When rats are offered palatable fluids to drink, excess free water is excreted without major changes in plasma sodium concentration (Rolls et al. 1978). However, if renal free water excretion is reduced, for example with exogenous AVP administration, drinking of the palatable fluids persists despite significant hyponatraemia and acute morbidity (Rolls et al. 1978). In fact, the profound effect of palatability and water intake associated with a liquid diet overriding the satiating effect of hyponatraemia on thirst has led to the establishment of an animal model of chronic hyponatraemia (Verbalis and Drutarosky 1988) and is seen to occur in patients with the syndrome of inappropriate antidiuretic hormone secretion and dialysis-dependent renal failure who cannot excrete excess water. Hyponatraemia and fluid overload associated with either excess thirst or intake of palatable liquids can be a significant problem in these situations. Although the acute availability of palatable fluids does not seem to play a major role in improving water deprivation-induced dehydration in the elderly, water intake association with food, habit or under social conditions, such as meals, may be a source of "inappropriate" non-homeostatic excess water intake in the elderly. Even in healthy young subjects ad libitum water intake occurs mainly in association with meals and before any body water deficits occur (Phillips et al. 1984b). Whether oropharyngeal satiety mechanisms are normal in the elderly is unknown. However, since oropharyngeal symptoms of thirst are reduced in the elderly (Phillips et al. 1984a; Rolls and Phillips 1990), and other oropharyngeal senses such as taste change with age (Cowart 1981) it may be that oropharyngeal satiety mechanisms are also diminished with age. This combined with reduced ability to excrete excess water, could predispose the elderly to hyponatraemia and overhydration. This remains to be tested experimentally.

Hyponatraemia may also be associated with hypovolaemia (Anderson et al. 1985), when extracellular fluid sodium losses exceed water losses. Diuretics are a common cause of this clinical problem. Recently, Friedman et al (1989) noted that in 11 patients (56–87 years old) with diuretic-induced hyponatraemia, recurrence of hyponatraemia in the hours following acute rechallenge with the diuretic was associated with significant weight gain presumably due to excess water intake. This finding requires further clarification as to whether any excess water intake was due to thirst or other factors as hypovolaemia per se does not seem to be a potent thirst stimulus in man (see above and Chapter 25).

Conclusions

Awareness of reduced thirst in the elderly has led to advice that the aged should drink extra fluids during hot weather or during periods of fluid loss to prevent dehydration and hypernatraemia. Similarly, the recognition of the reduced capacity of the elderly to excrete excess water has led to recommendations of caution when giving exogenous fluids to the elderly (as with intravenous therapy postoperatively) so as to prevent overhydration and hyponatraemia. With this increasing recognition that the elderly have reduced homeostatic capacity, earlier detection of dehydration or overhydration should occur. However, the mechanisms leading to this reduced homeostatic capacity still require further clarification. Only through such studies might new preventive strategies and therapeutic interventions be developed to prevent these causes of significant morbidity and mortality in the ageing population.

Acknowledgements. The authors gratefully acknowledge support from the National Health and Medical Research Council of Australia and the Sandoz Foundation for Gerontological Research.

References

Anderson RJ, Chung HM, Kluge R, Schrier R (1985) Hyponatremia: a prospective study of its epidemiology and the pathogenetic role of vasopressin. Ann Intern Med 102:164–168

Beck N, Yu BP (1982) Effect of aging on urinary concentrating mechanism and vasopressin dependent cAMP in rats. Am J Physiol 243:F121–F125

Bevilacqua M, Norbiato G, Chebat E et al. (1987) Osmotic and non-osmotic control of vasopressin release in the elderly: effect of metoclopramide. J Clin Endocrinol Metab 65:1243–1247

Cleroux J, Giannattasio C, Grassi G et al. (1988) Effects of ageing on the cardiopulmonary receptor reflex in normotensive humans. J Hypertens 6 (Suppl 4):S141–S144

Cowart BJ (1981) Development of taste perception in humans: sensitivity and preference throughout the life span. Psychol Bull 90:43–73

Crowe MJ, Forsling ML, Rolls BJ, Phillips PA, Ledingham JGG, Smith RF (1987) Altered water excretion in healthy elderly men. Age Aging 16:285–293

Epstein M (1979) Effects of aging on the kidney. Fed Proc 38:168–172

Fitzsimons JT (1979) The physiology of thirst and sodium appetite. Cambridge University Press, Cambridge

Follenius M, Candas V, Bothorel B, Brandenberger G (1989) Effects of rehydration on atrial natriuretic peptide release during exercise in the heat. J Appl Physiol 66:2516–2521

Friedman E, Shadel M, Halkin H, Farfel Z (1989) Thiazide-induced hyponatremia: reproducibility by single dose rechallenge and an analysis of pathogenesis. Ann Intern Med 110:24–30

Geelen GL, Keil LC, Kravik SE et al. (1984) Inhibition of plasma vasopressin after drinking in dehydrated humans. Am J Physiol 247:R967–R971

Gribbin B, Pickering TG, Sleight P, Peto R (1971) Effects of age and high blood pressure on baroreflex sensitivity in man. Circ Res 29:424–431

Helderman JH, Vestal RE, Rowe JW, Tobin JD, Andres R, Robertson GL (1978) The response of arginine vasopressin to intravenous ethanol and hypertonic saline in man: the impact of aging. J Gerontol 33:39–47

Ledingham JGG, Growe MJ, Forsling ML, Phillips PA, Rolls BJ (1987) Effects off aging on vasopressin secretion, water excretion and thirst in man. Kidney Int 32:S90–S92

Lindeman RD, Lee TD, Yiengst MJ, Shock NW (1966) Influence of age, renal disease, hypertension, diuretics and calcium on the antidiuretic responses to suboptimal infusions of vasopressin. J Lab Clin Med 68:206–233

Miescher E, Fortney SM (1989) Responses to dehydration and rehydration during heat exposure in young and older men. Am J Physiol 257:R1050–R1056

Miller JH, Shock NW (1953) Age differences in the renal tubular response to antidiuretic hormone. J Gerontol 8:446–450

Miller PD, Krebs RA, Neal BJ, McIntyre DO (1982) Hypodipsia in geriatric patients. Am J Med 73:354–356

Norbiato G, Bevilacqua M, Carella F et al. (1988) Alterations in vasopressin regulation in Alzheimer's disease. J Neurol Neurosurg Psychiatry 51:903–908

Phillips PA, Rolls BJ, Ledingham JGG et al. (1984a) Reduced thirst following water deprivation in healthy elderly men. N Engl J Med 311:753–759

Phillips PA, Rolls BJ, Ledingham JGG, Morton JJ (1984b) Body fluid changes, thirst and drinking in man during free access to water. Physiol Behav 33:357–363

Phillips PA, Rolls BJ, Ledingham JGG, Forsling ML, Morton JJ (1985a) Osmotic thirst and vasopressin release in man: a double-blind cross-over study. Am J Physiol 248:R645–R650

Phillips PA, Rolls BJ, Ledingham JGG, Morton JJ, Forsling ML (1985b) Angiotensin II induced thirst and vasopressin release in man. Clin Sci 68:669–674

Phillips PA, Hodsman GP, Johnston CI (1990) Neuroendocrine mechanisms and cardiovascular homeostasis in the elderly. Cardiovasc Drugs Ther (in press)

Rolls BJ, Rolls ET (1982) Thirst. Cambridge University Press, Cambridge

Rolls BJ, Phillips PA (1990) Aging and distribution of thirst and fluid balance. Nutr Rev 48:137–144

Rolls BJ, Wood RJ, Stevens RM (1978) Palatability and body fluid homeostasis. Physiol Behav 20:15–19

Rolls RT, Wood RJ, Rolls ET, Lind H, Lind W, Ledingham JGG (1980) Thirst following water deprivation in humans. Am J Physiol 239:R476–R482

Rowe JW, Minaker KL, Sparrow D, Robertson GL (1982) Age-related failure of volume–pressure-mediated vasopressin release. J Clin Endocrinol Metab 54:661–664

Snyder NA, Feigal DW, Arieff AI (1987) Hypernatremia in elderly patients: a heterogenous, morbid and iatrogenic entity. Ann Intern Med 107:309–319

Verbalis JG, Drutarosky MD (1988) Adaptation to chronic hypoosmolality in rats. Kidney Int 34: 351–360
Yamamoto T, Harada H, Fukuyama J, Hayashi T, Mori I (1988) Impaired arginine vasopressin secretion associated with hypoangiotensinemia in hypernatremic dehydrated elderly patients. JAMA 259: 1039–1042

Commentary

Booth: Could it be that because the elderly have had longer experience of life than the middle-aged or young their drinking behaviour is much better learned and they are less enthused by new-fangled or adolescent beverages or indeed by the plain tapwater that few people drink very often?

If so, the lower sensitivity of fluid intake to hypohydration may not be a thirst deficit, at least not in the sense of the defect you imply in the neural networks between the sensors and the effectors. It could be the result of rated overall interest in drinking of interaction of dehydration-induced thirst with the effects of test-drink palatability to the individual, decline of interest in ingestive or other physical activity, or even sensory or motor defects. This is a clear example of the usefulness for physiological analysis of the psychological dissociation I outline in Chapter 4.

Phillips: We do not think that being less enthused about the style of beverage or tapwater accounts for the hypohydration of the elderly. Not only was the volume of water drunk after 24 hours' water deprivation less than in younger subjects, but thirst ratings were less in that original experiment and subsequent experiments. The nature of the mechanism remains unknown.

Chapter 27

The Consequences of Exercise on Thirst and Fluid Intake

J.E. Greenleaf

Introduction

Physical exercise, defined as body movement or body limb movement that increases
energy utilization above the resting (sitting) level, and accompanying physiological
responses can have a profound influence on fluid intake (the only practical means
for rehydration in humans). The degree of thirst sensation (the desire for fluid intake)
in exercising humans seems to be a poor indicator of body fluid content and may or
may not be indicative of fluid requirements, depending on the duration and intensity
of the exercise and on the ambient conditions (Hunt 1912; Adolph and Dill 1938;
Vernon and Warner 1932). In some situations humans drink excessively when
adequately hydrated, and in others they consume inadequate fluid volumes when
dehydrated. Most fluid–electrolyte problems arise from body-fluid deficits, but severe
hyponatraemia (water intoxication) occurs occasionally with forced intake of hypotonic
fluids, for example during intense exercise for many hours without food intake.
There is no evidence of short-term adaptation to dehydration in humans.

The term dehydration used herein means loss of body fluid (negative water balance);
rehydration means restoration of normal body water balance; and hypo- and
hyperhydration refer to steady-state conditions of reduced and increased body water
content, respectively.

This chapter, about the effect of exercise on thirst and fluid intake, will be presented
in three sections: fluid deficits and electrolyte balance, body temperature and fluid
shifts. These three areas of study are not mutually exclusive.

Recent reviews of some aspects of fluid intake during exercise have been prepared
by Greenleaf (1990), Hubbard et al. (1990) and Murray (1987), and Murray discusses
extensively various drink compositions used for rehydration.

Fluid Deficits and Electrolyte Balance

Classical theory holds that the urge to drink in humans is stimulated either separately
or simultaneously by reduction in normal body-water volume (dehydration), and by
increased sodium (osmolality) in the cellular and extracellular fluid spaces (Adolph
and Dill 1938). One model suggests two pathways, involving changes in plasma
osmolality and angiotensin II, that can stimulate drinking (Oatley 1973). The stimuli
are probably additive and include dehydration and administration of hypertonic saline

(osmotic), and haemorrhage, dehydration and administration of polyethylene glycol or hypertonic saline which can affect extracellular fluid volume and stimulate angiotensin II and drinking. Most of the mammalian species studied restore moderate water deficits within minutes, but normal humans and rats may require many hours to do so depending on food (osmotic) consumption. The extended presence of this deficit (termed voluntary dehydration by Adolph and Wills (1947)) is called involuntary dehydration (or involuntary hypohydration) because, presumably, the person voluntarily drinks to assuage thirst or stops drinking because of gastrointestinal discomfort before the deficit is restored. Thus the remaining deficit occurs "involuntarily".

Combined Stress Factors

To determine the relative effects of heat exposure, chronic dehydration, and moderate exercise on the level of involuntary dehydration, fluid intake of four exercise-heat-acclimated young men was measured during a 4-h exposure to all eight possible combinations of heat–cool, hydrated–dehydrated and rest–exercise treatments (Fig. 27.1). The fluid deficits were not similar in each of the eight separate experiments, but these dissimilarities were "factored out" when each factor, exercise, heat, and

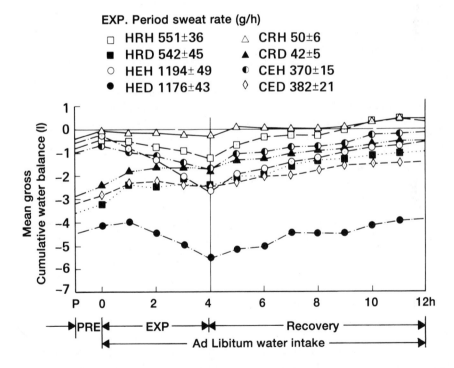

Fig. 27.1. Mean (±SE) sweat rates and cumulative water balances in four men during eight experiments: HRH, heat (Ta=49°C), rest (sitting), hydrated; CED, cool (Ta=24°C), exercise (treadmill 6.4 km/h), dehydrated (4% body wt). Meals were eaten at hours 4 and 10. (Modified from Greenleaf and Sargent (1965).)

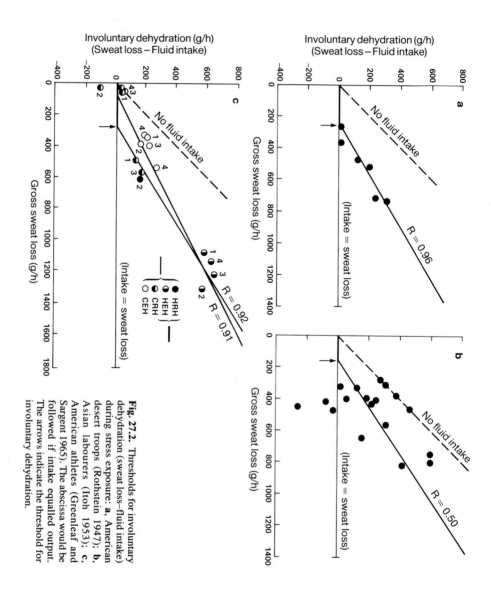

Fig. 27.2. Thresholds for involuntary dehydration (sweat loss–fluid intake) during stress exposure: **a**, American desert troops (Rothstein 1947); **b**, Asian labourers (Itoh 1953); **c**, American athletes (Greenleaf and Sargent 1965). The abscissa would be followed if intake equalled output. The arrows indicate the threshold for involuntary dehydration.

dehydration, was combined mathematically. It was concluded that voluntary cool-water consumption increased 2.5 times in heat versus cool temperature, 2.1 times when subjects were dehydrated versus hydrated, but only 1.4 times during exercise versus rest (Greenleaf and Sargent 1965). Thus, factors resulting from an increase in metabolism (exercise) appear to inhibit drinking more than those from heat exposure or prior dehydration. During recovery after the most stressful experiment (heat–exercise–dehydration) the subjects stated that they felt comfortable and were not particularly thirsty even though they had body-water deficits of about 5 l (Fig. 27.1); note that the slopes of the recovery curves are similar regardless of the level of the 4 h fluid deficits.

Thresholds for Involuntary Dehydration

Thresholds at which involuntary dehydration (loss from sweating minus water intake) becomes positive, i.e. water intake does not equal water output, have been calculated from data from two field studies and one laboratory study (Fig. 27.2). One field study was done with 31 American Air Force ground crew members working in the hot, dry Mohave desert (Rothstein et al. 1947) (Fig. 27.2a), and the other with 11 Japanese plus seven Chinese labourers working in hot, humid tropical conditions (Itoh 1953) (Fig: 27.2b); the laboratory study was done with four physically fit young athletes in various controlled laboratory environmental conditions (Greenleaf and Sargent 1965) (Fig. 27.2c). Despite the wide variability in time, place and environmental conditions, there was good agreement in the thresholds (arrows in figures) for fluid loss (sweating) calculated for the subjects, who were normally hydrated at the beginning of the experiments and were resting and exercising in hot environments: a, 268 g/h, b, 155 g/h, and c, 260 g/h.

In hydrated subjects, the daily induction of involuntary dehydration is normalized by the following day if proper rest and meals are provided. But if hypohydration occurs over a few days by withholding fluids, fluid recovery after a stressful exercise exposure often takes more than 24 h (Greenleaf and Sargent 1965; Greenleaf et al. 1967) (Fig. 27.1), probably because of slower cellular rehydration as a result of decreased cellular potassium and particularly because of water loss associated with glycogen oxidation (Nose et al. 1985; Nielsen et al. 1986). Kozlowski and Saltin (1964) reported greater cellular dehydration when exercise was performed to reduce body water, compared to equivalent weight loss from resting in heat. But the greater cellular dehydration with exercise alone should stimulate more rather than less drinking (Greenleaf 1982). A concomitant increase in plasma osmolality should also stimulate drinking unless some factor associated with exercise retards or eliminates detection of the hyperosmolality by sodium osmoreceptors.

Sodium–Osmotic Hypothesis

Involuntary dehydration is not confined to situations involving only sweating, but sweating is associated with greater inhibition of drinking, especially when it accompanies exercise. Arden (1934) was one of the first to suggest that thirst may be governed or stimulated by sodium alone, because ingestion of potassium chloride or bicarbonate produced no drinking. Gilman (1937) substantiated this finding after injecting subjects with equal osmotic concentrations of sodium chloride and urea: drinking was greater after sodium chloride injection. Similar increases in cellular osmolality were caused by diffusion of urea into the cell, but since sodium chloride could not enter the cell effectively, cellular water was shifted into the extracellular space. The result was a decreased cellular volume which stimulated fluid intake even though cellular osmolality was essentially the same. Later, cerebral osmoreceptors were proposed that control secretion of antidiuretic hormone (Verney 1947). These receptors might be located in a circumventricular organ (Thrasher 1982). Cerebral (anterior third ventricle) sodium-sensitive receptors (Andersson 1953), which might act separately or in co-ordination with osmoreceptors whose sensitivity might be influenced by a cerebral sodium–angiotensin interaction (Fitzsimons 1972; Leksell and Rundgren 1977; Andersson et al. 1980; Phillips et al. 1982), might also affect control of drinking and water balance.

Dill (1938) applied the sodium–osmotic hypothesis to explain the involuntary dehydration that occurs during exercise in a hot environment. During prolonged exercise in heat, previously hydrated, untrained, non-heat-acclimatized subjects had

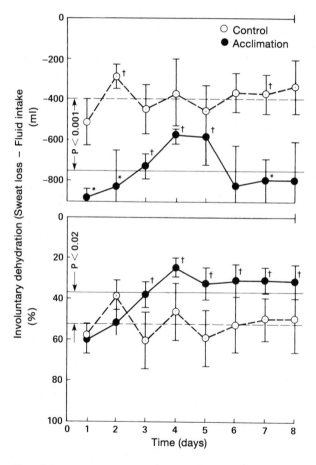

Fig. 27.3. Effects of exercise-heat acclimation on involuntary dehydration. Regimen was five men exercising at 75W for 2 h/day in control (Ta=23.8°C, 50% rh) and acclimation (Ta=39.8°C, 50% rh) experiments. (Modified from Greenleaf et al. (1983).)

a sweat sodium chloride concentration similar to that in their plasma. Thus, loss of isotonic sweat would reduce the extracellular volume with little change in plasma sodium concentration, would effect little change in cellular water content (volume), and would stimulate insufficient drinking (involuntary dehydration) (Rothstein et al. 1947; Greenleaf et al. 1983). After heat-exercise acclimatization, similar exercise in heat produces a significantly greater volume of hypotonic sweat, which also results in a reduced extracellular volume; but there are greater increases in extracellular sodium and osmotic concentrations, which would cause cellular dehydration, increased cellular osmolality, and greater fluid intake. Increased fluid intake during laboratory acclimation is characterized by a progressively shortened time to the first drink, an increase in the number of drinks per exercise session (Greenleaf et al. 1983), and an increase in the volume of each drink (Adolph and Wills 1947). In previously dehydrated and hyperosmotic subjects, access to water at the beginning of exercise results in a large immediate intake to replace the lost water (Fig. 27.3). Thus the sodium–osmotic hypothesis in conjunction with hypovolaemia seems well-established as a general explanation for drinking or lack of it in subjects in stressful situations (Dill 1938; Greenleaf et al. 1967; Nose et al. 1985; Hubbard et al. 1990). Similar data on the effect of angiotensin II on drinking during exercise are not available.

Osmometric analysis of water balance and drinking (Wolf 1950; Hubbard et al. 1990) assumes that, in general, cells behave like perfect osmometers, that body fluid compartments and their osmotic concentrations can be measured reasonably accurately, and that the normal level of body water can be determined. During stressful exposure to heat, altitude, $+G_z$ acceleration, head-up tilting, and submaximal and maximal exercise, mean corpuscular volume is unchanged within a range of plasma osmolality between -1 and $+13$ mosmol/kg of the normal level (Greenleaf 1979). Muscle cells might or might not respond similarly. Whereas plasma volume can be measured to within ± 25 ml (Greenleaf 1979), there are proportionately much greater errors in the measurements of extracellular and total body water volumes; interstitial and cellular volumes have not been measured accurately, if at all. Because of measurement imprecision, circadian factors, varying rates of water and osmotic losses (dermal, renal), intermittent food consumption, and variable non-physiological fluid consumption, like the fictitious "normal" body temperature, there can be no "normal" level of body water, only an average level over a specified time. Thus, osmometric analyses can be only moderately good estimates at best.

Free Circulating Water

Ladell (1955) has suggested that there are about 2.5 l of "free circulating" water in the extracellular fluid compartment that can be drawn upon before intracellular loss occurs. Hubbard et al. (1990), from osmometric calculations, have concluded that in a "fully hydrated" resting subject with a plasma osmolality of 280 mosmol/kg and 42 l of total body water, a 2.1 l water loss will be incurred before the thirst threshold of 295 mosmol/kg is reached. If this is also true during exercise, perhaps involuntary dehydration results partly from a normal delay in rehydration: it is not necessary to restore the 2.1–2.5 l immediately. This suggests that it is an intracellular fluid deficit and/or intracellular hyperosmolality that stimulates or inhibits drinking during exercise. The free circulating water might act as an "anatomical canteen", which is helpful in providing for fluid needs in stressful situations, but it does not seem reasonable that a portion of the largest and most important single compound in the body is controlled in such an apparently imprecise manner whereas the plasma volume is controlled day-to-day to at least ±25 ml (Greenleaf 1979). It is unclear where these 2.5 l are located in the vascular and interstitial fluid spaces and by what mechanism the fluid is transferred and restored.

Physical Fitness

There is ample evidence that voluntary fluid intake varies widely when men of various levels of physical fitness and heat acclimatization are subjected to many hours of strenuous hiking in a hot environment with water available either ad libitum during exercise (Brown 1947; Rothstein et al. 1947; Kuno 1956) (Fig. 27.2a) or only after exertion (Itoh 1953) (Fig. 27.2b). In the latter study some men drank nothing and some consumed more (by 500–700 ml) than they lost. On the other hand, men who are fully acclimatized to hard exercise in the heat can consume very large volumes of fluid. For example, Japanese steel workers consume 7–9 l of water in an 8-h shift (Kuno 1956), and desert troops and railroad repair crews up to 12 l per 24 h (Brown 1947). Reduction of involuntary dehydration can occur during one continuous 6-h exercise exposure in heat: from 34% replacement in the first hour to 68% replacement in hour 6 (Rothstein et al. 1947); and by 40% over 17 days in Polish steelworkers (Spioch and Nowara 1980). A similar reduction of involuntary dehydration occurred in laboratory experiments in which acclimating men exercised for 2 h/day for eight consecutive days in the heat. Fluid replacement was 40% (450 ml/h) on day 1 and increased to 1000–1200 ml/h on days 4–8 after only 8 h of intermittent exposure (Greenleaf et al. 1983) (Fig. 27.3). Increased drinking thus requires 6–8 hr of exposure and does not require intermittent rest and recovery periods.

Gastrointestinal Factors

One, perhaps the initial, inhibitory factor for involuntary dehydration could be the sensitivity of the mechanoreceptors in the stomach that govern the feeling of fullness and the normal, maximal rate of gastric emptying which varies from 900 to 1200 ml/h (Davenport 1982). As mentioned earlier, gastric emptying probably reaches about 1200 ml/h during acclimation (Greenleaf et al. 1983); as sweating increases, so does drinking. The rate of gastric emptying is fastest with water; and it is increased with exercise intensities between 28% and 65% of the maximal oxygen uptake (Vo_{2max}) in a neutral environment and inhibited beyond 65% of Vo_{2max} (Neufer et al. 1989b). Emptying is retarded during similar exercise in the heat (49°C) and at 35°C when subjects are hypohydrated (Neufer et al. 1989a). Gastric emptying is inversely proportional to its caloric content and to its osmolality. The rate of intestinal absorption is about ten times faster than gastric absorption.

Body Temperature

The proximity of sodium–osmotic and thermoregulatory centres in the brain suggests interactive control. Hypernatraemia is associated with increased body temperature, and hyponatraemia and hypercalcaemia with attenuated increases or with decreased temperature (Greenleaf 1979). Extracellular hyperosmolality induced in dogs with mannitol, with no change in plasma sodium concentration, results in excessive increase in rectal temperature during exercise but not at rest (Kozlowski et al. 1980). Hypervolaemia and increased body temperature suppress osmotic-induced thirst (Szczepanska-Sadowska 1979). Thus, interaction between sodium, other osmols, body temperature and thirst and drinking seems established.

In spite of copious evidence that dehydration and hypohydration adversely affect thermoregulation and physical performance (Adolph and Wills 1947; Greenleaf et al. 1979), the idea that drinking during stressful exertion may be harmful has not been refuted completely. Kuno (1956) had discussed the practice of Japanese workers and athletes avoiding "excessive" drinking during exertion because it hastens fatigue. He suggested that exercise-induced increases in body temperature cause vasodilation in muscle and skin with compensatory vasoconstriction in splanchnic vascular beds. This vasoconstriction may be effective, when there is some dehydration, in keeping "unused" organs at rest to retard heat production. Thus drinking during exercise should cause greater hyperthermia, but experimental evidence suggests the contrary (Pitts et al. 1944; Greenleaf et al. 1979). Drinking during exercise should require additional splanchnic blood flow which would shift blood from active muscles and skin, and this might impair thermoregulation somewhat.

Fluid Shifts

Performance of mild to intensive exercise initially causes isotonic hypovolaemia with hyperproteinaemia (Greenleaf et al. 1977), with a concomitant increase in the equilibrium level of core temperature which is directly proportional to absolute exercise intensity. Fluid leaving the vascular space is transferred into active muscle cells. Inhibition of thirst and drinking during exercise could be caused by the net effect of hypovolaemia which would tend to stimulate drinking, and hyperthermia plus the "exercise factor", which would inhibit thirst and drinking. Clearly the latter prevail. Perhaps the plasma volume portion of the free circulating water (20% of 2.5 l = 500 ml) must be depleted by an exercise-induced increase in hydrostatic pressure, causing fluid movement to interstitial and cellular spaces, or by sweating, before significant stimulation of drinking occurs. A vascular volume of 500 ml is transferred in an hour at an exercise intensity of about 75% of the Vo_{2max}. With exercise times greater than 1 h and intensities greater than 45% of Vo_{2max} (Greenleaf et al. 1977), hypernatraemia and hyperosmotaemia occur which would tend to stimulate thirst and drinking. Thus, hydrostatic and exercise factors would accentuate movement of plasma free circulating water.

Exercise-induced shift of vascular to interstitial isotonic fluid is a fast way to increase plasma total protein (enzymes and hormones) concentration, without changing the electrolyte concentrations, required to facilitate homeostasis during the sudden increase in metabolism and changes in flow and distribution of blood. Thus, it appears better to rehydrate voluntarily, not forcefully, during exercise, rather than to hyperhydrate before (Candas and Brandenberger 1989) or during (Frizzell et al. 1986) exercise. Cases of severe hyponatraemia (water intoxication) have been reported after ultramarathon races (> 8 h) in which there was forced intake of hypotonic fluids (Noakes et al. 1985; Frizzell et al. 1986).

Summary

Exposure to exercise and environmental stress causes intercompartmental fluid shifts, loss of body water and extended delay in fluid replacement by drinking (involuntary dehydration), especially when sweating occurs. Sodium–osmotic and volume-depletion stimuli induce thirst and drinking during and after exercise. Exercise-induced involuntary dehydration is associated with and may be caused by depletion of extracellular sodium and other osmols by sweating. Gastric capacity probably limits intake of large fluid volumes, and exercise-induced hyperthermia appears to affect CNS control of drinking. Some deficit in replacement may be due to "response competition", that is, people are too busy to make the effort to drink. Why most mammals restore fluid deficits quickly whereas humans do not is still unresolved.

References

Adolph EF, Dill DB (1938) Observations on water metabolism in the desert. Am J Physiol 123:369–378

Adolph EF, Wills JH (1947) Thirst. In: Adolph EF and associates (eds) Physiology of man in the desert. Interscience Publishers Inc, New York, pp 241–253

Andersson B (1953) The effect of injections of hypertonic NaCl-solutions into different parts of the hypothalamus of goats. Acta Physiol Scand 28:188–201

Andersson B, Olsson K, Rundgren M (1980) ADH in regulation of blood osmolality and extracellular fluid volume. JPEN 4:88–96

Arden F (1934) Experimental observations upon thirst and on potassium overdosage. Aust J Exp Biol Med Sci 12:121–122

Brown AH (1947) Fluid intakes in the desert. In: Adolph EF and associates (eds) Physiology of man in the desert. Interscience Publishers Inc, New York, pp 110–135

Candas V, Brandenberger G (1989) Hydration level during exercise: thermoregulatory and endocrine responses. Progress in biometeorology. SPR Academic Publishing bv, The Hague, The Netherlands, pp 129–141 (Milestones in environmental physiology, vol 7)

Davenport HW (1982) Physiology of the digestive tract, 5th edn. Yearbook Medical Publishers, Chicago

Dill DB (1938) Life, heat, and altitude. Harvard University Press, Cambridge, MA

Fitzsimons JT (1972) Thirst. Physiol Rev 52:468–561

Frizzell RT, Lang GH, Lowance DC et al. (1986) Hyponatremia and ultramarathon running. JAMA 255:772–774

Gilman A (1937) The relation between blood osmotic pressure, fluid distribution and voluntary water intake. Am J Physiol 120:323–328

Greenleaf JE (1979) Hyperthermia and exercise. In: Robertshaw D (ed) International Review of Physiology. University Park Press, Baltimore, pp 157–208 (Environ Physiol III, vol 20)

Greenleaf JE (1982) Dehydration-induced drinking in humans. Fed Proc 41:2509–2514

Greenleaf JE (1990) Environmental issues that influence intake of replacement beverages. In: The use of carbohydrate electrolyte solutions by soldiers in the field. National Academy of Sciences Press, Washington DC, pp XV-1–XV-30

Greenleaf JE, Sargent F II (1965) Voluntary dehydration in man. J Appl Physiol 20:719–724

Greenleaf JE, Douglas LG, Bosco JS et al. (1967) Thirst and artificial acclimatization in man. Int J Biometeor 11:311–322

Greenleaf JE, Convertino VA, Stremel RW et al. (1977) Plasma [Na$^+$], [Ca^{2+}], and volume shifts and thermoregulation during exercise in man. J Appl Physiol 43:1026–1032

Greenleaf JE, Convertino VA, Mangseth GR (1979) Plasma volume during stress in man: osmolality and red cell volume. J Appl Physiol 47:1031–1038

Greenleaf JE, Brock PJ, Keil LC et al. (1983) Drinking and water balance during exercise and heat acclimation. J Appl Physiol 54:414–419

Hubbard RW, Szlyk PC, Armstrong LE (1990) Influence of thirst and fluid palatability on fluid ingestion during exercise. In: Gisolfi CV, Lamb DR (eds) Fluid homeostasis during exercise. Benchmark Press, Indianapolis, IN, pp 39–95 (Perspectives in exercise science and sports medicine, vol 3)

Hunt EH (1912) The regulation of body temperature in extremes of dry heat. J Hyg (Lond) 12:479–488

Itoh S (1953) The water loss and blood changes by prolonged sweating without intake of food and drink. Jpn J Physiol 3:148–156

Kozlowski S, Saltin B (1964) Effect of sweat loss on body fluids. J Appl Physiol 19:1119–1124

Kozlowski S, Greenleaf JE, Turlejska E et al. (1980) Extracellular hyperosmolality and body temperature during physical exercise in dogs. Am J Physiol 239:R180–R183

Kuno Y (1956) Human perspiration. CC Thomas, Springfield, Illinois

Ladell WSS (1955) The effects of water and salt intake upon the performance of men working in hot and humid environments. J Physiol 127:11–46

Leksell LG, Rundgren M (1977) Cerebral sodium–angiotensin interaction demonstrated with "subthreshold" amounts of angiotensin II. Acta Physiol Scand 100:494–496

Murray R (1987) The effects of consuming carbohydrate–electrolyte beverages on gastric emptying and fluid absorption during and following exercise. Sports Med 4:322–351

Neufer PD, Young AJ, Sawka MN (1989a) Gastric emptying during exercise: effects of heat stress and hypohydration. Eur J Appl Physiol 58:433–439

Neufer PD, Young AJ, Sawka MN (1989b) Gastric emptying during walking and running: effects of varied exercise intensity. Eur J Appl Physiol 58:440–445

Nielsen B, Sjogaard G, Ugelvig J et al. (1986) Fluid balance in exercise dehydration and rehydration with different glucose-electrolyte drinks. Eur J Appl Physiol 55:318–325

Noakes TD, Goodwin N, Rayner BL et al. (1985) Water intoxication: a possible complication during endurance exercise. Med Sci Sports Exerc 17:370–375

Nose H, Yawata T, Morimoto T (1985) Osmotic factors in restitution from thermal dehydration in rats. Am J Physiol 249:R166–R171

Oatley K (1973) Stimulation and theory of thirst. In: Epstein AN, Kissileff HR, Stellar E (eds) The neuropsychology of thirst: new findings and advances in concepts. Winston, Washington, DC, pp 199–223

Phillips, MI, Hoffman WE, Bealer SL (1982) Dehydration and fluid balance: central effects of angiotensin. Fed Proc 41:2520–2527

Pitts GC, Johnson RE, Consolazio FC (1944) Work in the heat as affected by intake of water, salt and glucose. Am J Physiol 142:253–259

Rothstein A, Adolph EF, Wills JH (1947) Voluntary dehydration. In: Adolph EF and associates (eds) Physiology of man in the desert. Interscience Publishers Inc, New York, pp 254–270

Spioch FM, Nowara M (1980) Voluntary dehydration in men working in heat. Int Arch Occup Environ Health 46:233–239

Szczepanska-Sadowska E (1979) Neurohormonal control of thirst. Acta Physiol Pol 30 (Suppl 19):39–53

Thrasher TN (1982) Osmoreceptor mediation of thirst and vasopressin secretion in the dog. Fed Proc 41:2528–2532

Verney EB (1947) The antidiuretic hormone and factors which determine its release. Proc R Soc Lond [Biol] 135:25–106

Vernon HM, Warner CG (1932) The influence of the humidity of the air on capacity for work at high temperatures. J Hyg (Lond) 32:431–463

Wolf AV (1950) Osmometric analysis of thirst in man and dog. Am J Physiol 161:75–86

Chapter 28

Effects of Environmental Stresses and Privations on Thirst

M.J. Fregly and N.E. Rowland

Introduction

The adverse environment that is most closely associated with thirst in the minds of most laymen as well as scientists is a hot, dry environment, e.g. the desert. Descriptions of the tortures of thirst in such an environment when water is not available can be found in many novels and other publications. Although less well known, other adverse environments can also affect thirst and drinking, e.g. cold and hypoxia. An objective of this review is to attempt a comparison among these three environments with respect to their effects on fluid exchange and drinking in experimental animals. Where possible, the mechanisms contributing to the changes observed will be discussed.

Whether a single physiological mechanism can be invoked to induce water intake in mammals is unknown at present. At least six distinct stimuli have been used experimentally for this purpose: (a) hyperosmotic solutions; (b) beta-adrenergic agonists, especially isoproterenol; (c) the octapeptide, angiotensin II; (d) the hyperoncotic colloid, polyethylene glycol; (e) parasympathomimetic agents; and (f) dehydration and exposure to heat. The diversity of these stimuli and the differences in their responsiveness from different routes of administration suggest the difficulty in ascribing the induction of water intake to a single physiological mechanism. Indeed, with such a basic function as drinking, there is likely to be more than one underlying mechanism.

As a working model, Oatley (1973) proposed that experimentally induced thirst and drinking may arise in rats by either or both of two separate pathways: an osmoreceptor pathway and an angiotensin II (Ang II) pathway (Fig. 28.1). Stimuli that affect extracellular fluid (ECF) volume, such as haemorrhage, dehydration, and administration of polyethylene glycol can initiate drinking. Stimuli that affect ECF osmolality, such as dehydration and administration of hyperosmotic solutions, also affect the intracellular fluid (ICF) volume. The classical technique to induce drinking is water deprivation with its consequent dehydration, where the ECF volume decreases and plasma osmolality increases. An increase in the osmolality of ECF is an adequate stimulus to induce drinking by way of stimulation of osmoreceptors. Other pathways for induction of drinking most likely exist, as evidenced by the reports that nephrectomized rats drink following peritoneal dialysis with hyperoncotic solutions and certain other conditions (Fitzsimons 1979; Stricker 1978).

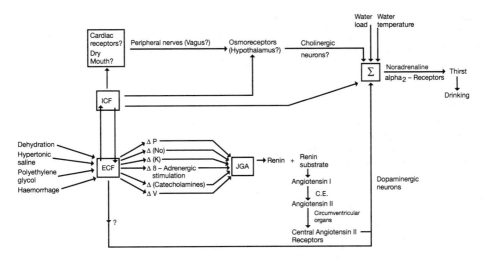

Fig. 28.1. A schema to characterize the potential pathways for the induction of thirst and drinking in rats. See text for explanation. (Adapted from Greenleaf and Fregly (1982).)

Other stimuli that induce drinking, such as a decrease in blood pressure (ΔP) and/ or volume (ΔV), a decrease in plasma sodium concentration ($\Delta[Na]$), an increase in plasma potassium concentration ($\Delta[K]$), or beta-adrenergic stimulation via adrenergic nerves or plasma, initiate release of renin from the juxtaglomerular apparatus (JGA) of the kidneys (Fig. 28.1). After the release of renin into the blood, Ang II is formed and gains access to receptors in the brain. Dopaminergic neurons, among others, may mediate this response (Fig. 28.1) (Block and Fisher 1975; Fitzsimons and Setler 1975; Rowland and Engle 1977; Fregly and Rowland 1988).

The relative importance of each pathway in the control of fluid intake is unknown. In some cases, e.g. dehydration and cold exposure, several pathways appear to be activated, as discussed below. As indicated in Fig. 28.l, all pathways are designated as terminating in a summing device, presumably located in the brain. This device is postulated because stimulation of both osmoreceptor and angiotensin II pathways at the same time always results in a summation of drinking rather than an interaction (Oatley 1973). Studies from this laboratory have probed the pathways beyond the summing device that may be responsible for mediating the drinking response (Fregly and Kelleher 1980; Fregly et al. 1984; Wilson et al. 1984; Fregly and Rowland 1986). The results of these studies indicate that noradrenaline is the neurotransmitter of the final common pathway, and that the alpha$_2$-adrenoceptor agonist, clonidine, administered either centrally or peripherally, can inhibit all types of experimentally induced drinking (Fregly and Kelleher 1980; Fregly and Rowland 1986).

Dehydration-Induced Drinking

When offered water at the end of a period of water deprivation, rats ingest the water immediately (Ramsay et al. 1977; Fitzsimons 1979). The amount ingested during the first hour of access to water is a function of the length of time the animals were deprived of water (Adolph et al. 1954). There appear to be at least two components mediating deprivation-induced water intake: an intracellular component relating to changes in cellular osmolarity/sodium concentration/volume (osmoreceptor pathway) and an extracellular component relating to changes in plasma volume (Ang II pathway) (Fig. 28.1) (Ramsay et al. 1977; Fitzsimons 1979; Greenleaf and Fregly 1982). During water deprivation, both osmolality of the plasma, plasma renin activity, and the concentration of Ang II in plasma increase (Yamaguchi 1981; Barney et al. 1983). Activation of both pathways most likely contributes to drinking under these conditions. However, the sequence of activation and relative contribution of each pathway to the drinking response required elucidation.

One approach to a determination of the contribution of each of the pathways discussed above to dehydration-induced drinking is to study the effect of administration of either Ang II antagonists or inhibitors of the Ang I converting enzyme (ACE) on the drinking response of dehydrated animals. Thus, saralasin, an Ang II antagonist, has been shown to inhibit the water intake normally induced by administration of Ang II (Ramsay and Reid 1975; Abraham et al. 1976; Lee et al. 1981). However, administration of saralasin, either centrally or peripherally, to dehydrated rats, dogs, sheep and goats appears to have no effect on water intake (Olsson 1975; Ramsay and Reid 1975; Abraham et al. 1976; Severs et al. 1977; Lee et al. 1981), although there is one report that saralasin decreased dehydration-induced drinking (Malvin et al. 1977). Experiments with centrally administered teprotide, one of the early ACE inhibitors, also failed to show any depression of dehydration-induced water intake (Lehr et al. 1973; Severs and Klase 1978).

More potent ACE inhibitors, such as captopril and enalapril, have since been developed, and peripheral administration of these agents partially suppresses dehydration-induced drinking (Barney et al. 1980; Fregly et al. 1982). These agents also have no effect on either hypertonic saline- or Ang II-induced drinking, but completely suppress Ang I-induced drinking, attesting to their pharmacological and behavioural specificity in this regard. It is clear, however, that converting enzyme inhibitors not only inhibit the conversion of Ang I into Ang II, but inhibit the metabolism of bradykinin as well. This peptide is dipsogenic under these conditions (Fregly et al. 1981). The resulting increased half-life of bradykinin would be expected to increase water intake, and thereby counteract the antidipsogenic effect of captopril in dehydrated rats. Since the net effect observed is an inhibition of drinking, it can be assumed that the inhibition of the formation of Ang II is more potent than the dipsogenic effect of bradykinin. Thus, with the caveat that pharmacological agonists and antagonists are seldom specific, it would appear that Ang II contributes partially to dehydration-induced drinking.

However, the picture is slightly more complex because the relative contribution of the extracellular (Ang II) and intracellular (osmoreceptor) components of thirst in the control of water intake in water-deprived rats appears to vary with the duration of dehydration (Barney et al. 1983). Thus, when rats were deprived of water, but not food, for 12, 24, 36 and 48 h, administration of captopril (50 mg/kg body weight) 1 h prior to return of water, reduced water intake significantly after 24, 36 and 48 h of dehydration, but not after 12 h. In subsequent tests, measurement of plasma osmolality and plasma renin activity (PRA) in similarly treated rats revealed a significant ($P<0.01$) increase in plasma osmolality within the first 12 h of water deprivation, but a significant ($P<0.05$) increase in PRA only after 24 h of deprivation (Fig. 28.2a,b). These results suggest that the drinking response following the initial 12 h of water deprivation is more dependent on changes in plasma osmolality and ICF volume (osmoreceptor pathway) than on the renin–angiotensin system and the

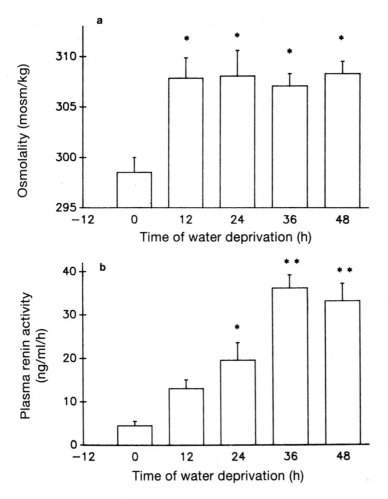

Fig. 28.2. Mean serum osmolality (**a**) and plasma renin activity (**b**) of rats deprived of water for 0, 12, 24, 36 and 48 h, $n = 6$ for each group. One SEM is set off at each mean. Asterisks represent significant ($P < 0.05$) differences from 0 h group. (Adapted from Barney et al. (1983).)

Ang II pathway. As the duration of water deprivation increases up to 48 h, the magnitude of the water intake attributable to an intracellular component (osmoreceptor pathway), as assessed by plasma osmolality, appears to remain at the level observed after 12 h of water deprivation. Thus, the magnitude of the osmoreceptor-dependent (i.e., captopril-insensitive) component of the deprivation-induced water intake did not change with the duration of water-deprivation; at least in the range of 12–48 h.

Effect of Certain Pharmacological Agents on Dehydration-Induced Drinking

The schema in Fig. 28.1 predicts that drugs that inhibit the release of renin from the kidneys should inhibit drinking mediated by the Ang II pathway, provided that the

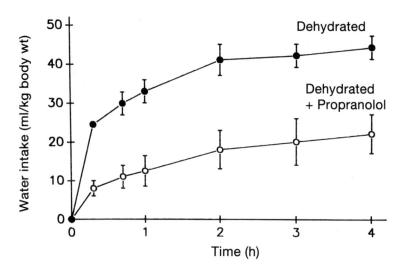

Fig. 28.3. Effect of acute administration of the beta-adrenoceptor antagonist, propranolol (6 mg/kg, i.p.), on water intake of rats dehydrated for 24 h. One SEM is set off at each mean.

dipsogenic stimulus acts to initiate release of renin. Since dehydration-induced drinking is mediated both by the osmoreceptor pathway and the Ang II pathway, it would be expected that dehydration-induced drinking should at least be reduced by administration of a beta-adrenoceptor antagonist known to inhibit the release of renin from the kidneys. That such is the case has been confirmed (Fig. 28.3). The results of this study show clearly that acute administration of propranolol resulted in a reduction in water intake that persisted throughout the 4-h experiment. Water intake of the propranolol-treated group was 40%–50% of that of controls throughout the test. This level of reduction of dehydration-induced drinking by propranolol is comparable to that induced by captopril after a 24-h dehydration (Barney et al. 1980). This suggests that the contribution of the Ang II pathway to drinking after a 24-h dehydration represents about 40% of the total stimulus for drinking. The remaining 60% is apparently contributed by the osmoreceptor pathway.

    Since the osmoreceptor pathway is presumed to use cholinergic neurons to initiate drinking (Fig. 28.1), it was of interest to determine the effect of the cholinergic antagonist, atropine sulphate, on dehydration-induced drinking (Table 28.1). Administration of atropine at 6 and 12 mg/kg, i.p., 30 min prior to return of water inhibited significantly ($P<0.05$–0.01) water intake measured at 1, 2, and 3 h after return of water to the dehydrated rats. At the higher dose used (12 mg/kg), water intake was reduced by approximately 50% of that of the untreated, dehydrated group. This suggests that the osmoreceptor pathway may account for about 50% of the total fluid ingested during the first hour of access to water following a 24-h dehydration. When considered in combination with the results of the experiment above in which propranolol was administered, these results suggest that there was roughly a 50% contribution to drinking after a 24-h dehydration by the osmoreceptor and Ang II pathways, respectively. Barney et al (1983), showed that plasma osmolality after 24 h of dehydration was maximal while plasma renin activity had achieved slightly more than half its maximal value (Fig. 28.2). It would be of interest to measure the

Table 28.1. Effect of the cholinergic antagonist, atropine sulphate, on dehydration-induced water intake by female rats[a]

Experimental group	No. of rats	Mean body wt. (g)	Cumulative water intake (ml/kg body wt.) during:		
			1h	2h	3h
Control	6	248 ± 6[b]	1.2 ± 0.4[d]	1.6 ± 0.4[d]	1.6 ± 0.4[d]
Dehydrated	6	258 ± 8	29.4 ± 3.6	29.6 ± 3.7	32.5 ± 4.8
Dehydrated + atropine (6 mg//kg, i.p.)	6	255 ± 11	21.9 ± 2.2[c]	22.0 ± 2.2[c]	23.0 ± 2.3[c]
Dehydrated + atropine (12 mg/kg, i.p.)	6	252 ± 6	14.5 ± 1.4[d]	15.2 ± 1.6[d]	17.5 ± 1.8[d]

[a] Dehydrated for 25 h. Atropine administered 30 min prior to availability of water.
[b] 1 SEM.
[c] Significantly different from dehydrated group ($P<0.05$).
[d] Significantly different from dehydrated group ($P<0.01$).

effect of the two inhibitors used above in rats dehydrated for 48 h, when both plasma osmolality and plasma renin activity are at their maximal levels.

Additional pharmacological and physiological studies support this "dual-depletion" model of dehydration-induced drinking. Administration of dopaminergic antagonists suppresses drinking to different degrees depending on the nature of the stimulus. Ang II-related drinking is particularly susceptible to blockade by dopaminergic antagonists (as well as lesions of central dopaminergic systems), yet, at an effective dose in this system, dehydration-induced drinking is only partially reduced (Block and Fisher 1975).

Rolls et al. (1980) used a relative rehydration model to assess the contribution of the two pathways of drinking after a 24-h dehydration in dogs. They found that intracarotid infusion of water, to normalize cerebral osmolality, suppressed drinking by some 70%; conversely peripheral administration of isotonic saline to restore vascular volume inhibited drinking by some 30%. These relative numbers compare reasonably with those observed in the pharmacological studies in rats, discussed above.

It is also likely that substantial species differences exist. Although dehydration is a potent stimulus to drinking in all species studied, the relative depletion of the various compartments may differ (Rolls et al. 1980). It would be of interest to assess by pharmacological means the relative contributions of the two pathways of drinking, especially in those species that are behaviourally unresponsive to exogenous administration of Ang II.

With respect to humans, we are not aware of any studies that have attempted to analyse in a similar fashion the components contributing to dehydration-induced drinking.

Effect of Water Temperature on Dehydration-Induced Drinking

After a period of water deprivation, the presentation of water results in different rates of rehydration among the various animal species. Thus, dogs, cats, rabbits, and burros rehydrate within approximately 15–30 min after presentation of water, but

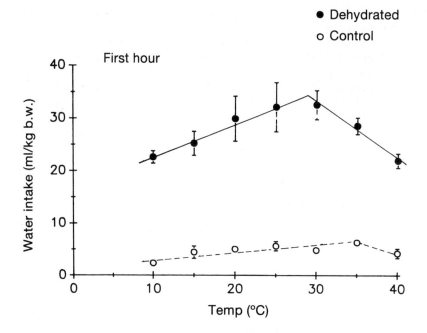

Fig. 28.4. Effect of temperature of water on water intake of rats dehydrated for 24 h and controls. Maximal intake of water appears to be ingested at approximately body temperature. One SEM is set off at each mean.

rats, hamsters, guinea pigs and man require approximately 1–2 h to rehydrate to predehydration levels (Adolph 1950, 1957). It is noteworthy that the temperature of the water presented to dehydrated rats influences significantly the amount of water ingested (Kapatos and Gold 1972). A wide range of temperatures was studied and it was found that maximal water intake occurred when the temperature of the water was nearly at body temperature (30–35°C) (Fig. 28.4). This study also shows that the preference-aversion relationship of water temperature on water intake is not a special aspect of dehydration, but is a normal function of water intake by euhydrated as well as dehydrated rats.

Others have also reported a relationship between water intake and water temperature (Nelson et al. 1974; Fregly et al. 1979). It seems clear that water temperature may be an important factor in the rate at which rehydration occurs and may be of special significance to those species requiring 1–2 h to rehydrate. The question also arises as to whether the preference–aversion relationship between water temperature and water intake in those animals that rehydrate quickly after dehydration is different from those that do not. This remains for further study. Of particular interest is the possibility that a species requiring 1–2 h to rehydrate after dehydration might rehydrate more quickly if water at an optimal temperature was made available, rather than water at colder temperatures. In the case of man, studies on French subjects have been carried out by Boulze et al. (1983) suggesting that the optimum temperature needed to maximize both intake and preference is 15°C. Another study by Sandick et al. (1984), carried out on dehydrated Americans, suggests that most water was drunk when the temperature was 5°C. Thus, there may be cultural differences in the temperature of water preferred by dehydrated subjects.

The mechanism by which water temperature affects water intake is incompletely understood. Kapatos and Gold (1972) suggested that ingestion of water below body temperature cools the tongue and signals satiation. Mendelson and Chillag (1970) showed that thirsty rodents licked cold, dry metal in preference to metal maintained at either room or body temperature. Water was not available to the animals during the sessions. These investigators concluded that cooling of the tongue was a primary reward for thirsty rodents. Additional study is needed to understand this behaviour more fully.

Heat-Induced Drinking

One of the principal physiological means by which mammals maintain thermal homeostasis in the heat is by evaporative cooling. This involves loss of body fluid via either panting, sweating, or spreading of saliva, and is dependent upon an adequate initial level of hydration (Hainsworth et al. 1968). It is also evident, from long-term balance studies, that thirst is stimulated rather precisely to replace these fluid losses (Adolph 1957).

Hainsworth et al. (1968) reported the effects of exposing rats to a temperature of 40°C for up to 6 h without water. Animals that survived this full period of exposure spread saliva and maintained a rectal temperature of 41 ± 1°C. Their evaporative water loss was about 13 mg/g per h, or 19.5 ml/5 h per 300 g rat. This volume approximates 6.5% of body weight and exceeds the initial plasma volume. The plasma became hyperosmotic and hypovolaemic, with the most rapid changes occurring early in the exposure. Water was displaced from tissues such as muscle as the duration of exposure to heat increased.

If water is available during exposure to heat, rats start to drink only after several hours (i.e., when they are already severely dehydrated), but, by the end of 6 h at 40°C, they have drunk 17–18 ml (Hainsworth et al. 1968). Water loss was not measured in this study, but because the evaporative losses decrease with increasing duration of dehydration, it is surmised that rats that have access to water during exposure to heat might actually lose more water than they ingest. In this regard, Paque (1980) noted that desert-dwelling humans, who, of necessity, economize their fluid intake, sweat less and have a lower fluid requirement than visitors to the desert.

The relative contribution of osmolar, Ang II and other pathways will depend on the principal mode(s) of fluid loss, and therefore on the species under study. We know of no published studies that have probed this aspect, although an unpublished study from this laboratory, to be discussed below, suggests a partial involvement of the Ang II pathway. However, it is likely that, as with dehydration by deprivation, heat exposure produces a stimulus to thirst that is of both intra- and extracellular origin.

For humans, it is clear that the salt lost in sweat during exposure to heat must be replaced. Exposure to extreme heat and/or exercise in moderate heat are accompanied by both a need and a preference for salt, either in the form of salt tablets or as electrolyte-balanced beverages. Interestingly, exposure of rats to 35°C for 10 days failed to increase their spontaneous intake of 0.15 M sodium chloride solution, but did increase their intake of water offered simultaneously when compared with their intakes during a previous 10-day period at 25°C (Fregly 1954).

Miescher and Fortney (1989) reported a study in which young and older men reclined in 45°C dry heat for 4 h. Water was available for only the last hour of exposure. It was found that young men maintained their plasma osmolality and had a 4.9% loss of plasma volume and a 1.5% loss of body weight in the first 3 h. They showed high subjective levels of thirst and, when water was available, drank approximately half of their 3 h weight loss, recovered their plasma volume deficit, and abolished their thirstiness. Relative to the young men, older men showed a larger increase in plasma osmolality, a larger decrease in plasma volume, but a similar loss of body weight. Relative to young men, older men reported less sensation of thirst, but still ingested a comparable amount of water to that of young men and partially reversed their reduction in plasma volume.

Thus, there may be age-related differences in water compartments from which fluids are mobilized for sweating and panting, but in any event, it is the dehydration per se that stimulates thirst. Studies in the rat clearly show that dryness of the mouth is not correlated with thirst during exposure to heat (Hainsworth et al. 1968). As with dehydration subsequent to water deprivation, the failure to make up body weight (in humans) in the short term can be considered a voluntary dehydration (Rothstein et al. 1947). Studies with access to salty water are needed to clarify further both the physiological and behavioural processes of restoration.

Effect of Certain Pharmacological Agents on Heat-Induced Drinking

Recent unpublished studies from our laboratories have shown that administration of the beta-adrenoceptor antagonist, d, l-propranolol (12 mg/kg, i.p.), to rats undergoing exposure to heat (30°C) for 0.5, 1.0 or 2.0 h, 0.5 h prior to presentation of water to them, inhibited their drinking response during the first hour after removal from heat. Administration of propranolol also reduced the deficit of body weight during exposure to heat compared to the untreated control group. However, the water intake for a given deficit of body weight by the control group was significantly greater than that for the treated group. Although water intake was correlated significantly with deficit of body weight for both groups, the intercepts, but not the slopes, of the lines were significantly different. Thus, propranolol did not change the basic relationship between deficit of body weight and water intake, it affected the deficit at which water intake was first initiated. This suggests a role for the Ang II pathway in the induction of drinking following exposure to heat; it also suggests that another factor, i.e. the osmoreceptor pathway, may be the initial mediator of drinking after a smaller deficit of body weight than is required for a contribution by the Ang II pathway. Additional studies, however, are needed to interpret these data more clearly.

Much less research has been done on the various factors affecting heat-induced drinking in rats than has been done on dehydration-induced drinking. The general assumption is that both procedures induce drinking by essentially the same mechanisms, and that the volume of water ingested during the first hour that water becomes available is a function of the deficit of body weight incurred. However, it is clear that the extent of drinking following a 24-h dehydration cannot be mimicked by as much as a 2-h exposure to 30°C. Shorter periods of dehydration are difficult to compare with exposure to heat because of the nycthemeral variation in fluid intake. Hence, it would be worthwhile to determine whether a similar relationship exists

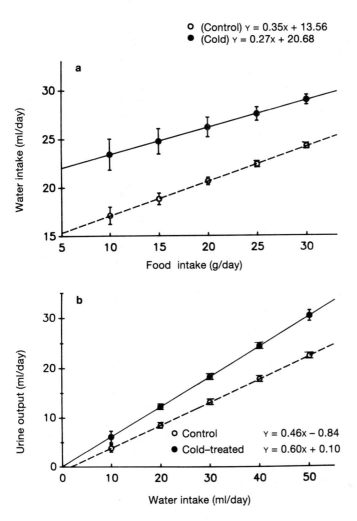

Fig. 28.5. The relationship between **a** food intake and water intake and **b** water intake and urine output of control and cold-treated rats is shown. The equations of the lines are given. (Adapted from Fregly (1968).)

between the volume of water ingested when water is made available to the animals and the deficits of body weight induced by exposure to heat and to dehydration.

Thermogenic (Post-Cold Exposure) Drinking

Nearly 25 years ago, the chance observation was made that rats removed to air at 26°C after continuous exposure to cold (5–6°C) for 3 weeks, manifested a striking thirst (Fregly and Waters 1966a). Water intake began within 15 min after removal from cold and persisted for at least one hour. The rate of water intake by the cold-treated group was significantly ($P<0.01$) greater than that of the control group at all times measurements were made (0.5, 1.0 and 2.0 h). The volume of water ingested within one hour after removal from cold was approximately 10% of the total daily water intake of these rats. Additional details of this experiment have been described by Fregly and Waters (1966a).

The time of exposure to cold required to initiate a thermogenic drink was as little as 6 h of exposure to 5°C. Additional studies showed that it persists during at least 120 days of exposure to cold (Fregly et al. 1978).

Effect of Exposure to Cold on Food and Water Intake

The thermogenic drinking response following removal from cold prompted studies to characterize food and water intakes, as well as urine output, accompanying exposure to cold. The results indicated that at a given food intake, water intake during exposure to cold was significantly less than prior to exposure to cold (Fig. 28.5a). This occurred in spite of the increase in food intake initiated by exposure to cold and the well-known direct linear relationship between food and water intakes in the rat (Cizek and Nocenti 1965). This suggests that the cold-treated rats may be dehydrating themselves voluntarily. Indeed, a regression analysis of urine output on water intake for these same animals indicated that more urine was excreted at a given water intake by cold-treated rats than by controls (Fig. 28.5b). The deficit of water intake may be related to cold-induced changes in mechanisms regulating thirst. Thus, Sobocinska and Kozlowski (1987) reported that dogs acutely exposed to cold had a significant elevation of their osmotic thirst threshold.

An additional possibility existed that the kidneys of cold-treated rats compensated for the reduced water intake by concentrating urine to a greater extent than controls. To test this possibility, male rats were exposed to cold for 10 days and urine osmolality of individual rats was measured daily. At the end of this time, the rats were dehydrated without food for 24 h. Four days later each rat was injected with 500 mU vasopressin tannate in oil, s.c. Their urine outputs and osmolalities were then compared with those of controls maintained at 26°C. Cold-treated rats increased urine output during the 10 days of exposure to cold, but urine osmolality was unchanged. Cold-treated rats failed to decrease urine volume and increase urine osmolality to the level of controls after either a 24-h dehydration or administration of vasopressin (Fregly and Tyler 1972). The extra solute load resulting from the increased food intake by cold-treated rats appeared to be excreted in urine of similar osmolal concentration to that of control rats. The extra solutes thus appeared to be eliminated in a greater volume of urine rather than by concentration of urine to a greater extent.

Itoh (1954) showed that the concentration of antidiuretic substance in the plasma of acutely (2 h) cold-exposed rats was reduced below that of controls. Itoh et al. (1959) also showed that the urine output of acutely cold-exposed rats administered vasopressin was reduced to a lesser extent than that of control rats. Further, Bray (1965) reported that the urinary volume and osmolar responses to daily injection of 500 mU vasopressin were negated when the rats were exposed to air at 3°C. These results suggest that exposure to cold may reduce both production of, and response to, endogenous antidiuretic hormone in rats. Other mechanisms may also play a role. Thus, it is known that the renal tubular response of rats to administered vasopressin is blunted when either glucocorticoid hormones (Gaunt et al. 1957; Sadowski et al. 1972) or catecholamines (Klein et al. 1971) are administered either separately or simultaneously (Fregly and Nelson 1989). Both of these hormones are elevated in the rat chronically exposed to cold (Munday and Blane 1960; Leduc 1961; Straw and Fregly 1967) and could also account for the failure of cold-treated rats to concentrate their urine.

It also appears that dehydration occurs in cold-treated rats as judged by serum osmolality (Fregly et al. 1972, 1974; Fregly 1982). Within 1 day of exposure to cold (either 7.5 or 5.0°C), serum osmolality increased significantly. Increasing the time of exposure to either temperature did not appear to increase serum osmolality further (Fregly 1982).

Thermogenic drinking was not thwarted by preventing access to water for either 1 or 2 h after transfer from 5 to 26°C. (Nelson et al. 1975). These results are similar to those reported by Adolph et al (1954) and Barker et al (1953) for thirst induced in rats and dogs, respectively, by administration of a hypertonic saline load. A further similarity is that delayed access to water did not modify the urge to drink whereas intragastric administration of a water load before removal from cold inhibited drinking (Nelson et al. 1974).

These results raise the question of the mechanism by which the thermogenic drinking response may arise; i.e., absolute dehydration (osmoreceptor pathway), or extracellular dehydration and involvement of the renin–angiotensin pathway (Oatley 1973; Greenleaf and Fregly 1982). The results of studies described above favour dehydration and activation of osmoreceptors as the mechanism in view of the following: (a) a greater urine output compared to water intake; (b) a smaller water intake at a given food intake; (c) increased serum osmolality; (d) increased evaporative water loss; (e) abolition of thermogenic drinking by prior administration of a water load, and (f) persistent thirst when access to water is delayed for 2 h after return of cold-treated rats to air at 26°C.

Effect of Certain Pharmacological Agents on Thermogenic Drinking

Additional studies were carried out to determine whether the renin–angiotensin system might play a role in thermogenic drinking. A possibility existed that at least one factor stimulating the secretion of Ang II and drinking might be the increased rate of secretion of catecholamines induced by either acute or chronic exposure of rats to cold (Leduc 1961). To assess this possibility, the beta-adrenergic antagonist, d, l-propranolol (6 mg/kg body weight, 0.5 h prior to removal from cold) was administered to determine its effect on thermogenic drinking (Fregly et al. 1978). It significantly inhibited thermogenic drinking in cold-exposed rats to about one-third that of untreated, cold-exposed controls. This dose of propranolol did not affect significantly the drinking response of control rats, presumably because water intake was already quite low. d-Propranolol (6 mg/kg, i.p.), administered in the same manner, failed to influence the thermogenic drink (Fregly et al. 1978).

Since beta-adrenergic receptors can be subdivided into $beta_1$- and $beta_2$-subtypes, additional studies were carried out to characterize further the receptor subtype concerned with thermogenic drinking (Fregly et al. 1978). It appears that the thermogenic drinking response arises, in part at least, as a result of stimulation of $beta_2$-adrenergic receptors during transfer of rats from a cold to a warm environment. Whether $beta_2$-adrenergic receptors may also mediate the release of renin from the juxtaglomerular cells of the kidney is debatable at present (Weber et al. 1974; Capponi et al. 1977). If it can be shown ultimately that they do, it will strengthen the possibility that the renin–angiotensin system might play a role in the thermogenic drinking response. To test whether the renin–angiotensin system became activated under these conditions, plasma renin activity (PRA) was measured in control rats maintained at 26°C; in cold-adapted rats maintained at 5°C for at least 6 weeks, and in similarly treated cold-adapted rats 15 min after removal from 5 to 26°C (Katovich et al. 1979). The results of this study revealed that a 6-week cold exposure per se did not affect PRA significantly. However, within 15 min after removal from cold, PRA increased fourfold. An additional study showed that administration of propranolol prior to removal from cold prevented the increase in PRA. This indicates that removal from cold is associated with an increase in the formation of Ang II which could account, in part at least, for the thermogenic drink.

To test this possibility further, the Ang I converting enzyme inhibitor, captopril, was administered to rats 15 min before transfer from the cold to a neutral environment (Katovich et al. 1979). These studies revealed a graded reduction in the thermogenic drinking response with graded increases in the dose of captopril administered.

These results suggest that thermogenic drinking may be mediated, in part at least, by activation of the beta-adrenergic system. This, in turn, induces an increase in PRA and the formation of Ang II. The latter is the dipsogenic agent responsible, at least in part, for the induction of drinking. Support for a role of Ang II in thermogenic drinking is obtained from the elevated PRA of rats removed from the cold to a neutral environment. The time-course for the increase in PRA also coincides with the time-course for induction of a thermogenic drinking response. Additional evidence that Ang II may be important in the thermogenic drinking response is found in the dose-dependent blocking effect of captopril which inhibits conversion of Ang I into Ang II (Katovich et al. 1979).

Experimental results to date suggest that thermogenic drinking is mediated both by osmoreceptors and Ang II receptors. Present results do not show clearly the extent to which each may contribute to the thermogenic drink. Thus, a possibility exists that the thermogenic drinking response of the cold-treated rat is of sufficient physiological importance to have redundancy in the mechanisms initiating it. While thermogenic drinking has been studied only in the rat, the studies by Conley and Nickerson (1945) and Spealman et al. (1948) indicate that cold-exposed man may have a similar response.

A number of factors may contribute to the cold-induced dehydration. Evaporative water loss has been measured and is nearly doubled during exposure of rats to air at 5°C (Fregly 1967). However, the increased evaporative water loss of cold-treated rats may be counterbalanced by their increased food intake since Radford (1959) estimated that evaporative water loss approximated the water present in the food ingested by his rats. Proof that this is the case during exposure to cold would require additional study. The relationship between food intake and water intake is also changed such that less water is ingested at a given food intake by cold-exposed than by control rats (Fregly 1968). A possibility exists that this food–water imbalance results in an accumulation of solutes and thereby contributes to the increased serum osmolality and the dehydrating effect of cold. In addition, the altered relationship between urine output and water intake induced by cold suggests changes in either the thirst mechanism or in the mechanisms influencing renal water loss. Thus, the ability both to produce and to respond to endogenous antidiuretic hormone, or both, may be reduced in rats exposed to cold.

Inasmuch as the cold-treated rat does not appear to make up for its reduced water intake by concentrating its urine to a greater extent than controls, a problem of continuing dehydration with no apparent stimulation of thirst appears to exist during exposure to cold. It is possible that a portion of the deficit of body water may be made up by the preferential metabolism of lipids that would yield more water of oxidation per g than carbohydrates or proteins. It is known that both the utilization and turnover of lipids are increased by exposure to cold (Kodama and Pace 1964), and both body fat content (Pagé and Babineau 1953; Young and Cook 1955) and respiratory quotient are reduced (Pagé and Chenier 1953). The reduction in body weight accompanying exposure to cold can be attributed, in part at least, to the metabolism of body fat stores. Further, Pagé and Babineau (1953) reported that high fat diets promote growth of cold-exposed rats. In addition, high fat diets may enhance the ability of rats to maintain their body temperatures on exposure to cold (Leblanc 1957), and to adapt to a cold environment (Pagé and Babineau 1953). Dietary self-selection studies also suggest that an enhanced adaptive resistance to cold accompanies increased ingestion of fat (Dugal et al. 1945). Although the importance of fat in the maintenance of energy balance of animals exposed to cold has been well studied, its importance and contribution to the maintenance of water balance require further study.

With respect to thermogenic drinking, the results seem clear that abrupt removal of rats from a cold to a warm environment results in a striking thermogenic drinking response. However, little is known about the failure to drink adequate amounts of water during exposure to cold, at least up to 12 days. A possibility exists that a blunting of the mechanisms for induction of thirst occurs during exposure to cold. This is not a function of the fact that only cold water is available to the rats to drink since studies in which the rats were supplied with warm water while in the cold failed to affect their thermogenic drink (Fregly et al. 1976). At present, it is possible only to speculate why exposure to cold, with its consequent increase in food intake and solute load, fails to induce an increased water intake. On the other hand, the possibility must also be considered that relative dehydration is a necessary facet of the process of adaptation to cold.

In addition to the factors mentioned above, it is possible that the following sequence of events occurs which may explain the thermogenic drinking response. Exposure to cold increases peripheral vasoconstriction which, in turn, increases central or "effective" blood volume and decreases water intake, antidiuretic hormone and PRA. This results in increased excretion of salt and water with the consequent restoration of effective blood volume to normal. On removal from cold there is a decrease both in peripheral vasoconstriction and control or effective blood volume accompanied by an increased water intake, PRA, and antidiuretic hormone secretion.

Effect of Water Temperature on Thermogenic Drinking

Studies were also carried out to assess the effect of water temperature on the thermogenic drinking response of rats that had been exposed to cold (5°C) for 24 h by Nelson et al. (1974). Water intakes of the cold-treated rats during the first hour after removal from cold were greater than those at all water temperatures tested. As was the case with dehydrated rats (Fig. 28.4), both cold-treated and control rats ingested the greatest volume when water temperature was at approximately body temperature. Of additional importance is the revelation that the thermogenic drinking response is not due merely to the desirability of warm water by cold-exposed rats, but is also present when water at 5°C is given (Nelson et al. 1974).

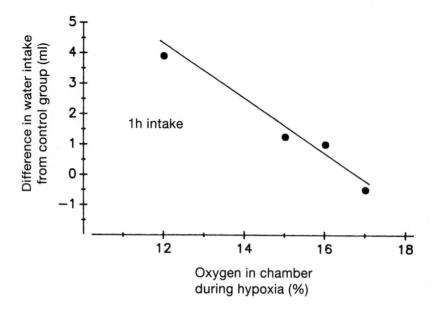

Fig. 28.6. The difference in water intake of hypoxia-treated rats from controls during the first hour after return from hypoxia to normoxia is shown at graded percentages of oxygen to which the rats were exposed. Each point represents six rats. (Adapted from Fregly (1971).)

Post-Hypoxic Drinking

While the induction of the thermogenic drinking response of rats is related both to excessive secretion of Ang II and to increases in the osmolality of plasma, a possibility existed that thermogenic drinking might generalize to other "stressful" environments. A question naturally arose as to whether a similar drinking response might occur following return from exposure to hypoxia.

To assess this possibility, four separate experiments were performed (Fregly 1971). Each used 12 male Sprague–Dawley rats. A 2-week control period preceded each 8-day exposure to hypoxia. At the end of this time, half the rats in each experiment were exposed to hypoxia ($17.0\pm0.2\%$, $16.0\pm0.2\%$, $15.0\pm0.2\%$ or $12.0\pm0.2\%$ oxygen). At the end of the 8-day exposure to hypoxia, the experimental rats were returned to the normoxic (20.9% oxygen) control environment and spontaneous water intake of individual control and treated rats was measured for 1.0 h thereafter.

The difference in water intake between control and treated groups during the first hour after removal from hypoxia increased in a linear fashion with decreasing oxygen percentage to which the rats were previously exposed (Fig. 28.6). These results suggest that the difference reached zero at 16.5% oxygen.

An additional study was carried out to determine whether the post-hypoxic drinking response continued throughout a much longer period of exposure to hypoxia (Fregly and MacArthur 1990). Rats were exposed to an atmosphere of 12% oxygen in nitrogen

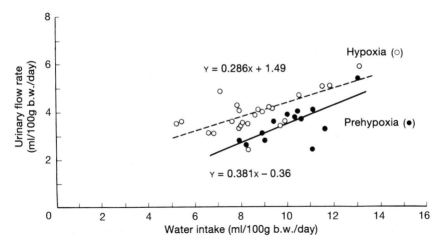

Fig. 28.7. The relationship between urinary flow rate and water intake of rats exposed to hypoxia (12% oxygen) and controls is shown. The equations for the lines are shown in the figure. (Adapted from Fregly and Waters (1966b).)

for 72 days. On days 14, 22, 29, 35, 41, 49 and 71 of the study, the treated rats were removed from their chambers, placed in individual metabolic cages without food, and given a preweighed water bottle containing water at 26°C. The first significant increase in water intake following removal from hypoxia occurred after 14 days of exposure, that is, the first time the measurement was made. Post-hypoxic drinking remained significantly greater than that of the control group throughout the 71 days of the experiment.

When return of water to the hypoxia-treated rats was delayed for 1 h after return to 20.9% oxygen, the magnitude of the post-hypoxic drink was not affected significantly (Fregly and MacArthur 1990).

Since the studies described above showed that the post-hypoxic drinking persisted throughout the duration of exposure to hypoxia, two experiments were performed to study water intake and urinary output during chronic exposure to hypoxia (11.0% and 12. 5% oxygen, respectively) (Sisson and Fregly 1955; Fregly and Waters 1966b).

Water intake was depressed by hypoxia whereas urinary output remained at the pre-hypoxic control level. A plot of urinary flow rate against water intake revealed that at a given water intake, urinary flow rate was greater during exposure to hypoxia than prior to it (Fig. 28.7). This suggests, as was the case for cold-exposure, that rats exposed to hypoxia might suffer dehydration. The increase in plasma osmolality of the hypoxia-treated rats also supports this possibility (Fregly and Waters 1966b).

It is apparent that the changes in water metabolism induced by acute exposure to hypoxia, including increased urinary flow (Fregly and Waters 1966b), increased insensible water loss (Fregly 1967), and no increase in water intake (Fregly and Waters 1966b), could not be maintained during chronic exposure to hypoxia. The question as to whether the changes occurring during acute exposure to hypoxia, even discontinuous exposures over many days' duration, are representative of those taking place in animals exposed chronically and continuously is worthy of consideration. Picon-Reategui et al. (1953), using tritiated water, measured body water content of rats exposed chronically to a simulated altitude of 15 000 feet. The results suggested that body water depletion was greatest from days 4 to 7 of exposure to this altitude. Measurements made during this period indicated that 20% of the body water content had been lost. With time at altitude, both body weight loss and body water loss diminished such that by 30 days of exposure to altitude, about 5 g of water was lost/100 g original body weight.

Sundstroem and Michaels (1942), in their comprehensive study of the physiological effects of exposure to hypoxia, reported that plasma specific gravities of rats increased during the first day of exposure to 14 000, 24 000 and 26 000 feet simulated altitude. Plasma specific gravity returned to control levels after 7 days of exposure. Earlier studies from this laboratory reported increases in plasma specific gravity and osmolality after 32 days of exposure to 12.0% oxygen (Fregly and Waters 1966b). This suggests that the relative dehydration induced by hypoxia may continue for longer periods than were observed by Sundstroem and Michaels (1942). The relative contributions of renal and extrarenal water losses to the dehydration occurring during hypoxia cannot be assessed at present. Nor can one assess the contribution to dehydration of the apparent failure to drink sufficient water to maintain serum osmolality at control levels. Further study will be needed to determine these.

A manifestation of the relative dehydration occurring during exposure to hypoxia is the drinking response which occurs immediately after return to normoxia. Experimental rats increased significantly their spontaneous water intake above that of controls within the first hour after returning to control environment (20.9% oxygen). The difference in water intake between treated and control groups during the first hour after removal from hypoxia increased linearly with decreasing oxygen concentration (Fig. 28.6). whatever factors are responsible for the post-hypoxic thirst in rats, they do not appear to be activated by exposure to 16.5% oxygen or higher for 8 days. These results are the first to suggest that the extent of the post-hypoxic thirst is an inverse linear function of the degree of hypoxia to which the rats were exposed. The results also suggest, but do not prove, that the extent of dehydration induced in rats may also be an inverse linear function of the degree of hypoxia to which the animals were exposed, at least within the period of exposure used in this study.

It is difficult to explain the thirst following removal from hypoxia. In spite of the increase in serum osmolality and specific gravity during exposure to hypoxia reported earlier (Fregly and Waters 1966b), the rats failed to ingest sufficient water to prevent a relative dehydration from occurring. Experiments of others show that the extent of water ingested following dehydration can be correlated with serum osmolality and deficit of body water prior to ingestion of fluid (Adolph 1957). The failure of the usual response to a normally adequate stimulus highlights the interesting question posed by Swann and Collings (1943) concerning the lack of thirst during exposure to hypoxia. The point at which the stimulus–response system fails is unknown and could involve afferent and efferent nervous pathways, receptors, as well as the areas of the brain mediating thirst and/or satiation. Additional experiments will be required to understand whether one or all of these may be affected.

Effect of Certain Pharmacological Agents on Post-Hypoxic Drinking

Administration of d, l-propranolol (6 mg/kg body weight, i.p.) immediately after removal from hypoxia, and 30 min prior to return of water to each animal, failed to affect the post-hypoxic drinking response (Fregly and MacArthur 1990). In this important respect, post-hypoxic drinking differs from thermogenic drinking.

Summary

Return to control conditions after exposure to any one of three adverse environments was accompanied by thirst, the characteristics of which were very similar. Water intake began immediately and was maximal within 0.5–1.0 h after return to control conditions. In the cases of dehydration, heat exposure, and exposure to cold, the drinking response observed after return to control conditions appears to be mediated both by the osmoreceptor and angiotensin II pathways. The relative contribution of each pathway to the drinking response may differ in each situation and with time following return to the control condition. Much work remains to be done in this regard. The drinking response following removal from hypoxia may have only a minor contribution from the Ang II pathway, but this conclusion remains tentative until further studies can be carried out.

Acknowledgement. Supported by grant N00014-88-J-1221 with the Office of Naval Research.

References

Abraham SF, Denton DA, McKinley MJ, Weisinger RS (1976) Effect of an angiotensin antagonist, Sar[1]-Ala[8]- Angiotensin II on physiological thirst. Pharmacol Biochem Behav 4:243–247

Adolph EF (1950) Thirst and its inhibition in the stomach. Am J Physiol 161:374–386

Adolph EF (1957) Regulation of water metabolism in stress. Brookhaven Symp Biol 10:147–161

Adolph EF, Barker JP, Hoy PA (1954) Multiple factors in thirst. Am J Physiol 178:538–562

Barker JP, Adolph EF, Keller AD (1953) Thirst tests in dogs and modifications of thirst with experimental lesions of the neurohypophysis. Am J Physiol 171:233–238

Barney CC, Katovich MJ, Fregly MJ (1980) The effect of acute administration of an angiotensin converting enzyme inhibitor, captopril (SQ 14,225), on experimentally induced thirsts in rats. J Pharmacol Exp Ther 212:53–57

Barney CC, Threatte RM, Fregly MJ (1983) Water deprivation-induced drinking in rats: role of angiotensin II. Am J Physiol 244 (Reg Integ Comp Physiol 13):R244–R248

Block ML, Fisher AE (1975) Cholinergic and dopaminergic blocking agents modulate water intake elicited by deprivation, hypovolemia, hypertonicity and isoproterenol. Pharmacol Biochem Behav 3:251–262

Boulze D, Montastruc P, Cabanac M (1983) Water intake, pleasure and water temperature in humans. Physiol Behav 30:97–102

Bray GA (1965) Rhythmic changes in renal function in the rat. Am J Physiol 209:1187–1192

Capponi AM, Gourjon M, Vallotton MB (1977) Effect of beta-blocking agents and angiotensin II on isoproterenol-stimulated renin release from rat kidney slices. Circ Res 40:89–93

Cizek LJ, Nocenti MR (1965) Relationship between water and food ingestion in the rat. Am J Physiol 208:615–620

Conley CL, Nickerson JL (1945) Effects of temperature change on the water balance in man. Am J Physiol 143:373–384

Dugal LP, LeBlond CP, Therien M (1945) Resistance to extreme temperatures in connection with different diets. Can J Res 23:244–258

Fitzsimons JT (1979) The physiology of thirst and sodium appetite. Cambridge University Press, Cambridge, pp 128–265; 383–391

Fitzsimons JT, Setler PE (1975) The relative importance of central nervous catecholaminergic and cholinergic mechanisms in drinking in response to angiotensin and other thirst stimuli. J Physiol (Lond) 250:613–631

Fregly MJ (1954) Effect of extremes of temperature on hypertensive rats. Am J Physiol 176:275–281

Fregly MJ (1967) Effect of exposure to cold on evaporative loss from rats. Am J Physiol 213:1003–1008

Fregly MJ (1968) Water and electrolyte exchange in rats exposed to cold. Can J Physiol Pharmacol 46:873–881

Fregly MJ (1971) Thirst immediately following removal of rats from graded levels of hypoxia. Proc Soc Exper Biol Med 138:448–453

Fregly MJ (1982) Thermogenic drinking: mediation by osmoreceptor and angiotensin II pathways. Fed Proc 41:2515–2519

Fregly MJ Kelleher DL (1980) Antidipsogenic effect of clonidine on isoproterenol-induced water intake. Appetite 1:279–289

Fregly MJ, MacArthur SA (1990) Some characteristics of post-hypoxia-induced drinking in rats. Aviat Space Environ Med (in press)

Fregly MJ, Nelson EL Jr (1989) Hormonal interaction in maintenance of fluid and electrolyte homeostasis in cold-exposed rats. In: Mercer JB (ed) Thermal physiology. Elsevier, Amsterdam, pp 593–599

Fregly MJ, Rowland NE (1986) Role for alpha$_2$- adrenoceptors in experimentally-induced drinking in rats. In: de Caro G, Epstein AN, Massi M (eds) The physiology of thirst and sodium appetite. Plenum Press, New York, pp 509–519

Fregly MJ, Rowland NE (1988) Augmentation of isoproterenol-induced drinking by acute treatment with certain dopaminergic agonists. Physiol Behav 44:473–481

Fregly MJ, Tyler PE (1972) Renal response of cold-exposed rats to pitressin and dehydration. Am J Physiol 222:1065–1070

Fregly MJ, Waters IW (1966a) Water intake of rats immediately after exposure to a cold environment. Can J Physiol Pharmacol 44:651–662

Fregly MJ, Waters IW (1966b) Posthypoxic drinking response of rats. Fed Proc 25:1220–1226

Fregly MJ, Lutherer LO, Tyler PE (1972) Variation of both ambient temperature and duration of cold exposure on water intake following removal from cold. In: Smith RE, Hannon JP, Shields JL, Horwitz BA (eds) International symposium on environmental physiology: bioenergetics. FASEB, Washington, DC, pp 96–100

Fregly MJ, Lutherer LO, Tyler PE (1974) The effect of exposure to cold, hypoxia and both combined on water exchange in rats. Aerospace Med 45:1223–1231

Fregly MJ, Kaplan BJ, Brown JG, Nelson EL Jr, Tyler PE (1976) Effect of water temperature during cold exposure on thermogenic drinking in rats. J Appl Physiol 41:497–501

Fregly MJ, Katovich MJ, Tyler PE, Dasler R (1978) Inhibition of thermogenic drinking by beta-adrenergic antagonists. Aviat Space Environ Med 49:861–867

Fregly MJ, Barney CC, Katovich MJ, Miller EA (1979) Effect of water temperature on isoproterenol-induced water intake. Proc Soc Exp Biol Med 160:359–362

Fregly MJ, Lockley OE, Simpson CE (1981) Effect of the angiotensin converting enzyme inhibitor, captopril, on development of renal hypertension in rats. Pharmacology 22:277–285

Fregly MJ, Fater DC, Greenleaf JE (1982) Effect of the angiotensin I converting enzyme inhibitor, MK-421, on experimentally induced drinking. Appetite 3:309–319

Fregly MJ, Rowland NE, Greenleaf JE (1984) A role for presynaptic alpha$_2$- adrenoceptors in angiotensin II-induced drinking in rats. Brain Res Bull 12:393–398

Gaunt R, Lloyd CW, Chart JJ (1957) The adrenal neurohypophyseal interrelationship. In: Heller H (ed) The neurohypophysis. Butterworths, London, pp 233–250

Greenleaf JE, Fregly MJ (1982) Dehydration-induced drinking: peripheral and central aspects. Fed Proc 41:2507–2508

Hainsworth FR, Stricker EM, Epstein AN (1968) Water metabolism of rats in the heat: dehydration and drinking. Am J Physiol 214:983–989

Itoh S (1954) The release of antidiuretic hormone from the posterior pituitary body on exposure to heat. Jpn J Physiol 4:185–190

Itoh S, Toyomasu Y, Konno T (1959) Water diuresis in cold environment. Jpn J Physiol 9:438–443

Kapatos G, Gold RM (1972) Tongue cooling during drinking: a regulator of water intake in rats. Science 176:685–686

Katovich MJ, Barney CC, Fregly MJ, Tyler PE, Dasler R (1979) Relationship between thermogenic drinking and plasma renin activity in the rat. Aviat Space Environ Med 50:721–724

Klein LA, Liberman B, Laks M, Kleeman CR (1971) Interrelated effects of antidiuretic hormone and adrenergic drugs on water metabolism. Am J Physiol 221:1657–1665

Kodama AM, Pace N (1964) Effect of environmental temperature on hamster body fat composition. J Appl Physiol 19:863–867

Leblanc J (1957) Prefeeding of high fat diet and resistance of rats to intense cold. Can J Biochem Physiol 35:25–30

Leduc J (1961) Catecholamine production and release in exposure and acclimatization to cold. Acta Physiol Scand 53 (Suppl 183):1–101

Lee M, Thrasher TN, Ramsay DJ (1981) Is angiotensin essential in drinking induced by water deprivation and caval ligation? Am J Physiol 240 (Reg Integ Comp Physiol 9):R75–R80

Lehr D, Goldman HW, Casner P (1973) Renin angiotensin role in thirst: paradoxical enhancement of drinking by angiotensin converting enzyme inhibition. Science 82:1031–1032

Malvin RL, Mouw D, Vander AJ (1977) Angiotensin physiological role in water-deprivation-induced thirst of rats. Science 197:171–173

Mendelson J, Chillag D (1970) Tongue cooling: a new reward for thirsty rodents. Science 170: 1418–1420

Miescher E, Fortney SM (1989) Responses to dehydration and rehydration during heat exposure in young and older men. Am J Physiol 257 (Reg Integ Comp Physiol 26):R1050–R1056

Munday KA, Blane GF (1960) Changes in electrolytes and 17-oxysteroids in the rat subjected to a cold environment. J Endocrinol 20:266–275

Nelson EL, Fregly MJ, Tyler PE (1974) Effects of water temperature on post-cold exposure drinking response of rats. Am J Physiol 227:977–980

Nelson EL Jr, Fregly MJ, Tyler PE (1975) Factors affecting thermogenic drinking in rats. Am J Physiol 228:1875–1879

Oatley K (1973) Stimulation and theory of thirst. In: Epstein AN, Kissileff HR, Stellar E (eds) The neuropsychology of thirst: new findings and advances in concepts. Winston, Washington, DC, pp 199–223

Olsson K (1975) Attenuation of dehydrative thirst by lowering the CSF [Na^+]. Acta Physiol Scand 94: 536–538

Pagé E, Babineau LM (1953) The effects of diet and cold on body composition and fat distribution in the white rat. Can J Med Sci 31:22–40

Pagé E, Chenier L (1953) Effects of diet and cold environment on respiratory quotient of the white rat. Rev Can Biol 12:530–541

Paque C (1980) Saharan Bedouins and the salt water of the Sahara: a model for salt intake. In: Kare MR, Fregly MJ, Bernard RA (eds) Biological and behavioral aspects of salt intake. Academic Press, New York, pp 31–47

Picon-Reategui E, Fryers CR, Berlin NI, Lawrence JH (1953) Effect of reducing the atmospheric pressure on body water content of rats. Am J Physiol 172:33–36

Radford EP Jr (1959) Factors modifying water metabolism in rats fed dry diets. Am J Physiol 196: 1098–1108

Ramsay DJ, Reid IA (1975) Some central mechanisms of thirst in the dog. J Physiol (Lond) 253:517–525

Ramsay DJ, Rolls BJ, Wood RJ (1977) Body fluid changes which influence drinking in the water deprived rat. J Physiol (Lond) 266:453–469

Rogers TA, Setliff JA, Klopping JC (1964) Energy cost, fluid and electrolyte balance in subarctic survival situations. J Appl Physiol 19:1–8

Rolls BJ, Wood RJ, Rolls ET (1980) Thirst: the initiation, maintenance and termination of drinking. Prog Psychobiol Physiol Psychol 9:263–321

Rothstein A, Adolph EF, Wills JH (1947) Voluntary dehydration. In: Adolph EF and associates (eds) Physiology of man in the desert. Interscience, New York, pp 254–270

Rowland N, Engle DJ (1977) Feeding and drinking interactions after acute butyrophenone administration. Pharmacol Biochem Behav 7:295–301

Sadowski J, Nazar K, Szczepanska-Sadowska E (1972) Reduced urine concentration in dogs exposed to cold: relation to plasma ADH and 17-OHCS. Am J Physiol 222:607–610

Sandick BL, Engle DB, Maller O (1984) Perception of drinking water temperature and effects for humans after exercise. Physiol Behav 32:851–855

Severs WB, Klase PA (1978) Drinking in water-deprived rats after combined central angiotensin receptor and converting enzyme blockage. Pharmacol Biochem Behav 9:259–260

Severs WB, Kapsha JM, Klase PA, Keil LC (1977) Drinking behavior in water deprived rats after angiotensin receptor blockade. Pharmacology 15:254–358

Sisson GM, Fregly MJ (1955) A simplified apparatus for chronic exposure of rats to low oxygen tension. J Appl Physiol 8:128–131

Sobocinska J, Kozlowski S (1987) Osmotic thirst suppression in dogs exposed to low ambient temperature. Physiol Behav 40:171–175

Spealman CR, Yamamoto W, Bixby EW, Newton M (1948) Observations on energy metabolism and water balance of men subjected to warm and cold environments. Am J Physiol 152:233–241

Straw JA, Fregly MJ (1967) Evaluation of thyroid and adrenal-pituitary function during cold stimulation. J Appl Physiol 23:825–830

Stricker EM (1978) The renin-angiotensin system and thirst: some unanswered questions. Fed Proc 37:2704–2710

Sundstroem ES, Michaels G (1942) The adrenal cortex in adaptation to altitude, climate, and cancer. Mem Univ Calif 12:1–409

Swann HG, Collings WD (1943) The extent of water loss by rats at lowered barometric pressures. J Aviat Med 14:114–118

Weber MA, Stokes FS, Gain JM (1974) Comparison of the effects on renin release of beta adrenergic antagonists with differing properties. J Clin Invest 54:1413–1419

Wilson KM, Rowland N, Fregly MJ (1984) Drinking: a final common pathway? Appetite 5:31–38

Yamaguchi K (1981) Effects of water deprivation on immunoreactive angiotensin II levels in plasma, cerebroventricular perfusate and hypothalamus of the rat. Acta Endocrinol 97:137–144

Young DR, Cook SF (1955) Body lipids in small mammals following prolonged exposures to high and low temperatures. Am J Physiol 181:72–74

Chapter 29

Effect of Changes in Reproductive Status on Fluid Intake and Thirst

P.H. Baylis

Introduction

The purpose of this chapter is to review the effects of the menstrual cycle, pregnancy, lactation and the menopause on thirst and osmoregulation in humans. Over the last decade there has been considerable increase in our knowledge of factors that influence osmoregulation, which can be attributed to advances in methods of assessment of thirst and vasopressin, the two principal regulators of water balance. The development of a series of specific and highly sensitive radioimmunoassays capable of measuring the low physiological plasma concentrations of vasopressin has allowed the characterization of the functional properties of the vasopressin osmoreceptor (for review see Baylis and Thompson 1988). Accurate assessment of thirst is difficult but the pioneering approach of Rolls and her colleagues who documented a variety of sensations associated with thirst using visual analogue scales, has led to reliable and reproducible methods to quantitate this sensation (Rolls et al. 1980; Phillips et al. 1984).

Studies on healthy resting adults by Robertson et al. (1976) have clearly defined a simple linear correlation between plasma vasopressin concentration and plasma osmolality which can be defined by the function, $PVp = m\,(Pos-c)$ where PVp represents plasma vasopressin, Pos, plasma osmolality, m, the slope of the regression line and c, the abscissal intercept. The slope of the line is a measure of the sensitivity of the vasopressin secreting system and the abscissal intercept denotes the osmotic threshold for vasopressin release. Since there is also a linear increase in thirst in response to rises in plasma osmolality a similar equation for thirst can be derived, $Th = m\,(Pos-c)$, where Th represents thirst (Robertson 1984). Some workers have demonstrated a remarkable similarity between the functional osmoregulatory characteristics of thirst and vasopressin secretion in man (Thompson et al. 1986).

In the present review, the analytical approach to osmoregulated thirst and vasopressin release described above will be used to define the changes in osmoregulation observed in various reproductive states of women. Major emphasis will be put on the osmoregulation of thirst but mention will be made of the closely related plasma vasopressin alterations.

Menstrual Cycle

Normal Individuals

Many healthy women in their reproductive years notice changes in body fluid distribution and possibly fluid retention which are related to their menstrual cycle (Greene and Dalton 1953), the causes of which are unknown. Recent studies have indicated that plasma osmolality is lower in the luteal than the follicular phase of the ovulatory cycle (Mira et al. 1984: Spruce et al. 1985; Vokes et al. 1988), which suggests that cyclical variation in osmoregulation may occur. However, the fall in plasma osmolality is small, of the order of 3–5 mosmol/kg. Interestingly, there is little or no significant weight gain associated with the reduced plasma tonicity.

To investigate the cause of the relative plasma hypotonicity, Spruce et al. (1985) assessed the response of thirst sensation and vasopressin release to infusion of hypertonic saline in the mid-follicular and mid-luteal phases of the cycle in a group of healthy young women. They showed a fall in the osmotic threshold for thirst of 4 mosmol/kg in the luteal phase, and a similar fall of 5 mosmol/kg in the vasopressin osmotic threshold. Thus, it appeared that the osmostat for thirst and vasopressin release had been reset at a lower level in the luteal phase. They also demonstrated a small but significant reduction in the slope of the vasopressin osmoregulatory line, indicating a decrease in sensitivity of the vasopressin secreting system.

More comprehensive studies have been reported recently in which healthy young women were studied at five points throughout their cycles and underwent dynamic osmoregulatory tests (water loading and hypertonic saline infusion) in the early follicular (days 4–6), ovulatory (days 13–15), and advanced luteal (days 23–25) phases of their cycles (Vokes et al. 1988). With the use of a visual analogue scale to assess thirst, these workers confirmed that there was a fall in the thirst osmotic threshold in the luteal phase of 5 mosmol/kg (Table 29.1 and Fig. 29.1). There was no difference between the thirst thresholds at the follicular and periovulatory timepoints. The sensitivity of the thirst osmostat did not vary during the menstrual cycle (Table 29.1). Similar alterations in the vasopressin osmoregulatory line were observed, with a reduction in the osmotic threshold of 5 mosmol/kg in the luteal phase but no change in the slope of the line (Table 29.1).

Water loading of the same young women at the three phases in their cycles clearly indicated that there were no differences in their responses, but that plasma osmolality was lower throughout the study in the luteal phase. The changes in plasma

Table 29.1. Osmotic characteristics for thirst and vasopressin release during the menstrual cycle

Phase of menstrual cycle	Thirst			Vasopressin		
	Osmotic threshold (mosmol/kg)	Slope	Regression coefficient	Osmotic threshold (mosmol/kg)	Slope	Regression coefficient
Follicular	289	0.4	0.94	285	0.78	0.94
Ovulatory	287	0.5	0.97	284	0.97	0.95
Luteal	284*	0.5	0.95	280**	0.77	0.95

* $P<0.01$ Luteal vs follicular
** $P<0.02$ Luteal vs follicular
From Vokes et al. (1988), with permission.

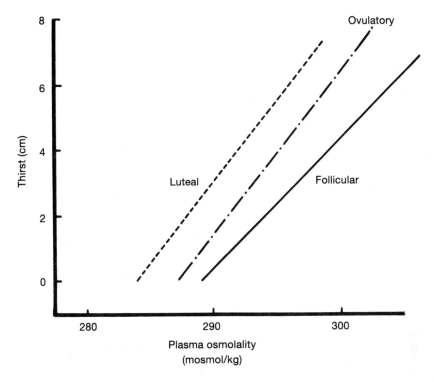

Fig. 29.1. The relationship between thirst ratings and plasma osmolality in the follicular, ovulatory and luteal phases of the human menstrual cycle. (Adapted from Vokes et al. (1988), with permission.)

vasopressin, urine osmolality, and free water clearance were all entirely normal throughout the cycle.

So, Vokes' studies have confirmed the downward resetting of osmotic thresholds for both thirst and vasopressin secretion, which readily explains the observation of the lowered plasma osmolality in the luteal phase of the menstrual cycle in healthy women. As yet, however, the cause for the resetting has to be elucidated. The cyclical variations in plasma concentrations of oestrogens, progesterone and gonadotrophins are obvious potential causes but the evidence to date fails to confirm, in humans, that any of them account for the resetting. No correlation was found between the degree of resetting and the magnitude of the changes in plasma sex steroid hormones or luteinizing hormone (Vokes et al. 1988). Studies designed specifically to document the effect of exogenous oestrogen and progesterone on thirst and vasopressin osmoregulation in a small group of young women with primary ovarian failure indicate that these steroid hormones are without any influence on osmoregulation (Baylis 1988). Not all workers agree with these findings. The administration of physiological quantities of oestrogen to postmenopausal women caused an increase in basal plasma vasopressin concentrations whereas progesterone decreased the values (Forsling et al. 1982). These workers, however, did not test osmotically stimulated vasopressin release and did not assess thirst.

Lowering the osmotic threshold for vasopressin release occurs following moderate degrees of hypovolaemia or hypotension (Robertson and Athar 1976). Whether a

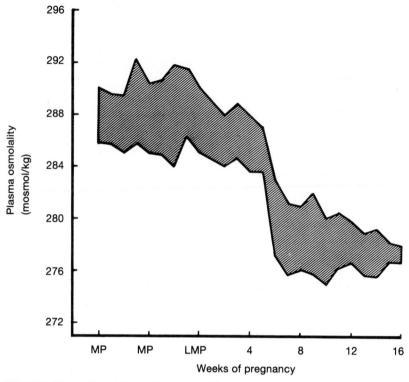

Fig. 29.2. The decline in basal plasma osmolality associated with pregnancy. MP, menstrual period; LMP, last menstrual period. (From Davison et al. (1981), with permission.)

similar fall occurs in the thirst threshold is not known. No changes were observed in the indices measured to assess blood volume and pressure in the luteal phase (Vokes et al. 1988) which militates against baroregulatory influences being responsible for the lowered vasopressin osmostat. Furthermore, there are data to suggest that blood volume increases in the late luteal phase (Turner and Fortney 1984). Thus, the downward resetting of the osmostat in the luteal phase of the menstrual cycle appears to be due to an unidentified factor.

Cyclical Oedema

The osmoregulatory changes in thirst and vasopressin release described above have also been documented in a small group of women suffering from cyclical oedema (Thompson et al. 1988). These individuals who all demonstrated dependent oedema and weight gain of at least 3 kg during the luteal phase of two consecutive cycles had a 4 mosmol/kg reduction in plasma osmolality in the mid-luteal phase compared to the mid-follicular phase. The osmotic thresholds for thirst and vasopressin release fell by 3 and 4 mosmol/kg respectively, in the luteal phase, values that are identical to healthy young women. Thus it can be concluded that there are no significant alterations in osmoregulation in cyclical oedema.

Table 29.2. Osmoregulatory characteristics for thirst and vasopressin release in pregnancy

	Thirst			Vasopressin		
	Osmotic threshold (mosmol/kg)	Slope	Regression coefficient	Osmotic threshold (mosmol/kg)	Slope coefficient	Regression coefficient
Preconception	290	0.52	0.97	285	0.61	0.87
5–8 weeks pregnant	280*	0.49	0.97	278*	0.59	0.89
10–12 weeks pregnant	280*	0.45	0.97	276*	0.53	0.93
28–33 weeks pregnant	279*	0.47	0.97	276*	0.23*	0.89
Postpartum	290	0.45	0.97	286	0.61	0.90

* $P<0.001$, pregnant vs non-pregnant
From Davison et al, (1988a), with permission.

Pregnancy

Far more profound changes in body fluids occur in pregnancy than during the menstrual cycle. One of the earliest changes to occur is a considerable fall in plasma osmolality, of the order of 10 mosmol/kg, in the first few weeks of gestation (Fig. 29.2). The lowest plasma osmolality values are attained by week 10 of gestation and they remain low until term (Davison et al. 1981).

Detailed studies on osmoregulation were hampered by the circulating placental enzyme, vasopressinase, which avidly degrades vasopressin in vitro. Following the development of techniques to inhibit completely the activity of this enzyme during venepuncture, sample preparation and assay of vasopressin, substantial falls were clearly demonstrated in the osmotic thresholds for vasopressin secretion and thirst of 6 and 9 mosmol/kg, respectively, in the latter part of pregnancy compared to 8–10 weeks postpartum (Davison et al. 1984).

Subsequent studies performed prior to conception, on three occasions during pregnancy (at 5–8, 10–12 and 28–33 weeks' gestation) and postpartum have defined the changing osmoregulatory characteristics associated with pregnancy (Davison et al. 1988a). By 5–8 weeks' gestation there was a 10 mosmol/kg reduction in the thirst osmotic threshold (Table 29.2). The slope of the thirst osmoregulatory line failed to alter during pregnancy and was the same as preconceptual and postpartum values (Fig. 29.3).

Similar but not identical changes occurred to vasopressin osmoregulation. The reduction in the vasopressin osmotic threshold was slightly less at 7 mosmol/kg by 5–8 weeks' gestation, but the major difference between thirst and vasopressin osmoregulation was seen at 28–33 weeks' gestation, when the slope of the vasopressin osmoregulatory line was reduced significantly by over 50% of the non-pregnant value (Table 29.2). These changes in functional characteristics of thirst and vasopressin osmoregulation account entirely for the large fall in basal plasma osmolality observed in pregnancy (Lindheimer et al. 1989).

Thus, the changes in osmoregulation documented in pregnancy are very similar to those associated with the luteal phase of the menstrual cycle, but they are more profound. The mechanisms responsible for the decreases in thirst and vasopressin osmotic thresholds are not yet clearly understood. Although it has been proposed that there is a reduction in "effective" blood volume or blood pressure in human pregnancy, there are few data to substantiate this view. Certainly studies in the rat

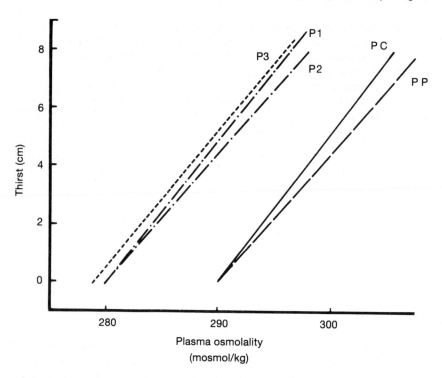

Fig. 29.3. The relationship between thirst ratings and plasma osmolality in the pregnant (P1–P3) and non-pregnant (PC, PP) states. P1 data obtained at 5–8 weeks' gestation; P2, 10–12 weeks; P3, 28–30 weeks. PC, preconception and PP, 10–12 weeks postpartum. (Adapted from Davison et al. (1988a), with permission.)

suggest strongly that volume regulation in pregnancy cannot account for these findings (Barron et al. 1984). It would seem more likely that one or more circulating pregnancy-associated hormones or hormonal changes are responsible. Again, some studies in the rat have failed to implicate oestrogens, progesterone, prolactin, opioids or aspects of the renin–angiotensin system (Lindheimer et al. 1987), whereas others have documented differences in drinking during oestrus and have noted variations in plasma vasopressin concentrations with exogenous oestrogens and oestrus (Findlay et al. 1979; Skowsky et al. 1979).

However, the serendipitous observation that a patient with a pregnancy due to a hydatidiform mole, who had circulating human chorionic gonadotrophin concentrations about 4000 U/ml, demonstrated plasma hypotonicity and reductions in osmotic thresholds for thirst and vasopressin, suggested that chorionic gonadotrophin may be responsible for the osmoregulatory changes of pregnancy, at least in part (Davison et al. 1988a). Injection of this hormone into women wishing to conceive does cause a reduction of basal plasma osmolality and osmotic thresholds, but the decrement is only about half that seen in pregnancy. However, the circulating concentrations of human chorionic gonadotrophin achieved after injection were substantially less than those observed in pregnancy. It thus appears that human chorionic gonadotrophin may play an important role in lowering the osmotic thresholds for thirst and vasopressin during human pregnancy, but there may be other factors yet to be identified.

The reason(s) for the reduction in slope of the vasopressin osmoregulatory line at 28–33 weeks' gestation (Table 29.2) are unclear. One possible explanation is that hypervolaemia associated with pregnancy has decreased osmotic sensitivity in a manner similar to that described in primary hyperaldosteronism (Ganguly and Robertson 1980). Another potential reason might be the enhanced metabolic clearance rate of vasopressin that has recently been demonstrated in human pregnancy (Davison et al. 1989). Further studies are required to elucidate the cause of this subtle change in osmoregulation of vasopressin in late pregnancy.

It has been realized for the past few years that drinking rapidly abolishes thirst and vasopressin secretion in dehydrated man (Rolls et al. 1980; Geelen et al. 1984) and after infusion of hypertonic saline (Thompson et al. 1987), despite continued plasma hypertonicity. An oropharyngeal reflex appears to mediate the effect of drinking on vasopressin secretion and thirst (Seckl et al. 1986). A similar oropharyngeal neuroendocrine reflex occurs in human pregnancy (Davison et al. 1988b).

Lactation

No formal studies on osmoregulation have been undertaken in lactating women. However, information is available on water intake in this group.

Fluid intake in healthy adults is estimated to be 1000–1600 ml/day under normal circumstances (Goodhart and Shils 1980; Pike and Brown 1975), although there is a large variation between individuals, as great as 600–6000 ml per day (Holmes 1964). Lactating women are generally advised to increase their fluid intake (Nichols and Nichols 1981), but until recently there were no data on which to base these recommendations. It has been reported that lactating women have a mean total daily intake of about 2860 ml (see Table 29.3 for details; Stumbo et al. 1985). It was found that there was a large variability in fluid intake both within subjects and between subjects, but little variation with the seasons. Thus it appears that lactating women require about twice the fluid intake of non-lactating adults. Enhanced fluid intake is also seen in the lactating rat.

Postmenopause

Few data, if any, are available which show whether there are any major changes in osmoregulation in the immediate postmenopausal period. As age advances into the sixth and seventh decades significant alterations in osmoregulation do occur. Two independent groups have documented that the sensitivity of the vasopressin osmostat increases with age, but there appears to be no variation in the osmotic threshold for vasopressin release (Helderman et al. 1978; Bevilacqua et al. 1987); in contrast, thirst appreciation declines with age (Phillips et al. 1984). These topics are considered in detail in Chapter 26.

Forsling et al. (1982) showed that the basal plasma vasopressin concentration of postmenopausal women did not appear to be particularly dissimilar to basal values reported in other studies on the menstrual cycle (Forsling et al. 1981), suggesting perhaps no large changes in vasopressin secretion. Precise evaluation of osmoregulated thirst and vasopressin secretion was not undertaken.

Table 29.3. Daily water intake of 26 lactating women

	Water intake		
	From food (ml)	From drinking (ml)	Total (ml)
Mean (SEM)	648 (48)	2220 (110)	2860 (108)
Range: min	299	1360	1920
max	1515	3139	3957

From Stumbo et al. (1985), with permission.

Summary

Careful studies on osmoregulation of thirst and vasopressin secretion have defined subtle changes in the menstrual cycle of healthy women. There is a reduction of the osmotic thresholds for both thirst and vasopressin secretion by about 4 mosmol/kg, in the luteal phase of the cycle. The observations account for the fall in basal plasma osmolality which occurs in the luteal phase. The mechanisms responsible are not known.

Similar but more profound alterations are seen in pregnancy. Within a few weeks of gestation basal plasma osmolality falls by about 10 mosmol/kg again due to a reduction in osmotic thresholds for both thirst and vasopressin. The sensitivity of the vasopressin osmostat may be reduced in the third trimester or the increased metabolic clearance of vasopressin at this time in pregnancy appears to account for the reduced increments in osmotically stimulated plasma vasopressin concentrations. Circulating human chorionic gonadotrophin may be responsible for a degree of the shifts in the osmotic thresholds.

There are no clear data on osmoregulation during lactation or in the immediate postmenopausal period, but fluid intake increases twofold in lactating women.

References

Barron WM, Stamoutsos BA, Lindheimer MD (1984) Role of volume in the regulation of vasopressin secretion during pregnancy in the rat. J Clin Invest 73:923–932

Baylis PH (1988) Vasopressin secretion during the menstrual cycle: an overview. In: Cowley AW, Liard J-F, Ausiello DA (ed) Vasopressin: cellular and integrative functions. Raven Press, New York, pp 273–280

Baylis PH, Thompson CJ (1988) Osmoregulation of vasopressin secretion and thirst in health and disease. Clin Endocrinol 29:549–576

Bevilacqua M, Norbiato G, Chebat E et al. (1987) Osmotic and non-osmotic control of vasopressin release in the elderly: effect of metoclopramide. J Clin Endocrinol Metab 65:1243–1247

Davison JM, Vallotton MB, Lindheimer MD (1981) Plasma osmolality and urinary concentration and dilution during and after pregnancy: evidence that lateral recumbency inhibits maximal urinary concentrating ability. Br J Obstet Gynaecol 88:472–479

Davison JM, Gilmore EA, Durr J, Robertson GL, Lindheimer MD (1984) Altered osmotic thresholds for vasopressin secretion and thirst in human pregnancy. Am J Physiol 247:F105–F109

Davison JM, Shiells EA, Philips PR, Lindheimer MD (1988a) Serial evaluation of vasopressin release and thirst in human pregnancy: role of human chorionic gonadotrophin in the osmoregulatory changes of gestation. J Clin Invest 81:798–806

Davison JM, Shiells EA, Philips PR, Lindheimer MD (1988b) Suppression of AVP release by drinking despite hypertonicity during and after gestation. Am J Physiol 254:F588–F592

Davison JM, Shiells EA, Barron WM, Robinson AG, Lindheimer MD (1989) Changes in the metabolic clearance of vasopressin and of plasma vasopressinase throughout human pregnancy. J Clin Invest 83:1313–1318

Findlay ALR, Fitzsimons JT, Kucharczyk J (1979) Dependence of spontaneous and angiotensin-induced drinking in the rat upon the oestrous cycle and ovarian hormones. J Endocrinol 82:215–255

Forsling ML, Akerlund M, Stromberg P (1981) Variations in plasma concentrations of vasopressin during the menstrual cycle. J Endocrinol 89:263–266

Forsling ML, Stromberg P, Akerlund M (1982) Effect of ovarian steroids on vasopressin secretion. J Endocrinol 95:147–151

Ganguly A, Robertson GL (1980) Elevated threshold for vasopressin release in primary hyperaldosteronism. Clin Res 280:33A (Abstr)

Geelen G, Keil LC, Kravik SE et al. (1984) Inhibition of plasma vasopressin after drinking in dehydrated humans. Am J Physiol 247:R968–R971

Goodhart RS, Shils ME (1980) Modern nutrition in health and disease, 6th edn. Lea and Febiger, Philadelphia

Greene R, Dalton K (1953) The premenstrual syndrome. Br Med J i:1007–1014

Helderman JH, Vestal RE, Rowe JW, Tobin JD, Andres R, Robertson GL (1978) The response of arginine vasopressin to intravenous ethanol and hypertonic saline in man: the impact of aging. J Gerontology 33:39–47

Holmes JH (1964) Thirst and fluid intake problems in clinical medicine. In: Wayne MJ (ed) Thirst. Macmillan, New York

Lindheimer MD, Barron WM, Durr J, Davison JM (1987) Water homeostasis and vasopressin release during rodent and human gestation. Am J Kidney Dis 9:270–275

Lindheimer MD, Barron WM, Davison JM (1989) Osmoregulation of thirst and vasopressin release in pregnancy. Am J Physiol 257:F159–F169

Mira M, Stewart PM, Gebski V, Llewellyn-Jones D, Abraham SF (1984) Changes in sodium and uric acid concentrations in plasma during the menstrual cycle. Clin Chem 30:380–381

Nichols BL, Nichols VN (1981) Human milk: nutritional resources. In: Tsang RG, Nichols BL (eds) Nutrition and child health: perspectives for the 1980's. Alan R Liss, New York, pp 109–146

Phillips PA, Rolls BJ, Ledingham JGG et al. (1984) Reduced thirst after water deprivation in healthy elderly man. N Engl J Med 311:753–759

Phillips PA, Rolls BJ, Ledingham JGG, Forsling ML, Morton JJ (1985) Osmotic thirst and vasopressin release in humans: a double-blind crossover study. Am J Physiol 248:R645–R650

Pike RL, Brown ML (1975) Nutrition: an integrated approach, 2nd edn. John Wiley, New York

Robertson GL (1984) Abnormalities of thirst regulation. Kidney Int 25:460–469

Robertson GL, Athar S (1976) The interaction of blood osmolality and blood volume in regulating plasma vasopressin in man. J Clin Endocrinol Metab 42:613–620

Robertson GL, Shelton RL, Athar S (1976) The osmoregulation of vasopressin. Kidney Int 10:25–37

Rolls BJ, Wood RJ, Rolls ET, Lind H, Lind W, Ledingham JG (1980) Thirst following water deprivation in humans. Am J Physiol 239:R476–R482

Seckl JR, Williams TDM, Lightman SL (1986) Oral hypertonic saline causes transient fall of vasopressin in humans. Am J Physiol 251:R214–R217

Skowsky RW, Swan L, Smith P (1979) Effects of sex steroid hormones on arginine vasopressin in intact and castrated male and female rats. Endocrinology 104:105–108

Spruce BA, Baylis PH, Burd J, Watson MJ (1985) Variation in osmoregulation of arginine vasopressin during the human menstrual cycle. Clin Endocrinol 22:37–42

Stumbo PJ, Booth BM, Eichenberger JM, Dusdieker LB (1985) Water intakes of lactating women. Am J Clin Nutr 42:870–876

Thompson CJ, Bland J, Burd J, Baylis PH (1986) The osmotic thresholds for thirst and vasopressin release are similar in healthy man. Clin Sci 71:651–656

Thompson CJ, Burd JM, Baylis PH (1987) Acute supression of plasma vasopressin and thirst after drinking in hypernatraemic humans. Am J Physiol 252:R1138–R1142

Thompson CJ, Burd JM, Baylis PH (1988) Osmoregulation of vasopressin secretion and thirst in cyclical oedema. Clin Endocrinol 28:629–635

Turner C, Fortney S (1984) Daily plasma volume changes during the menstrual cycle. Fed Proc 43:718 (Abstr)

Vokes TJ, Weiss NM, Schreiber J, Gaskill MB, Robertson GL (1988) Osmoregulation of thirst and vasopressin during normal menstrual cycle. Am J Physiol 254:R641–R647

Commentary

Szczepanska-Sadowska: The author postulates that the changes in osmoregulation of thirst and vasopressin release cannot be related to baroregulatory influences because of lack of changes in blood volume and pressure. However, the latter does not exclude blood flow redistribution, changes in the central haemodynamics (especially in pregnancy) or in the sensitivity of the baroreflex. The observation that the blood volume may be increased in the luteal phase may in fact disclose some inefficiency of baroregulation. Inadequate inhibitory function of baroreflex could result in downward shift of osmoregulation of thirst and vasopressin.

Baylis: Dr Szczepanska-Sadowska is correct in introducing an alternative explanation for the lowering of osmotic thresholds for thirst and vasopressin, particularly in pregnancy. It is well recognized from both animal and human studies that moderate hypovolaemic/hypotensive stimuli will lower the vasopressin osmotic threshold. Controversy has existed in earlier years about the role of non-osmotic baroregulatory changes in resetting the osmostat in pregnancy. The Denver group certainly have proposed that the changes in pregnancy could be accounted for by relative underfilling of the vascular compartment despite the increase in extracellular fluid volume in pregnancy. Consensus of opinion, however, is against this view at present.

Robertson: The observation that lactating women have increased rates of fluid intake raises the following questions; (a) is all the extra fluid needed to replace losses in milk and; if not, is the surplus intake a form of primary polydipsia or an appropriate response to increased renal and/or insensible losses? (b) What is the stimulus to drink?

Baylis: In response to Dr Robertson's questions, I am unaware of any careful studies in lactating humans which clearly define fluid balance. No data are available on thirst appreciation or osmotic thresholds during lactation. It is certainly an area that warrants investigation.

Chapter 30

Disorders of Thirst in Man

G.L.Robertson

Introduction

Water Balance

In humans and other terrestrial mammals, water is lost continuously from the skin, lungs and kidneys (Adolph 1969). The rate of loss varies considerably depending on ambient temperature, physical activity, antidiuretic function and solute load. On average, however, even a healthy, sedentary adult eating a normal diet and living in a temperature-controlled environment, loses an average of at least 1500 ml/day from renal and extrarenal sites. These obligatory losses are offset partially by water taken with meals or generated from the metabolism of dietary fat. However, the amounts derived from these sources are usually only about 1000–1500 ml/day. Hence, the maintenance of normal water balance requires the existence of some mechanism to ensure that the rate of water ingestion will always be sufficient to replace obligatory losses. This indispensable homeostatic role is normally played by the thirst mechanism (Robertson et al. 1982; Robertson 1984; Robertson and Berl 1986).

Definition of Terms

For physiological purposes, thirst is defined as a conscious sensation of a need for water and a desire to drink. The sensation must be of a kind that normally motivates water seeking behaviour because that is the only condition which is meaningful for the physiology of water balance. The term thirst should not be equated with the absence of satiety since the latter really corresponds to a neutral condition in which the individual is not motivated either to avoid or to seek the ingestion of fluid. Although the two sensations often go together, thirst also is not equated with the sensation of "dry mouth" because clinical observations suggest that the latter can occur independently as a result of various activities, drugs or diseases that reduce salivary flow or increase evaporation from oropharyngeal mucous membranes.

Thirst Independent Influences on Fluid Intake

It is important to note that thirst is not the only or even the major determinant of the rate of water intake under ordinary conditions. In many cultures, there are ritualistic behaviours like the coffee break in which the ingestion of fluids is not necessarily motivated by thirst and does not seem to serve a physiological so much as a social or psychological need (see Section V). The same may be said of the clinical disorder known as psychogenic or compulsive water drinking which is almost always explained by a motive other than thirst and may have its origin in the same abnormal psychobiological drives as other compulsive behaviours. Not infrequently, water is also drunk from a belief in its therapeutic values. In rats and most humans, the acts of eating and drinking seem to be inextricably connected but it is unclear whether the link is true thirst, taste or another oropharyngeal sensation (see Section IV). Likewise, the absence of water intake does not necessarily signify a lack of thirst because the behavioural response to this potent biological drive can be diverted or thwarted by a variety of physical, psychological or environmental obstacles.

Measurement of Thirst

Despite its crucial and powerful influence on water homeostasis, true thirst is difficult to measure reliably. A variety of linear rating scales have been used with some success to quantitate changes in thirst in humans during controlled stimuli of various types (Rolls et al. 1980; Thompson et al. 1981, 1986, 1987; Robertson et al. 1982; Robertson 1984, 1985). This approach seems to give fairly reproducible values that correlate significantly not only with stimulus strength but also with subsequent measures of ad libitum water intake. However, as a subjective measure, it is totally dependent on patient reliability and cannot be used to quantitate thirst in long-term or uncontrolled situations. An objective measure of water intake can obviate some of these difficulties and is the only approach available for studies in experimental animals. It can also be shown to correlate well with stimulus strength in many situations and has the advantage of providing quantitative values of direct physiological relevance. However, this approach also has major limitations because drinking itself alters the strength of dipsogenic stimuli and, as noted above, it does not always reflect the sensation of thirst. This lack of a simple, safe and consistently reliable method for measuring the sensation of thirst remains a major obstacle to a full understanding of the physiology and pathophysiology of this vital homeostatic mechanism. Despite these limitations, however, a clearer picture of its role in various clinical disorders of water balance has begun to emerge.

Physiology

Osmoregulation of Thirst

In healthy people, thirst appears to be regulated primarily by the "effective" osmotic pressure of extracellular fluid (Wolf 1950; Robertson et al. 1982; Robertson 1984). This control appears to be mediated by an osmoregulatory mechanism very similar to that which regulates secretion of the antidiuretic hormone, arginine vasopressin (Fig. 30.1). Thus, there appears to be a level of plasma osmolality below which thirst is absent. Above this "threshold", or "set point", thirst increases very rapidly in direct proportion with small increases in plasma osmolality. The sensitivity of this osmoregulatory mechanism is such that a rise in effective plasma osmolality of as little as 2%–3% is sufficient to trigger a very powerful desire to drink. The

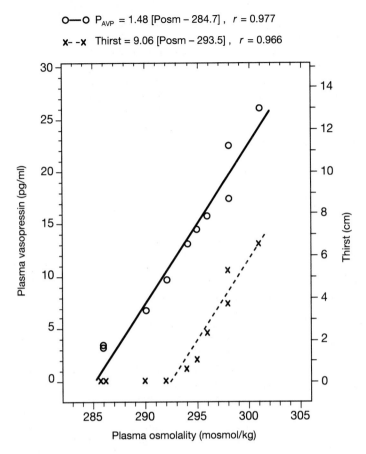

o—o P_{AVP} = 1.48 [Posm − 284.7] , r = 0.977

x- -x Thirst = 9.06 [Posm − 293.5] , r = 0.966

Fig. 30.1. The relationship of thirst and plasma vasopressin to plasma osmolality during the infusion of hypertonic saline in a healthy adult. Thirst values are expressed in centimetres from the starting point of an analogue rating scale. (From Robertson (1984), with permission.)

intensity of this feeling seems to vary considerably from subject to subject but, in a given person, it almost always correlates closely with the magnitude of the increase in plasma osmolality. Thus, the system that osmoregulates thirst is very precise as well as sensitive.

Relationship to Osmoregulation of Vasopressin

The only important difference between the osmoregulation of thirst and of vasopressin secretion appears to be the "threshold" or "set point" of the two systems. Although each of these properties varies considerably from person to person (see below) the osmotic threshold for thirst is slightly higher than that for vasopressin release (Wolf 1950; Robertson et al. 1982; Robertson 1984). Consequently, thirst does not begin until plasma osmolality reaches a level well above that required to stimulate vasopressin secretion. This arrangement is highly advantageous because it enables the antidiuretic mechanism to be used as the first line of defence against dehydration.

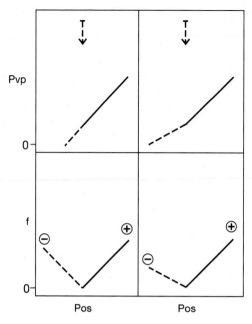

Fig. 30.2. Schematic representation of the operation of a bimodal osmoregulatory system. *Top panels*, the expected relationship of plasma vasopressin (Pvp) and onset of thirst (T) to plasma osmolality (Pos) under two hypothetical systems in which the inhibitory (–––) and stimulatory (——) components are of equal potency (*left*) or unequal potency (*right*). *Bottom*, frequency of firing (f) for the two components as a function of plasma osmolality in the two systems.

Thus, a person is free to range a considerable distance from a supply of fresh water and is not "driven to drink" until the intake of salt and/or the obligatory renal and extrarenal losses of water raise plasma osmolality by at least 2%–3% (6–9 mosmol/kg).

It has been reported that the osmotic thresholds for thirst and vasopressin secretion are similar in healthy adults (Thompson et al. 1986). In this study, however, thirst was measured with a rating scale that resulted in positive values under basal conditions. Since the water intake of the subjects had not been restricted, these ratings do not represent thirst as defined above, i.e. conscious desire for water that motivates one to drink. Rather, they probably represent the absence of a feeling of satiation. This interpretation is consistent with the fact that the "thirst" ratings did not increase immediately during hypertonic saline infusion but remained "flat" until plasma osmolality had risen a few percent above basal levels. Hence, the disparity between these findings and other reports is probably more semantic than real.

Hypothetical Model of Osmoregulatory System

A difference between the osmotic thresholds for thirst and vasopressin secretion does not necessarily mean that the two regulatory systems are "set" to begin firing at different levels of plasma osmolality. Evidence from patients with complete destruction of their osmoregulatory system strongly suggests that it is composed of inhibitory as well as stimulatory components (Robertson 1980). According to this theory, at the normal "basal" level of plasma osmolality both components are inactive and vasopressin is secreted tonically at a rate sufficient to maintain a state of half maximum antidiuresis (Fig. 30.2). When plasma osmolality falls below the "normal" level, the inhibitory arm of the osmoregulatory system is activated and vasopressin secretion is suppressed sufficiently to permit urinary dilution. On the other hand, when plasma osmolality rises above the normal level, the stimulatory arm of the control system is activated and vasopressin secretion rises to levels that permit maximum antidiuresis. In all probability, the osmoregulation of thirst is similarly constituted. Thus, at normal basal levels of plasma osmolality neither arm of the control system is active and the individual is in a basal state in which neither thirst nor satiety are present. Thirst begins to occur only when the stimulatory arm is activated by a rise in plasma osmolality. Since this response occurs at a level well above that at which vasopressin secretion is maximally suppressed, the "thresholds" of the two systems appear to differ. However, if "threshold" is redefined as the level of plasma osmolality at which both the inhibitory and stimulatory arms of the two control systems are inactive, then they may be very similar if not identical.

The existence of such a bimodal osmoregulatory system is consistent with prior electrophysiological studies which have shown that sodium sensitive neurons in the anterior hypothalamus respond in a symmetrical way to hyper and hypo-osmolar

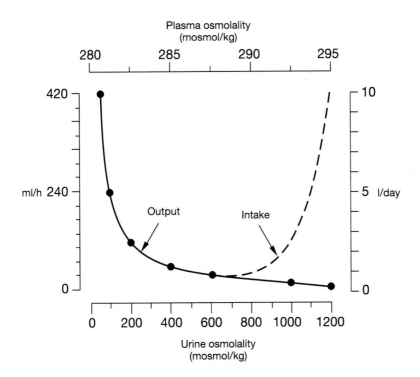

Fig. 30.3. Schematic diagram of the relationship between plasma osmolality, urine osmolality, the rate of urine output (——) and the rate of water intake (– – –) in a typical healthy adult.

changes (Nicolaïdis 1969). Thus, the firing rate of some neurons is decreased by hyperosmolality whereas others respond in the opposite way. This bimodal response pattern is observed when the osmolar changes are induced either by intracarotid injection or application to the oral cavity, indicating that the recordings were probably not obtained from the osmoreceptors themselves but from interneurons in the efferent projections to the neurohypophysis and/or cortical thirst areas.

Relation Between Thirst and Fluid Intake

The degree of thirst perceived during a stimulus such as hypertonic saline infusion correlates closely not only with the levels of plasma osmolality achieved but also with the amount of water drunk spontaneously at the end of the test. Thus, it is possible to derive a quantitative relationship between plasma osmolality and the rate of water intake (Fig. 30.3). The exponential curve that described this relationship is almost a mirror image of that which describes the relationship between plasma osmolality and urine output. These two curves cross near their bottoms at a plasma osmolality which approximates the normal basal level. Thus, any deviation above or below this "resting" point is counteracted by an increase in water intake or output. Since the lower portion of each curve is relatively shallow, small deviations from the norm are counteracted relatively slowly and may persist for long periods of time. However, larger deviations are opposed much more briskly. For example, a rise in plasma osmolality of 2.5% (7 mosmol/kg) increases water intake to a level of about 6 ml/min (Fig. 30.3). Under conditions of maximum antidiuresis, this rate of intake would be sufficient to expand body water and lower plasma osmolality at the rate of about 1% an hour (0.006 l/min x 60 min - 0.5 x 70 kg body weight = 0.01) for a total correction time of 2.5–3 hours.

Integration of Thirst and Antidiuretic Mechanisms

The frequency with which the thirst mechanism is called upon to defend against dehydration varies considerably, depending upon dietary and environmental circumstances. Assuming total body sodium does not change and obligatory losses of water are minimum (about 20 ml/kg body weight per day), a rise in plasma osmolality sufficient to stimulate thirst (1%–2%) would occur every 7 to 14 hours. In reality, however, the typical human diet contains enough free (1000 ml) and metabolically generated (300 ml) water to offset most if not all the daily obligatory loss. In addition, as noted above and discussed elsewhere (see Section V), water intake may be further augmented by various psychological or cultural influences. Generally speaking, therefore, the thirst mechanism is called into play only when the intake of sodium or the rate of water loss is unusually high. This situation probably occurs frequently among primitive peoples because greater physical activity and exposure to the elements predisposes to high rates of insensible water loss (Adolph 1969). In technically developed societies, however, this kind of stress occurs infrequently and then only as a result of certain occupational, recreational or environmental factors. Consequently, patients who lack a thirst mechanism can maintain a normal level of plasma osmolality for weeks or even months without specific treatment (see below).

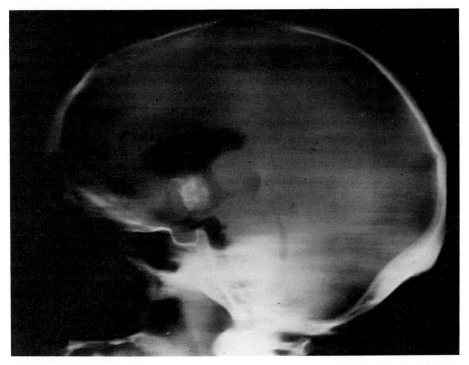

Fig. 30.4. Pneumoencephalogram from a 41-year-old male with adipsic hypernatraemia due to selective destruction of osmoreceptor neurons. Note the calcified tumour anterior to the third ventricle.

Pathophysiology

Hypodipsia

Anatomy

The osmoreceptor neurons that regulate thirst and vasopressin secretion appear to be located in the anterior hypothalamus, near but distinctly separate from the magnocellular neurons of the supraoptic nuclei where vasopressin is produced (see Section II). Thus, tumours and a variety of pathological lesions involving this area of the brain are frequently associated with diminished or absent thirst in response to osmotic stimulation (Table 30.1; Fig. 30.4). One of the lesions that can produce this syndrome is surgery on the anterior communicating artery (Robertson 1980). Since this artery is part of a large collateral system that circles the base of the brain, the only area likely to be infarcted by this procedure is the portion of the anterior hypothalamus supplied by small perforating end-arteries that arise from the back of the anterior communicating artery. Hence, either the osmoreceptors themselves or their efferent projections would appear to be concentrated in the small midline area of the anterior hypothalamus supplied by these short perforators. Interruption of these arteries does not infarct the more laterally situated supraoptic nuclei (Robertson 1980) since they receive their blood supply from separate branches of the internal carotid arteries.

Table 30.1 Causes of osmoreceptor destruction

Tumours
  Craniopharyngioma
  Pinealoma
  Meningioma
  Metastatic

Granuloma
  Histiocytosis
  Sarcoidosis

Vascular
  Occlusion of the anterior communicating artery

Trauma
  Penetrating
  Closed

Other
  Hydrocephalus
  Developmental defects of corpus callosum and other
  midline structures
  Cysts
  Inflammation
  Degenerative (Alzheimer's disease)

Idiopathic

Clinical Manifestations/Natural History

Destruction of the thirst osmoreceptors results in a clinical syndrome characterized by chronic or recurrent hypernatraemia in association with inadequate or absent thirst (hypodipsia or adipsia). The hypernatraemia can vary greatly in magnitude (plasma sodium concentrations as high as 200 mEq/l have been observed) and appears to be due largely if not totally to a severe depletion of body water. Thus, these patients almost always have evidence of volume depletion such as tachycardia, postural hypotension, prerenal azotaemia, hyperreninaemia and secondary hyperaldosteronism. Possibly as a consequence of the resultant hypokalaemia, they may also develop severe hyperglycaemia, rhabdomyolysis and paralysis. The severe hypertonic dehydration that characterizes this syndrome is due largely if not totally to a failure to ingest enough water to replenish obligatory renal and insensible losses. Although most patients with hypodipsia also have deficiencies of vasopressin secretion (see below), their basal urine outputs are not increased. Hence, urinary losses do not contribute significantly to the genesis of the dehydration. However, polyuria sometimes develops when large volumes of fluid are given in an effort to correct dehydration (see below). Hence, it can be a significant problem in the treatment of this condition.

In virtually every patient with hypodipsia or adipsia, the hypertonic dehydration eventually recurs if the patient is not taught to drink prophylactically in response to a loss of weight or some other indicator of dehydration. In the absence of such a regimen, plasma osmolality and sodium may remain relatively normal for weeks or months if the obligatory renal and insensible losses of water are low enough to be covered by the amount of water ingested with food and other thirst-independent

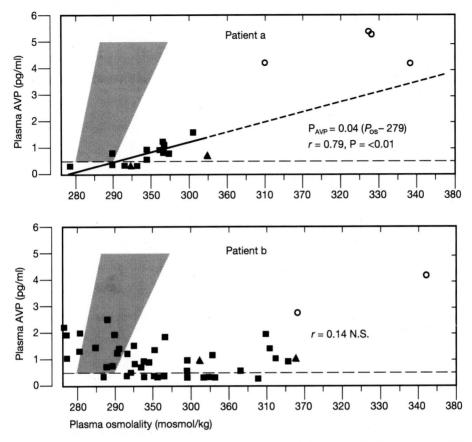

Fig. 30.5. The relationship of plasma vasopressin to corrected plasma osmolality in two patients with adipsic hypernatraemia. There is a significant positive relationship in patient a but not in patient b and plasma vasopressin was subnormal for plasma hypertonicity in both patients. Shaded areas indicate the range of normal.

activities. Sooner or later, however, an increase in physical activity and/or temperature prompts a rise in respiratory and perspiratory losses. When this occurs, ordinary rates of intake are inadequate to offset losses and dehydration ensues, sometimes with amazing rapidity. In one patient, for example, plasma sodium was documented to rise from 145 to 175 mEq/l during a hot summer's day on the beach. This kind of abrupt and rapid descent into dehydration is readily explained by lack of a dipsogenic response to the marked increase in insensible losses that are known to result from increases in activity and environmental temperature (Adolph 1969).

Vasopressin Deficiencies

Destruction of the thirst osmoreceptors in humans is almost always associated with deficiencies in the osmoregulation of vasopressin. These deficiencies manifest as absent or markedly subnormal increases of plasma vasopressin under hypertonic conditions (Fig. 30.5). They appear to be due to selective destruction of the vasopressin

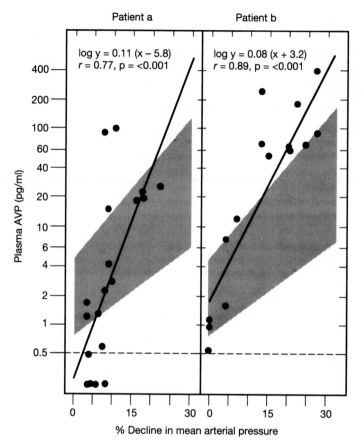

Fig. 30.6. The relationship of plasma vasopressin to percentage decrease in mean arterial pressure induced by infusion of a ganglionic blocker in the two patients with adipsic hypernatraemia shown in Fig. 30.5. Each patient showed a significant response which equalled or exceeded that found in healthy adult controls (shaded area).

osmoreceptors rather than the neurohypophysis because the plasma vasopressin response to a non-osmotic stimulus is usually normal or even supranormal (Fig. 30.6). These observations are consistent with much other evidence that the osmoreceptors are distinct from the neurosecretory neurons themselves. They also indicate that the systems for thirst and vasopressin are either the same or, more likely, are located in close if not overlapping areas of the brain. A deficiency of osmotic dipsogenesis without a deficiency of osmotically mediated vasopressin secretion has been described (Hammond et al. 1986) but this disassociation is rare and does not necessarily indicate completely separate systems because it could result from disruption of efferent projections rather than of the osmoreceptors themselves.

In patients with hypodipsia, the deficiencies in the osmoregulation of vasopressin secretion can be complete or partial (Fig. 30.5). In the former, plasma vasopressin is completely refractory to changes in plasma osmolality and remains relatively constant within the normal, basal range even under conditions of hypotonic over-hydration (Fig. 30.5b). This inability osmotically to suppress or stimulate vasopressin secretion is not attributable to interference by non-osmotic influences and appears to be due to a complete lack of osmoregulatory input to the neurohypophysis. This implies that the osmoregulatory system is probably bimodal, that is, it normally contains inhibitory as well as stimulatory components, and when both influences are absent, vasopressin is secreted tonically in amounts which approximate the normal basal level. In other cases, plasma vasopressin varies in close correlation with changes in plasma osmolality but the slope or sensitivity of this relationship is markedly subnormal (Fig. 30.5a). Thus, plasma vasopressin is low or undetectable under conditions of normal hydration and does not reach physiologically significant levels unless plasma osmolality and sodium concentration are abnormally high. The preservation of a limited capacity to suppress as well as stimulate vasopressin secretion implies that the deficiency of osmoregulation is incomplete and may primarily involve the stimulatory component.

It is noteworthy that the rather subtle difference between complete and partial deficiency of osmoregulation results in marked differences in the clinical characteristics of the disorder. The patients with a complete deficiency can never dilute their urine and will develop overhydration if their rate of water intake exceeds the rate of obligatory renal and extrarenal losses. Hence, they are vulnerable to episodes of severe hypo- as well as hypernatraemia. In contrast, the patients with the partial defect tend to dilute their urine at inappropriately high levels of plasma osmolality. This abnormality protects against hyponatraemia but also impairs efforts to correct hypernatraemia by water loading alone (see below). It resembles the concentrating defect of partial neurogenic diabetes insipidus or upward resetting of the osmostat. However, the latter two entities can be distinguished from a partial osmoreceptor defect by demonstrating that the patient's plasma vasopressin response to a non-osmotic stimulus is normal whereas that to osmotic stimulation is subnormal due solely to a marked reduction in the slope or sensitivity of the relationship to plasma osmolality.

Treatment

The treatment of the hypodipsic syndromes is based on correction and prevention of the dehydration by educating the patient on the need to drink in response to indicators other than thirst. During hospitalization for acute episodes, the physician can prescribe oral or intravenous intake based on frequent measures of body weight and plasma sodium concentration. After discharge, however, the patient must be taught to drink a certain daily minimum (based on estimates of his or her basal urine output and insensible loss) and also to increase intake during periods of increased insensible losses. These requirements are met most easily by providing the patient with a scale or schedule which specifies the amount of fluid intake required each day as a function of changes in body weight (Robertson 1984). In some patients with adipsic hypernatraemia, the administration of clofibrate or chlorpropamide decreases hypertonicity and hypernatraemia by promoting the retention of water (Fig. 30.7). However, it is not clear whether this improvement is due to increased intake, decreased output or both factors since neither variable changes significantly even when measured in a metabolic unit. Most likely, the drugs act by preventing the water diuresis that would otherwise develop when plasma osmolality, sodium and vasopressin decline with increasing hydration (Fig. 30.5a). This antidiuretic effect is not mediated by stimulation of vasopressin secretion since plasma levels of the hormone actually decrease to undetectable levels during administration of the drugs (Fig. 30.7).

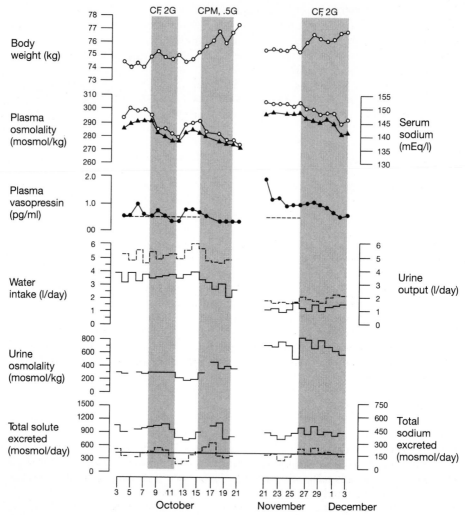

Fig. 30.7. Effect of clofibrate (CF) or chlorpropamide (CPM) on various indicators of water balance and antidiuretic function in a patient with partial osmoreceptor deficiency (Figs. 30.5a and 30.6a). The patient was studied twice, once while on a regimen of forced water loading of 5 l a day (October) and again when he was allowed to drink ad libitum (November–December).

These drugs or dDAVP itself are sometimes helpful in preventing recurrence of dehydration in patients with partial osmoreceptor deficiencies who tend to dilute their urine at inappropriately high levels of plasma osmolality. However, they are of little or no value in patients with complete osmoreceptor destruction and never replace the need for the basic regimen of weight-controlled water intake.

Polydipsia

Clinical disorders characterized by increased intake of fluids can be classified as either secondary or primary (Table 30.2).

Table 30.2. Causes of polydipsia

Primary
Dipsogenic
Idiopathic
Neurosarcoid
Tuberculous
Meningitis
Multiple sclerosis (?)
Psychogenic
Schizophrenia
Compulsive disorders (?)
Secondary
Dehydration
Diabetes insipidus
Insensible loss
Ingestion of salt
Hyperglycaemia (insulin-deficient diabetes mellitus)
Hypokalaemia
Hyperreninaemia (?)

In secondary polydipsia, the increased water intake is a physiologically appropriate response to dehydration or some other hypertonic stimulus. It is always associated with thirst and is exemplified most clearly by patients with neurogenic or nephrogenic diabetes insipidus.

Secondary Polydipsias

Diabetes Insipidus. In this disorder, thirst and polydipsia are induced by the rise in plasma osmolality that results when body water is reduced by the excretion of abnormally large volumes of dilute urine. The degree of dehydration required to stimulate polydipsia is relatively small (1%–2%) because basal plasma osmolality is normally maintained only slightly below the thirst threshold. Once this threshold is reached, the resultant increase in drinking is sufficient to offset even maximum rates of polyuria of 20 l/day. Hence, plasma osmolality and sodium concentration do not rise above the normal range unless the thirst mechanism is also defective or drinking is prevented by some other factor. This fact attests to the singular capacity of the thirst mechanism to maintain water balance even in the complete absence of antidiuretic function.

In patients with neurogenic diabetes insipidus, thirst and polydipsia rapidly subside when the polyuria and dehydration are corrected by antidiuretic therapy (Fig. 30.8). This reduction in water intake correlates closely with the reduction in plasma osmolality that results from the retention of water (Fig. 30.9). The relationship of these two variables differs considerably from patient to patient but, in almost every case, is sufficient to restore balance between intake and output at plasma osmolalities well within the normal range. Hence, hypo-osmolaemia or other signs of overhydration do not develop unless the patient also has abnormal thirst. This fact provides another instructive illustration of the capacity of the normal thirst mechanism precisely to regulate water balance even when antidiuretic function is no longer responsive to

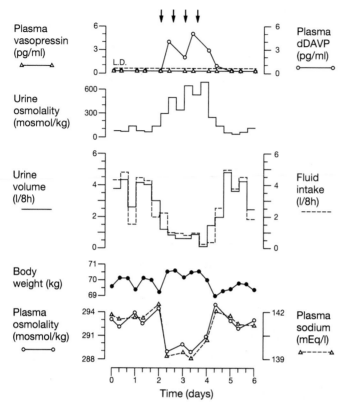

Fig. 30.8. The effect of the vasopressin analogue, desmopressin (dDAVP) on water balance in a patient with neurogenic diabetes insipidus. dDAVP was given intranasally at a dose of 25 μg every 8 hours.

osmotic variables. Indeed, the experiments of nature described here and in the section on hypodipsia clearly show that, in humans at least, a normal thirst mechanism is both a necessary and a sufficient condition for the maintenance of normal water balance. The antidiuretic mechanism serves only to free the individual from the need to remain close to a supply of fresh water. This freedom was undoubtedly of great survival value among primitive peoples but is little more than a convenience in modern societies.

Other. Secondary polydipsia also occurs when plasma osmolality is raised by increased rates of sodium intake or extrarenal losses of water due to perspiration and/or hyperventilation. The polydipsia that occurs in uncontrolled insulin-deficient diabetes mellitus is another example since it appears to be due at least in part to stimulation of the thirst osmoreceptor by the increased levels of plasma glucose (Vokes and Robertson, unpublished). Finally, thirst and water intake can also be stimulated by severe haemorrhage or other acute reductions in blood volume. The degree of hypovolaemia required is unknown but appears to be relatively large. Controlled studies in healthy volunteers indicates that reducing "effective" blood volume 20%–30% by a combination of frusemide diuresis and upright posture significantly lowers the osmotic threshold for thirst as well as for vasopressin secretion

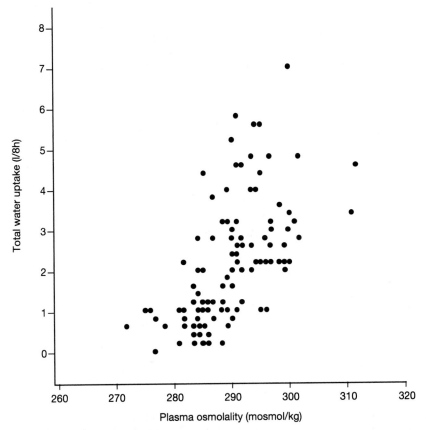

Fig. 30.9. Water intake as a function of plasma osmolality before and during dDAVP therapy in 13 patients with neurogenic diabetes insipidus. The two variables showed a significant positive correlation ($r = 0.53$, $P<0.001$) described by the equation water intake (l/8h) = 0.29 (plasma osmolality − 280).

(Weiss et al. 1984). Thus, polydipsia secondary to hypovolaemic stimulation probably does contribute to the hyponatraemia that often develops in patients with diuretic abuse, mineralocorticoid deficiency (Robertson et al. 1982), cholera and other forms of severe gastroenteritis.

Primary

In primary polydipsia, the drinking is a physiologically inappropriate response to some abnormal stimulus. It may be subdivided into two categories; the dipsogenic variety in which the patient reports an abnormal increase in thirst; and the psychogenic variety in which thirst is denied and the cause appears to be a more generalized cognitive defect that leads to a fixed, irrational belief in the therapeutic value of a high water intake.

Dipsogenic. The clinical presentation of dipsogenic diabetes insipidus is very similar to that of partial neurogenic or partial nephrogenic diabetes insipidus (Robertson 1987). Thus, the patients complain of constant thirst, polydipsia and polyuria with

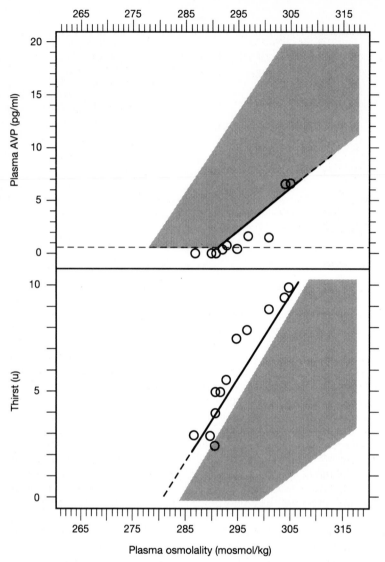

Fig. 30.10. Relationship of plasma vasopressin (*top*) and thirst ratings (*bottom*) to plasma osmolality during the infusion of hypertonic saline in a 37-year-old woman with dipsogenic diabetes insipidus and a 12-year history of multiple sclerosis. Her daily urine volume and osmolality were 6 l and 160 mosmol/kg, respectively. Basal plasma osmolality and sodium concentration were 290 mosmol/kg and 140 mEq/l, respectively. The shaded areas indicate the range of values found during the same procedure in 30 healthy adults.

no obvious abnormality in basal plasma osmolality or sodium concentration. Although their polydipsia and polyuria tend to lessen at night, they are still awakened often by thirst and/or a need to urinate. Also like partial neurogenic or nephrogenic diabetes insipidus, their urine is dilute (osmolality less than 300 mosmol/kg) under basal conditions of ad libitum intake but usually becomes concentrated when their drinking is restricted.

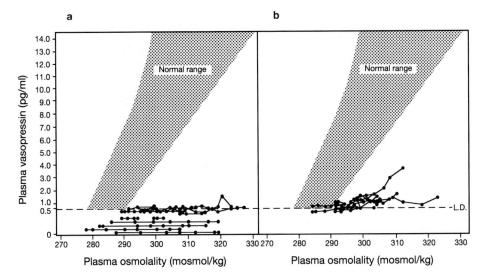

Fig. 30.11. The relationship of plasma vasopressin to plasma osmolality during the infusion of hypertonic saline in patients with **a** severe or **b** partial neurogenic diabetes insipidus. The shaded areas indicate the range of values found in 30 healthy adults studied under identical conditions.

Despite the many clinical similarities between dipsogenic, neurogenic and nephrogenic diabetes insipidus, the three disorders are easily distinguished by their characteristic differences in vasopressin function (Robertson 1988). Thus, in patients with dipsogenic diabetes insipidus, plasma vasopressin rises normally with plasma osmolality during fluid deprivation or hypertonic saline infusion tests (Fig. 30.10). This result excludes neurogenic diabetes insipidus which is distinguished by a markedly deficient vasopressin response to the same stimulus (Fig. 30.11). Renal resistance to the antidiuretic effect of vasopressin (nephrogenic diabetes insipidus) can also be excluded since the relationship of urine osmolality to plasma vasopressin is also normal in patients with dipsogenic diabetes insipidus (maximum concentrating capacity may be blunted but this abnormality occurs to varying degrees in all forms of chronic polyuria).

The lack of any significant defect in the secretion or action of vasopressin in these patients indicates that their thirst, polydipsia and polyuria must be due to the presence of some abnormal dipsogenic stimulus. This deduction is supported by the estimates of thirst reported by the patients during the hypertonic saline infusion (Fig. 30.10). Analysis of these ratings indicate that their feelings of thirst continue to be influenced by changes in plasma osmolality but the "set" of the osmostat appears to be lower than normal. As a consequence, thirst is experienced at levels of plasma osmolality below those required to stimulate vasopressin secretion. This inversion of the normal relationship results in a vicious cycle in which the patient drinks excessively in an effort to quench thirst but is unable to do so because the resultant fall in plasma osmolality suppresses vasopressin secretion and produces a counterbalancing polyuria before the requisite degree of water retention and hypo-osmolaemia are achieved. In many cases, this problem is aggravated by a slight upward resetting of the vasopressin "osmostat" that may be caused by the same pathological process that resets the thirst osmostat in the opposite direction.

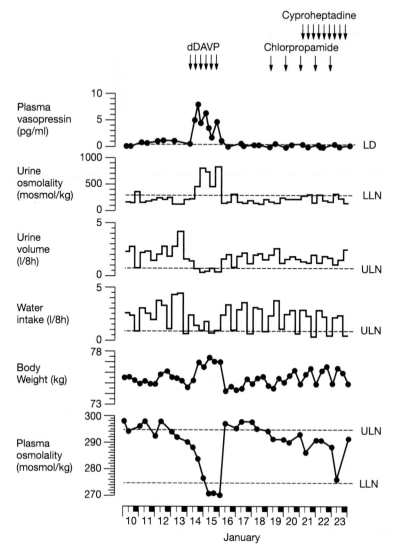

Fig. 30.12. The effect of the vasopressin analogue, desmopressin (dDAVP) on water balance in a woman with dipsogenic diabetes insipidus. The patient is the same as shown in Fig. 30.10. dDAVP was given intranasally at a dose of 25 μg every 8 hours. Intake and output are shown for successive 8-hour periods. The period of sleep from 11:00 p.m. to 8:00 a.m. is indicated by darkened areas on the horizontal time scale. LD, ULN and LLN refer, respectively, to limit of detectability, upper limit of normal and lower limit of normal for the variables indicated. The patient was also given a therapeutic trial of chlorpropamide which had little or no effect and of cyproheptadine (a histamine and serotonin antagonist) which produced a fall in plasma osmolality and sodium almost as great as that produced by the dDAVP.

Further evidence that dipsogenic diabetes insipidus results from an abnormal reduction in the osmotic threshold for thirst is provided by the response of these patients to antidiuretic therapy. As noted above, the administration of antidiuretic hormone to a patient with neurogenic diabetes insipidus promptly abolishes thirst

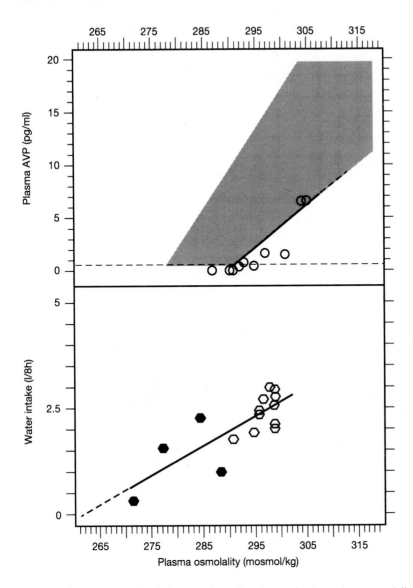

Fig. 30.13. The relationship of plasma vasopressin and water intake to plasma osmolality in a woman with dipsogenic diabetes insipidus. The patient is as shown in Figs. 30.10 and 30.12. In the bottom panel, the open and closed symbols signify the values obtained before and during dDAVP treatment. The shaded area indicates the range observed under identical conditions in 13 patients with neurogenic diabetes insipidus. (Fig. 30.9). The top panel is identical to Fig. 30.10.

and polydipsia as well as polyuria without causing excessive retention of water (Fig. 30.8). When the same therapy is given to a patient with dipsogenic diabetes insipidus, the polyuria and polydipsia also cease (Fig. 30.12). However, intake declines more slowly than output and does not reach normal levels until enough water has been retained to suppress plasma osmolality below the normal range. The nature of this

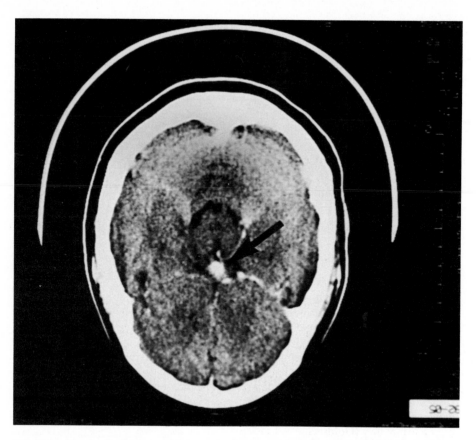

Fig. 30.14. Computerized tomographic scan showing a top-down view of the brain of a 38-year-old woman with dipsogenic diabetes insipidus and neurosarcoid. The *arrow* indicates a large, white-appearing granuloma typical of neurosarcoid surrounding the pituitary stalk.

delayed response can be appreciated most easily when the rate of water intake before and during treatment is plotted as a function of the concurrent plasma osmolality (Fig. 30.13). Compared to the relationship observed in patients with uncomplicated neurogenic diabetes insipidus (Fig. 30.9), water intake as well as thirst are clearly osmoregulated around an abnormally low "set point". These behavioural observations confirm the results obtained with the subjective rating scale and illustrate how the clinical consequences of abnormal thirst are radically altered by eliminating the protection normally afforded by the ability osmotically to suppress vasopressin.

In our experience, the syndrome of dipsogenic diabetes insipidus is not rare and can result from a variety of pathology in or around the hypothalamus. Of the 16 patients studied, organic lesions were found in about half. The most common association was with neurosarcoid, a disease of unknown aetiology that results in the formation of granulomas particularly in the meninges around the base of the brain (Fig. 30.14). Neurosarcoid can also destroy the posterior pituitary and in two patients with this disease a gradual progression from "pure" dipsogenic diabetes insipidus to a combination of dipsogenic and neurogenic diabetes insipidus has been observed. Other diseases associated with dipsogenic diabetes insipidus include tuberculous meningitis and multiple sclerosis. Unfortunately, all these disorders including neurosarcoid produce multiple lesions in the central nervous system, making it impossible to trace the abnormal thirst to destruction of a particular area. Indeed, it is remarkable and possibly significant that the syndrome of dipsogenic diabetes insipidus has been observed only in association with diseases that produce diffuse lesions of the brain.

Inappropriate Antidiuresis. Polydipsia due to an abnormal reduction in the osmotic threshold for thirst has also been observed in some patients with the syndrome of inappropriate antidiuresis (Robertson et al. 1982). In these patients, hyponatraemia and inappropriate urinary concentration rather than polyuria are the presenting abnormalities. This picture is identical to that seen after the administration of antidiuretic hormone to patients with dipsogenic diabetes insipidus and has virtually the same pathophysiology. The only difference is that the "inappropriate antidiuresis" is endogenous in the former and exogenous in the latter. The aetiology of the abnormal thirst in this condition is unknown but presumably derives from the same defect that produces inappropriate release of vasopressin.

Psychogenic. Psychogenic polydipsia exhibits several clinical features that differ significantly from dipsogenic and other types of diabetes insipidus (Illowsky and Kirch 1988). In addition to signs and symptoms of schizophrenia or other chronic severe mental disorder, patients with psychogenic polydipsia often present with moderate to severe reductions in basal plasma osmolality and sodium concentration. In addition, they do not report thirst as a reason for their sometimes astronomically high rates of water intake. Instead, they ascribe it to other motives usually of a therapeutic nature. However, these claims may reflect a defective interpretation or description of abnormal thirst because these patients often misinterpret other information and appear to vary their desire for water in accordance with changes in plasma osmolality (Goldman et al. 1988).

As in dipsogenic diabetes insipidus, patients with psychogenic polydipsia do not have a deficiency of either the secretion or action of vasopressin (Goldman et al. 1988). If anything, both properties of the hormone are enhanced since they exhibit a slight shift to the left in the osmotic threshold for vasopressin release as well as in the stimulus response relationship between urine osmolality and plasma vasopressin. These changes probably explain their tendency to hypo-osmolaemia/hyponatraemia and indicate that the slight upward resetting of the osmostat seen in many patients with dipsogenic diabetes insipidus is probably not due to chronic polydipsia per se.

Because of their high rate of water intake, patients with psychogenic polydipsia rapidly develop severe, sometimes fatal degrees of water intoxication if treated with antidiuretic hormone or any other drug that reduces the rate of urinary free water excretion (Illowsky and Kirch 1988). The same complication occurs not infrequently as a result of inappropriate secretion of endogenous vasopressin. This complication can be truly devastating and underscores again the great importance of a normal renal diluting and diuresing mechanism in protecting against the more severe consequences of the primary polydipsias.

The cause of psychogenic polydipsia is unknown. The neuroleptic drugs are not a major factor because the syndrome was observed long before these drugs were available (Sleeper and Jellinek 1936) and anecdotal evidence suggests an association between the severity of the psychosis and the polydipsia (Illowksy and Kirch 1988). These patients do not have any demonstrable anatomic abnormalities of the brain other than occasional cortical atrophy and ventricular enlargement which are thought to be consequences of chronic recurrent hyponatraemia (Weinberger et al. 1979). However, psychogenic polydipsia is so common in schizophrenia (up to 50% of all institutionalized patients are affected), the two disorders may share a common biochemical aetiology.

Conclusion

Thirst and other physiological mechanisms for regulating water intake play the pre-eminent role in regulating salt and water balance in humans.

Current methods for measuring thirst are not completely satisfactory because they lack objectivity and/or specificity. Nevertheless they have provided a revealing if somewhat fuzzy picture of how water intake is regulated in health and disease.

In healthy adults, thirst appears to be regulated primarily by the "effective" osmotic pressure of extracellular fluid.

Abnormal decreases or increases in thirst can occur either as a secondary manifestation of some other disease or as a result of some primary defect in the osmoregulatory system (Sleeper and Jellinek 1936).

The clinical manifestations of the thirst disorder vary considerably depending on the type of defect present and the presence of associated abnormalities in antidiuretic function. Generally speaking, however, primary hyperdipsia results in polyuria and hyponatraemia whereas hypodipsia results in hypernatraemia.

The causes of primary hypodipsia and hyperdipsia are not completely clear but either type of abnormality can apparently be caused by a variety of destructive lesions of brain.

At present there is no effective treatment for any of the primary thirst disorders.

References

Adolph EF (1969) Physiology of man in the desert. Hafner Publishing Company, New York

Goldman MB, Luchins DJ, Robertson GL (1988) Mechanisms of altered water metabolism in polydipsic, hyponatremic psychotic patients. N Engl J Med 318:397–403

Hammond DN, Mole GW, Robertson GL, Chelmicka-Schorr E (1986) Hypodipsic hypernatremia with normal osmoregulation of vasopressin. N Engl J Med 14:433–436

Illowsky BP, Kirch DG (1988) Polydipsia and hyponatremia in psychiatric patients. Am J Physiol 145:6

Nicolaïdis S (1969) Early systemic responses to orogastric stimulation in the regulation of food and water balance; functional and electrophysiologic data. Ann NY Acad Sci 157:1176–1203

Robertson GL (1980) Control of the posterior pituitary and antidiuretic hormone secretion. In: Berlyne GM et al. (eds) Contributions to nephrology. Karger, Basel, pp 33–40

Robertson GL (1984) Abnormalities of thirst regulation (nephrology forum). Kidney Int 25:460–469

Robertson GL (1985) Osmoregulation of thirst and vasopressin secretion: functional properties and their relationship to water balance. In: Schrier RW (ed) Vasopressin. Raven Press, New York, pp 203–212

Robertson GL (1987) Dipsogenic diabetes insipidus: a newly recognized syndrome caused by a selective defect in the osmoregulation of thirst. Trans Assoc Am Physicians C:241–249

Robertson GL (1988) Differential diagnosis of polyurias. Ann Rev Med 39:425–442

Robertson GL, Berl T (1986) Water metabolism. In: Brenner BM, Rector FC (eds) The kidney. WB Saunders, Philadelphia, pp 385–388; 398–399

Robertson GL, Aycinena PR, Zerbe RL (1982) Neurogenic disorders of osmoregulation. Am J Med 72:339–353

Rolls BJ, Wood RJ, Rolls ET, Lind H, Lind W, Ledingham JGG (1980) Thirst following water deprivation in humans. Am J Physiol 239:R476–R482

Sleeper FH, Jellinek EM (1936) A comparative physiologic, psychologic and psychiatric study of polyuric and nonpolyuric schizophrenic patients. J Nerv Ment Dis 83:557–563

Thompson DA, Campbell RG, Lilivivat V, Welle SL, Robertson GL (1981) Increased thirst and plasma arginine vasopressin levels during 2-deoxy-D-glucose induced glucoprivation in humans. J Clin Invest 60:1083–1093

Thompson CJ, Bland J, Burd J, Baylis PH (1986) The osmotic thresholds for thirst and vasopressin are similar in healthy man. Clin Sci 71:651–656

Thompson CJ, Burd JM, Baylis PH (1987) Acute suppression of plasma vasopressin and thirst after drinking in hypernatremic humans. Am J Physiol 252:R1138–R1142

Weinberger DR, Torrey EF, Neophytides AN, Wyatt RJ (1979) Lateral cerebral ventricular enlargement in chronic schizophrenia. Arch Gen Psychol 36:735–739

Weiss NM, Robertson GL, Byun K (1984) The effect of hypovolemia on the osmoregulation of thirst and AVP. Clin Res 32 (4):786A

Wolf AV (1950) Osmometric analysis of thirst in man and dog. Am J Physiol 161:75–86

Commentary

Baylis: In response to the points made by Dr Robertson on the visual thirst rating scale, I would like to emphasize the important differences between his method of assessing thirst and my own, and discuss the effects these two methods have on the results thus generated. With regard to the methodology, the first difference rests with the type of information on thirst that we are trying to obtain from individuals undergoing osmotic stimulation. In my method, the individual is asked to place a mark on a 10-cm undivided line, which represents the degree of thirst sensation appreciated at that moment in time; the ends of the line represent the extremes of thirst, at one end "no desire to drink at all" and the other "the most severe thirst you can imagine". At regular time intervals during the osmotic challenge (e.g. infusion of hypertonic saline) the process is repeated on a separate line, but no reference is made to previous ratings. All data points of thirst rating measured from one end of the line and concomitant plasma osmolality are submitted to simple linear regression analysis which then provides a slope to the line (the sensitivity of gain of the thirst osmoreceptor) and an abscissal intercept (the osmotic threshold of thirst). This type of analysis gives two determinants which can be used to describe thirst, but I should emphasize that they ought to remain theoretical and not necessarily absolute physiological concepts. They are reproducible in healthy individuals and are useful in defining disorders of thirst and abnormalities of water balance.

In contrast to the results obtained by Robertson's method which normalizes basal resting values to zero, the above method gives basal values in the lower range of the scale. Consequently, regression analysis of my data results in a lower abscissal intercept ("osmotic thirst threshold") than that derived from Robertson's method.

The method described above clearly uses all the data points from an osmotic study and does not arbitrarily discard data. One objection that has been raised is that some healthy individuals fail to record a rise in thirst rating at low plasma osmolalities (i.e. during the first 15–30 minutes of a hypertonic saline infusion), but this is certainly not the case in all healthy individuals. Furthermore, oral water loading causes a rapid and profound reduction in thirst sensation from basal values. My method of thirst assessment can readily accommodate a decrease in thirst sensation, whereas normalization of basal ratings to zero results in negative ratings for thirst.

Robertson: Dr Baylis' summary of the differences between his method and my method for characterizing the osmoregulation of thirst is correct except for two points; we do not "normalize basal resting values to zero" or "arbitrarily discard data". We accept the ratings of basal thirst made by the subject (usually between 0 and 2 on a scale of 10) but perform regression analysis of the relation to plasma osmolality only on those ratings which show an increase above basal levels. This decision is not arbitrary but is controlled by fixed objective criteria which can be defined mathematically. This approach is necessary to avoid the downward bias in both the slope and threshold values which results from including the very low ratings from the flat or unchanging portion of the curve (see Fig. 30.3). Whether these very low ratings represent true thirst (as Dr Baylis believes) or simply an absence of satiety (as I believe) cannot be resolved at present. This particular issue is not very important since we both agree that, whatever the nature of the sensation, it does not result in drinking. However, it seems clear to me that these low basal ratings must be excluded from the calculation of osmoregulatory thirst since Dr Baylis' observations as well as my own indicate that they are not appreciably increased by osmotic stimulation until plasma osmolality rises significantly above basal levels. Hence, these basal levels are probably not affected by plasma osmolality.

Phillips: I disagree with the idea that an individual senses thirst as an all or nothing phenomenon. It is a common experience that one can be a "little bit thirsty" or "extremely thirsty". If one is going to attempt to assess this then, in my opinion, the best method is by visual analogue scales as developed initially by Barbara Rolls for thirst research. It has become endemic that by correlating thirst ratings with plasma osmolality/sodium then information about changes in the physiological control mechanisms can be obtained. In using such correlation analysis I think it is better to use Dr Baylis' method (i.e. utilizing all the data in the analysis) rather than Dr Robertson's, which relies on an arbitrary decision of when thirst is increasing and only using the data above this point in the correlation analysis. Obviously, the plasma osmolality "threshold" when thirst is prominent and the primary conscious sensation will be higher than the "threshold" when thirst is mild and distractable. Robertson's method tends toward the former, Baylis' method tends to the latter. Personally, I have steered away from these analyses and used the ratings less quantitatively, but more qualitatively, in combination with water intake, to assess alterations in thirst and water intake mechanisms.

Robertson: Dr Phillips' argument that thirst is not an all or none phenomenon and is best assessed by visual analogue scales is one with which I fully agree. Indeed, it is the basis for all of the findings described in Chapter 30. However, for the reasons summarized above, I must take issue with his suggestion that we make "an arbitrary decision of when thirst is increasing". The decision of which data points to include in the regression analysis is dictated by purely objective mathematical criteria. By applying these criteria, we are able to obtain an unbiased estimate of the level of plasma osmolality at which thirst begins to be influenced by osmotic stimuli. Thus, the difference between the two approaches is not a matter of whether thirst is strong and prominent or mild and distractable. It is a matter of the best way to determine the "threshold" level, i.e. the point at which it begins to be stimulated appreciably by the increases in plasma osmolality. As evidenced by our findings in patients with neurogenic or dipsogenic diabetes insipidus, this quantitative approach provides a singularly effective way of understanding the pathophysiology as well as the pathogenesis of clinical thirst disorders. A more qualitative approach which ignores the relationship to osmolality can also be useful for some purposes but, we believe, it is necessarily less informative in explaining certain aspects of the problem.

Section VIII
A Concluding View

Chapter 31

Thirst and Salt Intake: A Personal Review and Some Suggestions

A.N. Epstein

Introduction

We are fortunate to be working on thirst. It's a phenomenon of wonderful complexity that will continue to fascinate us throughout our scientific lifetimes. After all, in order to have a complete understanding of thirst we need to know how the brain creates the urge to drink water, how it generates an anticipation of water, how it emits the behaviours that lead an animal to water and allow it to be selected from among other ingestible commodities, how it makes thirst and water drinking inherently reinforcing and memorable, and how it controls the consumption of it until satiation is achieved, all with accompanying hedonic states that include the distress of water depletion, the excitement of its anticipation and the pleasure of thirst satiation.

Think for a moment about how little was said about so many of these issues in the preceding chapters and you will understand why this volume is not a eulogy for an exhausted field but, is, instead, a summary of a largely unfinished research venture that is diverse, exciting and vigorous.

Second, we are fortunate in our choice of research problem because thirst is an important natural phenomenon. It is necessary for the successful life of most terrestrial animals. Whatever we learn will be interesting and can be useful for human and animal welfare.

Third, we are fortunate because thirst lends itself to biological analysis. In dealing with thirst experimentally we deal with an innate motivated behaviour; a behaviour, in other words, that is governed by a preformed neural apparatus which is perfected epigenetically and then mediates the search for and consumption of water as part of the biological machinery of the body.

The Nature of Thirst in Mammals and Man

The subject matter of this book is thirst in mammals including man. How, then, should we think about thirst in mammals? I suggest that we characterize it as the central and specific, motivational state of readiness to seek and consume water.

By central I mean, of course, that it goes on in the brain. It is modified, certainly, by events in the periphery, but what we are seeking to understand is essentially a problem of brain function, and we will not know what thirst is until we have understood its neuroscience.

By specific I mean two things. First, I mean that when thirst arises, water (its sensory qualities and its location in the environment) is somehow represented in the brain as the specific object of the animal's intentions, and that this neural representation or image of water governs the behaviour until the animal contacts a commodity and accepts it as sufficiently water-like for ingestion. And, second, by specific I mean that the behaviours of thirst are biologically specific. They are specific in the form that they take and can typify a species as uniquely as do its body parts and their arrangement. Think of how differently and characteristically rats, pigeons and blowflies drink. Moreover, the behaviours of thirst are biologically specific in their links to other freely emitted behaviours (feeding, reproduction and sleep–waking cycles in particular), and in their accommodation to the animal's ecological setting (nature of its diet, the availability of water, and the necessities of defence and predation). And they are biologically specific because they are expressions of the operation of a preformed neural apparatus that was assembled under genetic instruction and then modified by complementary epigenetic events that usually occur in the neonatal period. These events (use and practice, hormones, experience) complete and perfect the apparatus which then mediates optimal adult behaviours that occur in phases of appetite and consumption and have the other characteristics of motivation.

By motivational I again mean two things. First, thirst often arises in the absence of water, and drinking behaviour often occurs in the absence of depletion signals. It cannot therefore be merely reflexive. And, second, the behaviour is complex. It is divided into appetitive and consummatory phases, is guided by an anticipation of water, and is accompanied by affect. The appetitive behaviours of thirst will be individuated in ways that depend on the animal's past history of drinking behaviour and on the ecological problems that it must solve in order to reach the water. Even the caged animal must move to its water reservoir and must emit behaviours that are appropriate to the problem of bringing the water into its mouth. The animal, in other words, uses instrumental behaviours to gain access to the water (Teitelbaum 1977). And, remember, from its onset, which often occurs in the absence of water, the behaviour is guided by an image of water, by an expectancy of it that is satisfied (or frustrated) only when the behavioural sequence is completed and consumption (or rejection) begins. And all of this is accompanied by affect. Feelings and emotional expressions are an integral part of thirst as they are of all motivated behaviours (Epstein 1982a).

If this concept of the nature of thirst is accepted we can understand why we are not just investigating drinking behaviour. Drinking is not thirst. Although licking and swallowing are the common consummatory behaviours for water ingestion there is no mandatory fixed action pattern for thirst in mammals. A mammal can "eat" water if it is a succulent plant (Milgram et al. 1974) or, more demonstrably, a solid mixture of non-nutritive cellulose and water (Hamilton 1969), and it can "drink" its food if it is a liquid. Licking and swallowing are often an expression in behaviour of the central state of thirst and we use them to study the central state, but they are not thirst. They can, in fact, be dispensed with if the thirsty animal is provided with an alternate means for delivery of water into its stomach (Epstein 1960) or veins (Rowland and Nicolaïdis 1974).

We can also understand why we are not investigating just a sensation. Thirst generates a percept that is represented in consciousness, but it is only the sensory aspect of thirst and can only be studied in humans. It is not the major issue of our science.

And, lastly, we can understand why we should not limit ourselves just to the study of the signals that control the arousal and satiation of drinking behaviour. The brain does indeed process signals of several kinds to begin and to terminate water drinking, and the study of these signals (neural afferents, hydrational deficits, hormones) is one of the most feasible and informative of our current research strategies. But the brain is more than a processor of information. It is a self-activating organ that issues commands to somatic, autonomic, and endocrine action systems. These are, of course, modified by signals, but the signals do not produce the behaviour. The brain does, while signals are integrated into neural programmes for action that are both innate and acquired. If we limit our concept of thirst and the other motivated behaviours to the signal-processing idea we risk misunderstanding them as kinds of Sherringtonian reflexes occurring in an essentially reactive brain.

Sleep, grooming and locomotion are better models for our thinking. They are the most common behaviours of most mammals and their neuroscience suggests a better programme for the future of our work. What we know about them illustrates the brain's self-activating capabilities and its command functions. They are not imposed on the brain by afferent inputs or other signals. They are generated by the brain and modulated by signals and I am suggesting that thirst is also a product of the brain's self-activation, that the state of thirst is inherent to the brain, modulated, of course, by signals, but essentially endogenous.

We are confident of sleep's endogenous nature because the EEG and other forms of electrical recording give us its signature (Kleitman 1963). The brain sleeps both behaviourally and electrically, and our colleagues in sleep research use the electrical signs of sleep to do their work. We do not have this advantage. There is no recordable index of the brain activities of thirst. In order to obtain it we will need a technology that we do not have, one that reveals multifocal activities that are both chemical and electrical and that occur dynamically in neuron assemblies throughout the neuraxis over extended periods of time. But I am confident that there is a specific state of activity underway inside the brain which is active episodically when animals seek water and drink it, and I believe that one of the major turning points in the future of thirst research will be the discovery of its signature.

The Biological Diversity of Thirst

Thirst has evolved in the fully terrestrial animals – the insects, and the vertebrates beginning with the reptiles. But its evolution has not been simple. Among the vertebrates water swallowing occurs in eels and other fishes (Hirano 1974; Balment and Carrick 1985), and, because it is a form of water ingestion, may represent an evolutionary prelude to true thirst. Where thirst is surely present, there is biological diversity in the behaviours of water drinking. Lapping, sucking or pecking are all used to ingest water, and some animals bring it to their mouths with their forepaws or hands. The temporal patterns of drinking behaviour vary. Some animals are frequent small draught drinkers, others drink large draughts infrequently, and the behaviour may occur either nocturnally or diurnally. There is even diversity in whether or not thirst is included in the animal's behavioural repertoire. There are desert rodents that do not drink even when water is available (Schmidt-Nielsen 1979), and the fully marine vertebrates (the marine turtles, birds and mammals) have virtually no access to fresh water and rely on salt glands or on the isotonic flesh of their prey for water balance. Other vertebrates that live in fresh water or prey on freshwater species (turtles, many birds and mammals like the beaver and otter) must solve the problem, not of water ingestion but of reducing their water intake and of excreting the excess water that enters through permeable membranes.

With all this diversity we should not be surprised by species differences in the neurological mechanisms of thirst among animals, even among those that are closely related. There are, for example, differences in the responsiveness of rodents to the dipsogenic and natriorexigenic actions of angiotensin II (Wright et al. 1987; Rowland and Fregly 1988). Differences will be even greater across taxons as shown by the fact that, when given into the brain ventricles, the tachykinins are dipsogenic in rats but antidipsogenic in pigeons (de Caro 1986).

We still have much to learn about species differences in the neurobiology of thirst, and it is safe to predict that they will become even greater in number and will be even more surprising as we study thirst and its brain mechanisms in more animal species. After all, most of our research uses only a handful of species (rat, sheep, dog, and to some extent the pigeon, monkeys and man), and even among this small number we are confronted with the following facts: (a) sheep (McKinley et al. 1986), unlike rats, dogs, the opossum and the pigeon, continue to drink to intravenous angiotensin after ablation of the subfornical organ (Simpson 1981), and (b) the organum vasculosum of the lamina terminalis must be intact in sheep and dogs in order for them to drink normally to cellular dehydration (Thrasher et al. 1982), but in the rat the necessary tissue is in the lateral preoptic area and nearby portions of the anterior hypothalamus (Blass and Epstein 1971; Peck and Novin 1971; Almli and Weiss 1974).

Biological Diversity Within Species

There are three other kinds of biological diversity that are not as readily acknowledged for thirst, in particular, and for other areas of behavioural neuroscience in general. All three occur within rather than across species and all of them have been only poorly studied. These are the diversities that arise, first, from differences in developmental stage and ageing, second, from differences in sex and, third, from differences in the states of thirst themselves.

Development and Diversity

Animals that drink do so differently as they progress through their ontogeny and the brain mechanisms for the behaviour change as they mature (Epstein 1984). We know something about this problem from studies of the onset of water drinking in rats. Young rats begin to drink in their third postnatal week (Babicky et al. 1972) and then drink more frequently during weaning, but without the careful segregation of feeding and drinking that is characteristic of the adult (Kissileff 1969a, 1971). These are the only published studies of the natural onset of drinking and both use the rat, and there are none of drinking in the aged animal.

Changes in the brain mechanisms for thirst with development are illustrated by work on the ontogeny of the dipsogenic effect of angiotensin (Leshem and Epstein 1988; Leshem et al. 1988). Very young rat pups drink in response to both systemic and intracranial angiotensin, but they accept either water or milk. Leshem showed that the selective water drinking that is diagnostic of the adult behaviour begins at 8 days of age when the pups begin to drink more water than milk in response to intracranial angiotensin, and reaches its developmental climax at 16 days when they drink only water. Throughout this sequence of developmental stages angiotensin arouses drinking only when the pups are expressing nascent forms of adult ingestion while they are away from their dams and not suckling. When they are reunited with her and treated with angiotensin, they do not respond with increased consumption. They do not increase their intake of suckled milk despite its high water content, as if the thirst evoked by angiotensin were somehow uncoupled from suckling and, therefore, not expressed in their behaviour (for further discussion, see pp 491–494). Again the work has been done only in the rat, and longitudinal developmental studies of the other determinants of thirst have not been published.

Sex Differences

In the rat, differences in ingestive behaviours between the sexes are common and conspicuous. They occur for water and food (Tarttelin and Gorski 1971; Vijande et al. 1978) intake, for salt consumption (Krecek 1973), and for the intake of other sapid solutions (Zucker 1969). They are often expressions of gender and can be altered by castration or hormone administration in the neonatal period and in adulthood. For example, female rats drink more salt than males, but the difference disappears if the males are castrated shortly after birth (Krecek 1973). And the decreased water intake that occurs during oestrous in the adult female is eliminated by ovariectomy and reinstated by oestrogen administration (Findlay et al. 1979).

Except for Jonklass and Buggy's intriguing work, the neurobiological aspects of this problem have not attracted the interest they deserve. They first confirmed their earlier work (Jonklass and Buggy 1984) in which intracranial implants of oestrogen reduced the water intake that is evoked by central injection of angiotensin II, and then showed (1985) that this occurs only in female rats and is most effective when the oestrogen is administered to the medial preoptic area. They also demonstrated, with regional binding studies, that angiotensin II binding was decreased in the medial preoptic region of oestrogen treated animals. And, lastly, they found that the antidipsogenic effect of oestrogen occurs only in animals whose neonatal hormonal history was normal. That is, genetic females that were masculinized by neonatal androgen did not reduce their water intake when treated with oestrogen in adulthood, and males that were castrated at birth were feminized and drank less while receiving oestrogen. Sex differences in angiotensin-induced water drinking and in salt intake (see section on "Need-free salt intake" below) appear, therefore, to be expressions in adulthood of the sexual differentiation of the brain that is produced in the perinatal period by the central actions of the gonadal steroids.

The Several States of Thirst

It should not be necessary to insist on the diversity of the states or kinds of thirst, but much of our published work implies otherwise. We often write about thirst as if it was qualitatively singular, one kind of activity in the brain that underlies all water drinking behaviour and that does not vary except quantitatively among episodes of drinking. I do not believe this. The drinking behaviour of our most frequently studied animal argues against it. Rats that have free access to water and are eating dry food drink most often just before or just after they eat (food-associated drinking occurring pre- or postprandially), but they also drink, albeit rarely, between meals, usually in association with grooming (Kissileff 1969b). When their salivary production is impaired they drink repeatedly within meals (prandial drinking), often excessively (Epstein et al. 1964). They also drink, of course, after they have been deprived of water for periods in excess of 6–8 hours, and when they are defending themselves against heat stress by saliva-spreading (Hainsworth et al. 1968) or when they are moved abruptly from the cold to a neutral environment (Katovich et al. 1979). They can also be required to use water drinking as an operant, entirely outside of its natural context (Williams and Teitelbaum 1956), and will drink water in excess when they are forced to wait while hungry for the delivery of successive morsels of dry food (Falk 1967).

This diversity in the kinds of water drinking behaviour warns us that there are kinds of thirst just as there are kinds or states of sleep. An animal that is drinking preprandially cannot be expressing a thirst that is the same in its biological determinants and neurological mechanism as one that drinks after eating a meal, and both of these must be different from the drinking that is done after a bout of grooming or saliva-spreading. Conditions within the oral cavity will be major determinants of the drinking that is associated with grooming and thermolysis, whereas water depletions and expectancies of the need for water are the most likely determinants of the drinking that precedes a meal. This kind of drinking must be different, again in its biological determinants and neurological mechanisms, from the drinking that occurs postprandially. After a meal, but not before, isotonic serum will have shifted into the gut without net water loss, and meal-taking has neurochemical and hormonal consequences that simply cannot operate to control preprandial drinking.

Most of us assume the singularity of the neural mechanisms of thirst because we provoke it experimentally and study drinking behaviour out of its natural context. Kraly's recent analysis of meal-associated water drinking in the rat is a welcome counterexample (Kraly 1989, and Chapter 18). In a careful analysis of the drinking done by the rat before, during, and after eating, he finds that water intake that is food-associated (done while the animal is eating) is reduced by one-third when either angiotensin or histamine actions are blocked, whereas water intake before or after meals is unaffected by the same treatments. In the sham-feeding rat, in which pregastric food-associated events predominate, histamine or angiotensin blockade abolishes the animal's robust drinking.

We should not let the sameness of water drinking behaviour deceive us into believing that there is only one kind of thirst. We should, instead, be asking how many kinds there are? Or, more precisely, we should be asking how many qualitatively different states of brain activity are there that can generate drinking behaviour in response to different sets of biobehavioural determinants?

Angiotensin as a Hormone of Thirst

There is an impressive number of dipsogenic agents including carbachol, noradrenaline, serotonin, histamine, ADH (in dogs), neurotensin, and neuropeptide Y, but angiotensin is the best understood. Kraly's demonstration of a role for angiotensin in the drinking done while rats are eating (pregastric food-associated drinking) is the most recent evidence of a role for the hormone in thirst. From the first demonstrations of the effects of its injection it was clear that angiotensin was an exceptionally powerful dipsogen (Fitzsimons 1969). Now, more than 20 years later, we know that the hormone is potent when infused intravenously (Fitzsimons and Simons 1969; Hsiao et al. 1977; Evered and Robinson 1981), and that the circulating levels of the hormone that are achieved by its intravenous infusion match the endogenous levels that are produced by several dipsogenic treatments including those that occur naturally (Johnson et al. 1981). It is also now known that several experimental thirsts are entirely angiotensin dependent (Fitzsimons 1964; Fitzsimons and Elfont 1982; Katovich et al. 1979; Evered and Robinson 1981), and it is becoming increasingly likely that the hormone plays a role in deprivation-induced and food-associated drinking behaviour.

Angiotensin remains the most potent dipsogenic agent known when given into the anterior cerebral ventricles and especially when it is injected directly into the subfornical organ where it can arouse drinking behaviour in the satiated rat at doses in the femptomole range (Simpson et al. 1978). Its potency as a blood-borne hormone of thirst is also impressive. This was underestimated in the original studies of the problem (Fitzsimons and Simons 1969; Hsiao et al. 1977) in which artifacts of the method of intravenous infusion were not avoided. Studies that have abolished the pressor effect of intravenous angiotensin II (Ang II) (Evered and Robinson 1981) demonstrate a 100% increase in the dipsogenic potency of the blood-borne hormone across a broad range of doses (1–100 ng/min), and they show that the threshold for elicitation of drinking is the same as that for the pressor response. Other studies that suppress prostaglandin production with oral indomethacin yield the same results (Kenney and Moe 1981). A similar enhancement of the dipsogenic potency of systemic Ang II is produced by parabrachial (Ohman and Johnson 1986) or area postrema (Edwards and Ritter 1982) lesions. And, lastly, infusions of the hormone into the carotid circulation of the dog, which give it direct access to the brain and which do not provoke a marked pressor response, elicit drinking at low physiological doses (Fitzsimons et al. 1978).

The work of Johnson and his colleagues (Mann et al. 1980; Johnson et al. 1981) is especially relevant. They showed that thirst-provoking treatments such as 48 h of water deprivation, caval ligation, moderate doses of isoproterenol and hyperoncotic colloid dialysis all produce blood levels of endogenous Ang II that are well above the amounts (approx. 200 fmol/ml of plasma) that are produced by intravenous infusion of the hormone at the dipsogenic threshold (approx 10 ng/min) that was determined by Hsaio et al. (1977). This threshold was measured before any of us were aware of the artifacts of the inhibitory effect of the pressor response and suppression of drinking by the prostaglandins. When the dipsogenic threshold for circulating Ang II is re-evaluated with precautions against these antidipsogenic artifacts it will undoubtedly be lower and closer to, if not below, the plasma Ang II levels that are produced by everyday dipsogenic conditions such as consumption of a dry meal and 12–24 hours of water deprivation.

It must also be remembered that judging the dipsogenic potency of blood levels of infused angiotensin places an abnormal burden on it as a cause of drinking. Angiotensin never acts alone during instances of physiological thirst. After water deprivation, for example, angiotensin is always accompanied by other dipsogenic preconditions, most frequently hypovolaemia and cellular dehydration (Hatton and Almli 1969; Ramsay et al. 1977; Epstein 1982b). It is remarkable that exogenous angiotensin is as potent as it is even when it is taken out of its normal physiological context and is administered as the sole dipsogen.

Lastly, angiotensin's role in thirst has been confirmed by use of specific blockers (Malvin et al. 1977; Barney et al. 1983), either analogues of the natural octapeptide that are competitive inhibitors of receptor occupancy or other compounds such as captopril that inhibit the angiotensin-converting enzyme and thereby prevent the biosynthesis of endogenous angiotensin II. Experimental thirsts such as those which are induced by isoproterenol (Houpt and Epstein 1971) or by caval ligation (Fitzsimons 1964) are completely abolished by angiotensin receptor blockers and by captopril (Rettig et al. 1981), and the water intake that is produced by 48 hours of water deprivation is markedly reduced by the same treatments (Barney et al. 1983).

Remarkable progress has been made in understanding the neural circuit that mediates the thirst that is aroused by angiotensin. It derives from two discoveries. These were (a) the discovery by Ganten et al. (1978) of the cerebral renin–angiotensin system and their demonstration of its independence from the blood-borne system of renal origin, and (b) the identification by Simpson (1981) and Simpson and Routtenberg (1973) of the subfornical organ (SFO) as the essential site at which blood-borne angiotensin II acts to arouse thirst.

These new findings set the stage for two concurrent lines of work that focused attention on the nucleus medianus of the preoptic area (MnPO) as the crucial site within the brain for the dipsogenic action of angiotensin. These were the mapping of the neural connections of the SFO by Miselis (1981) and Miselis et al. (1979), which includes a reciprocal pathway between SFO and MnPO, and the findings of Johnson and Buggy (1977), Johnson and Wilkin (1987), Lind et al. (1985) and Lind (1988) of the important role of the tissues of the ventral lamina terminalis (the "AV3V"), which includes the MnPO, in thirst, ADH release, and control of blood pressure. The work of Lind and Johnson (1982) and Lind et al. (1985) was especially important because it demonstrated, on the one hand, that the MnPO was rich in both angiotensinergic terminals and angiotensin-sensitive cells (Mendelsohn et al. 1984), and that it is, on the other hand, essential for thirst aroused by both blood-borne angiotensin and angiotensin that acts within the brain.

As a result of this remarkable series of investigations, it is now known that both peripherally and centrally generated angiotensin II contribute to the water drinking that is evoked by the hormone. They do so through an angiotensin sensitive circuit that includes the SFO, which lacks a blood–brain barrier, and the MnPO which is within the barrier. Drinking is evoked by activation of angiotensin receptors at both sites and they are linked by angiotensinergic neurons of the SFO that project to the MnPO. It is also known that noradrenergic projections to the MnPO from the brainstem are necessary for its role as the mediator of the dipsogenic effect of angiotensin (Bellin et al. 1987).

Much still needs to be learned about the role of angiotensin in the generation of drinking behaviour. We need to know how the two renin–angiotensin systems (the peripheral renin–angiotensin–aldosterone system and the renin–angiotensin system that is endogenous to the brain) co-operate to produce the behaviour, how the hormone is integrated into the complex of conditions that lead to normal drinking, and how it may participate in thirst satiety. However, despite what still needs to be understood, the evidence that we now have justifies the prediction made 20 years ago by Fitzsimons (1969) when he discovered a renal dipsogen and suggested that angiotensin II is a hormone of thirst.

Antidipsogens

There is also an impressive list of agents that suppress thirst including the prostaglandins (especially those of the E series), vasoplegic drugs (nitroprusside, papaverine), bombesin and the tachykinins. The latter include substance P, and the neurokinins in mammals and a group of non-mammalian analogues (elidoisin, physalaemin, kassinin and others) that have been studied extensively by our colleagues in Camerino (see de Caro et al. 1988). They are of special interest to us first because the mammalian brain is well provided with tachykinins and with a variety of tachykinin receptors, and these occur in areas like the amygdala, the anterior forebrain, and the brainstem, and second because they are potent suppressors of thirst in mammals when they are injected directly into the brain. Lastly, they are of interest because of their specificity. None of the tachykinins suppresses feeding behaviour (Massi et al. 1986), and some of them have remarkable specificity for dipsogenic state. Kassinin, for example, is a powerful suppressor of drinking induced by cellular dehydration, but has no effect on angiotensin-induced thirst even at high intracerebroventricular doses (de Caro et al. 1988). The mammalian tachykinins (the neurokinins in particular) may be active during the neural mediation of natural thirst satiety, and we can look forward to investigating this possibility when specific inhibitors of them are found.

The Neural Substrates of Thirst

Progress is being made in this important problem, details of which are given in previous chapters. We do know that thirst is mediated by the forebrain: the chronic decerebrate rat does not seek water and will not ingest it in response to a variety of thirst challenges even when it is infused directly into its mouth (Grill and Miselis 1981). Where thirst determinants such as cellular dehydration and angiotensin have been identified, neural circuits for their action are being described in the circumventricular organs and the anterior forebrain (Miselis 1981; Johnson and Wilkin 1987). Global deficits in several kinds of thirst are produced by ablations of the lateral hypothalamus (Epstein 1971), of the ventral lamina terminalis (the "AV3V") (Buggy and Johnson 1977; Johnson 1985) and of the zona inserta (Grossman 1984), suggesting that they are convergence areas for information in the thirst circuit. And Mogenson's (1987) synthesis of mechanisms for motivated behaviour is a reminder of the equal importance of brainstem circuitry for the organization of efferent commands and sequential action.

Much remains to be done and it will be a formidable task. First, because, as emphasized above, we are not searching for the neural substrate of behaviours like jaw and tongue movements or of swallowing, which are problems of motor control. We are instead attempting to describe a neural apparatus that uses the motor programmes for jaw and tongue action and for swallowing while it mediates a cluster of central states whose common property is the intention to drink water. Moreover, we will have to understand how the several states of thirst arise within the neural apparatus, and how the special combination of biological determinants of each kind of thirst gain control of it while it governs the expression of drinking behaviour.

Second, the task is formidable because all kinds of thirst are complex behaviours. They occur in appetitive sequences, are characterized by expectancy and expression of affect, and have endocrinological, perceptual, motivational, cognitive, memorial and hedonic aspects whose mediation requires coherent activity throughout the cerebrum, brainstem and cord. We should expect the neural apparatus for thirst to be distributed throughout the anterior forebrain, the limbic system, and the amygdala, and we should expect it to use the brainstem and cord to generate not only the motor sequences of the appetitive and consummatory behaviours of water ingestion but also the more evanescent expectancies and affects that are equally important aspects of the behaviour.

And, lastly, we should acknowledge that we know very little about one of the most powerful controls of thirst, the biological clock. Thirst is strongly influenced by the day–night cycle and by the free-running clock when animals are in conditions of continuous dim illumination or darkness. We know that the clock for thirst is in the suprachiasmatic nucleus in the rat because ablation of it (Stephan and Zucker 1972) or paralysis of it with local tetrodotoxin (Schwartz et al. 1987) stops the circadian rhythm of drinking without disrupting the behaviour itself, but this is all that is known about the biological basis of the rhythmicity of thirst.

Spontaneous or Freely Emitted Drinking Behaviour

Thanks to the work of Ramsay et al. (1977) (and see Rolls et al. 1980 for a review and species comparisons) we now have a physiological account of the causes of the drinking that is induced by water deprivation. It is aroused by the additive effects of deficits in both the cellular and extracellular water compartments. Earlier work had shown that the dipsogenic effect of cellular dehydration depends on osmosensors in the lateral preoptic area and immediately adjacent portions of the anterior hypothalamus (Blass and Epstein 1971; Almli and Weiss 1974), and that the drinking evoked by hypovolaemia is mediated by a combination of neural afferents from the great vessels of the low pressure circulation and of angiotensin II acting on the subfornical organ (Fitzsimons 1979). In the experiments by Ramsay et al. (1977) rats, dogs, and monkeys that had been prevented from drinking overnight were found to be dehydrated by water losses from both their cellular and extracellular water compartments, and their thirsts were satiated, by selective and partial restoration of both deficits. Rehydration of their forebrain osmosensors was achieved with intravascular infusion of small volumes of water, and their extracellular volume was restored with intravenous isotonic saline. When offered access to water after these treatments animals that had been deprived of it overnight drank only trivial volumes, despite the fact that they were still in negative water balance. These experiments were a convincing confirmation of the double-depletion hypothesis which predicted that thirst would be understood as the sum of the dipsogenic effects of deficits in both water compartments (Epstein 1973), and they have the added importance of making clear that the thirst that is aroused by double depletion is the thirst of water deprivation.

We also have a satisfactory account of the drinking that is done, at least by rats, while they eat. Kraly's work demonstrates that water intake that is associated with active food ingestion is controlled by an angiotensin mediated histaminergic mechanism (Kraly and Corneilson 1988), but there is still no account of the mechanism of the drinking that is done by non-deprived animals just prior to meals or at times that are unrelated to eating. We simply do not know why animals drink at these times, and the several names that have been invented for this kind of drinking are symptomatic of our ignorance. It has been called secondary, non-homeostatic (or non-regulatory), or spontaneous. I prefer to call it "freely emitted" drinking which does not imply, as does spontaneous, that it has no cause; which does not carry the implication that it is somehow less important than primary or deficit-induced drinking; and which does not define it as what remains when homeostatic drinking is accounted for.

The only thing we can say with certainty about freely emitted drinking (and, again, the reference animal is our most common subject, the domestic rat) is that it occurs most frequently at night when the animal is performing its other freely emitted behaviours (locomotion and investigation, elimination of urine and faeces, yawning and stretching, grooming and scratching, and, of course, eating). These occur in bouts that interrupt periods of sleep that are briefer than in the daytime. The behaviours are episodic and associated with each other (de Castro 1989). The animal wakes and performs several of these freely emitted behaviours in a cluster or bout that often includes water drinking, and this fact suggests how freely emitted drinking may be generated.

Could these bouts of freely emitted behaviours be habitual domestic routines that have been acquired over the lifetime of the animal? Could they be based on a combination of causes that have recurred on a daily (or, rather, nightly) basis since the animal was weaned, and that have provided the comfort of relief from minor distresses and peripheral irritants, the pleasure of the performance of benign behaviours, and the opportunity to forestall nutrient deficits by the consumption of small amounts of water (and food) before it is needed in the homeostatic sense? Freely emitted drinking behaviour may, in other words, be one of the behaviours in the well-practised domestic routines of behaviour that afford the rat comfort, pleasure, and the avoidance of nutrient deficits.

This suggestion rejects the idea that freely emitted drinking behaviour can be understood as an instance of classic, feed-back homeostasis. It predicts that water deficits will not be found in animals expressing this behaviour, and that they are not performing it to restore water losses. Behaviour serves homeostasis but is not its slave. There is a richness in the natural behaviours of animals and a complexity in the capabilities of their nervous systems that goes beyond homeostasis. We should be looking elsewhere for our hypotheses about the mechanism of phenomena like freely emitted drinking, and should take more seriously the suggestion, made by the work of Fitzsimons and Le Magnen (1969), that drinking can anticipate water need. Although they did not use the term, they studied freely emitted feeding and drinking behaviours in rats that were switched from a high carbohydrate to a high protein diet that required more water for its metabolism. The animals increased their water intakes during the first day of high protein feeding, but they did not consume the excess water while they ate until several days later. That is, the animals adjusted to the greater need for water in two stages. First, they responded to it by drinking after they had eaten, but then they reorganized the extra water intake into temporal register with their food intake thereby anticipating the greater need for it that was created by the metabolism of the added protein.

The Ontogeny of Thirst, Especially in Mammals

Everything that is alive, both plants and animals, has had an individual developmental history, from the most lowly of creatures to animals like ourselves, and no aspect of biology can be completely understood without an understanding of its ontogeny. Research on the ontogeny of thirst began only in the early and mid 1970s (Babicky et al. 1972; Almli 1973; Wirth and Epstein 1976), and it has already taught us several things of value.

First, we know from research on the suckling rat that thirst is precocious. Neonatal rat pups that are still entirely dependent on mother's milk for all of their food and water will drink water that is delivered directly into their mouths or is available from puddles at their feet when they are made thirsty by cell dehydration, deprivation (removal from the dam), hypovolaemia, or by systemic or intracranial angiotensin, and they do so before the end of the first week of postnatal life (Wirth and Epstein 1976). We also know that they do not begin to ingest water freely until they are 16 or 17 days old (Babicky et al. 1972; Almli 1973), demonstrating that the neural mechanisms for water drinking and for thirst are mature well before the animal begins to drink water.

Second, we know that the water drinking that is elicited from the neonatal rat by all of these dipsogens is an early expression of adult-like ingestive behaviour that is mediated by thirst mechanisms which are nascent in the brain of the pup. It occurs only when they are away from their dam, and not while they are suckling. In fact, dipsogens either have no effect on pups that are suckling from their dam or they decrease their intake of mother's milk despite its high water content (Leshem and Epstein 1988). Moreover, pups that are expressing this precocious form of adult thirst make the movements of drinking (repeated mouth opening, lapping and swallowing) and do not attempt to suck the water they are offered (Almli 1973; Wirth and Epstein 1976).

This distinction between suckling on the one hand and feeding and drinking (or independent ingestion) on the other and their underlying neurological mechanisms arose from studies of fluid intake in neonatal rats that were either sucking mother's milk from their dam or were being fed milk or water, usually by direct infusion into their mouths, while they were away from her. It was first made by Drewett (1978), and has been discussed recently by several authors (Hall and Williams 1983; Epstein 1984). It is illustrated by Leshem's recent work, which describes the developmental time-table for the maturation of the brain mechanisms for angiotensin-induced water intake (Leshem et al. 1988). Brain angiotensin was activated by injecting pups intracranially with renin and the selective intake of water, which is characteristic of adult angiotensin-induced drinking behaviour, was demonstrated to mature in the neonatal rat at 16 days of age. This was shown in pups that had been removed from their dam and were offered water or milk that was infused directly into their mouths while they rested in a heated and humidified chamber. Activation of their brain angiotensin resulted in an increase in both water and milk intake in pups that were between 8 and 15 days of age. Pups that were 16 days old and older increased only their water intake. But when the experiment was repeated in pups that were reunited with their dam and were suckling in a natural litter, intracranial renin had no effect on their intake of mother's milk even when they were less than 16 days old and could be induced by the brain renin injections to consume both water and milk while they were in the test chamber and away from their dam.

The newborn rat, and presumably newborn mammals in general, is born with several separate neurological systems for ingestion; one is for suckling, which is the most recent ingestive behaviour to have evolved among vertebrates and which is expressed when the pup is with its dam, and the others are for independent feeding and drinking. These latter mechanisms are nascent in the neonate's brain and are somehow suppressed whenever the pup is with its dam. At weaning suckling is suppressed and the neurological systems for independent feeding and drinking gain complete control of ingestive behaviour. From the ontogenetic point of view, thirst (and feeding) in neonatal mammals is a behaviour in waiting.

To understand this, think of locomotion and of how common it is among animals like amphibians and birds to have two kinds of locomotory behaviour, each with its own neurological mechanism and each with its own developmental calendar. Many birds, for example, both walk and fly. They usually walk before they fly, and different neurological mechanisms are obviously employed for the two behaviours. Similarly, the mammalian brain contains neurological mechanisms for different kinds of ingestive behaviour and much remains to be done in future research to understand both the nature of these neural mechanisms and how their succession is controlled.

A third way in which this work on the ontogeny of the ingestive behaviours has been valuable is the clarity with which it demonstrates that these behaviours are innate. Animals that have never encountered water and that have not previously been deprived of it (recall that the high water content of milk and the pups' frequent ingestion of it assure that they are not dehydrated) drink it selectively when treated for the first time with a variety of dipsogens and this is true of birds (Stricker and Sterritt 1967) as well as mammals (Wirth and Epstein 1976). The neural mechanisms for thirst are products of the animal's genome that have been realized in its normal development and are preformed in its brain. Thirst and the acts of water ingestion are not acquired. Learning, especially of the places in its environment at which fluids will be found and of the operants that can be employed to gain access to it, is, of course, important once weaning has begun (Kriekhaus and Wolf 1968; Teitelbaum 1977; Paulus et al. 1984), but the urge to drink water and the motor programmes for its ingestion once it has passed the lips are part of the animal's phenotype.

Fourth, the ontogenetic research has revealed that the behaviours of independent ingestion mature abruptly and sequentially (Wirth and Epstein 1976; Ellis et al. 1984; Leshem et al. 1988). The animal does not become competent for feeding and drinking slowly over several days and it does not express all aspects of both behaviours at once. Instead they are either absent from or, once a critical age has been reached, present in the animal's behavioural repertoire. For example, there is a clear sequence in the ages at which the several kinds of thirst make their debuts during the first week of the rat pup's life. Immediately after birth the pup can not be made to drink water by any known dipsogen, but at 3 days of age cellular dehydration thirst matures, followed two days later by that of hypovolaemia, and then by the drinking that is elicited by angiotensin which, as discussed above, becomes selective for water abruptly at 16 days of age.

The same principles are illustrated by the ontogeny of feeding. Distension of the upper gastrointestinal tract inhibits milk intake at birth (Houpt and Epstein 1973), but decreases in intracellular fuel utilization (glucoprivation and ketoprivation) increase food intake only after weaning (Houpt and Epstein 1973; Gisel and Henning 1980; Leshem et al. 1990). Again, there is an invitation here for future work. What is happening in the brain of the pup to provide for these successive steps toward fully mature ingestive behaviours? What are the developmental events in growth of neural circuitry or in maturity of receptor systems or effector mechanisms that occur in what could be a matter of hours suddenly to increase the animal's behavioural competence?

And lastly, what should we make of the precocity of independent ingestive behaviours in general, and of thirst in particular? Is the fact that the neonatal rat has neural mechanisms for responses to dipsogens and for the drinking of water long before they are required, nothing more than an expression of nascent adult behaviours? Or are these early capabilities for adult ingestion important for the optimal development of adult ingestive behaviours? There are at least two ways in which their precocity may be important. First, the arousal of the several kinds of thirst and the performance of the behaviours of water ingestion early in life may be necessary for their full development as is the case for pattern vision (Wiesel and Hubel 1963). As was pointed out earlier mammals are endowed by their genome with the programmes for the development of the appropriate neural mechanisms for innate motivated behaviours, and the optimal ontogeny of the mechanisms may require complementary epigenetic events especially in the neonatal period. For the ingestive behaviours these events may be performance of the behaviours and the use of their neural mechanisms. We have an indication that a process of this kind may operate in the development of suckling. Newborn rat pups that are deprived of contact with their dam for the first 12 hours after birth never suckle as effectively as their normally reared peers (Dollinger et al. 1978).

Second the early performance of the independent ingestive behaviours may be important for their insertion into the routines of freely emitted behaviours that the animal will express later in life. The rat pup suckles, locomotes, grooms, scratches, urinates and defaecates before it eats and drinks. When weaning begins it nibbles at food and samples water. It does so while performing its other behaviours, and these may be added to the clusters of behaviours it expresses while it is awake because, as discussed above, they are comforting and pleasurable, and because they forestall nutrient deficits. Later, when weaning is completed, meals and draughts of water may occur as parts of habitual domestic routines of behaviours that, as I suggested above, are the settings in which freely emitted drinking (and eating) will be understood.

This makes weaning a crucial period for future research. How do animals like rats make the transition from suckling (which occurs most often in daytime because the dam is away from her litter more often at night) to adult feeding and drinking? And how do they organize the bouts or clusters of activities in which their freely emitted domestic behaviours will occur for the rest of their lives? We know that in the rat the process of weaning begins at the end of the second week of suckling and continues for at least another week or ten days. We know something about the decline in the dam's attractiveness to her pups as her production of ceacotroph declines (Leon 1974), about the adoption by the pups of their nocturnal way of life (Levin and Stern 1975), and about the tutoring the pups receive from adult females in the location and acceptability of foods (Galef and Clark 1977). However, I know of only one study (Kissileff 1971) of both the feeding and drinking behaviours of pups at weaning (and it shows, interestingly, that they have a high incidence of prandial drinking), and of no studies of the ontogeny of all the freely emitted behaviours as they are expressed by the developing mammal.

Salt Intake

Salt intake (or natriorexia) is similar to thirst in important respects. It is innate, it serves homeostasis, and it can also occur when the animal is replete and is therefore, on occasion, another of the freely emitted behaviours. Salt appetite or salt hunger is the form of the behaviour that serves homeostasis. It is the increase in the intake of salty commodities that occurs when the animal is sodium deficient. The freely emitted form of the behaviour has been called salt preference, but I prefer to think of it as "need-free" salt intake. It is common among mammals and occurs in birds that have access to dilute salty solutions, and, of course, is one of man's most interesting and medically relevant ingestive behaviours (Fregly and Kare 1982).

Salt Appetite

Future research on salt appetite will be shaped by the two contrasting proposals for its physiological mechanism that have been made in the past decade. Denton and his colleagues in Australia, on the one hand (Weisinger et al. 1982; Blair-West et al. 1987), have demonstrated in sheep and cows that salt appetite can be aroused and satiated by decreases or increases in brain sodium, and they believe that the behaviour is governed by sodium sensors in the anterior forebrain. My colleagues and I, on the other hand (Epstein 1985), can arouse or satiate the behaviour in the rat and pigeon by either administering or by pharmacologically blocking angiotensin II and aldosterone, which are the hormones of renal sodium conservation, and we propose that they are also the hormones of salt appetite. We believe that they generate the behaviour by acting in synergy on the brain. In both proposals, sodium deficiency in the periphery triggers events in the brain (activity in brain sodium sensors in the Australian proposal, activation of brain angiotensin and release of aldosterone and its subsequent action on the brain in ours) that then generate the behaviour, but the mechanisms by which these are produced are unknown.

The two proposals may be the result of real species differences (see Chapter 8, especially the evidence for a role for blood-borne angiotensin in the arousal of the sheep's salt intake which is contrary to the dependence of the rat's intake on angiotensin of cerebral origin). But a species difference is less interesting than the possibility that they are extremes of a common mechanism. Increased salt intake can be aroused in sheep by pharmacological doses of DOCA (a mimic of aldosterone) (Hamlin et al. 1988) and by angiotensin II especially if it is given at high doses into the cerebral ventricles (Coghlan et al. 1981). But the hormones have not been given at the same time, and at lower doses, which would test for the operation of a synergy mechanism. And the rat has not been fully evaluated for the possibility that sodium deficiency acts directly on the brain to trigger the activation of cerebral angiotensin in parallel with its well-known direct action on the kidney. Decreases in brain sodium may activate brain angiotensin in both sheep and rats, and when this is accompanied by peripheral sodium deficit (as it will certainly be during instances of the arousal of the appetite by natural causes) aldosterone will be released and a synergy of the two hormones in the brain may cause the behaviour.

Increases in salt intake can also be aroused by pregnancy (Pike and Yao 1971), and by the hormones of reproduction (Schultes et al. 1972) and by ACTH (Blaine et al. 1975; Denton 1984), and these have been demonstrated in the rabbit and mouse. Intracranial carbachol which is a pharmacological mimic of acetylcholine, suppresses salt intake (Fitts et al. 1987), as do the tachykinins (Massi et al. 1986) and the atrial natriuretic peptides (Fitts et al. 1985) when they are given into the cerebral ventricles. The amphibian peptide kassinin, which is a ligand for tachykinin-like receptors in the mammalian brain, is an especially interesting agent. It suppresses salt intake (both salt appetite and need-free intake) at very low doses without interfering with food intake or with angiotensin-induced drinking (Massi et al. 1988). And some consequence of uraemia, whether produced by nephrectomy, ureteric ligation, or bladder puncture, produces a profound aversion to salty commodities that has not yet been analysed (Fitzsimons and Stricker 1971). There may be an interesting and complex set of endogenous agents that arouse and satiate salt appetite. We can look forward to learning more about them and their interactions, as well as about the role that sodium deficiency plays in triggering their actions.

Additional Aspects of the Hormonal Synergy

The angiotensin/aldosterone synergy proposal assumes that in the intact, sodium-deficient animal the hormones act on separate receptor systems within the brain for each hormone. This seems likely. Salt appetite can be produced by each hormone acting alone (Rice and Richter 1943; Braun-Menendez and Brandt 1952; Chiaraviglio 1976; Avrith and Fitzsimons 1980; Bryant et al. 1980), pharmacological blockade of each hormone alone does not reduce the appetite aroused by administration of the other hormone (Sakai et al. 1986; Sakai and Epstein 1989), and medial amygdala damage to the rat brain selectively abolishes the salt appetite aroused by the mineralocorticoids (Schulkin et al. 1989). Such animals do not increase their salt intake in response to aldosterone or DOCA, but remain responsive to the natriorexigenic effect of angiotensin.

Research on the physiology of salt appetite began with Richter's discovery of the increase in salt intake that is produced by adrenalectomy (Richter 1936), and with his and Braun-Menendez's demonstrations that pharmacological doses of a mineralocorticoid also produce the behaviour in the sodium-replete rat (Rice and Richter 1943; Braun-Menendez and Brandt 1952). Both of these classic phenomena can now be understood as instances of the abnormal operation of the synergy mechanism. The mechanism is activated by both hormones when they are acting within their physiological ranges, but it can also be activated by each hormone acting alone if it is in excess. The increased salt intake that is induced by large doses of DOCA or aldosterone is an obvious example. Adrenalectomy is its counterpart. Removal of the adrenal removes the source of aldosterone. In its absence sodium loss in the urine cannot be controlled. This leads to release of renal renin and, somehow, to activation of brain angiotensin which is the cause of the salt appetite of the adrenalectomized rat. Activation of brain angiotensin with intracranial renin or renin substrate (Fitzsimons 1979) leads to increases in sodium chloride as well as water intake, and blockade of angiotensin receptors within the brain of the adrenalectomized rat with competitive analogues of angiotensin II completely suppresses the adrenalectomized rat's avidity for salt (Sakai and Epstein 1989) without interfering with its other ingestive behaviours.

Lastly, studies of the synergy mechanism have revealed what may be an organizational effect of the hormones on the brain mechanisms for salt appetite. Rats that have had prior sodium depletions or that have been exposed to brain angiotensin and aldosterone while sodium replete, drink salt solutions more rapidly and in greater volume during subsequent expressions of salt appetite (Sakai et al. 1987). This results in a lifelong enhancement of depletion-induced salt intake. The enhancement is not gradual but quantal, and it does not depend on the drinking of the salt solution during the initial expression of the appetite. It appears that the hormones have two effects during the first occasion of their synergistic action. They arouse an innate brain circuit for salt intake, and they somehow produce irreversible changes in the brain that make the animal thereafter more responsive to the combined actions of angiotensin and aldosterone.

Need-Free Salt Intake

Rats and other animals drink salt that they do not need. That is, if sodium chloride solutions are available in addition to water and commercial diets (which usually contain ten times more sodium than is required by the rat for its daily nutrition) rats that are in good health will consume small volumes of it nightly even if its concentration is as high as 3%. This need-free intake is greater in females, which may be an expression of a suppressive effect of androgen (Krecek 1973). This kind of salt intake is also enhanced by prior sodium depletions. Several such depletions are required for maximum effect, but fewer are needed in females and their intakes escalate to higher asymptotic levels (Sakai et al. 1989). Four prior depletions at weekly intervals produce final levels of average daily 3% sodium chloride intake that are similar in males to those that are consumed by many adrenalectomized rats (10–12 ml/night) and that approach 20–25 ml/night in females. This is voluntary salt intake by rats that have restored their prior deficits, that have no renal pathology, and that do not have elevated levels of angiotensin and aldosterone. It is, therefore, a model of human salt overconsumption.

It should attract research interest in the future. First, because of its medical relevance. Second, because of its ontogenetic implications: can the enhancement of need-free intake be produced in adult animals that have had sodium depletions in infancy? Third, it will attract interest because of its biological implication. What role does it play in the life of the rat and why have the brain mechanisms for it evolved? Does it promote high avidity for salt in animals that do not need it in order to assure that future needs will be avoided? Is it greater in females because they must donate sodium to their offspring?

And, lastly, this phenomenon of enhanced daily need-free salt intake in the multidepleted rat will attract attention because it, like the related phenomenon of increased need-induced salt intake in animals with similar depletion histories, is an irreversible change in brain function that is somehow linked to actions of hormones. This means that all the power and precision of modern biological technology is available for its cellular, molecular and genetic analysis, and this analysis can be done with continuous reference to an innate, specific, sexually dimorphic and medically relevant motivated behaviour.

Acknowledgements. This is a revised and shortened version of an essay entitled *Prospectus: Thirst and Salt Intake.* Stricker E (ed) Handbook of Behavioral Neurobiology, volume 10, pp 489–512 Plenum, New York, 1990.

The writing of this chapter and the personal research discussed here were supported by NS 03469, HD 25857, and MH 43787.

References

Almli RC (1973) Ontogeny of onset of drinking and plasma osmotic pressure regulation. Dev Psychobiol 6:147–158

Almli RC, Weiss CR (1974) Drinking behaviors: effects of lateral preoptic and lateral hypothalamic destruction. Physiol Behav 13:527–538

Avrith D, Fitzsimons JT (1980) Increased sodium appetite in the rat induced by intracranial administration of components of renin angiotensin system. J Physiol (Lond) 301:349–364

Babicky A, Pavlik L, Pavlik J, Ostadalova I, Kolar J (1972) Determination of the onset of spontaneous water intake in infant rat. Physiol Bohemoslov 21:467–471

Balment RJ, Carrick S (1985) Endogenous renin–angiotensin system and drinking behavior in flounder. Am J Physiol 248:R157–R160

Barney CC, Threatte RM, Fregly MJ (1983) Water deprivation-induced drinking in rats: role of angiotensin II. Am J Physiol 244:R244–R248

Bellin SI, Bhatnagar RK, Johnson AK (1987) Periventricular noradrenergic systems are critical for angiotensin-induced drinking and blood pressure responses. Brain Res 403:105–112

Blaine E, Covelli M, Denton D, Nelson J, Shulkes A (1975) The role of ACTH and adrenal glucocorticoids in the salt appetite of wild rabbits (*Oryctolagus cuniculus* (L)). Endocrinology 97:793–801

Blair-West JR, Denton DA, Gellatly DR, McKinley MJ, Nelson JF, Weisinger RS (1987) Changes in sodium appetite in cattle induced by changes in CSF sodium concentration and osmolality. Physiol Behav 39:465–469

Blass EM, Epstein AN (1971) A lateral preoptic osmosensitive zone for thirst in the rat. J Comp Physiol Psychol 76:378–394

Braun-Menendez E, Brandt P (1952) Augmentation de l'appetitspecifique pour le chlorure de sodium provoguee par le desoxycorticosterone: caracteristiques. C R Soc Biol 146:1980–1982

Bryant RW, Epstein AN, Fitzsimons JT, Fluharty SJ (1980) Arousal of a specific and persistent sodium appetite in the rat with continuous intracerebroventricular infusion of angiotensin II. J Physiol (Lond) 301:365–382

Buggy J, Johnson AK (1977) Preoptic-hypothalamic periventricular lesions: thirst deficits and hypernatremia. Am J Physiol 233:R44–R52

Chiaraviglio E (1976) Effect of renin–angiotensin system on sodium intake. J Physiol (Lond) 255:57–66

Coghlan J, Considine P, Denton D et al. (1981) Sodium appetite in sheep induced by cerebral ventricular infusion of angiotensin: comparison with sodium deficiency. Science 214:195–197

de Caro G (1986) Effects of peptides of the "gut-brain-skin triangle" on drinking behavior of rats and birds. In: de Caro G, Epstein AN, Massi M (eds) The physiology of thirst and sodium appetite. Plenum Press, New York, pp 213–226

de Caro G, Perfumi M, Massi M (1988). Tachykinins and body fluid regulation. In: Epstein AN, Morrison AR (eds) Progress in psychobiology and physiological psychology. Plenum Press, New York, 13: 31–61

de Castro JM (1989) The interactions of fluid and food intake in the spontaneous feeding and drinking patterns of rats. Physiol Behav 45:861–810

Denton DA (1984) The hunger for salt. Springer-Verlag, New York

Dollinger MJ, Holloway WR, Denenberg VH (1978) Nipple attachment in rats during the first 24 hours of life. J Comp Physiol Psychol 92:619–626

Drewett RF (1978) The development of motivational systems. Prog Brain Res 48:407–417

Edwards G, Ritter R (1982) Area postrema lesions increase drinking to angiotensin and extracellular dehydration. Physiol Behav 29:943–950

Ellis S, Axt K, Epstein AN (1984) The arousal of ingestive behaviors by chemical injection into the brain of the suckling rat. J Neurosci 4:945–955

Epstein AN (1960) Water intake without the act of drinking. Science 131:497–498

Epstein AN (1971) The lateral hypothalamic syndrome: its implications for the physiological psychology of hunger and thirst. In: Stellar E, Sprague JM (eds) Progress in physiological psychology. Academic Press, New York, pp 263–311

Epstein AN (1973) Epilogue: retrospect and prognosis. In: Epstein AN, Kissileff HR, Stellar E (eds) The neuropsychology of thirst. Winston, New York, pp 315–332

Epstein AN (1982a) Instinct and motivation as explanations for complex behavior. In: Pfaff DW (ed) Physiological mechanisms of motivation. Springer-Verlag, New York, pp 25–55

Epstein AN (1982b) The physiology of thirst. In: Pfaff DW (ed) Physiological mechanisms of motivated behavior. Springer-Verlag, New York, pp 315–332

Epstein AN (1984) The ontogeny of neurochemical systems for feeding and drinking. Proc Soc Exp Biol Med 175:127–134

Epstein AN (1985) The dependence of the salt appetite of the rat on the hormonal consequences of sodium deficiency. J Physiol (Paris) 79:496–498

Epstein AN, Spector D, Samman A, Goldblum C (1964) Exaggerated prandial drinking in the rat without salivary glands. Nature 201:1342–1343

Evered MD, Robinson MM (1981) The renin–angiotensin system in drinking and cardiovascular responses to isoprenaline in the rat. J Physiol (Lond) 316:357–362

Falk JL (1967) Control of schedule-induced polydipsia: type, size, and spacing of meals. J Exp Anal Behav 10:199–206

Findlay ALR, Fitzsimons JT, Kucharczyk J (1979) Dependence of spontaneous and angiotesin-induced drinking upon the estrous cycle and ovarian hormones. J Endocrinol 82:215–225

Fitts DA, Thunhorst RL, Simpson JB (1985) Diuresis and reduction of salt appetite by lateral ventricular infusions of atriopeptin II. Brain Res 348:118–124

Fitts DA, Thunhorst RL, Simpson JB (1987) Modulation of salt appetite by lateral ventricular infusions of angiotensin II and carbachol. Brain Res 346:273–280

Fitzsimons JT (1964) Drinking caused by contraction of the inferior vena cava in the rat. Nature 204: 479–480

Fitzsimons JT (1969) The role of a renal thirst factor in drinking induced by extracellular stimuli. J Physiol (Lond) 201:349–368

Fitzsimons JT (1979) The physiology of thirst and sodium appetite. Cambridge University Press, Cambridge

Fitzsimons JT, Elfont RM (1982) Angiotensin does contribute to drinking induced by caval ligation in rat. Am J Physiol 243:R558–R562

Fitzsimons JT, Le Magnen J (1969) Eating as a regulatory control of drinking in the rat. J Comp Physiol Psychol 67:273–283

Fitzsimons JT, Simons BJ (1969) The effect on drinking in the rat of intravenous angiotensin, given alone or in combination with other stimuli of thirst. J Physiol (Lond) 203:45–57

Fitzsimons JT, Stricker EM (1971) Sodium appetite and the renin–angiotensin system. Nature New Biol 231:58–60

Fitzsimons JT, Kucharczyk J, Richards G (1978) Systemic angiotensin-induced drinking in the dog: a physiological phenomenon. J Physiol (Lond) 276:435–448

Fregly MJ, Kare MR (1982) The role of salt in cardiovascular hypertension. Academic Press, New York

Galef BG Jr, Clark MM (1977) Mother's milk and adult presence: two factors determining initial dietary selection by weanling pups. J Comp Physiol Psychol 78:220–225

Ganten D, Fuxe K, Phillips MI, Mann JFE, Ganten V (1978) The brain isorenin–angiotensin system: biochemistry, localization, and possible role in drinking and blood pressure regulation. In: Ganong D, Martini L (eds) Frontiers in neuroendocrinology, vol 5. Raven Press, New York pp 61–99

Gisel EG, Henning SJ (1980) Appearance of glucoprivic control of feeding behavior in the developing rat. Physiol Behav 24:313–318

Grill HJ, Miselis RR (1981) Lack of ingestive compensation to dehydrational stimuli in decerebrates. Am J Physiol 240:R81–R86

Grossman SP (1984) A reassessment of the brain mechanisms that control thirst. Neurosci Biobehav Rev 8:95–104

Hainsworth FR, Stricker EM, Epstein AN (1968) The water metabolism of the rat in the heat: dehydration and drinking. Am J Physiol 214:983–989

Hall WG, Williams CL (1983) Suckling isn't feeding, or is it? A search for developmental continuities. In: Rosenblatt J, Beer C, Hinde R, Busnel MC (eds) Advances in the study of behavior, vol 13: 219–254

Hamilton CL (1969) Problems of refeeding after starvation in the rat. Ann NY Acad Sci 157:1004–1017

Hamlin MN, Webb RC, Ling WD, Bohr DF (1988) Parallel effects of DOCA on salt appetite, thirst, and blood pressure in sheep. Proc Soc Exp Biol Med 188:46–51

Hatton GI, Almli CR (1969) Plasma osmotic pressure and volume changes as determinants of drinking thresholds. Physiol Behav 4:207–214

Hirano T (1974) Some factors regulating water intake by the eel Anguilla japonica. J Exp Biol 61: 737–747

Houpt KA, Epstein AN (1971) The complete dependence of beta-adrenegic drinking on the renal dipsogen. Physiol Behav 7:897–902

Houpt KA, Epstein AN (1973) The ontogeny of the controls of food intake in the rat: GI fill and glucoprivation. Am J Physiol 225:58–66

Hsiao S, Epstein AN, Camardo JS (1977) The dipsogenic potency of peripheral angiotensin II. Horm Behav 8:129–140

Johnson AK (1985) The periventricular anteroventral third ventricle (AV3V): its relationship with the subfornical organ and neural systems involved in maintaining body fluid homeostasis. Brain Res Bull 15:595–601

Johnson AK, Buggy J (1977) A critical analysis of the site of action for the dipsogenic effect of angiotensin II. In: Buckley J, Ferrario CM (eds) Central actions of angiotensin and related hormones. Pergamon Press, New York pp 357–386

Johnson AK, Wilkin LD (1987) The lamina terminalis. In: Gross PM (ed) Circumventricular organs and body fluids, vol III. CRC Press, Boca Raton, pp 125–141

Johnson AK, Mann JEF, Rascher W, Johnson JK, Ganten D (1981) Plasma angiotensin II concentrations and experimentally induced thirst. Am J Physiol 240:229–234

Jonklass J, Buggy J (1984) Angiotensin–estrogen interaction in female brain reduces drinking and pressor responses. Am J Physiol 247:R167–R172

Jonklass J, Buggy J (1985) Angiotensin–estrogen central interaction: localization and mechanism. Brain Res 326:239–249

Katovich MJ, Barney CC, Fregly MJ, Tyler PE, Dasler R (1979) Relationship between thermogenic drinking and plasma renin activity in the rat. Aviat Space Environ Med 50:721–724

Kenney NJ, Moe KE (1981) The role of endogenous prostaglandin E in angiotensin II-induced drinking. J Comp Physiol Psychol 95:383–390

Kissileff H (1969a) Food associated drinking in the rat. J Comp Physiol Psychol 67:284–300

Kissileff H (1969b) Oropharyngeal control of prandial drinking. J Comp Physiol Psychol 67:309–319

Kissileff H (1971) Acquisition of prandial drinking in weaning rats and in rats recovering from lateral hypothalamic lesions. J Comp Physiol Psychol 77:97–109

Kleitman N (1963) Sleep and wakefulness. University of Chicago Press, Chicago

Kraly S (1989) Drinking elicited by eating. In: Epstein AN, Morrison AR (eds) Progress in psychobiology and physiological psychology. Winston, New York, pp 315–332

Kraly S, Corneilson R (1988) Angiotensin II mediates drinking elicited by histamine in rats. Soc Neurosci Abstr 14:196

Krecek J (1973) Sex differences in salt taste: the effect of testosterone. Physiol Behav 10:683–688

Kriekhaus EE, Wolf G (1968) Acquisition of sodium by rats: interaction of innate mechanisms and latent learning. J Comp Physiol Psychol 65:197–201

Leon M (1974) Maternal pheromone. Physiol Behav 13:441–453

Leshem M, Epstein AN (1988) Thirst-induced anorexias and the ontogeny of thirst in the rat. Dev Psychobiol 21:651–662

Leshem M, Boggan B, Epstein AN (1988) The ontogeny of drinking evoked by activation of brain angiotensin in the rat pup. Dev Psychobiol 21:63–75

Leshem M, Flynn FW, Epstein AN (1990) The ontogeny of the metabolic controls of ingestion: does brain energy privation control ingestion in the rat pup? Am J Physiol 258:R365–375

Levin R, Stern JM (1975) Maternal influences on ontogeny of suckling and feeding rhythms in the rat. J Comp Physiol Psychol 89:711–723

Lind, WR (1988) Angiotensin and the lamina terminalis: illustrations of a complex unity. Clin Exp Hypertens [A] 10 (Suppl.1):79–105

Lind WR, Johnson AK (1982) Central and peripheral mechanisms mediating angiotensin-induced thirst. In: Ganten D, Priuta M, Phillips MI, Scholkens A (eds) The renin angiotensin system in the brain. Springer-Verlag, New York, pp 353–364

Lind WR, Swanson LW, Ganten D (1985) Organization at angiotensin II immunoreactive cells and fibers in the rat central nervous system. Neuroendocrinology 40:2–24

Malvin RL, Mouw D, Vander AJ (1977) Angiotensin: physiological role in water-deprivation-induced thirst in rats. Science 197:171–173

Mann JFE, Johnson AK, Ganten D (1980) Plasma angiotensin II: dipsogenic levels and angiotensin-generating capacity of renin. Am J Physiol 238:R372–R378

Massi M, Epstein AN (1990) Angiotensin/aldosterone synergy governs the salt appetite of the pigeon. Appetite 14:181–192

Massi M, Micossi LG, de Caro G, Epstein AN (1986) Suppression of drinking but not feeding by central eledoisin and physalaemin in the rat. Appetite 7:63–71

Massi M, Perfumi M, de Caro G, Epstein AN (1988) Inhibitory effect of kassinin on salt intake induced by different natriorexigenic treatments in the rat. Brain Res 440:232–242

McKinley MJ, Denton DA, Park RG, Weisinger RS (1986) Ablation of subfornical organ does not prevent angiotensin-induced water drinking in sheep. Am J Physiol 250:R1052–R1059

Mendelsohn FAO, Quirion R, Saavedra JM, Aguilera G, Catt KJ (1984) Autoradiographic localisation of angiotensin II receptors in rat brain. Proc Natl Acad Sci USA 81:1575–1579

Milgram NW, Krames L, Thompson R (1974) Influence of drinking history on food-deprived drinking in the rat. J Comp Physiol Psychol 87:126–133

Miselis RR (1981) The efferent projections of the subfornical organ of the rat: a circumventricular organ within a neural network subserving water balance. Brain Res 230:1–37

Miselis RR, Shapiro ER, Hand PJ (1979) Subfornical organ efferents to neural systems for control of body water. Science 205:1022–1025

Mogenson GL (1987) Limbic-motor integration. In: Epstein AN, Morrison AR (eds) Progress in psychobiology and physiological psychology , vol 12. Academic Press, New York, pp 117–158

Ohman L, Johnson AK (1986) Lesions in lateral parabrachial nucleus enhance drinking to angiotensin II and isoproterenol. Am J Physiol 251:R504–R511

Paulus RA, Eng R, Schulkin J (1984) Preoperative latent place learning preserves salt appetite following damage to the central gustatory system. Behav Neurosci 98:146–151

Peck JW, Novin D (1971) Evidence that osmoreceptors mediating drinking in rabbits are in the lateral preoptic area. J Comp Physiol Psychol 74:134–147

Pike RL, Yao C (1971) Increased sodium chloride appetite during pregnancy in the rat. J Nutr 101: 169–176

Ramsay DJ, Rolls BJ, Wood RJ (1977) Thirst following water deprivation in dogs. Am J Physiol 232:

R93–R100

Rettig R, Ganten D, Johnson AK (1981) Isoproterenol-induced thirst: renal and extrarenal mechanisms. Am J Physiol 241:R152–R159

Rice KK, Richter CP (1943) Increased sodium chloride and water intake in normals rats treated with desoxycorticosterone acetate. Endocrinology 33:106–115

Richter CP (1936) Increased salt appetite in adrenalectomized rats. Am J Physiol 115:155–161

Rolls BJ, Wood RJ, Rolls ET (1980) Thirst: the initiation maintenance, and termination of drinking. In: Sprague JM, Epstein AN (eds) Progress in psychobiology and physiological psychology, vol 9. Academic Press, New York, pp 263–315

Rowland NE, Fregly M (1988) Sodium appetite: species and strain differences and role of renin–angiotensin–aldosterone system. Appetite 11:143–178

Rowland N, Nicolaïdis S (1974) Periprandial self-intravenous drinking in the rat. J Comp Physiol Psychol 87:16–25

Sakai RR, Epstein AN (1989) The dependence of adrenalectomy-induced sodium appetite on the action of angiotensin II in the brain of the rat. Behav Neurosci 104:167–176

Sakai RR, Nicolaïdis S, Epstein AN (1986) Salt appetite is completely suppressed by interference with angiotensin II and aldosterone. Am J Physiol 251:R762–R768

Sakai RR, Fine WB, Frankmann SP, Epstein AN (1987) Salt appetite is enhanced by one prior episode of sodium depletion in the rat. Behav Neurosci 101:724–731

Sakai RR, Frankmann SP, Fine WB, Epstein AN (1989) Prior episodes of sodium depletion increase the need-free sodium intake of the rat. Behav Neurosci 103:186–192

Schmidt-Nielsen K (1979) Desert animals. Dover, New York

Schulkin J, Marini J, Epstein AN (1989) A role for the medial region of the amygdala in mineralocorticoid-induced salt hunger. Behav Neurosci 103:724–731

Schultes AA, Covelli MD, Denton DA, Nelson JF (1972) Hormonal factors influencing salt appetite in lactation. Aust J Exp Biol Med Sci 50:819–826

Schwartz WJ, Gross RA, Morton MT (1987) The suprachiasmatic nuclei contain a tetrodotoxin-resistant circadian pacemaker. Proc Natl Acad Sci USA 84:1694–1698

Simpson JB (1981) The circumventricular organs and the central actions of angiotensin. Neuroendocrinology 32:248–256

Simpson JB, Routtenberg A (1973) The subfornical organ: site of drinking elicitation by angiotensin II. Science 818:1172–1174

Simpson JB, Epstein AN, Camardo JS (1978) Localization of dipsogenic receptors for angiotensin in subfornical organ. J Comp Physiol Psychol 92:768–795

Stephan FK, Zucker I (1972) Circadian rhythms in drinking behavior and locomotor activity are eliminated by hypothalamic lesions. Proc Natl Acad Sci USA 69:1583–1698

Stricker EM, Sterritt GM (1967) Osmoregulation in the newly hatched domestic chick. Physiol Behav 2:117–119

Stricker EM, Verbalis JG (1988) Hormones and behavior: the biology of thirst and sodium appetite. Am Sci 76:261–267

Tarttelin MF, Gorski RA (1971) Variations in food and water intake in the normal and acyclic female rat. Physiol Behav 7:847–852

Teitelbaum P (1977) Levels of integration of the operant. In: Honig WK, Staddon JER (eds) Handbook of operant behavior. Prentice-Hall, Englewood Cliffs, NJ, pp 67–83

Thrasher TN, Keil LC, Ramsay DJ (1982) Lesions of organum vasculosum of the lamina terminalis (OVLT) attenuate osmotically-induced drinking and vasopressin secretion in the dog. Endocrinology 110:1837–1845

Vijande M, Costales N, Schiaffini O, Marin B (1978) Angiotensin-induced drinking: sexual differences. Pharmacol Biochem Behav 8:753–755

Weisinger RS, Considine P, Denton DA et al. (1982) Role of sodium concentration of the cerebrospinal fluid in the salt appetite of sheep. Am J Physiol 242:R51–R63

Wiesel TN, Hubel DH (1963) Single-cell responses in striate cortex of kittens deprived of vision in one eye. J Neurophysiol 26:1003–1017

Williams DR, Teitelbaum P (1956) Control of drinking behavior by means of an operant-conditioning technique. Science 124:1294–1296

Wirth JB, Epstein AN (1976) Ontogeny of thirst in the infant rat. Am J Physiol 230:188–198

Wright JW, Morseth SL, Fairley PC, Petersen EP, Harding JW (1987) Angiotensin's contribution to dipsogenic additivity in several rodent species. Behav Neurosci 101:361–370

Zucker I (1969) Hormonal determinants of sex differences in saccharin preference, food intake, and body weight. Physiol Behav 4:595–602

Subject Index

Printing: COLOR-DRUCK DORFI GmbH, Berlin
Binding: Buchbinderei Lüderitz & Bauer, Berlin